AF667470

Fernando D. Dip • Takeaki Ishizawa
Norihiro Kokudo • Raul J. Rosenthal
Editors

Fluorescence Imaging for Surgeons

Concepts and Applications

Editors
Fernando D. Dip, M.D., M.A.A.C.
Section of Minimally Invasive Surgery
Department of General
and Vascular Surgery
The Bariatric and Metabolic Institute
Cleveland Clinic Florida
Weston, FL, USA

Oncological Surgical Division
Division of Surgical Research
Department of Surgery
Hospital de Clinicas Buenos Aires
University of Buenos Aires
Buenos Aires, Argentina

Norihiro Kokudo, M.D., Ph.D., F.A.C.S.
Professor and Chairman
Hepato-Biliary-Pancreatic Surgery
Division
Department of Surgery
Graduate School of Medicine
University of Tokyo
Bunkyo-ku, Tokyo, Japan

Takeaki Ishizawa, M.D., Ph.D., F.A.C.S.
Department of Gastroenterological
Surgery
Cancer Institute Hospital
Japanese Foundation for Cancer Research
Koto-ku, Tokyo, Japan

Hepato-Biliary-Pancreatic Surgery
Division
Department of Surgery
Graduate School of Medicine
University of Tokyo
Bunkyo-ku, Tokyo, Japan

Raul J. Rosenthal, M.D., F.A.C.S.
Chairman
Department of General Surgery
Section of Minimally Invasive Surgery
Department of General and Vascular
Surgery
The Bariatric and Metabolic Institute
Cleveland Clinic Florida
Weston, FL, USA

ISBN 978-3-319-15677-4 ISBN 978-3-319-15678-1 (eBook)
DOI 10.1007/978-3-319-15678-1

Library of Congress Control Number: 2015939720

Springer Cham Heidelberg New York Dordrecht London

Printed on acid-free paper

Springer International Publishing AG Switzerland is part of Springer Science+Business Media (www.springer.com)

Foreword

Half of all cancer patients, over 700,000 annually in the USA, undergo surgery to remove their tumors. Motivation for this surgical resection derives from the fact that the single most important predictor of patient survival is complete resection of the primary tumor, draining lymph nodes, and metastatic lesions. Historically, surgeons have depended on two tools to guide such surgical resection—their eyes to look for suspicious tissues and their hands to feel for abnormal masses. Unfortunately, up to 40 % of cancer patients experience disease recurrence at the site of surgery, and the majority of these individuals eventually die of the disease. Thus, there are urgent unmet needs to develop new and innovative technologies that can assist a surgeon in ensuring complete tumor resection by delineating tumor margins, highlighting metastatic lesions, and informing on the presence or absence of malignant cells in draining lymph nodes. Moreover, whenever disease relapse is suspected, minimally invasive methods such as fluorescence-guided endoscopy are needed to verify such suspicions and then characterize the nature and degree of dissemination of any recurrent disease.

The above challenges all suggest a need for new technologies that can delineate tumor margins, identify micrometastases, and establish whether tumor resection has indeed been quantitative before the patient's surgery is concluded. Moreover, because time in the operating room is expensive, there is also a need for new methods that will reduce the time a surgeon spends searching for malignant lesions and looking for draining lymph nodes. While intraoperative ultrasound, PET, and radioimaging have all been proposed as tools to guide the surgeon's knife during tumor resection, the methods that have drawn the greatest attention have all involved use of tumor-specific fluorescent probes to localize malignant lesions during fluorescence-guided surgery. In these strategies, either (1) a tumor-targeted fluorescent dye is administered systemically and allowed to concentrate in the malignant tissue, thereby illuminating diseased lesions for their facile resection, or (2) or a tumor-activated fluorescent prodrug is sprayed onto surgically exposed tissues and allowed to convert to its fluorescent form in response to the activity of a tumor-specific enzyme, or (3) a nontargeted fluorescent dye is injected into the vicinity of a tumor mass and permitted to passively accumulate in the diseased tissue and draining lymph nodes.

While great progress has been made in the construction of many tumor-specific fluorescent probes, progress must still be achieved in the design and optimization of both tumor-specific fluorescent dyes and intraoperative imaging cameras/endoscopes. Regarding the former, improvements will still be important in the: (1) tumor specificity, (2) tissue transparency, (3) brightness, (4) in vivo stability, (5) toxicity, and (6) ease of synthesis of the current tumor-specific fluorophores. And concerning the latter, modifications are still required in camera: (1) sensitivity, (2) background/autofluorescence suppression, (3) wavelength discrimination, and (4) ergonomic compatibility with limitations imposed by the structural and biological constraints of each specific surgery.

The chapters in this volume summarize many of the highlights of a conference held in Coral Gables, Florida, on February 15th of 2014 devoted to fluorescence-guided surgery of foregut and related cancers. The many speakers addressed a diversity of topics related to their experiences with the use of tumor and lymph node specific fluorescent dyes during localization and resection of foregut cancers. Successes and failures of specific systems and strategies were all honestly and objectively discussed, resulting in a wealth of clinical experience and associated recommendations. This compendium summarizes much of this clinical knowledge and should consequently aid any cancer surgeon interested in exploring the use of fluorescence in his/her surgical practice. Indeed, because fluorescence-guided surgery could well revolutionize the practice of surgery, this compendium is recommended for general reading by all practicing surgeons.

West Lafayette, IN, USA Philip S. Low, Ph.D.

Preface

Fluorescence Imaging for Surgeons: Concepts and Applications represents the experiences of leading researchers and surgeons worldwide practicing different fluorescence methods. The chapters range from basic science knowledge of fluorescence to the most current clinical applications and new methods on the horizon.

The chapters in the first section focus on the historical evolution and physical principles of fluorescence and provide the foundation for the reader to understand the current scope and limits of its use in surgery.

The second section describes new frontiers including fluorescence probes, imaging systems, and applications to photodynamic therapy. The third section focuses on the clinical applications of intraoperative fluorescence imaging including subsections on fluorescence cholangiography, its application to hepatectomy, lymph node navigation, the GI tract, and pelvic surgery, and the identification of cancer tissues.

Authored by leaders in the development of fluorescent methods worldwide, *Fluorescence Imaging for Surgeons: Concepts and Applications* will have an impact on numerous medical specialists including general surgeons, colorectal and minimally invasive surgeons, and surgical oncologists. Researchers will find the book to be an invaluable resource on the latest advances in the utilization of nanoparticles and fluorescent probes.

There are currently limited publications available relative to fluorescence imaging for surgeons; this book will add to the growing number of resources available on this topic.

Buenos Aires, Argentina — Fernando D. Dip, M.D., M.A.A.C.
Tokyo, Japan — Takeaki Ishizawa, M.D., Ph.D., F.A.C.S.
Tokyo, Japan — Norihiro Kokudo, M.D., Ph.D., F.A.C.S.
Weston, FL, USA — Raul J. Rosenthal, M.D., F.A.C.S.

Contents

Contributors

Vincent Agnus, Ph.D. Department of General, Digestive and Endocrine Surgery, IRCAD-IHU, Institute of Minimally Invasive Image-Guided Surgery, University of Strasbourg, Strasbourg, France

Hirofumi Akita, M.D., Ph.D. Department of Surgery, Osaka Medical Center for Cancer and Cardiovascular Diseases, Osaka, Japan

Jarmo T. Alander, Ph.D. Department of Electrical and Energy Engineering, University of Vaasa, Vaasa, Finland

Takaaki Arigami, M.D., Ph.D. Department of Digestive Surgery, Breast and Thyroid Surgery, Graduate School of Medicine, Kagoshima University, Kagoshima, Japan

Ihab Atallah, M.D. INSERM-UJF U823, Institut Albert Bonniot, CHU-Grenoble, University Joseph Fourier, Grenoble, France

Bettina Ballardini, M.D. Department of Senology, European Institute of Oncology, Milan, Italy

Gabriele Barabino, M.D. Hopital nord CHU Saint Etienne, Avenue Albert Raimond, Saint Etienne, France

Mohamed Bekheit, M.B.Ch.B., M.Sc., M.R.C.S. Hepatobiliary and Liver Transplant, Paul Brousse Hospital, Villejuif, France

Georges Bettega, M.D., Ph.D. Plastic and Maxillofacial Surgery Department, Grenoble University Hospital, Grenoble, France

Michael Bouvet, M.D. Department of Surgery, Moores UCSD Cancer Center, La Jolla, CA, USA

Nicole D. Bouvy, M.D. Ph.D. Department of Surgery, Maastricht University Medical Center, Maastricht, Limburg, The Netherlands

Nicolas Brozzi, M.D. Department of Cardiothoracic Surgery, Cleveland Clinic Florida, Weston, FL, USA

Nicolas C. Buchs, M.D. Department of Surgery, Clinic for Visceral and Transplantation Surgery, University Hospital of Geneva, Faculty of Medicine, University of Geneva, Rue Gabriel-Perret-Gentil, Geneva, Switzerland

Jean-Luc Coll, Ph.D. INSERM-UJF U823, Institut Albert Bonniot, University Joseph Fourier, Grenoble, France

Ronald M. van Dam, M.D., Ph.D. Department of Surgery, Maastricht University Medical Center, Maastricht, Limburg, The Netherlands

Alejandro Damonte, M.D. Department of Surgery, Hospital de Clinicas Buenos Aires, University of Buenos Aires, Buenos Aires, Argentina

Division of Surgical Research, Hospital de Clinicas Buenos Aires, University of Buenos Aires, Buenos Aires, Argentina

Marcelo Damonte, M.D. Department of Surgery, Hospital de Clínicas, Jose de San Martin, University of Buenos Aires, Buenos Aires, Argentina

Daniel Debonis, M.D., F.A.C.S. Oncological Surgery Division, Hospital de Clinicas Buenos Aires, Buenos Aires, Argentina

Michele Diana, M.D. Department of General, Digestive and Endocrine Surgery, IRCAD-IHU, Institute of Minimally Invasive Image-Guided Surgery, University of Strasbourg, Strasbourg, France

Fernando D. Dip, M.D., M.A.A.C. Section of Minimally Invasive Surgery, Department of General and Vascular Surgery, The Bariatric and Metabolic Institute, Cleveland Clinic Florida, Weston, FL, USA

Oncological Surgical Division, Division of Surgical Research, Department of Surgery, Hospital de Clinicas Buenos Aires, University of Buenos Aires, Buenos Aires, Argentina

Paul Dorval Fluoptics, Grenoble, France

Lau Bik Ee, B.A. Graduate School of Engineering, Chiba University, Chiba, Japan

Jorge Falco, M.D. Section of Surgery, Department of General Surgery, Hospital de Clinicas Jose de San Martin, Buenos Aires, Argentina

Pedro Ferraina, M.D. Hospital de Clinicas Jose de san Martin, Buenos Aires, Argentina

Giampaolo Formisano, M.D. Department of General and Minimally Invasive Surgery, Misericordia Hospital, Grosseto, Italy

Muneharu Fujisaki, M.D. Department of Surgery, The Jikei University School of Medicine, Tokyo, Japan

Yasuhiro Fujisawa, M.D., Ph.D. Department of Dermatology, University of Tsukuba, Tsukuba, Ibaraki, Japan

David Fuks, M.D., Ph.D. Department of Digestive Diseases, Institut Mutualiste Montsouris, Universite Paris Descartes, Paris, France

Hideo Fukuhara, M.D., Ph.D. Department of Urology, Kochi Medical School, Nankoku, Kochi, Japan

Tao Gao, M.S. Department of Frontier Surgery, Graduate School of Medicine, Chiba University, Chiba Japan

Brice Gayet, M.D., Ph.D. Department of Digestive Diseases, Institut Mutualiste Montsouris, Universite Paris Descartes, Paris, France

Pascal Gayet, Ph.D. Fluoptics Society, Grenoble, France

Shoichi Goto, B.A. Faculty of Engineering, Chiba University, Chiba Japan

Kunihito Gotoh, M.D., Ph.D. Department of Surgery, Osaka Medical Center for Cancer and Cardiovascular Diseases, Osaka, Japan

Stéphanie Guillermet, Ph.D. Fluoptics, Grenoble, France

Cynthia Hamou, M.D. Plastic and Maxillofacial Surgery Department, Grenoble University Hospital, Grenoble, France

Kazuhiro Hanazaki, M.D., Ph.D. Department of Surgery, Kochi Medical School, Nankoku, Kochi, Japan

Henricus J.M. Handgraaf, M.D. Department of Surgery, Leiden University Medical Center, Leiden, The Netherlands

Yoshinori Harada, M.D., Ph.D. Department of Pathology and Cell Regulation, Graduate School of Medicine, Kyoto Prefectural University of Medicine, Kyoto, Japan

Ari L. Harjula, M.D., Ph.D. Heart and Lung Center, HUCH, Helsinki, Finland

Hideki Hayashi, M.D., Ph.D. Center for Frontier Medical Engineering, Chiba University, Chiba, Japan

Maxime Henry INSERM-UJF U823, Institut Albert Bonniot, University Joseph Fourier, Grenoble, France

Marine Hitier, M.D. Plastic and Maxillofacial Surgery Department, Grenoble University Hospital, Grenoble, France

Robert M. Hoffman, Ph.D. AntiCancer, Inc., San Diego, CA, USA

Department of Surgery, University of California San Diego, La Jolla, CA, USA

Merlijn Hutteman, M.D., Ph.D. Department of Surgery, Leiden University Medical Center, Leiden, The Netherlands

Yoshinori Inagaki, Ph.D. Hepato-Biliary-Pancreatic Surgery Division, Department of Surgery, Graduate School of Medicine, University of Tokyo, Tokyo, Japan

Keiji Inoue, M.D., Ph.D. Department of Urology, Kochi Medical School, Nankoku, Kochi, Japan

Sumiya Ishigami, M.D. Department of Digestive Surgery, Breast and Thyroid Surgery, Graduate School of Medicine, Kagoshima, Japan

Osamu Ishikawa, M.D., Ph.D. Department of Surgery, Osaka Medical Center for Cancer and Cardiovascular Diseases, Osaka, Japan

Morihiko Ishizaki, M.D., Ph.D. Department of Surgery, Hirakata Hospital, Kansai Medical University, Osaka, Japan

Takeaki Ishizawa, M.D., Ph.D., F.A.C.S. Department of Gastroenterological Surgery, Cancer Institute Hospital, Japanese Foundation for Cancer Research, Koto-ku, Tokyo, Japan

Hepato-Biliary-Pancreatic Surgery Division, Department of Surgery, Graduate School of Medicine, University of Tokyo, Bunkyo-ku, Tokyo, Japan

Véronique Josserand, Ph.D. INSERM-UJF U823, Institut Albert Bonniot, University Joseph Fourier, Grenoble, France

Ilkka S. Kaartinen, M.D., Ph.D. Department of Plastic Surgery, Tampere University Hospital, Tampere, Finland

Masaki Kaibori, M.D. Department of Surgery, Hirakata Hospital, Kansai Medical University, Osaka, Japan

Mako Kamiya, Ph.D. Laboratory of Chemical Biology and Molecular Imaging, Graduate School of Medicine, University of Tokyo, Tokyo, Japan

Junichi Kaneko, M.D., Ph.D. Hepato-Biliary-Pancreatic Surgery Division, Department of Surgery, Graduate School of Medicine, University of Tokyo, Tokyo, Japan

Artificial Organ and Transplantation Division, Department of Surgery, Graduate School of Medicine, University of Tokyo, Tokyo, Japan

Yoshikuni Kawaguchi, M.D. Hepato-Biliary-Pancreatic Surgery Division, Department of Surgery, Graduate School of Medicine, University of Tokyo, Tokyo, Japan

Michèle Keramidas INSERM-UJF U823, Institut Albert Bonniot, University Joseph Fourier, Grenoble, France

Hirotoshi Kikuchi, M.D., Ph.D. Second Department of Surgery, Hamamatsu University School of Medicine, Higashi-ku, Hamamatsu, Japan

Norihiro Kokudo, M.D., Ph.D. Hepato-Biliary-Pancreatic Surgery Division, Department of Surgery, Graduate School of Medicine, University of Tokyo, Tokyo, Japan

Artificial Organ and Transplantation Division, Department of Surgery, Graduate School of Medicine, University of Tokyo, Tokyo, Japan

Hiroyuki Konno, M.D., Ph.D. Second Department of Surgery, Hamamatsu University School of Medicine, Higashi-ku, Hamamatsu, Japan

Mitsuo Kusano, M.D., Ph.D. Department of Surgery, Kushiro Rosai Hospital, Japan Labor Health and Welfare Organization, Hokkaido, Japan

Masanori Kon, M.D., Ph.D. Department of Surgery, Hirakata Hospital, Kansai Medical University, Osaka, Japan

Martin Lehecka, M.D., Ph.D. Department of Neurosurgery, Helsinki University Central Hospital, Helsinki, Finland

Federico Coluccio Leskow, M.D., Ph.D. Departamento de Química Biológica, IQUIBICEN-CONICET, Facultad de Ciencias Exactas y Naturales, Universidad de Buenos Aires, Buenos Aires, Argentina

Jana L. Lewis, M.D. Department of Surgery, Maimonides Medical Center, New York, NY, USA

Germana Lissidini, M.D. Department of Senology, European Institute of Oncology, Milan, Italy

Emanuele Lo Menzo, M.D., Ph.D. Digestive Disease Institute, Cleveland Clinic Florida, Weston, FL, USA

Alessandra Marano, M.D. Department of General Surgery, Humanitas Clinical and Research Hospital IRCCS, Rozzano, MI, Italy

Jacques Marescaux, M.D., F.A.C.S. (Hon), F.R.C.S. (Hon), F.J.S.E.S. Research Institute Against Cancers of the Digestive System (IRCAD), Institute of Image-Guided Surgery (IHU-Strasbourg), Strasbourg, France

Shigeru Marubashi, M.D., Ph.D. Department of Surgery, Osaka Medical Center for Cancer and Cardiovascular Diseases, Osaka, Japan

Hisahiro Matsubara, M.D., Ph.D. Department of Frontier Surgery, Graduate School of Medicine, Chiba University, Chiba, Japan

Kosuke Matsui, M.D., Ph.D. Department of Surgery, Hirakata Hospital, Kansai Medical University, Osaka, Japan

Daisuke Matsushita, M.D. Department of Digestive Surgery, Breast and Thyroid Surgery, Graduate School of Medicine, Kagoshima, Japan

Guillermo Oscar Menéndez, M.D., Ph.D. Departamento de Quimica Orgánica, CIHIDECAR-CONICET, Facultad de Ciencias Exactas y Naturales, Universidad de Buenos Aires, Buenos Aires, Argentina

Clément Milet INSERM-UJF U823, Institut Albert Bonniot, University Joseph Fourier, Grenoble, France

Takeo Minamikawa, Ph.D. Department of Pathology and Cell Regulation, Graduate School of Medical Science, Kyoto Prefectural University of Medicine, Kyoto, Japan

Norio Mitsumori, M.D., Ph.D. Department of Surgery, The Jikei University School of Medicine, Tokyo, Japan

Manuel Montesinos, M.D. Section of Surgery, Department of General Surgery, Hospital de Clinicas Jose de San Martin, Buenos Aires, Argentina

Alejandro Daniel Moreira Grecco, M.D. Department of Colorectal Surgery, Hospital de Clinicas Jose de San Martin, Buenos Aires, Argentina

Yoshifumi Morita, M.D., Ph.D. Second Department of Surgery, Hamamatsu University School of Medicine, Higashi-ku, Hamamatsu, Japan

Toshiya Nakaguchi, Ph.D. Center for Frontier Medical Engineering, Chiba University, Chiba, Japan

Tsutomu Namikawa, M.D., Ph.D. Department of Surgery, Kochi Medical School, Nankoku, Kochi, Japan

Shoji Natsugoe, M.D., Ph.D. Department of Digestive Surgery, Breast and Thyroid Surgery, Graduate School of Medicine, Kagoshima, Japan

David Nguyen, M.D. Section of Surgery, Department of General Surgery, Hospital de Clinicas Jose de San Martin, Buenos Aires, Argentina

Hiroshi Nimura, M.D., Ph.D. Department of Surgery, The Jikei University School of Medicine, Tokyo, Japan

Takeo Nomi, M.D., Ph.D. Department of Surgery, Nara Medical University, Nara, Japan

Tomás G. Núñez, M.D. Thoracic Surgery Division, Hospital de Clinicas, Universidad de Buenos Aires, Buenos Aires, Argentina

Membrillar 112 14 “B”, Buenos Aires, Argentina

Clotilde Ochala, M.D. Plastic and Maxillofacial Surgery Department, Grenoble University Hospital, Grenoble, France

Seiji Ohtsubo, M.D., Ph.D. Department of Oral and Maxillofacial Surgery, Kushiro Rosai Hospital, Japan Labor Health and Welfare Organization, Hokkaido, Japan

Aki Oishi Faculty of Engineering, Chiba University, Chiba, Japan

Tommi Pätilä, M.D., Ph.D. Hospital for Children and Adolescents, Helsinki University Central Hospital, Helsinki, Finland

Sandy Perrins, M.D. Section of Minimally Invasive Surgery, Department of General and Vascular Surgery, The Bariatric and Metabolic Institute, Cleveland Clinic Florida, Weston, FL, USA

Tamara Portas, M.D. Thoracic Surgery Division, Hospital das Clinicas, Universidad de Buenos Aires, Buenos Aires, Argentina

Pablo Quadri, M.D. Section of Surgery, Department of General Surgery, Hospital de Clinicas Jose de San Martin, Buenos Aires, Argentina

Gaston Quiche, M.D. Department of Surgery, Hospital de Clinicas Buenos Aires, University of Buenos Aires, Buenos Aires, Argentina

Division of Surgical Research, Hospital de Clinicas Buenos Aires, University of Buenos Aires, Buenos Aires, Argentina

Christian Righini, M.D., Ph.D. INSERM-UJF U823, Institut Albert Bonniot, CHU-Grenoble, University Joseph Fourier, Grenoble, France

Philippe Rizo, Ph.D. Fluoptics, Grenoble, France

Eben L. Rosenthal, M.D. Department of Surgery, The University of Alabama at Birmingham (UAB), Birmingham, AL, USA

Division of Otolaryngology, The University of Alabama at Birmingham (UAB), Birmingham, AL, USA

Raul J. Rosenthal, M.D., F.A.C.S. Section of Minimally Invasive Surgery, Department of General and Vascular Surgery, The Bariatric and Metabolic Institute, Cleveland Clinic Florida, Weston, FL, USA

Fernando M. Safdie, M.D. Department of General Surgery, Cleveland Clinic Florida, Weston, FL, USA

Masayo Sakabe, Ph.D. Laboratory of Chemical Biology and Molecular Imaging, Graduate School of Medicine, University of Tokyo, Tokyo, Japan

Takanori Sakaguchi, M.D., Ph.D. Second Department of Surgery, Hamamatsu University School of Medicine, Higashi-ku, Hamamatsu, Japan

Masato Sakon, M.D., Ph.D. Department of Surgery, Osaka Medical Center for Cancer and Cardiovascular Diseases, Osaka, Japan

Luis Sarotto, M.D. Section of Surgery, Department of General Surgery, Hospital de Clinicas Jose de San Martin, Buenos Aires, Argentina

Rutger M. Schols, M.D., Ph.D. Department of Surgery, Maastricht University Medical Center, Maastricht, Limburg, The Netherlands

Danny A. Sherwinter, M.D. Department of Surgery, Maimonides Medical Center, Brooklyn, NY, USA

Fabien Stenard, M.D. Groupe Hospitalier Mutualiste de Grenoble, CS, Grenoble, France

Shingo Shimada, M.D. Department of Surgery, Kushiro Rosai Hospital, Japan Labor Health and Welfare Organization, Hokkaido, Japan

Taro Shuin, M.D., Ph.D. Department of Urology, Kochi Medical School, Nankoku, Kochi, Japan

Diego Sinagra, M.D. Section of Surgery, Department of General Surgery, Hospital de Clinicas Jose de San Martin, Buenos Aires, Argentina

Carla Cecilia Spagnuolo, M.D., Ph.D. Departamento de Quimica Orgánica, CIHIDECAR-CONICET, Facultad de Ciencias Exactas y Naturales, Universidad de Buenos Aires, Buenos Aires, Argentina

Giuseppe Spinoglio, M.D. Center of Robotic Surgery, Department of General Surgery, Humanitas Clinical and Research Hospital IRCCS, Rozzano, MI, Italy

Laurents P.S. Stassen, M.D., Ph.D. Department of Surgery, Maastricht University Medical Center, Maastricht, Limburg, The Netherlands

Taku Suzuki, Ph.D. Graduate School of Engineering, Chiba University, Chiba, Japan

Samuel Szomstein, M.D., F.A.C.S., F.A.S.M.B.S. Section of Minimally Invasive Surgery, Department of General and Vascular Surgery, The Bariatric and Metabolic Institute, Cleveland Clinic Florida, Weston, FL, USA

Naoto Takahashi, M.D., Ph.D. Department of Surgery, The Jikei University School of Medicine, Tokyo, Japan

Hidenori Takahashi, M.D. Department of Surgery, Osaka Medical Center for Cancer and Cardiovascular Diseases, Osaka, Japan

Tetsuro Takamatsu, M.D., Ph.D. Department of Pathology and Cell Regulation, Graduate School of Medical Science, Kyoto Prefectural University of Medicine, Kyoto, Japan

Taro Toyota, Ph.D. Department of Basic Science, University of Tokyo, Komaba, Tokyo, Japan

Valery Tuchin, Ph.D. Optoelectronics and Measurement Techniques Laboratory, University of Oulu, Oulu, Finland

Department of Optics and Biophotonics, Saratov State University, Saratov, Russia

Institute of Precise Mechanics and Control of the RAS, Saratov, Russia

Masaki Ueno, M.D., Ph.D. Second Department of Surgery, Wakayama Medical University, Wakayama, Japan

Yoshikazu Uenosono, M.D., Ph.D. Department of Digestive Surgery, Breast and Thyroid Surgery, Graduate School of Medicine, Kagoshima University, Kagoshima, Japan

Naoki Unno, M.D., Ph.D. Second Department of Surgery, Hamamatsu University School of Medicine, Higashi-ku, Hamamatsu, Japan

Yasuteru Urano, Ph.D. Laboratory of Chemical Biology and Molecular Imaging, Graduate School of Medicine, University of Tokyo, Tokyo, Japan

Alexander L. Vahrmeijer, M.D., Ph.D. Department of Surgery, Leiden University Medical Center, Leiden, The Netherlands

Cornelis J.H. van de Velde, M.D., Ph.D. Department of Surgery, Leiden University Medical Center, Leiden, The Netherlands

Floris P.R. Verbeek, M.Sc. Department of Surgery, Leiden University Medical Center, Leiden, The Netherlands

Paolo Veronesi, M.D. Department of Senology, European Institute of Oncology, Milan, Italy

Eric Vibert, M.D., Ph.D. Hepatobiliary and Liver Transplant, Paul Brousse Hospital, Villejuif, France

Centre Hépato-Biliaire, Hôpital Paul Brousse, Villejuif, France

Outi M. Villet, Ph.D. Heart and Lung Center, HUCH, Helsinki, Finland

Jason M. Warram, Ph.D. Department of Surgery, The University of Alabama at Birmingham (UAB), Birmingham, AL, USA

Division of Otolaryngology, The University of Alabama at Birmingham (UAB), Birmingham, AL, USA

Terumasa Yamada, M.D., Ph.D. Department of Surgery, Higashiosaka City General Hospital, Osaka, Japan

Suguru Yamashita, M.D. Department of Surgery, Graduate School of Medicine, University of Tokyo, Tokyo, Japan

Katsuhiko Yanaga, M.D., Ph.D. Department of Surgery, The Jikei University School of Medicine, Tokyo, Japan

Shigehiro Yanagita, M.D., Ph.D. Department of Digestive Surgery, Breast and Thyroid Surgery, Graduate School of Medicine, Kagoshima, Japan

Masahiko Yano, M.D., Ph.D. Department of Surgery, Osaka Medical Center for Cancer and Cardiovascular Diseases, Osaka, Japan

Section I

Background

Basic Concepts of Fluorescence and Fluorescent Probes

1

Guillermo Oscar Menéndez, Federico Coluccio Leskow, and Carla Cecilia Spagnuolo

Fundamental Photophysical Concepts

Fluorescence is a type of photoluminescence that involves light emitted by an atom or molecule after the absorption of electromagnetic energy. The absorbed light promotes an electronic transition between the ground state and a certain excited state. After a short period of time, in the nanosecond scale, this electron relaxes to its ground state by emitting part of this energy at a different wavelength. The phenomenon was formally reported for the first time in 1,565 with the observation of the emission of light by an infusion of wood Lignum Nephriticum [1]. But major achievements would have to wait until the first half of the twentieth century with the contributions made by Perrin, Jablonsky and Förster, among others, who mainly described the theoretical basis of the phenomenon [2].

The wavelengths associated with absorption/fluorescence are in the range of 200–1,000 nm which can be subdivided in: ultraviolet (200–400 nm), visible (400–600 nm) and near-infrared region (600–1,000 nm).

The molecules that yield fluorescence emission are called *fluorophores*. A wide variety of luminescent materials are currently available: minerals, organic molecules, transition metal complexes and nanoparticles. Small fluorescent molecules are the most widely available offering high versatility regarding optical properties and chemical reactivity. Typically, a fluorophore contains merged or conjugated aromatic groups in its chemical structure.

The emission of fluorescence is sensitive to the influence of many parameters of the microenvironment such as: polarity, pH, viscosity, pressure, temperature, quenchers, ions, etc. Therefore, any variation of this signal would provide valuable temporal and spatial information of the surroundings of the fluorophore. When such a molecule allows exploring the structure and dynamics of a system then is named as a *fluorescent probe*. This fact defines the success of fluorescence techniques as a tool for studying structure and dynamics of a given system with high sensibility.

G.O. Menéndez, M.D., Ph.D. •
C.C. Spagnuolo, M.D., Ph.D. (✉)
Departamento de Química Orgánica, CIHIDECAR-CONICET, Facultad de Ciencias Exactas y Naturales, Universidad de Buenos Aires, Ciudad Universitaria, C1428EGA, Intendente Güiraldes 2160, Buenos Aires, Argentina
e-mail: carlacs@qo.fcen.uba.ar

F. Coluccio Leskow, M.D., Ph.D. (✉)
Departamento de Química Biológica, IQUIBICEN-CONICET, Facultad de Ciencias Exactas y Naturales, Universidad de Buenos Aires, Ciudad Universitaria, C1428EGA, Intendente Güiraldes 2160, Buenos Aires, Argentina
e-mail: federico@fbmc.fcen.uba.ar

F.D. Dip et al. (eds.), *Fluorescence Imaging for Surgeons: Concepts and Applications*, DOI 10.1007/978-3-319-15678-1_1, © Springer International Publishing Switzerland 2015

Four fundamental parameters of a fluorescent molecule are exploited to get practical information of the microenvironment:

- Fluorescence intensity, which is directly associated with the fluorescence quantum yield (Φ) and the molar extinction coefficient (ε). The emission spectrum remains the same, regardless of the excitation wavelength. Only the intensity is affected by varying the energy of excitation. The quantum yield depends on the radiative and non-radiative relaxation processes.
- Emission wavelength (λ): the energy associated with emission is typically lower than the excitation light, i.e. the wavelength of the emission is higher than the absorption. The difference between emission (λ_{em}) and absorption (λ_{exc}) is called Stokes shift.
- Fluorescence lifetime (τ): is the average time that the molecule spends in the excited state before the emission of a photon. The decay time is very important because it defines the observation window of a given dynamic event. It also allows the observation of different fluorophores simultaneously or an individual fluorophore in different locations.
- Fluorescence anisotropy or polarization (r): describes the polarization of the emission compared to the absorption revealing the angular displacement change between excitation and emission due to diffusional rotation occurring during the excited state lifetime. This parameter gives information about viscosity of the medium and size of the molecule.

As an investigative tool, fluorescence provides information in a broad range of areas, from materials technology (polymers, surfaces, etc.) to biological research such as the study of biological membranes, interaction between biomolecules in different cellular events, observation of macroscopic tissue and immunoassays.

More recently, as it will be discussed within this book, fluorescence has reached the medical arena hand-by-hand with new technologies that allow the performance of fluorescence image-guided surgery. This very promising technique based on molecular navigation using fluorescent probes increases the efficacy of surgical tissue resection allowing surgeons to real-time visualization of the target structure.

For further reading on theory and methods of fluorescence, the reader is encouraged to seek out the excellent books by Bernard Valeur [3] and Joseph Lakowicz [4].

Steady-State Spectroscopy

The photophysical processes suffered by a molecule and its interaction with light are well described by a Jablonski diagram. The incident light absorbed by the material leads to an electronic transition from the fundamental (S0) to an excited state (S1) and this process is called *excitation*. One of the possible de-excitation processes S1 → S0 is the *emission* of a photon, called fluorescence (Fig. 1.1).

Both excitation and emission are observable by means of a spectrofluorometer. Within this equipment the sample is exposed to a light source emitting a constant photon flow and the excitation and emission spectra are recorded. The parameter obtained by this measure is called *steady-state fluorescence intensity* which is related to the fluorescence quantum yield (Φ) by the following expression:

$$I_F = \alpha I_0 F$$

where α is the number of photons absorbed per unit volume and I_0 is the intensity of incident light. The term steady-state refers to the condition of continuous illumination which renders a constant concentration of molecules in the excited state.

The fluorescence spectrum shows the intensity of the emitted photons correlated with the wavelength, reflecting the probability of various transitions from the lowest vibrational level S1 to different vibrational levels of S0.

Time-Resolved Spectroscopy

Fluorescence is a dynamic process developed over time after an initial electronic excitation. It decays as a function of time typically in the

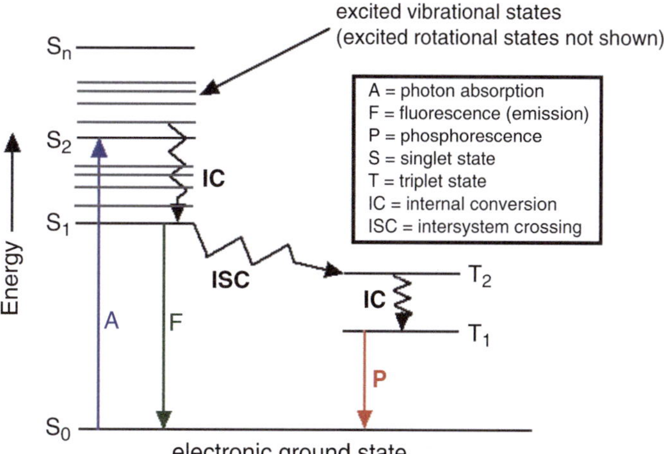

Fig. 1.1 Jablonski diagram

sub-nanosecond–nanosecond time range. Over this short period of time molecules could move, rotate, collide and participate in different reactions. The study of the dynamic of the fluorescence emission is a valuable tool to understand the mechanism of all these events. Moreover, changes in the duration of the fluorescence decay could be used to develop fluorescent sensors and markers, which could enhance the resolution and sensitivity of the conventional intensity-based fluorescence techniques.

Once a fluorophore molecule is excited it usually emits light independently, so the changes in the fluorescence intensity as a function of time are essentially concentration-independent. Moreover, the fluorescence decay is an intrinsic property of the dye, its interactions, dynamics and participation in reactions. Hence, in a simple model of identical independent fluorescence emitters, the fluorescence decay is a single exponential function of time:

$$F(t) = F_0 e^{-\frac{t}{\tau F}}$$

where F_0 is the initial emission intensity at time $t=0$. The constant τF is the excited-state (fluorescence) lifetime. Additionally, in case of a non-exponential fluorescence decay, it is often possible to describe it as a multiexponential decay and characterize the system through an averaged life-time $<\tau F>$.

The determination of fluorescence lifetimes of samples is usually performed by two time-resolved techniques: pulsed fluorometry and phase-modulation fluorometry. The first approach uses a short exciting pulse of light and the sample response is recorded as a function of time, convoluted by the instrument response. A deconvolution routine is applied in order to recover the fluorescence lifetime parameters of the sample.

In phase-modulation fluorometry, the sample is continuously excited by a high-frequency sinusoidally modulated radiation. The fluorescence response is also sinusoidally modulated at the same frequency but phase shifted and partially demodulated with respect to the excitation (Fig. 1.2). The phase shift (ϕ) and the modulation ratio M characterize the harmonic response of the system at a given frequency (ω), and the fluorescence lifetime could be calculated from both parameters.

$$\tau_\phi = \frac{1}{\omega}\tan\phi$$

$$\tau_m = \frac{1}{\omega}\sqrt{\frac{1}{m^2} - 1}$$

These two estimations yield the same lifetime result if the decay is truly monoexponential (i.e. a single fluorophore with amonoexponential decay). In most cases of interest, there will be multiple species with possible non-exponential decays. In such a case, these single-frequency

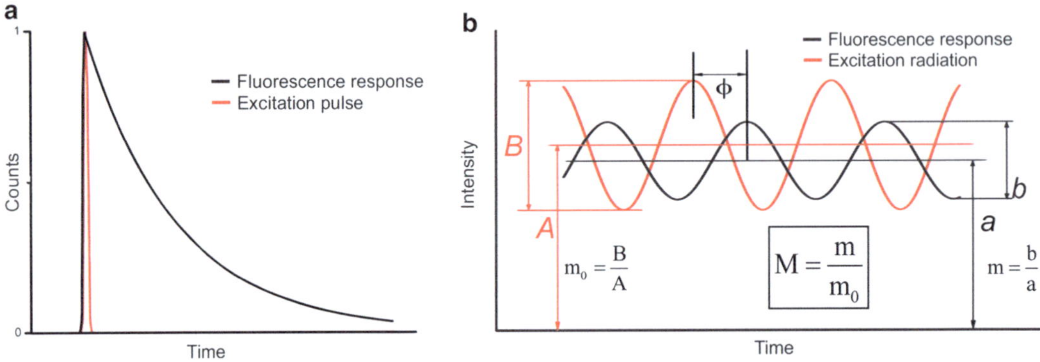

Fig. 1.2 Principles of (**a**) pulsed fluorometry and (**b**) time-resolved fluorometry

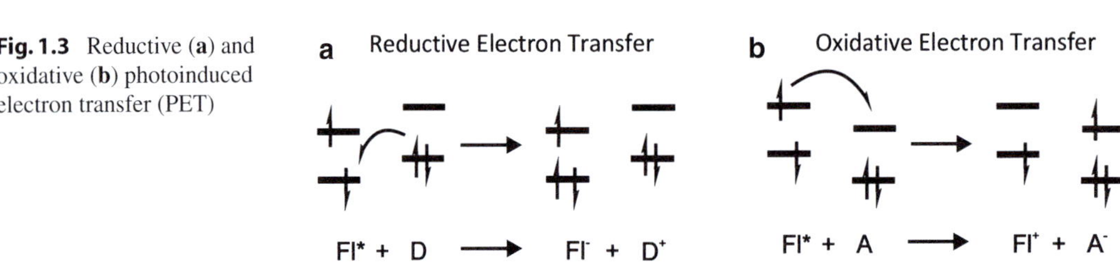

Fig. 1.3 Reductive (**a**) and oxidative (**b**) photoinduced electron transfer (PET)

lifetime estimations are a function of the lifetimes of the present species and the resolution of the complex system requires measurements at different modulation frequencies.

Relevant Biological Photophysical Processes

Over the last years, fluorescent molecules have been widely used as biomolecular labels, enzyme substrates, environmental indicators and cellular stains, constituting an indispensable tool in chemistry, physics, biology and medicinal sciences [5]. Owing to their high sensitivity, the detection of single fluorescent molecules and investigation of the interaction of these molecules with their local environment, the visualization of biochemical or biological process, have all became routinely possible through the use of appropriate instrumentation, such as near-field microscopy or confocal techniques. Moreover, the vast range of different types of fluorophores, the large variation in the photophysics and the chemical modifications available gives to fluorescent probe techniques an almost unlimited scope not just for in vitro and in vivo labeling of tissues, cells and intracellular structures of interest, but also for the detection of specific molecules and the investigation of intermolecular interactions on a molecular scale.

The fluorescence-based sensing is founded in the intermolecular interactions of dyes, both in fundamental and excited states, which change the optical properties of the fluorophores. These variations constitute the sensing signal of the fluorescent probes, which are reflected in changes of quantum yields, fluorescence lifetimes, and absorption and/or emission spectra of the fluorophore. Two of the most used processes in the designing of fluorescent sensors with biological relevance are *Photoinduced Electron Transfer* (PET) and *Förster Resonance Energy Transfer* (FRET).

Photoinduced electron transfer is a process frequently involved in many photochemical reactions and often responsible for fluorescence quenching. Moreover, PET processes play important roles during photosynthesis and in solar panels (Fig. 1.3).

When the fluorophore is excited by the absorption of a quantum of light (Fl*), the molecules gain energy, which allows its participation in different reactions. Therefore, the excited fluorophore molecules could be considered as redox sites that can donate or capture electrons from other species, being either oxidant or reductant. Photoinduced Electron Transfer is the process by which an electron is transferred from one excited site to specie. Two important requirements for this reaction are the closeness in space of the two sites and the matching of their redox potentials. PET can be facilitated by covalent bonding between donor and acceptor via a short spacer. Usually in PET systems just one of the states (reactant or product) is fluorogenic, consequently the reactions lead to a quenching or enhancement of the fluorescence emission of the sample.

Förster Resonance Energy Transfer (FRET) is a process that involves the non-radiative transfer of energy from a photoexcited specie, which is the *energy donor*, to another specie performing as the *energy acceptor*. FRET efficiency depends strongly upon the donor–acceptor distance and the spectral overlap integral between the emission spectrum of the donor and the absorption spectrum of the acceptor.

FRET has found many applications in sensing [6] and as a "*nanometric rule*" regarding the strong dependence of the energy transfer efficiency with the distance [7]. Additionally, FRET to a fluorescent acceptor allows the possibility of establish a two-channel self-calibrating system.

Fluorescence Microscopy Techniques

A fluorescence microscope is a type of optical microscope that uses the fluorescence properties of the samples instead of – or in addition to – the absorption and reflection. For this purpose, a given fluorescent or fluorescently labeled sample is illuminated with a wavelength-specific excitation radiation and the emitted light is separated using a defined array of filters and dichroic mirrors, known as "cube." The "cube" shown in

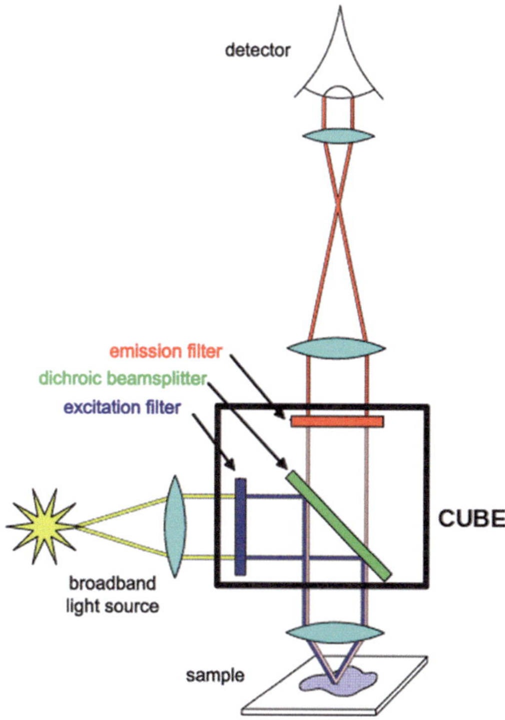

Fig. 1.4 Schematic representation of the working principle of a fluorescence microscopy showing the cube containing dichroic mirror, excitation, and emission filter

Fig. 1.4 is composed of an excitation filter, that allows only the right wavelength light to pass throw; a dichroic mirror displayed at a 45° angle that reflect the excitation light but permits the emitted light to pass; and an emission filter that allows to pass the desired wavelength light. Cubes are available or could be set up to match the spectral excitation and emission of the fluorophore used in the sample (Fig. 1.4).

Epifluorescence Microscopes

These are the simplest type and mostly used florescence microscopes, in which the excitation and emission light go through the same light path. The light sources used in these microscopes are usually xenon arc lamps, mercury-vapor lamps or more recently high power LEDs. These sources have discreet emission lines that allow to, with the right cube, excite the sample with monochromatic or near-monochromatic light.

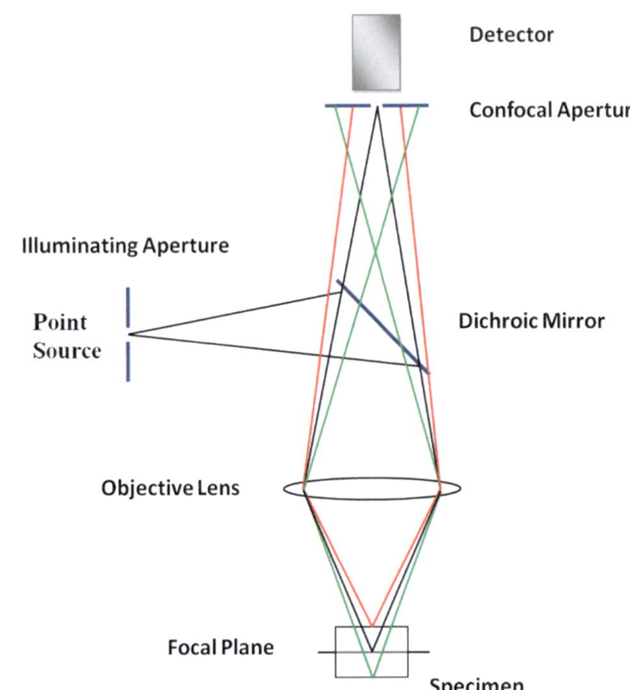

Fig. 1.5 Schematic representation of a confocal microscope. The incident light illuminates a single point on the focal plane of a specimen (*black*), the fluorescent light emitted passes through the dichroic mirror and the conjugated pinhole allows only the in-focus light from the region of interest on the sample to pass through to the detector. The light rays colored *red* and *green* in the figure represent the fluorescent emission from different focal planes

Confocal Microscopes

These are microscopes that achieve by a variety of methods the optical sectioning of the sample to get a better resolution of the image by eliminating out of focus emission. Confocality is reached by point illumination and detection using a pair of optically conjugate pinholes. For this purpose, the most widely used method is the confocal laser scanning microscope that combines a laser to excite point by point the sample and a fixed pinhole and detector. The scanning and descanning are achieved by using a set of computer-controlled mirrors and the image is reconstructed by a special software. Alternatives to the laser scanning microscope are the spinning disk microscope and the Programmable Array Microscope (PAM). The first one uses a set of moving pinholes on a spinning disk to scan the sample and the second one an array of micro-mirrors to generate a computer-controlled illumination pattern. Both microscopes usually use a CCD camera to acquired image and imaging frame rates are faster than the laser scanning systems (Fig. 1.5).

Two-Photon Excitation Microscopy

This microscopy technique uses a simple concept based on the fact that a fluorophore can be excited by two photons of carrying half of the energy needed. These lower energy photons are provided by infrared light (typically between 700 and 1,000 nm), allowing tissue imaging by an increased penetration depth since most of the tissues are transparent to this light. This excitation also minimized tissue damage and provides low background signal. Therefore, as appose to the conventional fluorescence techniques, the excited molecules emit fluorescence at a lower wavelength (higher energy) than the excitatory photons.

Since the fluorophore has to be excited simultaneously by two photons, this technique requires a high flux of excitation light that is provided by a laser source focused in one plane. This fact minimized out of focus fluorescence and provides a lower spread of the point-spread-function than single-photon excitation, resulting in an improved resolution along the z axis.

Molecular Fluorescent Probes for Imaging in the NIR Region

Imaging biological systems by fluorescence techniques became the most expanded tool to evidence and understand how a variety of biological events take place within a cell, tissue, or whole organism. Fluorescence determinations are especially sensitive and non-invasive and allow the observation of processes in real time with spatial resolution. Compared to absorbance measurements, the detection of fluorescence emission can be achieved at much lower probe concentrations and among imaging techniques fluorescence instrumentation is less expensive, simple and adaptable.

Some biomolecules, like proteins, contains intrinsic fluorophores like the amino acid tryptophan, but they are mainly not useful especially because of poor image contrast. Extrinsic molecular fluorophores ranging from visible to near infrared emission are widely preferred due to the versatility available when it comes to optical properties (absorption/emission, quantum yields, photostability) and the possibility to get them specifically attached to the target biomolecule. Compared to other extrinsic fluorophores like genetically encodable fluorescent proteins or fluorescent nanoparticles (Quantum dots), organic molecular probes have the advantage of being small in size, which leads to negligible perturbation of the structure or function of the labeled biomolecule.

Fluorescent probes are used for many different imaging applications using conventional fluorescence techniques [8] and more recently by advanced microscopic methodologies like single-molecule [9] or superresolution [10] techniques.

Ideally, a molecular fluorescent reporter should have high brightness that allows spatial and temporal resolution, high photostability in order to have a good observation window, and related to biological applications it should be non-toxic. Besides these basic requirements and foreseeing the application in image-guided surgery, the fact of being near infrared (NIR) emitters and with appropriate derivatization for specific labeling are high rated characteristics.

NIR fluorescent probes emit photons in the region between 650 and 900 nm. The main advantage of working in this range is that tissue autofluorescence in living systems are avoided (e.g. emission of elastin, collagen, hemoglobin). Therefore, a higher signal-to-noise ratio is accomplished leading to a better observation at the molecular level. Additionally, reduced light scattering and enhanced tissue penetration are achieved.

NIR emitting fluorophores have been mainly used in bioimaging applications such as pH fluctuations, ion sensing and multicolor imaging [11] in living cells with very successful results. The objective of translating these studies from the molecular biology to the medical field is in continuous growing since the last 5 years. Research made by different groups focus on the design of the chemical structure and optical properties of a fluorophore to meet the requirements for a probe to be successful as an imaging agent in surgery. Regarding clinical use, only Indocyanine Green (ICG) and Methylene Blue (MB) have received FDA approval (Fig. 1.6).

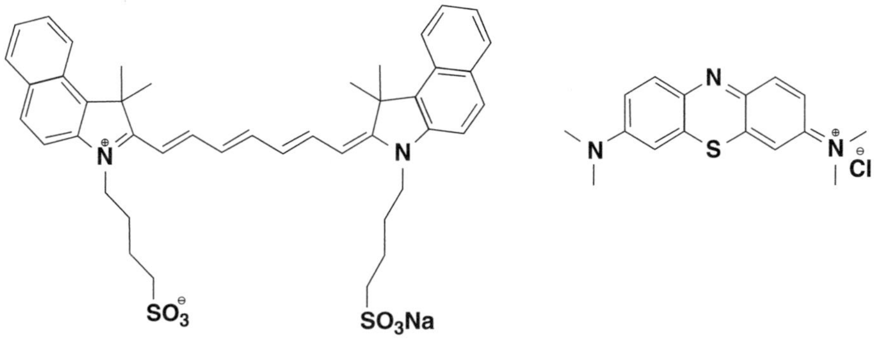

Fig. 1.6 Chemical structure of Indocyanine Green (*left*) and Methylene Blue (*right*)

MB is a phenotiazine with lower quantum yield and extinction coefficient than ICG. However, it has been used in surgery to identify ureters [12], extrahepatic bile ducts [13], among others.

ICG belongs to the heptamethine cyanines family and its use as surgical imaging agent is more expanded as will be discussed later on. The cyanine dyes are extensively studied and found application in a variety of research fields. One of the reasons of their widespread use is the chemical versatility they possess and that can be exploited to tune in a desired optical property or for further conjugation with a fragment of interest.

An important class of NIR dye are rhodamine and xanthene analogs, which have been chemically modified to shift their absorption/emission bands to the near infrared region. Although these dyes have excellent quantum efficiency they suffer from some disadvantages such as poor water solubility and short Stokes shifts. Recently, some progress has been made by Nagano et al. to overcome some of these issues [14].

There are other families of NIR dyes that are intensively used as sensors or probes for imaging cell cultures and tissues like: BODIPY analogs, Squaraine, Perylene derivatives, etc. [15]. Most of the research made on the design and synthesis of these molecules focus on the improvement of properties such as aqueous solubility, better photophysical parameters and convenient derivatization for specific labeling of biomolecules. Given the demand for new NIR probes coming from biomedical research is mandatory to study biocompatibility and pharmacokinetic of a probe intended to be applied as a diagnostic or treatment tool. Regarding imaging agents for a surgical procedure, according to Nguyen and Tsien, the probe is better to be administered topically or intravenously before operation and it should render high imaging contrast during the procedure or at the moment of tissue resection [16].

Although in its infancy, NIR imaging applied to surgical procedures will provide surgeons a real-time guidance to the target tissue for resection or to identify tissue that should be avoided. The reader is encouraged to seek up the review by Gioux et al. for further reading on the influence of key parameters of the imaging system for NIR fluorescence-guided surgery [17]. About light sources and hardware associated with these procedures, we recommend the reviews from Nguyen and Berg [16, 18].

NIR Molecular Probes for Imaging Guided Surgery

As mentioned before, ICG is actually the most widely used NIR imaging probe in intraoperative proceedings. Alander et al. have reviewed over 200 papers reporting on the use of this dye as a fluorescent probe, stressing the increase of publications related since 2009 [19]. Schaafsma et al. reported in the same year an extensive revision of the bibliography referred to the use of ICG in cancer-related surgery [20].

New successful applications of ICG were very recently published. For example, the use of this dye for assistance during a hepatic metastasectomy allows to identify superficially located colorectal liver metastases and in some patients it was possible to localize small lesions otherwise undetectable by typical procedures [21].

Here we present relevant potential imaging agents inspired by ICG which were reported the last 2 years, to illustrate how current disadvantages of ICG could be overcome following a rational chemical design.

Ebert & colleagues reported in 2011 a sugar derivative of ICG named SIDAG which foreseen in its design a superior hydrophilicity than ICG [22]. As a consequence SIDAG showed renal elimination, an acute tolerance 60-fold higher and strong tumor contrast in rat breast carcinomas.

We have been considering the use of a NIR probe which falls in the class of *non-targeted* dyes. This means that following administration, the pharmacokinetics of the molecule should be known in detail and it defines whether a probe is a good imaging agent for the observation of a given target according to its biodistribution, excretion ways, accumulation patterns, etc. In general, non-targeted probes have poor resolution between malignant and healthy tissue boundaries.

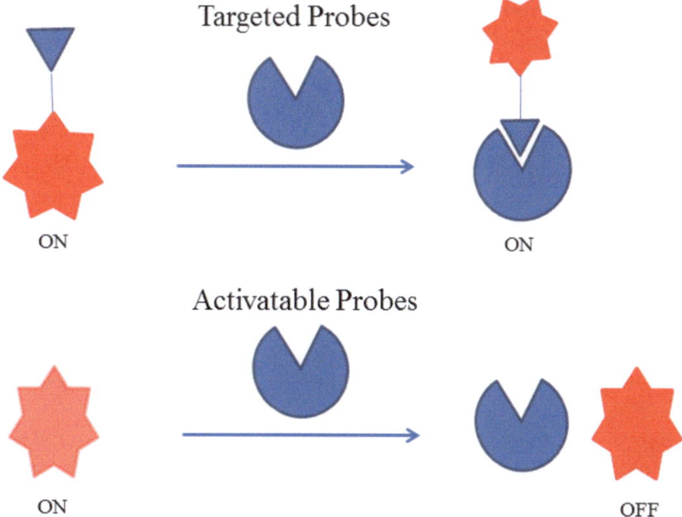

Fig. 1.7 General strategies to detect a target specifically

A key aspect of any imaging agent is the effective delivery to the tissue of interest. This requirement represents a major challenge from the chemical point of view because the probe has to be both a good fluorophore and a good ligand for its target receptor.

Targetability can be achieved by two distinct strategies (Fig. 1.7): attachment of a specific recognition fragment to the probe (*targeted probe*) or activation of the probe signal by means of a specific chemical reaction given in situ (act*ivatable probe*).

The common approach when using a targeted probe is to label specific cell-surface receptors such as the Folate receptor, integrins or EGFR among others. The corresponding ligands are small molecules, peptides, protein, antibodies, and aptamers [23].

In the following paragraphs, we will comment on recently reported potential imaging agents for use in this strategy.

Kelderhouse et al. developed NIR-specific probes by combining fluorophores like IR800, AlexaFluor® 647 and DyLight 680 with folic acid and DUPA as ligands for the folate receptor and the prostate-specific membrane antigen (PSMA), respectively [24]. They were able to image culture cancer cells at the nanomolar level and more important they could resect malignant tissue with minimal content of healthy tissue in tumor-bearing mice.

Cao et al. choose $\alpha_v\beta_3$ integrin as a target for solid tumor labeling. They synthesized a near-infrared derivative of ICG and dimeric cyclic peptide RGD as the ligand [25]. The authors claimed and showed excellent tumor activity accumulation when binding the probe to subcutaneous MDA-MB-231 and U87MG tumor cells in vivo.

With the aim of developing a probe for early gastric cancer detection, the group of Tian designed a series of pentamethine cyanine dyes conjugated to GX1 [26]. GX1 is a potential gastric tumor marker. They observed selective accumulation of the probe in the tumor site in xenograft models in vivo and were able to elaborate an efficient synthetic route of the fluorescent ligand.

Choi et al. developed the dye ZW800-1, a zwitterionic NIR-dye with improved signal-to-background ratio, low serum binding, ultralow non-specific tissue background and rapid elimination from the body through renal filtration [27]. Following this achievement they pursued a systematic study of this probe conjugated to different ligands (cyclic RGD peptide, fibrinogen or antibodies) and their performance to target the corresponding receptors in comparison with similar

Table 1.1 Chemical structures, ligands, and targets of some targetable probes

Structure	Ligand	Target	Reference
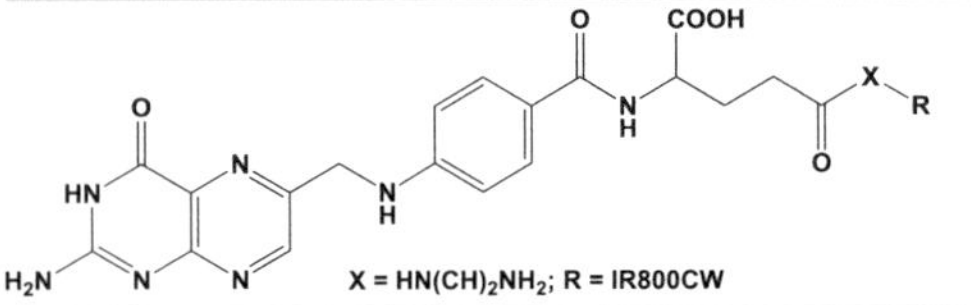 X = HN(CH)$_2$NH$_2$; R = IR800CW	Folate	Folate receptor	[25]
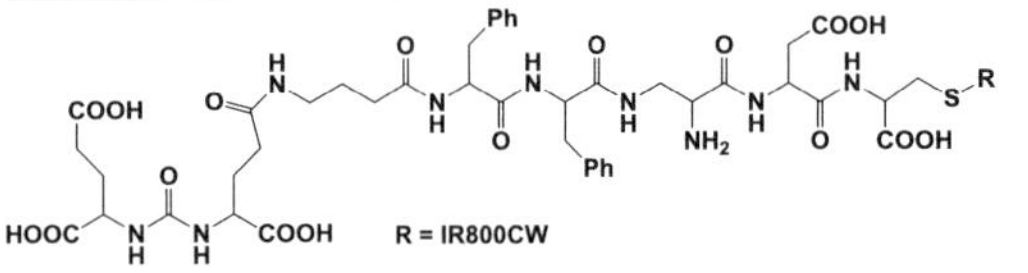 R = IR800CW	DUPA	PSMA receptor	[25]
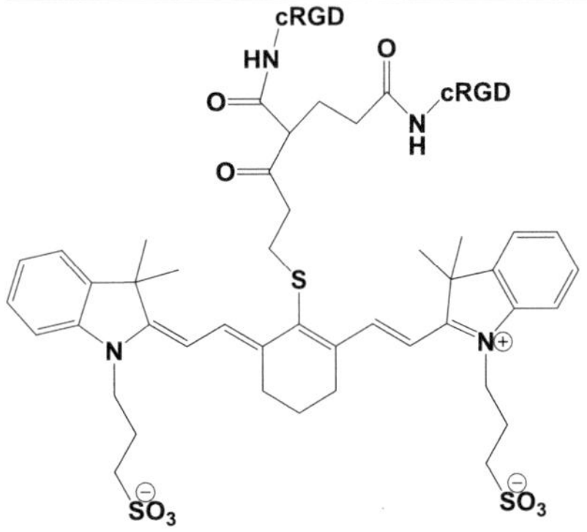 cRGD	cRGD	Integrin αvβ3	[26]

probes containing CV800 and Cy5.5 as fluorescent reporters [28]. They reached a significant increase of the tumor-to-background ratio for the ZW800-1 conjugate in tumor model systems and suggest that introducing zwitterionic properties would improve sensitivity in fluorescence-based imaging (Tables 1.1 and 1.2).

Alternatively, when using an activatable imaging probe, is expectable that only upon interaction with the target through any chemical transformation (oxidation, enzymatic cleavage) the probe switches from a dark to a bright state, thus generating a fluorescent signal. By following this strategy only the target tissue is turned ON and not the background. The activation of the quenched probe is given by an enzymatic reaction occurring in situ of the target tissue either, by reaction of the probe with oxygen species typically abundant in malignant cells, by chelation of metal ions present in the target tissue or by sensing significant pH changes between healthy and abnormal tissue.

Those activatable probes with enzymatic ON switches are the more exploited. One of the leading groups in the development of such imaging agents is the Hisataka Kobayashi's laboratory. They have been working on different probe designs based on the activatable galactosyl human serum albumin (hGSA)–fluorophore pair [29]. hGSA binds specifically to a lectin receptor overexpressed in the cell surface of peritoneal ovarian cancer metastases (POCM). Combining hGSA with the bacteriochlorin-based dye NMP1, an activatable unique probe is obtained because NMP1 has two absorption peaks, one in the green range and the other in the NIR range. This probe is self-quenched and generates a fluorescence signal after internalization by local enzymes at the target site. With this probe the authors were able to detect in vivo submillimeter POCM lesions with high sensitivity and specificity compared to other NIR–hGSA conjugates.

The same group also found that recently reported sililated rhodamines have superior performance

Table 1.2 Chemical structures, ligands, targets, and triggers of some targetable probes

Structure	Ligand/target	Trigger	Reference
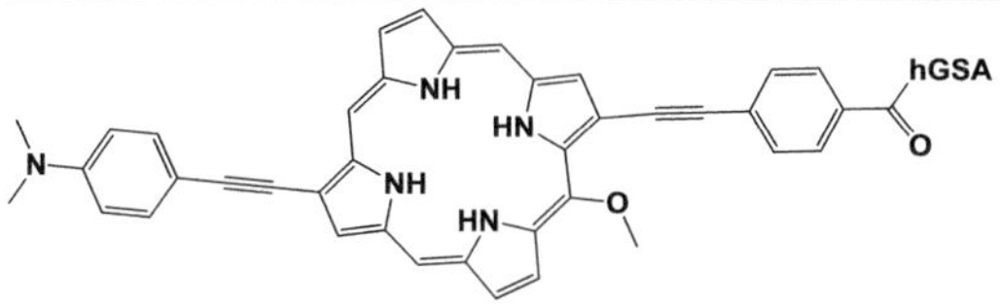 hGSA; NH; HN; NH; HN; N; O	hGSA/lectin receptor	Enzymatic cleavage upon internalization	[30]
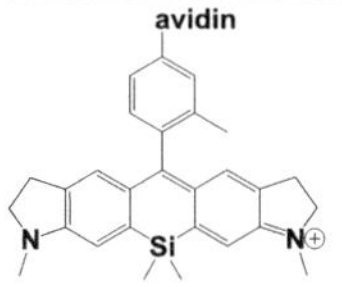 avidin; N; Si; N⊕	Avidin/-	Endocytosis	[31]
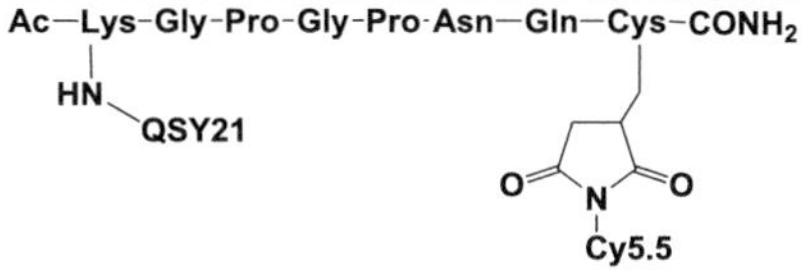 Ac–Lys-Gly-Pro-Gly-Pro-Asn–Gln–Cys–$CONH_2$; HN; QSY21; O; N; O; Cy5.5	Peptide/FAPα	Proteolitic cleavage	[32]
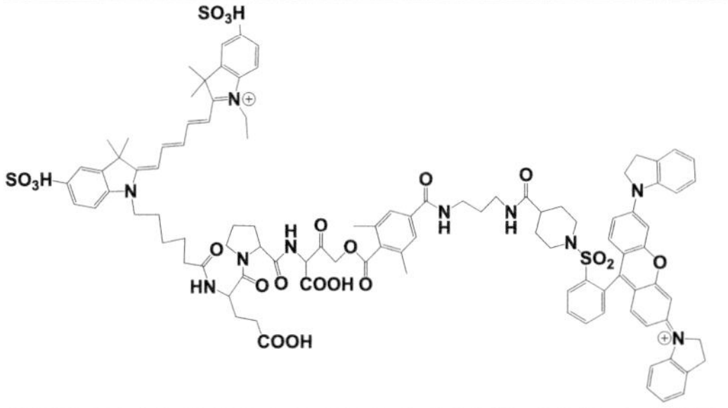 SO_3H; N⊕; SO_3H; N; HN; COOH; COOH; SO_2; O; N; ⊕N	Peptide/legumain	Proteolitic cleavage	[33]
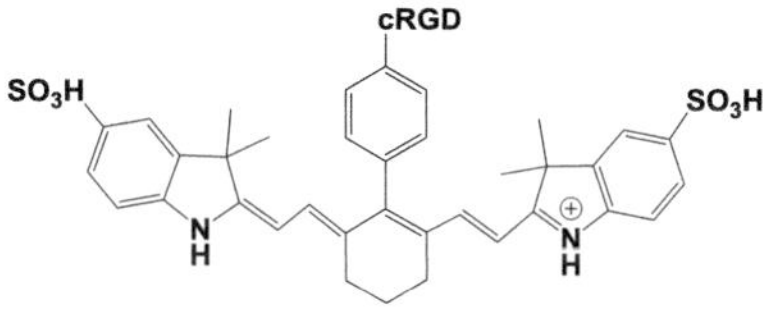 cRGD; SO_3H; SO_3H; NH; ⊕NH	cRGD/integrin αvβ3	pH change upon internalization	[34]

as potential activatable probes than cyanines or Alexa dyes in a mouse model of metastatic ovarian cancer during in vivo optical imaging with fluorophore–avidin conjugates [30]. The ability of these Si-rhodamines for self-quenching by forming H-dimers when conjugated to the protein was exploited. The quenching is disrupted after internalization by enzymes at the target site and the fluorescence signal is observed with a very low degree of non-specific binding.

Li et al. based their strategy on another glycosilated protein: Fibroblast activation protein-alpha (FAPα), a cell surface glycoprotein which is selectively expressed by tumor-associated fibroblasts in malignant tumors [31]. They synthesized ANPFAP consisting of a NIR dye (Cy5.5) and a quencher dye (QSY21), which are linked together by a short peptide sequence specific for FAPα cleavage. The authors showed fast tumor uptake as well as high tumor-to-background contrast in tumor models.

More recently, Bogyo's group developed an activatable probe highly selective to legumain, a very active protease in human cancers and inflammatory diseases [32]. This work represents an elegant example on how to chemically rationalize the design of the structure of probe following previous detailed information of the action mechanism of the enzyme to be used as the activatable trigger. Normally this information requires a deep study of the active of the enzyme. The probe LE28 consists on a dimer of a cyanine and a xanthene dye forming a FRET quenched pair linked by the target peptide of the enzyme.

A promising pH-activated probe was recently reported by Achilefu's group [33]. The probe combines actually both strategies: targeted and activatable probing. Target specificity is obtained by introducing a cRGD fragment in the structure of the pH-sensitive trycarbocyanine to achieve exclusive labeling at the $\alpha_v\beta_3$ integrin receptor, hence favoring accumulation in tumor cells. The tricarbocyanine is non-fluorescent at pH > 5 but becomes highly fluorescent in acidic environment typical of tumor organelles (lysosomes).

Given the need of the surgical field to distinct between different tissues in real operational time and the impact of having such possibility, many groups are focusing their research to satisfy this demand. Efforts toward success rely in a combination of different disciplines as: chemistry, molecular biology, medicine, and physics. An efficient interaction between them is mandatory for a correct molecular design and application of any probe.

Fluorescent Nanoprobes for In Vivo Imaging

Low molecular weight fluorescent dyes (typically with molecular weights up to 1,500 Da for NIR fluorophores) are currently the most widespread type of tracers. As mentioned before, their well-defined chemical structure could be modified in order to obtain the desired optical and chemical properties. However, small organic fluorescent dyes present some disadvantages when are applied to in vivo applications. Owing to their extended pi-conjugated system, these molecules are usually poorly soluble in aqueous solutions and their fluorescence quantum yield is generally lower than other fluorophores in the visible range. However, one of the most significant concerns in the development of potential fluorophores for in vivo imaging applications is their fast body clearance. This could be advantageous to limit dye toxicity, but also limits the temporal window in which the imaging can be performed, and the number of organs in which the fluorescent tracer can be distributed.

The evolution of nanoparticles, where fluorophore molecules can be encapsulated, adsorbed, or attached in the surface, has blossomed in recent years due to their excellent potential for use in biomedical, imaging, sensing, drug delivery and therapy systems [34, 35]. Biotechnological and medical advances relating to diagnosis and treatment are heavily dependent on an intensive understanding of biochemical processes. The judicious choice of suitable nanoparticle compositions and surface functionalization have enabled researchers to develop nanoparticle probes that produce enhanced fluorescent signals and exhibit increased selectivity and better reproducibility; which have opened up a new era of promising biomarker development.

Particularly, the native features of silica have been exploited in the development of numerous hybrid silica nanomaterials with different properties, making them good candidates for the construction of hybrid systems which can load and transport appropriate agents for applications in different fields [36]. The integration of fluorescent dye-doped silica nanoparticles with functional molecules using various surface modification techniques can bring about great improvements in bioapplications such as selective recognition, sensitive imaging, and reporting. Moreover, unlike other nanoparticles, such as metal nanoparticles or semiconductor nanocrystals, fluorescent silica nanoparticles exhibit higher hydrophilicity, biocompatibility, and stability under physiological conditions [37]. Additionally, the confinement of dye molecules in the silica matrix increases their fluorescence signal, making them significantly brighter than organic fluorophores [38]. This advantageous aspect becomes fluorescent silica nanoparticles especially suitable for highly sensitive bioapplications without additional amplification steps. Likewise, the high intensity and excellent optical stability of fluorescent silica nanoparticles facilitate ultrasensitive analyte determinations and monitoring of biological processes that are undetectable using conventional fluorescence labeling techniques [39].

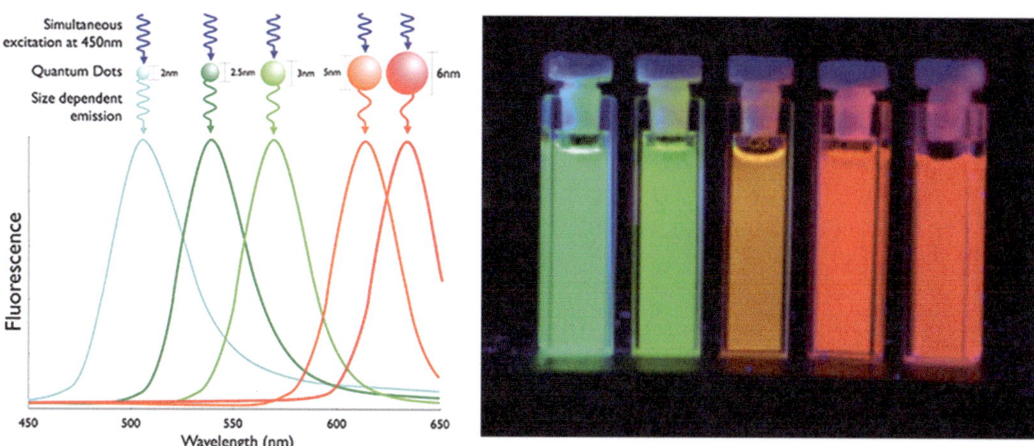

Fig. 1.8 Spectral characteristics of QDs

In addition to non-fluorescent nanoparticles which are doped with fluorophores, the intrinsically fluorescent nanoparticles known as quantum dots (QDs) appear as great alternative tools to dissect molecular processes, identify cellular targets or label discrete population of cells. A QD is a nanocrystal made of a semiconductor material that displays quantum mechanical properties related to the size and shape of the individual crystal. The emission wavelength increases with the size of the QD, resulting in a color shift from red to blue in the light emitted (Fig. 1.8). Because the size of the crystals can be controlled during synthesis, fluorescence emission can be carefully controlled. QDs provide unique advantages for cellular imaging due to their intense and uniform brilliance, permitting detection down to the single nanoparticle level and their photostability, allowing imaging over prolonged periods. QDs have wide excitation spectra in the ultraviolet range and narrow emission bands can be achieved by controlling their sizes, allowing tuning the emission band to fit any necessity. QDs are also biochemically stable and can be conjugated with a variety of molecules including antibodies, streptavidin, peptides, DNA, nucleic acid aptamers or small-molecule ligands (Fig. 1.8).

Biotinylated ligands in combination with streptavidin-conjugated QDs have been widely used for studying receptor tyrosine kinase dynamics in living cells, including the characterization of endocytosis and retrograde transport. For example, internalization of insulin—IR constitutes the major mechanism of insulin degradation and down-regulation of cell surface receptors. However, signal transduction by IR is not limited to its activation at the plasma membrane and endosome signaling showed to be important in mitogenic signaling. Internalization rates of the insulin receptor bound to insulin could be calculated using biotinylated human recombinant insulin in conjugation with QDs allowing tracking in real time the endocytosis and the receptor–ligand complex fate (Fig. 1.9). The activated ligand–receptor complex internalized into endosomes would activate substrates that are spatially distinct from those accessible at the plasma membrane. The dynamics of the activated insulin receptor depends on the cellular context and it is crucial for the correct signaling balance. Insulin-QD tracking in cells allowed a comparative and quantitative analysis of dynamics of insulin receptor signaling and its termination. This lead to the proposed hypothesis that internalization of activated receptor plays a crucial role in the signaling balance by turning off metabolic signaling by displacing the receptor away from the effectors molecules localized at the membrane and promoting mitogenic signaling from endosomes. Endosomal trafficking would finally lead to dissociation of insulin–receptor complexes in late endosomes and lysosomes switching off the signaling pathways (Fig. 1.9) [40].

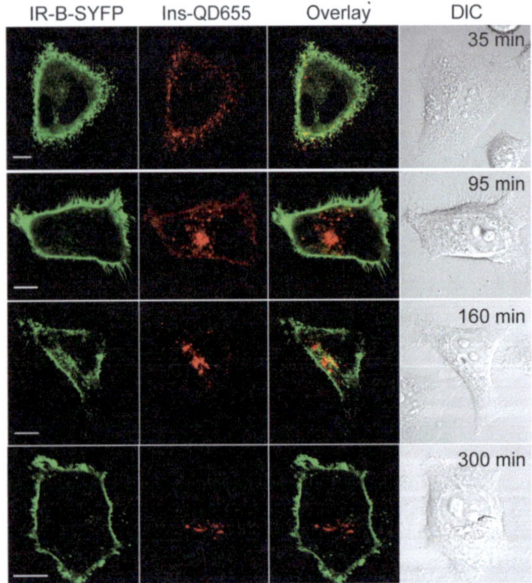

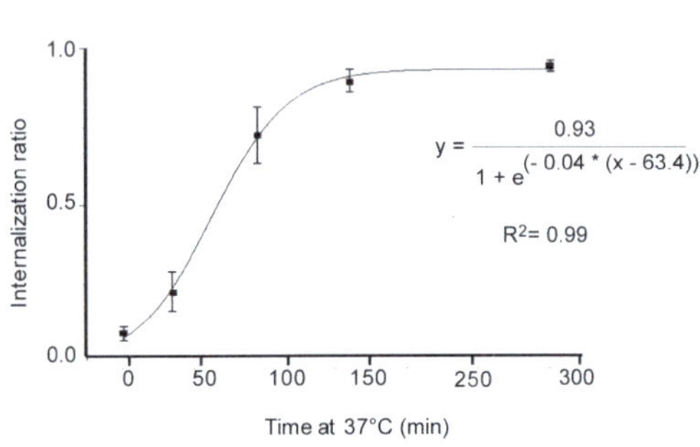

Fig. 1.9 Insulin-QDs internalization in living cells. HeLa cells expressing IR-B-SYFP labeled with 50 nM biotinylated human recombinant insulin and 1 nM QD655 were incubated at 37 °C for 35, 95, 160, and 300 min and visualized in living cells at RT by confocal microscopy (Olympus Fluoview FV 1000). Scale bars: 10 μm. Graph shows quantification of internalization ratio in individual cells [40]

Another example is the use of the epidermal growth factor (EGF) in complexed QDs to track by in vivo microscopy the binding of EGF to its receptors. EGF binds to the erbB family of tyrosine kinase receptors including erbB1 (the classical EGF receptor, EGFR), erbB2, erbB3, and erbB4. After ligand binding the receptors dimerized and the specificity of the response depends on the type of dimer formed. EGF-QDs are highly specific and potent in the binding and activation of the EGF receptor (erbB1), being rapidly internalized into endosomes that exhibit active trafficking and extensive fusion. When erbB2 or erbB3 were coexpressed with erbB1, the rates and extent of endocytosis of EGF-QD demonstrated that erbB2 but not erbB3 heterodimerizes with erbB1 after EGF stimulation, thereby modulating EGF-induced signaling [41]. ErbB1 receptors are situated on cellular filopodia and undergo systematic retrograde transport after binding of the epidermal growth factor (EGF). Using single molecule sensitivity by tracking the receptors labeled with EGF-QDs allowed to conclude that retrograde transport precedes receptor endocytosis, which occurs at the base of the filopodia [42].

EGF in conjugation with QDs were also used to label brain tumor cells, helping to identify and clearly demarcate between brain and tumor tissue at the cellular level with high contrast ratios by the fluorescence emission of the Qds. The strong and photostable fluorescence and rapid differential binding of these probes meet some of the criteria required by surgeons to distinguish tumor cells left in the resection cavity [43].

References

1. O'Haver TC. Development of luminescence spectrometry as an analytical tool. J Chem Educ. 1978; 55(7):423.
2. Nickel B. From the Perrin Diagram to the Jablonski Diagram. EPA Newsletter. 1996;58:9–38.
3. Valeur B, Berberan-Santos MN. Molecular fluorescence: principles and applications. Weinheim: Wiley-VCH; 2012.
4. Lakowicz JR. Principles of fluorescence spectroscopy. New York, NY: Springer; 2010.

5. de Silva AP, Gunaratne HQN, Gunnlaugsson T, Huxley AJM, McCoy CP, Rademacher JT, et al. Signaling recognition events with fluorescent sensors and switches. Chem Rev. 1997;97(5):1515–66.
6. Kudlacek O, Gsandtner I, Ibrisimovic E, Nanoff C. Fluorescence resonance energy transfer (FRET) sensors. BMC Pharmacol. 2008;8 Suppl 1:A44.
7. Stein IH, Schüller V, Böhm P, Tinnefeld P, Liedl T. Single-molecule FRET ruler based on rigid DNA origami blocks. Chem Phys Chem. 2011;12(3):689–95.
8. Demchenko AP, editor. Advanced fluorescence reporters in chemistry and biology III: Applications in sensing and imaging. New York: Springer; 2011.
9. Zheng Q, Juette MF, Jockusch S, Wasserman MR, Zhou Z, Altman RB, et al. Ultra-stable organic fluorophores for single-molecule research. Chem Soc Rev. 2014;43:1044–56.
10. Fernandez-Suarez M, Ting AY. Fluorescent probes for super-resolution imaging in living cells. Nat Rev Mol Cell Biol. 2008;9(12):929–43.
11. Escobedo JO, Rusin O, Lim S, Strongin RM. NIR dyes for bioimaging applications. Curr Opin Chem Biol. 2010;14(1):64–70.
12. Matsui A, Tanaka E, Choi HS, Kianzad V, Gioux S, Lomnes SJ, et al. Real-time, near-infrared, fluorescence-guided identification of the ureters using methylene blue. Surgery. 2010;148(1):78–86.
13. Matsui A, Tanaka E, Choi HS, Winer JH, Kianzad V, Gioux S, et al. Real-time intra-operative near-infrared fluorescence identification of the extrahepatic bile ducts using clinically available contrast agents. Surgery. 2010;148(1):87–95.
14. Koide Y, Urano Y, Hanaoka K, Terai T, Nagano T. Development of an Si-rhodamine-based far-red to near-infrared fluorescence probe selective for hypochlorous acid and its applications for biological imaging. J Am Chem Soc. 2011;133(15):5680–2.
15. Yuan L, Lin W, Zheng K, He L, Huang W. Far-red to near infrared analyte-responsive fluorescent probes based on organic fluorophore platforms for fluorescence imaging. Chem Soc Rev. 2013;42(2):622–61.
16. Nguyen QT, Tsien RY. Fluorescence-guided surgery with live molecular navigation [mdash] a new cutting edge. Nat Rev Cancer. 2013;13(9):653–62.
17. Gioux S, Choi HS, Frangioni JW. Image-guided surgery using invisible near-infrared light: fundamentals of clinical translation. Mol Imaging. 2010;9(5): 237–55.
18. van den Berg NS, van Leeuwen FWB, van der Poel HG. Fluorescence guidance in urologic surgery. Curr Opin Urol. 2012;22(2):109–20.
19. Alander JT, Kaartinen I, Laakso A, Pätilä T, Spillmann T, Tuchin VV, et al. A review of indocyanine green fluorescent imaging in surgery. Int J Biomed Imag. 2012;2012:26.
20. Schaafsma BE, Mieog JSD, Hutteman M, van der Vorst JR, Kuppen PJK, Löwik CWGM, et al. The clinical use of indocyanine green as a near-infrared fluorescent contrast agent for image-guided oncologic surgery. J Surg Oncl. 2011;104(3):323–32.
21. van der Vorst JR, Schaafsma BE, Hutteman M, Verbeek FPR, Liefers G-J, Hartgrink HH, et al. Near-infrared fluorescence-guided resection of colorectal liver metastases. Cancer. 2013;119(18):3411–8.
22. Ebert B, Riefke B, Sukowski U, Licha K. Cyanine dyes as contrast agents for near-infrared imaging in vivo: acute tolerance, pharmacokinetics, and fluorescence imaging. J Biomed Opt. 2011;16(6): 066003–9.
23. Hilderbrand SA, Weissleder R. Near-infrared fluorescence: application to in vivo molecular imaging. Curr Opin Chem Biol. 2010;14(1):71–9.
24. Kelderhouse LE, Chelvam V, Wayua C, Mahalingam S, Poh S, Kularatne SA, et al. Development of tumor-targeted near infrared probes for fluorescence guided surgery. Bioconjug Chem. 2013;24(6):1075–80.
25. Cao J, Wan S, Tian J, Chi X, Du C, Deng D, et al., editors. Synthesis of dimeric cyclic RGD based nearinfrared probe for in vivo tumor diagnosis. Proc SPIE. 2012;2:8224:82240T-1.
26. Xin J, Zhang X, Liang J, Xia L, Yin J, Nie Y, et al. In vivo gastric cancer targeting and imaging using novel symmetric cyanine dye-conjugated GX1 peptide probes. Bioconjugate Chem. 2013;24(7):1134–43.
27. Choi HS, Nasr K, Alyabyev S, Feith D, Lee JH, Kim SH, et al. Synthesis and in vivo fate of zwitterionic near-infrared fluorophores. Angew Chem Int Ed. 2011;50(28):6258–63.
28. Choi HS, Gibbs SL, Lee JH, Kim SH, Ashitate Y, Liu F, et al. Targeted zwitterionic near-infrared fluorophores for improved optical imaging. Nat Biotechnol. 2013;31(2):148–53.
29. Alexander VM, Sano K, Yu Z, Nakajima T, Choyke PL, Ptaszek M, et al. Galactosyl human serum albumin-NMP1 conjugate: a near infrared (NIR)-activatable fluorescence imaging agent to detect peritoneal ovarian cancer metastases. Bioconjugate Chem. 2012;23(8):1671–9.
30. McCann TE, Kosaka N, Koide Y, Mitsunaga M, Choyke PL, Nagano T, et al. Activatable optical imaging with a silica-rhodamine based near infrared (SiR700) fluorophore: a comparison with cyanine based dyes. Bioconjugate Chem. 2011;22(12): 2531–8.
31. Li J, Chen K, Liu H, Cheng K, Yang M, Zhang J, et al. Activatable near-infrared fluorescent probe for in vivo imaging of fibroblast activation protein-alpha. Bioconjugate Chem. 2012;23(8):1704–11.
32. Edgington LE, Verdoes M, Ortega A, Withana NP, Lee J, Syed S, et al. Functional imaging of legumain in cancer using a new quenched activity-based probe. J Am Chem Soc. 2013;135(1):174–82.
33. Lee H, Akers W, Bhushan K, Bloch S, Sudlow G, Tang R, et al. Near-infrared pH-activatable fluorescent probes for imaging primary and metastatic breast tumors. Bioconjugate Chem. 2011;22(4):777–84.
34. Tallury P, Payton K, Santra S. Silica-based multimodal/multifunctional nanoparticles for bioimaging and biosensing applications. Nanomedicine. 2008;3(4): 579–92.

35. Zhang L, Pornpattananangku D, Hu CM, Huang CM. Development of nanoparticles for antimicrobial drug delivery. Curr Med Chem. 2010;17(6):585–94.
36. Wang L, Wang K, Santra S, Zhao X, Hilliard LR, Smith JE, et al. Watching silica nanoparticles glow in the biological world. Anal Chem. 2006;78(3): 646–54.
37. Santra S, Dutta D, Walter G, Moudgil BM. Fluorescent nanoparticle probes for cancer imaging. Technol Cancer Res Treat. 2005;4:593–602.
38. Zhao X, Bagwe RP, Tan W. Development of organic-dye-doped silica nanoparticles in a reverse micro-emulsion. Adv Mater. 2004;16(2):173–6.
39. Santra S, Dutta D. Nanotechnologies for the Life Sciences. New Jersey: John Wiley & Sons Inc.; 2007. Nanoparticles for optical imaging of cancer; 44–85.
40. Giudice J, Jares-Erijman EA, Leskow FC. Endocytosis and intracellular dissociation rates of human insulin-insulin receptor complexes by quantum dots in living cells. Bioconjug Chem. 2013;24(3):431–42.
41. Lidke DS, Nagy P, Heintzmann R, Arndt-Jovin DJ, Post JN, Grecco HE, et al. Quantum dot ligands provide new insights into erbB/HER receptor-mediated signal transduction. Nat Biotechnol. 2004;22(2):198–203.
42. Lidke DS, Lidke KA, Rieger B, Jovin TM, Arndt-Jovin DJ. Reaching out for signals: filopodia sense EGF and respond by directed retrograde transport of activated receptors. J Cell Biol. 2005;170(4):619–26.
43. Kantelhardt SR, Caarls W, de Vries AH, Hagen GM, Jovin TM, Schulz-Schaeffer W, et al. Specific visualization of glioma cells in living low-grade tumor tissue. PLoS One. 2010;5(6):e11323.

Past, Present, and Future of Fluorescence

2

Daniel Debonis, Pablo Quadri, Manuel Montesinos, Jorge Falco, David Nguyen, Diego Sinagra, Fernando D. Dip, Raul J. Rosenthal, and Pedro Ferraina

Introduction

First, it is important to define fluorescence. There are several ways to describe this term: (1) the emission of electromagnetic radiation, especially of visible light, stimulated in a substance by the absorption of incident radiation and persisting only as long as the stimulating radiation is continued. (2) Emitting light during exposure to radiation from an external source. (3) Luminescence that is caused by the absorption of radiation at one wavelength followed by nearly immediate reradiation usually at a different wavelength and that ceases almost at once when the incident radiation stops. The common denominators for this term are a source of radiation, the absorption of this radiation, and the emission of a modified radiation that ends immediately as the source stops emitting radiation.

Chemically speaking, fluorescence is brought about by absorption of photons in the singlet ground state promoted to a singlet excited state. The spin of the electron is still paired with the ground state electron. As the excited molecule returns to ground state, it involves the emission of a photon of lower energy, which corresponds to a longer wavelength, than the absorbed photon and therefore lower energy, than the absorbed radiation (Fig. 2.1).

Application of Fluorescence in Medicine

By understanding what fluorescence consists on, we can now think on the uses that can be given in medicine. Fluorescence in the life sciences is used generally as a non-destructive way of tracking or analysis of biological molecules by means of the fluorescent emission at a specific frequency where there is no background from the excitation

D. Debonis, M.D., F.A.C.S. (✉)
Oncological Surgery Division, Hospital de Clinicas Buenos Aires, 2351 Córdoba Ave., Ciudad Autónoma de Bs. As, Buenos Aires C112OAAR, Argentina
e-mail: dl.debonis@intramed.net

P. Quadri, M.D. • M. Montesinos, M.D. • J. Falco, M.D. D. Nguyen, M.D. • D. Sinagra, M.D.
Section of Surgery, Department of General Surgery, Hospital de Clinicas Jose de San Martin, Buenos Aires, Argentina

F.D. Dip, M.D., M.A.A.C.
Section of Minimally Invasive Surgery, Department of General and Vascular Surgery, The Bariatric and Metabolic Institute, Cleveland Clinic Florida, 2950 Cleveland Clinic Boulevard, Weston, FL 33331, USA

Oncological Surgical Division, Division of Surgical Research, Department of Surgery, Hospital de Clinicas Buenos Aires, University of Buenos Aires, Buenos Aires, Argentina

R.J. Rosenthal, M.D., F.A.C.S.
Section of Minimally Invasive Surgery, Department of General and Vascular Surgery, The Bariatric and Metabolic Institute, Cleveland Clinic Florida, 2950 Cleveland Clinic Boulevard, Weston, FL 33331, USA

P. Ferraina, M.D.
Hospital de Clinicas Jose de San Martin, Buenos Aires, Argentina

F.D. Dip et al. (eds.), *Fluorescence Imaging for Surgeons: Concepts and Applications*, DOI 10.1007/978-3-319-15678-1_2, © Springer International Publishing Switzerland 2015

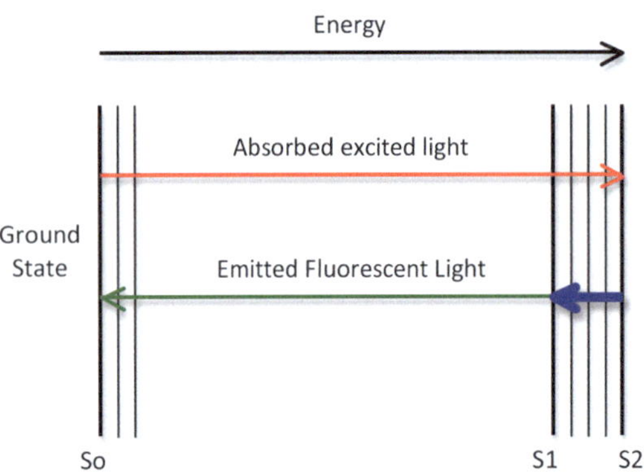

Fig. 2.1 Diagram representing fluorescence. The *red arrow* represents the absorption of light. The *blue arrow* represents vibrational relaxation from singlet excited state, S2 to S1. This process is a non-radiative relaxation in which the excitation energy is dispersed as vibrations or heat to the solvent, and no photon is emitted. The *green arrow* represents fluorescence to the singlet ground state, So

light. As relatively few cellular components are naturally fluorescent (autofluorescence), in fact, a protein or other component can be "labeled" with an extrinsic fluorophore, a fluorescent dye that can be a small molecule, protein, or quantum dot. These are the bases for most of the applications of fluorescence in medicine. The quantification of a dye is done with a spectrofluorometer and finds additional applications.

Some of the uses of fluorescence in medicine include:

Microscopy: when scanning the fluorescent intensity across a plane one has fluorescence microscopy of tissues, cells, or subcellular structures, which is accomplished by labeling an antibody with a fluorophore and allowing the antibody to find its target antigen within the sample. Labeling multiple antibodies with different fluorophores allows visualization of multiple targets within a single image. DNA microarrays are a variant of this.

Immunology: an antibody is first prepared by having a fluorescent chemical group attached, and the sites (e.g., on a microscopic specimen) where the antibody has bound can be seen, and even quantified, by the fluorescence.

Automated sequencing of DNA by the chain termination method; each of four different chain terminating bases has its own specific fluorescent tag. As the labeled DNA molecules are separated, the fluorescent label is excited by a UV source, and the identity of the base terminating the molecule is identified by the wavelength of the emitted light.

DNA detection: the compound ethidium bromide, in aqueous solution, has very little fluorescence, as it is quenched by water. Ethidium bromide's fluorescence is greatly enhanced after it binds to DNA, so this compound is very useful in visualizing the location of DNA fragments in agarose gel electrophoresis. Intercalated ethidium is in a hydrophobic environment when it is between the base pairs of the DNA, protected from quenching by water, which is excluded from the local environment of the intercalated ethidium.

Fluorescence image-guided surgery (FIGS) is a medical imaging technique that uses fluorescence to detect properly labeled structures during surgery. Its purpose is to guide the surgical procedure and provide the surgeon of real-time visualization of the operating field. When compared to other medical imaging modalities, FIGS is cheaper and superior in terms of resolution and number of molecules detectable. As a drawback, penetration depth is usually very poor (100 μm) in the visible wavelengths, but it can reach up to 1–2 cm when excitation wavelengths in the near infrared are used.

FIGS is performed using imaging devices with the purpose of providing real-time simultaneous information from color reflectance images (bright field) and fluorescence emission. One or more light sources are used to excite and illuminate the sample. Light is collected using optical filters that match the emission spectrum of the fluorophore. Imaging lenses and digital cameras are used to produce the final image.

Fluorescence excitation is accomplished using various kinds of light sources. Halogen lamps have the advantage of delivering high power for a relatively low cost. Using different band-pass filters, the same source can be used to produce several excitation channels from the UV to the near infrared.

During open surgery, handheld devices are usually preferred for their easy of use and mobility. A stand or arm can be used to maintain the system on top of the operating field. The disadvantage of these devices is that other lights in the operating room interfere in the fluorescence emission, which is simply solved by switching off the other lights. FIGS can also be performed using minimally invasive devices such as laparoscopes or endoscopes. In this case, a system of filters, lenses and cameras is attached to the end of the probe. Unlike open surgery, the background from external light sources is reduced.

The major limitation in FIGS is the availability of clinically approved fluorescent dyes. Indocyanine Green has been widely used as a non-specific agent to detect sentinel lymph nodes during surgery. Methylene Blue can also be used for the same purpose. First clinical applications using tumor-specific agents that detect deposits of ovarian cancer during surgery have been carried out.

History of Fluorescence

Bernardino de Sahagún and Nicolás Monardes described an early observation of fluorescence in the infusion known as lignum nephriticum (Latin for "kidney wood") in 1560 and 1565 respectively. It was derived from the wood of two tree species, Pterocarpus indicus and Eysenhardtia polystachya. The chemical compound responsible for this fluorescence is matlaline, which is the oxidation product of one of the flavonoids found in this wood.

In 1819, Edward D. Clarke, an English mineralogist and traveler who amassed valuable collections of minerals; and in 1822 René Just Haüy, a French mineralogist and one of the founders of the science of crystallography, described fluorescence in fluorites. Sir David Brewster described the phenomenon for chlorophyll in 1833 and Sir John Herschel did the same for quinine in 1845.

George Gabriel Stokes was a British physicist and mathematician noted for his studies of the behavior of viscous fluids, particularly for his law of viscosity, which describes the motion of a solid sphere in a fluid, and for Stokes's theorem, a basic theorem of vector analysis. In his 1852 paper on the "Refrangibility" (wavelength change) of light, he described the ability of fluorspar and uranium glass to change invisible light beyond the violet end of the visible spectrum into blue light. He named this phenomenon "fluorescence." The name was derived from the mineral fluorite (calcium difluoride), some examples of which contain traces of divalent europium, which serves as the fluorescent activator to emit blue light. In a key experiment he used a prism to isolate ultraviolet radiation from sunlight and observed blue light emitted by an ethanol solution of quinine exposed by it.

First Approaches of Fluorescence in Medicine

The first approaches of fluorescence remote to 1953, Lay and Randall, used a fluorescent staining for detection of cancer cells in vaginal smears. They used the affinity of the basic fluorochrome dyes (berberine sulfate, acid fuchsin, and acridine) for acid nucleic acids. Because neoplasic cells have shown an increase in the total nucleic acids and ribonucleic acids, showing an increase in protein synthesis, the cancerous cells were evidently seen. They used a fluorescent microscope to evidence this finding [1, 2].

In 1955 there was a publication of Peck and Mack for the use of hematoporphyrin fluorescence in biliary and cancer surgery. They investigated the hematoporphyrin and its increased affinity for malignant tissue and the biliary excretion of this component to evidence the biliary duct anatomy. They used the different fluorescent spectra of the hematoporphyrin to visualize malignant tumors and bile ducts. Now it is being used as an endobronchial tumor indicator to facilitate the bronchoscopic localization of early squamous cell carcinoma of the central tracheobronchial tree [1, 3–8].

Later, in 1963 Acherman and McFee investigated on tetracycline fluorescence in benign and malignant tissues. Tetracycline can be detected in the tissue by its bright yellow fluorescence in ultraviolet light. Using this method it has been shown that within a few minutes after parenteral administration of the drug, it is distributed to all the tissues of the body except the central nervous system. The highest concentrations are present at liver, kidneys, and bones. In 1957 Ral and his associates, while studying the effects of fluorescent riboflavin antagonist on metastases form Brest cancer, observed that patients that had received tetracycline some weeks before their study had other fluorescent material in tumor tissue. So they started investigating the detection of fluorescent tetracycline in tumors. Klinger and Katz reported that administering tetracycline orally to patients for 5 days and performing gastric washing for exfoliative cytology after 36 h, and placing the aspirate residue in a filter paper and examining under ultraviolet light could allow the visualization of cancerous cells. Fluorescence was seen in 17 of 18 patients with stomach cancer, and was not noted on 42 patients without cancer [9–11].

In 1965 Whitmore's group investigated on ultraviolet light cystoscopy following administration of tetracycline. This method was able to evidence tumor with more sensibility than light cystoscopy. It was able to see grade III tumors in patients with negative visible light cystoscopy but positive urinary sediments. No relationship was found between the occurrence of fluorescence and the grade of histologic malignancy or the type of bladder carcinoma. Ultraviolet cystoscopy can provide useful guidance for biopsy and thereby facilitate early detection of recurrence after treatment for bladder carcinoma in patients with persistent tumor cells in the urinary sediment but no tumor at visible light cystoscopy [12, 13].

Other lines of investigation opened in 1966 with the use of tetracycline for evaluation of ischemic changes in transplantations. Actually, there have been new publications on fluorescence spectroscopy in renal ischemia and reperfusion as a non-invasive evaluation of the organ viability. It is a fact that damage provoked by ischemia is difficult to quantify. So the bases of these studies consist on evaluating the findings of fluorescence spectroscopy with the histologic evidence of ischemia injury and organ viability, the gold standard. They concluded there was a strong correlation between fluorescence spectroscopy and histologic changes only in the reperfusion phase after renal ischemia, so this method was thus unable to assess the viability of organs before transplantation [14, 15].

In this way, fluorescence continued to be used in different studies including brain, rynolaringologic, and other tumors always trying to identify early tumors not evident with visible light. This included tetracycline fluorescence test in the surgery of rhinolaryngologic malignant tumors (1965) and fluorescence studies in brain tumors (1966) by Nessel E and Goldhahn WE respectively [16–20].

Another large line of investigations consisted on the evaluation of cardiac ischemia and reperfusion. This includes studies on the use of the fluorescent method for determining the degree of revascularization of the heart in the surgical treatment of experimental cardiac ischemia (1965), tetracycline fluorescence in experimental necrotizing cardiopathies (1969), and the value of fluorescence in the macroscopic delimitation of experimental acute myocardial ischemia before and after revascularization (1973) [16–18]. But the interesting things on these investigations is that there are several actual publications on all these topics that started on the 1950s, meaning that there were interesting topics that can gain important use in actual medicine and can be further developed.

Finally these lots of investigations of fluorescence lead to three Nobel Prizes: they received the Nobel Prize in Chemistry in 2008 for the discovery and development of green fluorescent protein (GFP). They are: Osamu Shimomura, born in 1928 in Japan, Kyoto. He is an organic chemist and marine biologist from Boston University and Marine Biological Laboratory. Martin Chalfie, he was born in Chicago in 1947. He is a scientist and Works at Columbia University. And Roger Yonchien Tsien, born in 1952, is a Chinese American biochemist. He is a professor at the Department of Chemistry and Biochemistry, University of California, San Diego.

To conclude, fluorescence has been studied since long time ago, but investigations still continue and can contribute to great advances in modern medicine, including biochemistry, biomolecular medicine, and anatomical localization in surgery.

References

1. Grubb C, Crabbe JGS. Fluorescence microscopy in exfoliative cytology. British J Cancer. 1961;15(3): 483–8.
2. Lay Cl, Randall LM. Fluorescent staining for detection of cancer cells in vaginal smears. Surg Forum. 1953;(38th Congress):321–7.
3. Raymond J. Lanzafame MD, David W. Rogers BS, John O. Nairn MS, Ralph P. Pennino MD, J. Raymond Hinshaw MD, DPhil, FACS1 and Jeffrey R. Blackman BS2 Hematoporphyrin derivative fluorescence: Photographic techniques for the localization of malignant tissue.Lasers in Surgery and Medicine "Volume 6, Issue 3, pages 328–335, 1986.
4. Rotomskis R, van de Meent EJ, Aartsma TJ, Hoff AJFluorescence spectra of hematoporphyrin and hematoporphyrin-diacetate aggregates in buffer solution Journal of Photochemistry and Photobiology. B, Biology 1989, 3(3):369–377.
5. Hackney JD, Collier CR. Autofluorescence and fluorochrome staining of alveolar linning. Med Thorac. 1965;22:77–88.
6. Peck GC, Mack HP. Use of hematoporphyrin fluorescence in biliary and cancer surgery. Annal Surg. 1955;21(3):181–8.
7. Ahlquist Jr RE, Figge FH. Fluorescence of the extrahepatic biliary system following parenteral hematoporphyrin administration. Surg Forum. 1956;6: 356–8.
8. Kato H, Cortese DA. Early detection of lung cancer by means of hematoporphyrin derivative fluorescence and laser photoradiation. Clin Chest Med. 1985;6(2): 237–53.
9. Ackermann NB, Mc Fee AS. Tetracycline fluorescence in benign and malignant tissues. Surgery. 1963;53:247–52.
10. Tapp E, Carroll R, Kovács K. Tetracycline fluorescence in experimental tumors. British J Cancer. 1965;19(3):538–46.
11. Sandlow LJ, Allen HA. The use of tetracycline fluorescence in the detection of gastric malignancy. Ann Intern Med. 1963;58(3):409–13. doi:10.7326/0003-4819-58-3-409.
12. Sørensen BL, Barlebo H. Ultraviolet light cystoscopy in patients with bladder cancer: one year's experience. Scand J Urol Nephrol. 1969;3(3):193–200. Departments of Surgery H and Pathology, Gentofte Hospital, Copenhagen, Denmark.
13. Whitmore Jr WF, Bush IM. Ultraviolet cystoscopy in patients with bladder cancer. Trans Am Assoc Genitourin Surg. 1965;57:149–55.
14. Málek P, Kolc J, Pultr V, Zástava V. Tetracycline fluorescence test in evaluation of ischemic changes in kidney transplantations. Rozhl Chir. 1966;45(4):252–6.
15. Cassini MF, da Costa MM. Fluorescence spectroscopy in renal ischemia and reperfusion: noninvasive evaluation of organ viability. Transplant Proc. 2013;45(5):1715–9.
16. Khutsishili TS, Milaeva MA. The use of the fluorescent method for determining the degree of revascularization of the heart in the surgical treatment of experimental cardiac ischemia. Grudn Khir. 1965;7(6):11–2. Russian.
17. Somogyi A, Kovács K. Tetracycline fluorescence in experimental necrotizing cardiopathies. Pathol Eur. 1969;4(3):257–64.
18. Loisance D, Sadony V. Value of fluorescence in the macroscopic delimitation of experimental acute myocardial ischemia before and after revascularization. Arch Mal Coeur Vaiss. 1973;66(6):701–7.
19. Nessel E, Wiggermann W. The tetracycline fluorescence test in surgery of malignant tumors; rhinolaryngologic experiences. HNO. 1965;13(12):346–9.
20. Goldhahn WE. Fluorescence studies of brain tumors. Beitr Neurochir. 1966;13:126–9.

Translation of Therapeutic Antibodies for Intraoperative Fluorescence Imaging

3

Eben L. Rosenthal and Jason M. Warram

Introduction

In oncologic imaging, actively targeted contrast agents have the power to overcome innate disadvantages that are associated with non-specific optical imaging. An intuitive approach that is currently advancing into human use involves the repurposing of previously approved therapeutic antibodies, which act as targeting vectors for disease-specific, for fluorescence enhancement. The use of therapeutic antibodies for imaging improves translational potential since the pharmacokinetic profile, biodistribution, and toxicity profile of these FDA-approved agents have been extensively described. Furthermore, recent data suggest that fluorescent labeling of the antibodies do not substantially alter this profile [1, 2] making antibody-based imaging a safe and cost-effective approach for translating targeted imaging agents into the clinic.

There are characteristics unique to antibodies that make their use as imaging agents controversial, such as a long half-life and limited number receptors on the cell surface. Preclinical and clinical data from recent studies have suggested that fluorescently labeled antibodies are highly specific for subclinical deposits of cancer and produce powerful tumor to background ratios in most types tested [3–6]. The use of antibodies for imaging has been rapidly expanding beyond preclinical models over the past several years and is being actively investigated for human use. The goal of this chapter is to provide an overview of the critical components necessary to adapt and transition fluorescently labeled antibodies for intraoperative imaging of cancer.

The Need for Cancer-Specific Imaging

Consistent with the themes of this book, there is a critical need for real-time imaging of subclinical disease in cancer. It is well known that failure to achieve negative margins in most cancer types leads to higher rates of recurrence and poor survival. Positive margins are associated with worse survival in most cancer types including breast, head and neck, prostate, melanoma, colon, and many others such that the presence of positive margins may actually necessitate the need to return the patient to the operating room and re-excise the positive tissue. Furthermore, in most cancer types adjuvant radiotherapy, with or without chemotherapy is indicated when surgical resection cannot achieve negative margins.

E.L. Rosenthal, M.D. (✉) • J.M. Warram, Ph.D.
Department of Surgery, The University of Alabama at Birmingham (UAB), BDB 563, 1720 2nd Ave S, Birmingham, AL 35294-0012, USA

Division of Otolaryngology, The University of Alabama at Birmingham (UAB), BDB 563, 1720 2nd Ave S, Birmingham, AL 35294-0012, USA
e-mail: oto@uab.edu

F.D. Dip et al. (eds.), *Fluorescence Imaging for Surgeons: Concepts and Applications*, DOI 10.1007/978-3-319-15678-1_3, © Springer International Publishing Switzerland 2015

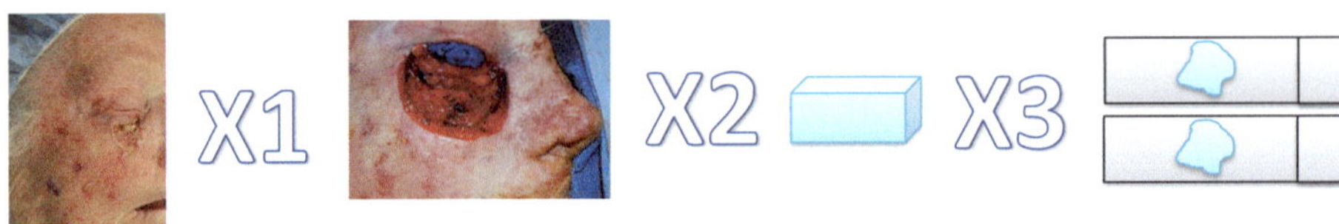

Fig. 3.1 Current methodology contains multiple possibilities of failure to accurately assess margin status. Pre-resection: The initial incision is based upon the surgeons' best assessment of the tumor margins, however overestimating the cancer results in loss of normal tissue while underestimation results in cutting through the cancer and leaving positive margins. Post-resection: the surgeon samples areas suspicious for cancer but rarely can sample the whole bed. Frozen specimen: the pathologist traditionally assesses only two cuts from a specimen, leaving at least 80 % of the tissue unsampled

Although each tumor is unique, the associated morbidity, patient burden and cost of adjuvant treatments can be greater than the surgery alone [7], suggesting that if tumor margins could be accurately determined intraoperatively, there would be a reduction in cost and morbidity.

Surgeons cannot detect or "see" subclinical disease at the time of resection. In most cancer types, the surgeon can use tactile feedback to estimate the gross extent of disease but requires frozen section analysis to determine the microscopic extension. Furthermore, there are some tumors such as glioma [8] or certain skin cancers [9] where the tumor–normal interface is nearly impossible to define by routine examination, which is detrimental to surgical success considering inaccurate resection leads to greater morbidity. In these cases, optical imaging using a cancer-specific contrast agent would direct where and when to obtain surgical margins. Preclinical and clinical studies have shown that detection of tumor volumes less than 1-mm is achievable using this technique [10–13].

Surgeons sample the wound bed blindly without real-time feedback. During conventional margin analysis, the wound bed is sampled randomly to determine the presence of residual disease. Using fluorescence guidance, areas suspicious for cancer can be biopsied or removed for definitive diagnosis by frozen section. The wound bed can then be re-imaged after resection repeatedly to confirm complete resection. Sampling of the wound bed after removal depends on the perceived closest margin and can be difficult to map in most wound cavities that are three-dimensional surfaced. Although gross disease is palpable, the microscopic extensions or skip lesions cannot be well visualized by any current technology in the operating room.

Frozen section analysis cannot adequately sample the entirety of the wound bed. Sampling a large wound bed by frozen section to identify positive margins is the current standard of care in oncologic resection. This methodology leads to sampling error that is exacerbated by wound bed dimension and size. However, this assessment remains critical to determine additional resection and adjuvant treatment. Because of the extensive surface area of many cancer types, or the three-dimensional nature of the wound cavities, sampling is limited to about 5 % of the wound bed. Furthermore, a positive frozen section can be difficult to map back to the wound cavity, and as a result, the process of re-excision often necessitates removal of large areas, leading to overcompensation and loss of normal tissue. Conversely, less aggressive sampling increases the risk of residual positive margins. The current methodology for margin assessment is hindered by multiple probabilities of failure due to human and sampling error (Fig. 3.1).

Components of a Successful Imaging Probe

Early fluorescent contrast agents consisted of small peptides. By their inherent physical structure, these agents contain some property of attenuation or photon emission that is detectable using a probe or camera. These early fluorescent probes used for diagnostic or localized cancer imaging have been single molecular agents, such as aminolevulinic acid (5-ALA) or indocyanine green (ICG), that localize to disease based on disease-specific characteristics rather than qualities inherent to the imaging agent. Early results using these probes have been promising, however the lack of an established standard for optimal fluorescence leaves the scientific community wondering if these probes are ideal.

Toward that end, a variety of molecular probes have been developed for cancer imaging. Variations that have been introduced and under investigation include receptor ligands [14, 15], nanoparticles [16, 17], receptor ligands [18, 19], antibodies/affibodies directed against cell surface molecules [20–23], and self-quenching probes that are activated by peri-tumoral enzymes or subtle changes in PH [24]. Although preclinical data support use of many of these agents, the translational hurdles posed by consistent manufacturing, costs associated with toxicology studies/IND submission, and safety concerns have severely limited the translation of these agents in humans. Importantly, industry is reluctant to support translation of imaging agents into the clinic because revenue associated with diagnostic agents is only a fraction of therapeutic agents with a similar initial investment for FDA approval [25]. Although some optical agents have been evaluated in phase I clinical trials [13], they are not readily applied to other tumor types or carry high costs of adoption.

Antibodies conjugated to fluorescent molecules can provide a viable approach for cancer imaging. FDA-approved antibodies that are repurposed for cancer localization can act as a vector for fluorescence imaging. These actively targeted complexes provide the specificity required to differentiate disease from normal tissue since they are specifically designed to target disease characteristics that are aberrantly expressed relative to normal tissue. In most imaging modalities, successful imaging is not solely dependent upon high tumor accumulation, but the tumor-to-normal ratio of accumulation. This ratio defines the amount of contrast generated between tissues during imaging, which determines the successful use of fluorescence imaging to assist surgical navigation. Antibodies previously approved for cancer therapy target these aberrantly expressed markers, which are innate to cancer but not normal tissue. When combined with fluorescent molecules, cancer-specific probes (Fig. 3.2) produce robust contrast between disease and normal tissue. Importantly, these combinations can be modified to incorporate advanced components as they become available to improve disease visualization.

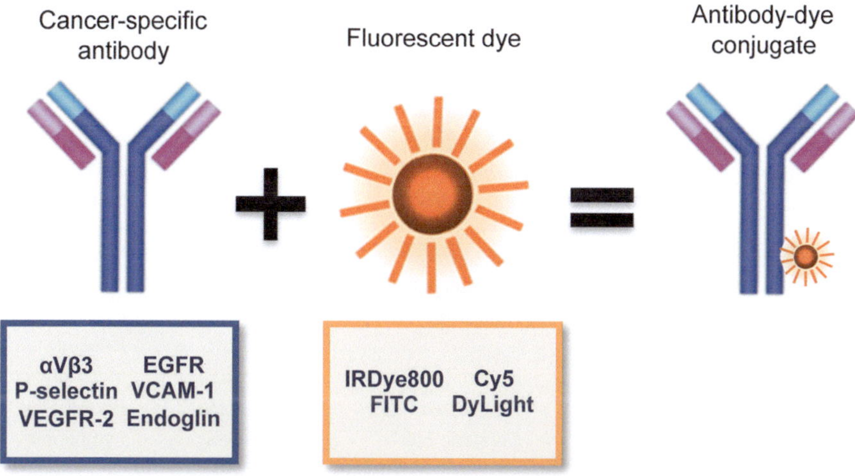

Fig. 3.2 Therapeutic antibodies that target cancer-specific markers can be conjugated to fluorescent molecules to yield a vector for fluorescence imaging of cancer

Cancer-Specific Targeting Using Monoclonal Antibodies

Currently, there are multiple antibodies which are approved for human use by the FDA [26] for the treatment of solid and blood-based cancer (Table 3.1), with many more in various stages of clinical trial. Although radiolabeled and fluorescently labeled antibodies have been used in early phase clinical trials, there is insufficient published data on which antigen will emerge as the optimal imaging target. There are several important properties of an antibody target that would allow accurate imaging of the tumor for detection or for surgical guidance. For example, antigens expressed on the cancer cell surface (EGFR or Her2) would permit greater tumor-specific localization compared to those within the extracellular milieu (VEGF). Realistically, the ideal agent will depend on factors specific to each cancer, but the ideal target would also be broadly applicable to a wide range of cancers. Antibodies that have been assessed in early stage clinical trials for imaging include cetuximab (ClinicalTrials.gov Identifier: NCT01987375) and bevacizumab (ClinicalTrials.gov Identifier: NCT01508572).

An additional consideration is that the fluorescent signal could be enhanced if the prolonged circulating time of the antibody leads to greater antibody–receptor complex internalization. This receptor cycling leads to greater cellular incorporation of the fluorescence molecule over time, leading to greater signal amplification in those tissues containing the targeted receptor, but not in normal tissues. Whole antibody conjugates would therefore have an inherent advantage over antibody fragments or small peptides due to their longer circulatory half-life, which in some cases can be many folds higher. Preliminary in vitro data and murine xenograft models suggest that the absolute expression level of the antigen may not be as important as the cellular incorporation or the permeability of the microvasculature [27, 28]. However, the conventional dogma has held that the targeted antigen itself should be widely expressed on tumor cells but not on normal tissues, should have equal distribution throughout the tumor, and expression levels relatively consistent between patients.

For use in fluorescence-guided surgery, it has commonly been assumed that the ideal imaging vector would allow tumor imaging immediately after systemic administration of the diagnostic

Table 3.1 Antibodies approved for the treatment of cancer

Antibody	Target	Isotype	Approval	Cancer
Alemtuzumab (Campath)	CD52	Humanized IgG1	2001 (US) 2001 (EU)	Chronic lymphocytic leukemia
Bevacizumab (Avastin)	VEGF	Humanized IgG1	2004 (US) 2005 (EU)	Glioblastoma, non-small cell lung cancer
Brentuximab (Adcetris)	CD30	Chimeric IgG1	2011 (US)	Lymphoma
Cetuximab (Erbitux)	EGFR	Chimeric IgG1	2004 (US) 2004 (EU)	Colorectal, head, and neck
Ibritumomab (Zevalin)	CD20	Murine IgG1	2002 (US) 2004 (EU)	Lymphoma
Obinutuzumab (Gazyva)	CD20	Humanized glycol-engineered	2013 (US)	Lymphoma
Ofatumumab (Arzerra)	CD20	Human IgG1	2009 (EU)	Chronic lymphocytic leukemia
Panitumumab (Vectibix)	EGFR	Human IgG2	2006 (US)	Metastatic colorectal
Pertuzumab (Perjeta)	Her2	Humanized IgG1	2012 (US)	Breast
Rituximab (Rituxan)	CD20	Chimeric IgG1	1997 (US) 1998 (EU)	Lymphoma
Tositumumab (Bexxar)	CD20	Murine IgG2a	2003 (US)	Lymphoma
Trastuzumab (Herceptin)	Her2	Humanized IgG1	1998 (US) 2000 (EU)	Breast

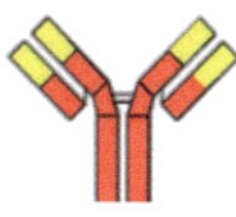

Cetuximab/Anti-EGFR Antibody

- Severe, potentially fatal allergic reactions in < 1%
- Transient infusion reaction during administration

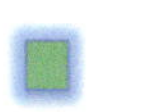

Fluorescent dye (IRDye800)

- Tested in rodents at supra-therapeutic doses without toxicity
- Chemically similar to ICG

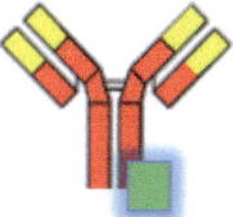

Cetuximab-Dye Conjugate (1:1.5 molar ratio)

- Possibility of infusion reaction
- Development of antibodies with repeated dosing
- Small but significant increase in QTc interval on ECG
- Half-life and pharmacology identical to cetuximab alone

Fig. 3.3 The addition of the fluorescent probe to an antibody is not associated with significant toxicity in non-human primate studies [1]

agent. However, the half-life of whole antibodies is usually measured in days when given as a therapeutic agent over many doses while the pharmacokinetics are non-linear such that when administered as a single dose the half-life is significantly shorter. Therapeutic antibodies administered at diagnostic levels have a half-life around 24 h, which will likely require administration several days before imaging to permit adequate tumor-specific accumulation and subsequent clearance from the circulation. This will produce the greatest amount of fluorescence contrast between tumor and normal tissue. Additionally, animal data of xenograft tumors suggest that the tumor to background ratio continues to increase over weeks as the background decreases [6].

The toxicity of the agent should also drive selection of the appropriate antibody for imaging. Because the acceptable range of toxicity for imaging agents is lower compared to anticancer agents, the antibody should have a limited toxicity profile when delivered in low doses. Several antibody-related toxicities have been reported at therapeutic dosing, especially after repeat dosing, however, the immediate infusion reactions and true anaphylaxis will occur after a limited administration of the drug. Infusion reactions are more commonly associated with chimeric antibodies as opposed to fully human antibodies. Recent studies have shown that at low molar ratios [2] liver clearance of antibodies is largely unchanged, and other studies in non-human primates have demonstrated that the antibody–dye conjugate has the same pharmacokinetics of the antibody alone [1, 2]. An antibody–dye agent currently being investigated in patients is shown in Fig. 3.3.

Selection of a Fluorescent Molecule

Most fluorescent molecules can be linked to an antibody, however the extent of modification will determine the immune response and clearance rates of the bioconjugate. There are a variety of fluorescent probes that can be conjugated to proteins through relatively simple covalent chemistry at consistent molar ratios (Fig. 3.2). Optical dyes are dependent on the properties of light and are therefore limited in depth of tissue penetration. While attenuation and scattering are consequential artifacts, optical imaging affords high resolution when imaged superficially. In the surgical setting, cancer-containing tissues are exposed and overlying tissue, which could attenuate fluorescence signal, is debulked or removed. Fluorescence agents are therefore well suited for the surgical arena, where high-resolution imaging can be applied in real-time. Furthermore, optical probes can be designed to emit fluorescence and toxic reactive oxygen species after light-based activation. This theranostic imaging strategy, referred to as photoimmunotherapy (PIT), can be used to optically resect the tumor and simultaneously treat residual microscopic disease.

There are several factors to consider before selecting the optimal dye for fluorescence-guided

Table 3.2 Fluorescent dyes currently under investigation

Dye	Manufacturer	Human use	Functional group	Available cGMP	Absorption: emission
Cy5/7	GE Healthcare	No	Yes	No	650:760 743:767
IcG	Various	Yes	No	Yes	810:830
Fluorescein (FITC)	Various	Yes	Yes	Yes	495:519
IRDye700/800	LI-COR Biosciences	Yes	Yes	Yes	689:700 773:792

surgical navigation. Examples of fluorescence molecules currently under investigation for use in image-guided surgery are listed in Table 3.2. The labeling procedure must be standardized so that the dye-to-protein (D:P) ratio remains consistent between conjugation procedures. Often, the optimal D:P ratio is discovered from a balance between unaltered pharmacokinetics and robust signal emission for each antibody conjugated. Anther factor to consider is the absorption and emission wavelengths. Excitation and emission wavelengths in the near infrared range (>700 nm) permit greater tissue penetration (1 cm) with less scattering and attenuation. In addition, properties of the fluorescent molecule will determine the appropriate hardware to use for imaging. This has significant impact on clinical translation and also the type of surgical wound bed that can be imaged (laparoscopic versus open field imaging). Certain nanoparticles and other novel substances have significant toxicity or antigenic properties and their application should be clearly justified prior to serious efforts toward clinical translation.

Most fluorescent dyes are easily conjugated to an antibody using a single step, amine-reactive cross-linker chemistry. The imaging probe is covalently linked to the antibody at very low molar ratios to prevent interference with the antigen-binding site and prevent a high rate of hepatic clearance. Furthermore, the imaging efficiency of the diagnostic probe can provide a strong signal even at these relatively low molar ratios. For the purposes of fluorescence-guided imaging, the ideal fluorescence molecule is one that is easily manufactured under cGMP conditions (for clinical use) and can be imaged using conventional intraoperative equipment.

Intraoperative Imaging Hardware

There are a growing number of FDA-approved imaging devices designed to visualize indocyanine green (ICG) fluorescence, including latest generation operative microscopes (Zeiss and Leica) as well as the LUNA and SPY fluorescence angiography systems manufactured by Novadaq [29–31]. Due to significant overlap with excitation sources and emission filters, clinically approved instruments can be repurposed to detect fluorescent dyes outside the specifications of the probe for which the instrument was designed. This approach is conceptualized in Fig. 3.4. An example of this repurposing strategy is demonstrated in a recent clinical trial (ClinicalTrials.gov Identifier: NCT01987375) using an FDA-approved instrument (LUNA), previously approved for intraoperative angiography of ICG perfusion, to image IRDye800-labeled material in the surgical setting. While this strategic approach is efficient and serves to demonstrate proof of principle without the initial investment of a custom instrument design, the pairing of discordant dyes and hardware may produce suboptimal results. For example, the fluorescence emission and absorption spectra of ICG and IRDye800 have some overlap, however the ICG fluorescence imaging platforms operate on the far edge of the fluorescence spectrum for IRDye800 (<10 % IRDye800 spectrum) and are not efficient in exciting the IRDye800 fluorescence. Because ICG is dosed in relatively large quantities for perfusion imaging applications, the fluorescence emission detection parameters of ICG imaging systems are not optimal for detecting very low optical emission that is required to detect a

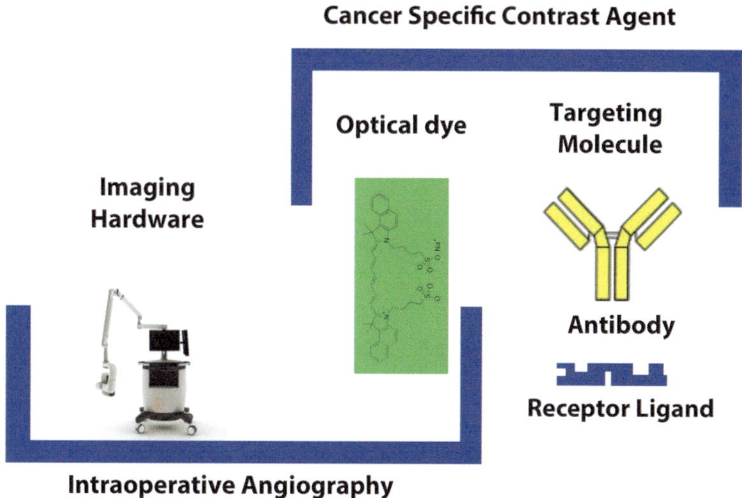

Fig. 3.4 Repurposing intraoperative imaging hardware for targeted cancer detection using fluorescence navigation

microscopic, subclinical disease. Furthermore, dedicated optical systems are currently designed for imaging cutaneous blood flow, and as a result cannot effectively measure body cavities created during oncologic surgery due to a lack of appropriate ergonomic design. Finally, the user adjustable parameters for ICG imaging systems, in particular, the operational software functions, have been designed for the intraoperative dynamics of perfusion imaging and not for imaging a wide dynamic range of possible fluorescence intensities encountered in microscopic disease detection. Although sufficiently effective at high doses in preclinical models, current optical platforms are designed for ICG imaging in milligram quantities, and not for the anticipated micro- and nanogram range of tumor-related uptake. Importantly, ICG is not chemically amendable to antibody labeling. Further, clinical translation will require industry partnerships to achieve platform optimization for effective imaging of subclinical disease in human trials.

Standardization of Probe, Hardware, and Quantification

There are several difficulties that need to be addressed as regulatory bodies, academia, and industry work together to implement antibody-based strategies for fluorescence-guided resection of cancer. In a recent meeting of the European Molecular Imaging Society, specific challenges to the safe and successful implementation of the technique were examined [32]. Notable discussions included issues regarding safety, quality control and manufacturing standards, instrument standardization, and regulatory affairs. The general conclusion of the meeting was that no one individual or group was responsible for solving these challenges. Instead, viable progress toward a solution will occur at the community level through education, awareness, and decisions formed by the consensus.

Another consideration centers on whether the fluorescently labeled contrast agents will be approved as an individual agent or paired with hardware. In the latter case, a specific imaging hardware (brand and model) will be approved for use only with a particular fluorescently labeled targeting probe. Nuclear medicine or computed tomography imaging devices have not required this coupling of the contrast agent and imaging hardware. The rationale for device–probe pairing is that surgeons will be making decisions based on fluorescence intensity, which is subject to a range of variations depending on the tissue properties, distance of measurement, shadowing from the wound cavity, and many other factors that cannot be easily standardized during an operative case. It is possible that clear standards developed for probes or hardware would significantly improve the ability to use these elements interchangeably, which would significantly reduce regulatory hurdles and promote competition in this highly technical field.

Another challenge to the routine use of fluorescence to specifically localize cancer is the subjective nature of fluorescence quantification. To date, pillar publications in the literature demonstrating the value and innovation of the technique use subjective, elementary classifications to describe the amount of fluorescence detected within tissue. For example, commonly reported terms have included "bright," "strong," and "vague" [12, 33, 34]. The literature has yet to introduce a standardized criterion for differentiating disease from normal tissue using cancer-targeted fluorescence imaging. These qualitative generalizations cannot accurately be used to determine specificity of the technique, which is an issue that delays attempts to show clinical benefit. Due to the natural artifacts of attenuation and scattering, combined with the difficulties associated with optimizing dye-instrument combinations, a standardized value of fluorescence signal to determine presence of cancer, similar to the standardized uptake value (SUV) used PET imaging, is a challenge that requires a great deal of investigation before true benefit of the technique can be shown.

Conclusion

There is tremendous potential for antibody-based fluorescence-guided imaging to routinely impact the field of surgical oncology. The use of an antibody to deliver cancer-specific fluorescence has the potential to produce robust optical localization of disease in contrast to normal tissue. The transition from bench to bedside can be streamlined with the incorporation of previously approved therapeutic antibodies, whose safety and pharmacokinetics are readily known. In addition, utilizing intraoperative hardware approved for similar procedures adds additional feasibility without relying on industry development, which can slow advancement of the technique. Before successful translation can occur, challenges remain to optimize dye/hardware interaction and standardize quality control and manufacturing processes. Finally, an imbalance between therapy-based reimbursement and imaging-based reimbursement is detrimental to the financial investment required for successful translation into the clinic. Once clearly defined parameters for measuring benefit to patient and cost are in place, the patient impact can be fully realized and reimbursement guidelines will be modified to reflect this patient benefit.

References

1. Zinn KR, Korb M, Samuel S, Warram JM, Dion D, Killingsworth C, et al. IND-directed safety and biodistribution study of intravenously injected Cetuximab-IRDye800 in Cynomolgus Macaques. Mol Imaging Biol. 2014;17:49–57.
2. Cohen R, Stammes MA, de Roos IH, Stigter-van Walsum M, Visser GW, van Dongen GA. Inert coupling of IRDye800CW to monoclonal antibodies for clinical optical imaging of tumor targets. EJNMMI Res. 2011;1(1):31.
3. Day KE, Beck LN, Heath CH, Huang CC, Zinn KR, Rosenthal EL. Identification of the optimal therapeutic antibody for fluorescent imaging of cutaneous squamous cell carcinoma. Cancer Biol Ther. 2013;14(3):271–7.
4. Heath CH, Deep NL, Beck LN, Day KE, Sweeny L, Zinn KR, et al. Use of panitumumab-IRDye800 to image cutaneous head and neck cancer in mice. Otolaryngol Head Neck Surg. 2013;148(6):982–90.
5. Heath CH, Deep NL, Sweeny L, Zinn KR, Rosenthal EL. Use of panitumumab-IRDye800 to image microscopic head and neck cancer in an orthotopic surgical model. Ann Surg Oncol. 2012;19(12):3879–87.
6. Korb ML, Hartman YE, Kovar J, Zinn KR, Bland KI, Rosenthal EL. Use of monoclonal antibody-IRDye800CW bioconjugates in the resection of breast cancer. J Surg Res. 2014;188(1):119–28.
7. Messali A, Villacorta R, Hay JW. A review of the economic burden of glioblastoma and the cost effectiveness of pharmacologic treatments. Pharmacoeconomics. 2014;32:1201–12.
8. Liu JT, Meza D, Sanai N. Trends in fluorescence image-guided surgery for gliomas. Neurosurgery. 2014;75(1):61–71.
9. McClain SE, Mayo KB, Shada AL, Smolkin ME, Patterson JW, Slingluff Jr CL. Amelanotic melanomas presenting as red skin lesions: a diagnostic challenge with potentially lethal consequences. Int J Dermatol. 2012;51(4):420–6.
10. Rosenthal EL, Kulbersh BD, Duncan RD, Zhang W, Magnuson JS, Carroll WR, et al. In vivo detection of

head and neck cancer orthotopic xenografts by immunofluorescence. Laryngoscope. 2006;116(9):1636–41.

11. Rosenthal EL, Kulbersh BD, King T, Chaudhuri TR, Zinn KR. Use of fluorescent labeled anti-epidermal growth factor receptor antibody to image head and neck squamous cell carcinoma xenografts. Mol Cancer Ther. 2007;6(4):1230–8.
12. Stummer W, Pichlmeier U, Meinel T, Wiestler OD, Zanella F, Reulen HJ, et al. Fluorescence-guided surgery with 5-aminolevulinic acid for resection of malignant glioma: a randomised controlled multicentre phase III trial. Lancet Oncol. 2006;7(5):392–401.
13. van Dam GM, Themelis G, Crane LM, Harlaar NJ, Pleijhuis RG, Kelder W, et al. Intraoperative tumor-specific fluorescence imaging in ovarian cancer by folate receptor-alpha targeting: first in-human results. Nat Med. 2011;17(10):1315–9.
14. Bailey KM, Wojtkowiak JW, Hashim AI, Gillies RJ. Targeting the metabolic microenvironment of tumors. Adv Pharmacol. 2012;65:63–107.
15. Zhan X, Lin Y, Gillies RJ. Tumor pH and its measurement. J Nucl Med. 2010;51(8):1167–70.
16. Gindy ME, Prud'homme RK. Multifunctional nanoparticles for imaging, delivery and targeting in cancer therapy. Expert Opin Drug Deliv. 2009;6(8):865–78.
17. Jiang S, Gnanasammandhan MK, Zhang Y. Optical imaging-guided cancer therapy with fluorescent nanoparticles. J R Soc Interface. 2010;7(42):3–18.
18. Adams KE, Ke S, Kwon S, Liang F, Fan Z, Lu Y, et al. Comparison of visible and near-infrared wavelength-excitable fluorescent dyes for molecular imaging of cancer. J Biomed Opt. 2007;12(2):024017.
19. Kramer-Marek G, Longmire MR, Choyke PL, Kobayashi H. Recent advances in optical cancer imaging of EGF receptors. Curr Med Chem. 2012;19(28):4759–66.
20. Andrades P, Bohannon IA, Baranano CF, Wax MK, Rosenthal E. Indications and outcomes of double free flaps in head and neck reconstruction. Microsurgery. 2009;29(3):171–7.
21. Newman JR, Gleysteen JP, Barañano CF, Bremser JR, Zhang W, Zinn KR, et al. Stereomicroscopic fluorescence imaging of head and neck cancer xenografts targeting CD147. Cancer Biol Ther. 2008;7(7):1063–70.
22. Shah N, Zhai G, Knowles JA, Stockard CR, Grizzle WE, Fineberg N, et al. (18)F-FDG PET/CT imaging detects therapy efficacy of anti-EMMPRIN antibody and gemcitabine in orthotopic pancreatic tumor xenografts. Mol Imaging Biol. 2012;14(2):237–44.
23. Withrow KP, Newman JR, Skipper JB, Gleysteen JP, Magnuson JS, Zinn K, et al. Assessment of bevacizumab conjugated to Cy5.5 for detection of head and neck cancer xenografts. Technol Cancer Res Treat. 2008;7(1):61–6.
24. Weissleder R. Molecular imaging in cancer. Science. 2006;312(5777):1168–71.
25. Agdeppa ED, Spilker ME. A review of imaging agent development. AAPS J. 2009;11(2):286–99.
26. Sliwkowski MX, Mellman I. Antibody therapeutics in cancer. Science. 2013;341(6151):1192–8.
27. Helman EE, Newman JR, Dean NR, Zhang W, Zinn KR, Rosenthal EL. Optical imaging predicts tumor response to anti-EGFR therapy. Cancer Biol Ther. 2010;10(2):166–71.
28. Netti PA, Roberge S, Boucher Y, Baxter LT, Jain RK. Effect of transvascular fluid exchange on pressure-flow relationship in tumors: a proposed mechanism for tumor blood flow heterogeneity. Microvasc Res. 1996;52(1):27–46.
29. van der Vorst JR, Schaafsma BE, Verbeek FP, Keereweer S, Jansen JC, van der Velden LA, et al. Near-infrared fluorescence sentinel lymph node mapping of the oral cavity in head and neck cancer patients. Oral Oncol. 2013;49(1):15–9.
30. Tanaka E, Choi HS, Humblet V, Ohnishi S, Laurence RG, Frangioni JV. Real-time intraoperative assessment of the extrahepatic bile ducts in rats and pigs using invisible near-infrared fluorescent light. Surgery. 2008;144(1):39–48.
31. Tanaka E, Ohnishi S, Laurence RG, Choi HS, Humblet V, Frangioni JV. Real-time intraoperative ureteral guidance using invisible near-infrared fluorescence. J Urol. 2007;178(5):2197–202.
32. Snoeks TJ, van Driel PB, Keereweer S, Aime S, Brindle KM, van Dam GM, et al. Towards a successful clinical implementation of fluorescence-guided surgery. Mol Imaging Biol. 2014;16(2):147–51.
33. Millesi M, Kiesel B, Woehrer A, Hainfellner JA, Novak K, Martínez-Moreno M, et al. Analysis of 5-aminolevulinic acid-induced fluorescence in 55 different spinal tumors. Neurosurg Focus. 2014;36(2):E11.
34. Stummer W, Novotny A, Stepp H, Goetz C, Bise K, Reulen HJ. Fluorescence-guided resection of glioblastoma multiforme by using 5-aminolevulinic acid-induced porphyrins: a prospective study in 52 consecutive patients. J Neurosurg. 2000;93(6):1003–13.

Review of Indocyanine Green Imaging in Surgery

4

Jarmo T. Alander, Outi M. Villet, Tommi Pätilä, Ilkka S. Kaartinen, Martin Lehecka, Toshiya Nakaguchi, Taku Suzuki, and Valery Tuchin

Abbreviations

AVM	Arteriovenous malformation
dAVM	Dural AVM
CABG	Coronary artery bypass grafting
CT	Computed tomography
DNIF	Dynamic near-infrared fluorescence (imaging)
DSA	Digital subtraction angiography
ICG	Indocyanine green
ICGA	ICG angiography
ICGI	ICG imaging
ICG-R15	ICG retention rate at 15 min
LED	Light-emitting diode
LP-ICG	Liposome-encapsulated ICG
LSCI	Laser speckle contrast imaging
M-LP-ICG	Mannosylated liposome-encapsulated ICG
NIR	Near-infrared
PA	Photoacoustic
PC	Personal computer
PCA	Principle component analysis
PL-PEG	Phospholipid-polyethylene glycol
SLN	Sentinel lymph node
SWNT	Single-walled carbon nanotube
TTFM	Transit-time flowmetry
USB	Universal serial bus

Introduction

Indocyanine green (ICG) is a nontoxic near-infrared fluorescent dye that has been used in medical diagnostics since late 1950s. It was first used clinically to evaluate liver function and later in cardiology.

J.T. Alander, Ph.D. (✉)
Department of Electrical and Energy Engineering, University of Vaasa, Yliopistonranta 10, 65101 Vaasa, Finland
e-mail: jarmo.alander@uva.fi; jal@uwasa.fi

O.M. Villet, Ph.D.
Heart and Lung Center, HUCH, Helsinki, Finland

T. Pätilä, M.D., Ph.D.
Hospital for Children and Adolescents, Helsinki University Central Hospital, Helsinki, Finland

I.S. Kaartinen, M.D., Ph.D.
Department of Plastic Surgery, Tampere University Hospital, Tampere, Finland

M. Lehecka, M.D., Ph.D.
Department of Neurosurgery, Helsinki University Central Hospital, Helsinki, Finland

T. Nakaguchi, Ph.D.
Center for Frontier Medical Engineering, Chiba University, Chiba, Japan

T. Suzuki, Ph.D.
Graduate School of Engineering, Chiba University, Chiba, Japan

V. Tuchin, Ph.D.
Optoelectronics and Measurement Techniques Laboratory, University of Oulu, Oulu, Finland

Department of Optics and Biophotonics, Saratov State University, Saratov, Russia

Institute of Precise Mechanics and Control of the RAS, Saratov, Russia

F.D. Dip et al. (eds.), *Fluorescence Imaging for Surgeons: Concepts and Applications*, DOI 10.1007/978-3-319-15678-1_4,

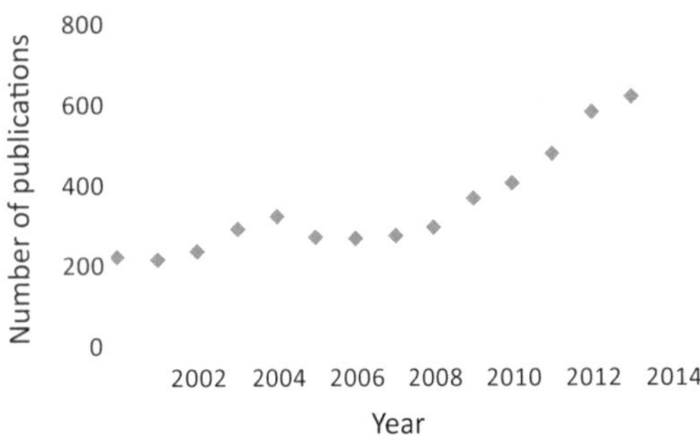

Fig. 4.1 The annual number of ICG publications according to PubMed (27.3.2014; "indocyanine green")

In ophthalmology, ICG angiography has been a standard measurement tool for the retinal and choroid vascularization for more than 20 years [1]. Intravenously injected ICG binds tightly to plasma proteins and becomes confined to the vascular system. The binding and spectral properties of ICG have enabled the development of new imaging systems for several surgical specialities. The introduction of new clinical applications has been especially rapid during the last few years (Fig. 4.1). References having "indocyanine green" in any field and published between 1st January 2012 and 11th March 2014 were collected from PubMed. In total 1,183 references were found. Naturally most of the references, nearly 400, deal with ophthalmological applications of ICG. Likewise the next most popular topic is related to liver and bile, their function, surgery, and oncology, about 150 references. The third most popular topic belongs to new application areas of ICG, namely to the use of ICG in the imaging of lymph nodes and lymphatic system in general. Next one is ICGA in neurosurgery, with about 100 references.

Here ICG fluorescence imaging is reviewed in the areas of plastic and reconstructive surgery, neurosurgery, cardiac, oncological, vascular, and hepatic surgery for the years 2012 and 2013 since our previous review of ICG imaging in surgery [2]. The large area of ICG imaging and angiography in ophthalmological applications has been excluded in this review. In addition to clinical applications, some technical aspects and potential future developments of ICG-based fluorescent imaging are discussed.

Diagnostic Applications

Liver function is assessed with several biochemical and other tests to detect the presence of liver disease, to evaluate the extent of liver damage and to follow the response to treatment. Especially preoperative assessment of liver function can help with the decision of the extent of a hepatic resection. These evaluations are important in preventing postoperative liver failure and mortality after hepatic resection.

Hoekstra et al. reviewed several clinically used liver function tests and concluded that due to the complexity of liver function, one single test does not stand for overall liver function [3]. Nevertheless, studies have shown that the ICG clearance test is a sensitive indicator to gauge liver function [4, 5]. After intravenous injection ICG bound in plasma proteins is taken up by hepatocytes and excreted in bile via an active transport system. Since ICG is not metabolized, its disappearance rate from plasma to bile reflects the excretory function of the liver [6]. ICG retention rate measured at 15 min (ICG-R15) has been reported to act as a significant predictor for postoperative outcome [7, 8]. In healthy liver ICG-R15 is usually less than 10 %, whereas 20 % is generally considered as a safety limit for extended hepatectomy in chronic liver disease [9, 10]. To aid the decision making process in surgical treatment of diseased liver, Makuuchi and Sano established an algorithm based on the following three parameters: the presence or absence of ascites, total bilirubin level, and the ICG-R15 [11]. Operability is assessed by the presence or absence of ascites and total bilirubin level, whereas the maximal resectable liver volume is assessed by ICG-R15. Though ICG-R15 test is very informative and widely accepted method, discrepancies still exist between ICG measurements and conventional liver function tests. Conditions such as hyperbilirubinemia and large port-systemic shunts interfere with the results of ICG tests [12, 13]. Additionally, a few patients are intolerant of ICG [14].

Orthotopic liver transplantations are often followed by early postoperative complications. Inconsistencies exist in graft evaluation results making the initial graft function problematic to determine. ICG plasma disappearance rate has been shown to predict absence of early postoperative complications, better and earlier than clinically used laboratory parameters [15, 16]; whereas, a lack of prognostic value to predict individual patient outcomes has also been shown [17].

ICG dynamics studies by Kang et al. suggest the possibility of ICG imaging for reliable diagnosis of Raynaud phenomenon [18]. A potential use of the ICG perfusion imaging as an effective evaluation tool to validate the vasoactive effect of acupuncture was also shown for normal healthy volunteers [19].

Surgical Applications

The binding and spectral properties of ICG have enabled the development of intraoperational imaging systems for several surgical specialities.

Reconstructive (Plastic) Surgery

There are three main purposes in reconstructive surgery, where indocyanine green imaging (ICGI) has been used and developed: For assessment of skin flap viability in elective surgery or traumatic skin lacerations; for intraoperative patency testing of microvascular anastomosis; and for postoperative flap monitoring.

Initially, ICGI was used for assessing viability of skin flaps formed by traumatic degloving injuries and depth of burn injuries [20, 21]. Although this method showed potential, it never gained wide popularity among clinicians. However, breast surgeons performing skin sparing mastectomies have adopted ICGI for evaluating viability of breast skin envelope before reconstruction [22–25]. Intraoperative use of ICGI might lead to decreased complication rates associated with skin necrosis [25].

Similarly, ICGI can be used to assess the perfusion of large cutaneous flaps, both microvascular and pedicled, used in reconstructive surgery. Using ICGI could reduce complications associated with partial flap necrosis due to insufficient perfusion to distal parts of the flap [26–35]. When combined with operational microscope, ICGA can be used to detect anastomotic failure after microvascular tissue transplantation [36]. The technical success of the anastomosis can easily be evaluated with ICGA through the operational microscope, and furthermore, the intristic transit time can be calculated [37]. This may prevent complications associated with anastomotic failure, which is a common cause of failures in reconstructive microsurgery.

After transplantation, ICGI can be used for flap monitoring. However, ICGI is more invasive than other commonly used methods and cannot be done real time. Instead, ICGI is more feasible in situations, where failure of the microvascular anastomoses is suspected based on clinical findings. When combined with endoscopy, ICGI can also be used to monitor flaps that are inside the oral cavity or in the upper digestive tract, and would otherwise be difficult to monitor [30, 38, 39].

Sentinel Lymph Node Biopsy

Sentinel lymph node (SLN) biopsy is a standard method for lymph node staging in breast cancer surgery. This avoids axillary lymphadenectomy in many patients, reducing associated morbidity. Technetium 99 m (^{99m}Tc) radiotracer is the most commonly used substance for locating the SLN(s) [40]. The isotope is injected next to the primary tumor, and drains through lymphatic vessels to the regional lymph nodes. The radiotracer is found using a gamma detector probe intraoperatively. Recently, ICGI has been used to detect SLNs in breast cancer patients. According to the first studies, ICGI is able to find the same lymph nodes as conventional radiotracer [40–51]. Also, ICG has been combined with ^{99m}Tc as a hybrid radiotracer, where it can be used to ease the detection of SLN, and to replace blue dye [52]. This is a promising aspect, since ^{99m}Tc is still believed to be a more reliable tracer, especially for deep sitting SLNs and in obese patients [47].

Skin Cancer Surgery

As well as in breast cancer surgery, SLN biopsy is used extensively in surgical treatment of skin malignancies. Melanoma and other forms of skin cancer, such as squamous cell carcinoma, some forms of basal cell carcinomas, and several other carcinomas can metastasize to regional lymph nodes. As well as in breast cancer, ^{99m}Tc has been the standard tracer. However, using ICGI for locating SLNs in skin cancer has some benefits. The fluorescence of ICG travelling from the injection site to the SLN can be traced real time, possibly enabling identification of SLNs in less common areas such as the cervical, popliteal, or parascapular areas [53]. ICGI is performed intra-operatively, and saves the patient and the staff from radiation associated with a radiotracer. The detection rate of SLNs using ICGI seems to be similar to using conventional tracers such as ^{99m}Tc, and the cost of ICG is also lower [54–58].

There are publications reporting the use of ICGI for SLN detection in vulvar carcinoma, cervical carcinoma, endometria carcinoma, and especially in head and neck carcinomas [59–65]. Yokoyama et al. also studied the optimal time between ICG injection and imaging, which seems to be in the interval from 30 min to 2 h [65]. In addition, they showed that ICG imaging is suitable for safe navigation in narrow spaces like the parapharyngeal space [66].

Lymphatics

Lymphastasis is a restriction of lymphatic flow out of an extremity, which may lead to chronic swelling of the affected extremity, lymphoedema. Although sometimes idiopathic, lymphastasis is most often secondary to surgical removal of lymphnodes in cancer surgery. Until recently, treatment of lymphoedema has mostly been conservative, but recent advancements in microsurgical techniques have made surgical treatment of lymphoedema possible.

Recently, ICGI has been used as a diagnostic tool to detect lymphastasis, both primary and secondary, where it seems to give equivalent results than lymphoscintigraphy, but with less invasiveness [67–71]. The findings of ICGI in lymphoedema patients also correlate with anatomical findings during surgery [72]. Early diagnosis of lymphastasis is valuable, because conservative treatment can be initiated already before onset of lymphoedema [73, 74].

An increasingly popular operation to treat upper and lower limb lymphoedema is connecting lymphatic vessels to subdermal veins (lymphaticovenous anastomosis). ICGI has become an integral part of planning and executing these procedures; ICGI is used to detect suitable lymphatics in the affected area before and during the operation [75–77]. The success of anastomosis is easy to assess by intraoperative ICGI through the operational microscope [78].

Neurosurgery

Over the last 10 years, intra-operative ICG angiography has become a standard tool for neurosurgeons operating on various vascular lesions of the brain. Microneurosurgical operations of the brain and spinal cord are an ideal area for the use of ICG in general due to three boundary conditions which are met: (1) All the surgeries are performed under the magnification of an operating microscope which can be fitted with all the necessary camera and filter systems; (2) The arteries and veins on the surface of the brain are exposed during the procedure and thus they can be seen directly by visual means; and (3) The ischemic time for brain tissue is very short so any method for providing information about the blood flow inside the cerebral vessels needs to be fast and easy to use.

Already in 1994 Wrobel et al. started to use the fluorescein for the intraoperative assessment of arterial branches and their patency during intracranial aneurysm surgery [79]. They worked with a complex and inconvenient camera and filter system which prevented the technique to become widespread at that time. In 2003 Raabe et al. published their first experience with the use of ICG video angiography in patients with various vascular lesions of the central nervous system [80], and later in 2005 the same group reported the use of ICG angiography integrated into a surgical microscope (Carl Zeiss Co., Oberkochen, Germany) [81]. A microscope-integrated light source containing infrared excitation light illuminated the operating field. The ICG dye was injected intravenously into the patient, and intravascular fluorescence from within the blood vessels was imaged using a video camera attached to the microscope. In 2005–2007 the ICG video angiography became commercially available and various groups around the world started to use it in clinical practice.

Before ICG angiography, the only rapid method to evaluate flow inside intracranial arteries was using a micro-Doppler flow probe which, unfortunately, is a rather unreliable method producing false-positive and false-negative results frequently and cannot be used for evaluating very small perforating vessels. On the other hand, intra-arterial digital subtraction angiography (DSA), the gold standard of brain vessel imaging, is a much more time-consuming method requiring significant infrastructure as well as experience to be readily available during neurosurgical operation. The introduction of ICG video angiography changed the game completely. Suddenly, there was a technique which was fast and easy to use with reliability close to that of DSA [82]. Nowadays, ICG video angiography is considered a must in centers performing high-quality microneurosurgery.

The most important clinical use for ICG video angiography in neurosurgery is in microsurgical treatment of different arterial and venous malformations, some congenital other acquired. ICG angiography has been mainly useful in: (a) evaluating intracranial aneurysm occlusion as well as patency of the surrounding branches [81, 82]; (b) treatment of brain arteriovenous malformation (AVMs) where ICG can be used to distinguish between arterial feeders and arterialized veins as well as for indirect evaluation of whether the whole AVM nidus has been removed [83]; (c) treatment of both cranial and spinal dural arteriovenous fistulas (dAVFs) for intraoperative identification of the fistulous region [84]; and (d) cerebral bypass surgery both for evaluation of anastomosis patency as well as for identification of recipient arteries [85]. In addition, the ICG video angiography has been also used for some less frequent purposes such as planning of approach to cortical tumors, identification and manipulation of important veins in their vicinity, and extracranial carotid and vertebral artery surgery.

Lately, a new fluorescent module has been introduced in neurosurgery, the Yellow 560 (Carl Zeiss, Germany). It visualizes fluorescent dyes in the wavelength range from 540 to 690 nm. It allows for highlighting the fluorescence-stained structures (e.g. with sodium fluoresceine) while viewing non-stained tissue in lifelike color. There is only limited experience with this application but main focus has been on intraoperative delineation of highly vascularized tissue of malignant brain tumors to allow for better distinction from the unaffected surrounding brain [86]. In general, ICG video angiography is an integral tool in modern vascular neurosurgery, but it has to be coupled with other imaging techniques for optimal visualization.

Adult Cardiac Surgery

At the moment, fluorescence imaging in cardiac surgery is an emerging technique for quality assessment namely in coronary artery bypass grafting (CABG). It possesses an opportunity to systemically improve the quality of the operational outcome. However, the outcome of CABG is already good, the 30-day survival rate being more than 99 % in relatively low-risk patients [87]. Thus the impact of fluorescence imaging in cardiac surgery is limited, but there clearly is a technical niche for its use. This suggestion is based on the fact that immediate graft failure occurs in up to 4 % of grafts, which relates to 8 % of the operated patients. At the time of the discharge after the operation, the graft occlusion rate is reported to be as high as 5–20 %. At 1 year after the operation, up to 30 % of the grafts are occluded. Some of these occlusions have no impact to heart function or long-term survival, but some probably will. There are several possible technical problems emerging in CABG. The most commonly used intraoperative method for graft functioning is transit-time flowmetry (TTFM). TTFM measures the mean flow of the bypass graft and calculates an index of the amount of diastolic flow relatively to systolic flow. The reliability of the TTFM is good but there are no standard values and in several patients it might lead to unnecessary graft revision [88].

The direct imaging of coronary arteries includes the location of the coronary arteries and anastomotic evaluation. Intraoperative ICG imaging is especially useful in the location of coronary vessels in coronary artery re-operations, but it would be helpful in any case, when difficulties are encountered in locating a desired vessel. It might also be useful to determine the location of the correct anastomosis site. Fluorescence imaging has inherent limitations of its use, which in cardiac surgery is mainly the low penetration of the light in tissues. This limits the use of the imaging method on the topical parts of the heart. The location of the conduit parts of coronary vessels is situated on top of the heart muscle. This is also the area, where the coronary heart disease is developed and eventually the area, where the anastomoses of the grafts are sutured in CABG operations. This location is optimal for fluorescent imaging. The evaluation of the anastomotic anatomy and function preoperatively is useful. However, direct evaluation of anastomoses requires a high-quality imaging device and maybe even from simultaneous angles to evaluate the technical quality of the anastomoses. This kind of device is not available in the market at the time being. The anastomoses evaluation with the devices in the market is more indirect and is estimated by the filling of the distal vessel.

Clinical studies of intraoperative direct imaging of the grafts by ICG angiography are increasing slowly but steadily. Twenty patients were assessed intraoperatively by ICG angiography by Rubens et al. in one of the earliest clinical studies and one patient was reported to eventually have a graft revision according to the imaging, which resembles 5 % [89]. Taggart et al. reported a study of 213 grafts with a revision rate of 4 grafts (1.9 %) due to findings in ICG angiography [90]. Reuthebuch and colleagues studied use of fluorescence imaging in 107 patients and reported a graft revision rate of 3.7 % [91]. In a larger study, Balacumaraswami et al. assessed 533 conduits in 200 patients. ICG imaging confirmed technical failure in 8 (1.5 %) conduits in 8 (4 %) patients, which eventually lead to graft revision [92]. Takahashi et al. studied 290 grafts, in which four grafts (1.9 %) were revised due to ICG angiographic findings [93]. Desai et al. reported that 348 grafts were studied by intraoperative ICG angiograms [94]. Patients who had their graft revised were 4.2 %. The evidence in these studies is rather constant. ICG imaging leading to graft revisions that would have otherwise gone unrecognized is roughly between 2 and 4 %. These numbers refer practically to the technical failures met in the operations. These graft revisions might have a long-term effect to the patients involved.

ICG imaging offers also a method for indirect measuring of coronary perfusion imaging. It can be assessed during the operation either before the coronary artery grafting or after the coronary artery grafting. Myocardial tissue perfusion is an indirect method to evaluate the coronary flow. Region of interest is the perfusion area of the measured coronary artery [95, 96]. In a study by Ferguson et al. in 2013, 167 patients underwent CABG with traditional angiography-based revascularization [97]. Coronary perfusion imaging was performed both before and after the bypass grafting and 359 grafts (53 % arterial) were assessed. After grafting, 24 % of the arterial and 22 % of the saphenous vein grafts showed no regional myocardial perfusion change in response to bypass grafting. In 165 in situ internal mammary artery grafts, 40 had no change in regional myocardial perfusion, and 32 out of the 40 were estimated to have competitive flow. This study showed that the anatomic evaluation of stenoses might not be enough and these physiological responses are possible to image with fluorescence perfusion method.

Pediatric Cardiac Surgery

In congenital heart surgery, the use of fluorescence imaging is rare. However, there is a slowly increasing amount of papers around the subject. Most of the heart operations in children are corrections within the heart itself, where the low penetration of the near-infrared light naturally inhibits use of fluorescence imaging. However, when there is an increasing tendency to correct the malformations in an early infancy and the structures are small, there certainly is a place for its use especially in certain defects. On the other hand, imaging of vascular structures including coronary arteries is feasible in children.

In a report of Kogon et al., 30 patients were undergoing the following repairs: (1) coronary artery reimplantation, (2) coarctation, (3) palliative shunts, and (4) pulmonary artery reconstruction [98]. After the repairs, imaging of the hearts was performed in the operating room. In these patients, adequate image quality was reached in 18 of 30 (60 %) patients. Image adequacy was highest for Blalock-Taussig shunts (100 %), coarctation repairs (86 %), and coronary reimplantations (66 %). The lowest quality was reached for the hemi-Fontan (0 %), Fontan (40 %), and pulmonary artery reconstructions (33 %). The authors report several reasons for the unsuccessful image quality, including small incision with difficult exposure, the overlying structures on top of the region of interest, indocyanine dosing and extravasation and the low penetration of the prosthetic material (PTFE) used in the corrections. In this study, the imaging leads to no corrections or changes to the planned surgical procedure. There were no adverse events related to the imaging. The authors concluded that indocyanine green fluorescence imaging may provide an important intraoperative imaging modality for congenital heart surgery. The method shows successful intraoperative images of the target area, which correlate to echocardiograms and cardiac cineangiograms.

Peripheral Arteriovenous Surgery

Even though the amount of scientific studies is limited in the area, ICGI has been used in peripheral arteriovenous surgery. Igari et al. propose a method for the quantitative measurement for assessing peripheral perfusion. According to their study comprising of 21 patients with peripheral arterial disease (PAD) the time to half maximum intensity ($T_{1/2}$) was the strongest parameter in their parameter set. They conclude that the outcomes of revascularization procedures can be assessed by the value $T_{1/2}$ at the distal region of the first metatarsal bone [99, 100]. Also Terasaki et al. used $T_{1/2}$ for assessing the severity of ischemia [101]. In addition to intraoperational use in peripheral arteriovenous surgery, ICG has also been used in the laser therapy of telangiectatic leg veins [102].

Liver Surgery

In oncologic liver surgery, ICG has been used in preoperative assessment as described in Chap. 2. Recently, Asian researchers have developed imaging techniques for ICG fluorescence, enabling visualization of bile ducts [103–105] and hepatic cancer during liver surgery [106, 107]. Lim et al. have very recently presented an excellent review of ICG imaging in surgical procedures of liver cancers [108].

The use of ICG fluorescence imaging has been studied in resection of colorectal liver metastases. Van der Vorst et al. concluded that the optimal minimal dose is 10 mg of ICG administered 24 h prior to the resection [49]. Compared to preoperative CT scan, intraoperative ultrasound, and palpation, ICG-based method is able to better found and discern small superficially located metastases [49]. This ability is based on the architecture of ICG fluorescing tissue, where ICG is actually surrounding the metastases and looks like a ring in the side view of the lesion. The reason for this ICG accumulation is unclear. It is assumed that the immature hepatocytes of the tumor have impaired transport of ICG. In any case, the spatial shape of the fluorescence is a clear indication of metastases and could be used for further image processing to increase the sensitivity of this approach.

Tanaka et al. compared the ability of ICG fluorescence navigation in detection of liver malignancies between cirrhotic and non-cirrhotic livers [109]. According to their study, ICG imaging was significantly better in detection of malignant liver tumors in non-cirrhotic livers. In any case, the positive prediction value of ICG fluorescence was small for pathologically diagnosed tumors. Obviously ICG is quite good in detection of near surface located and large enough tumors, but it seems that more advanced imaging is needed for further tumor classification.

Yokoyama et al. have used ICG imaging for detection of hepatic micrometastases from pancreatic cancer using the method developed by Ishizawa et al. [106] and Aoki et al. [110] for identifying small hepatic tumors during surgery [111]. In this method, the intravenous ICG injection is done on the day before surgery. Some of the ICG leaks from metastases thus revealing them while the ICG in blood circulation has already been excreted out of the body. Similarly, Ishizuka et al. have successfully detected small liver metastasis from colorectal cancer [112].

Imaging Engineering

The basic ICG fluorescence imaging system is small and easy to construct [113]. Replacing a standard computer monitor by goggles may provide a really useful aid for clinical work [114, 115]. Another display mode is to project the fluorescence image on the patient to aid navigation [116]. Optical devices already used in surgery such as endoscopes provide a tempting framework for further ICG imaging enhancements [117]. ICG can also be used as fiducial marker for aiding navigation in endoscopic or laparoscopic operations [118].

Towle et al. have compared ICG and laser speckle contrast imaging (LSCI) for the assessment of vasculature perfusion using a rodent model. Their conclusion is that the methods complement each other and LSCI provides accurate and continuous measurement of blood flow without the need of an external contrast agent. Both methods used same laser diode at 785 nm [119]. In some cases ICG imaging can compete or even supercede traditional imaging like CT as Ohba et al. have shown for intra-arterial chemotherapy for oral cancer, where ICGA was used to find the tumor supplying arteries [120, 121].

Resolution is not considerably reduced with the longer than visible light wavelength used in ICG fluorescence imaging. Therefore, microscopy or microendoscopy is possible with ICG up to resolutions that are in theory about half of that of traditional visual microscopy [122, 123]. Actually even super-resolution microscopy is possible with ICG as with other fluorescence dyes [124]. On the other extreme case, even whole-body ICG fluorescence imaging has been demonstrated in humans [125].

Lasers having wavelength around 800 nm can be used as monochromatic light sources in ICG fluorescence imaging. Another use of lasers with ICG is more active: radiation absorption can be used, e.g. in vascular repair. Using a rabbit model, Esposito et al. have shown how to repair blood veins by ICG-infused chitosan patches using laser tissue soldering. The laser was a AlGaAs diode laser having 810 nm wavelength [126]. Similar diode lasers have been used for ICG-assisted photothermal/photodynamic fat cell destruction [127–130]. The data obtained are useful for safe layer-by-layer dosimetry of laser illumination of ICG-stained adipose tissue for treatment of obesity and cellulite and support the hypothesis that photodynamic and photothermal effects induce fat cell lipolysis during tenths of minutes after treatment. Klein et al. have tested ICG in the laser therapy of cutaneous capillary malformations. The rapid clearance of ICG and other difficulties such as having a proper pulse length in controlling the laser-induced coagulation process of blood veins still remains to be solved in order to make this method more competitive with the alternative therapies. The diode laser had 808 nm wavelength [131].

A multichannel time-resolved optical monitoring system for measurements of diffuse reflectance allows for assessment of inflow and washout of the ICG in human brain tissue up to very large depths corresponding to source-detector separation up to 9 cm at intravenous dye injection [132]. Photoacoustic (PA) in vivo imaging of dynamic biological processes at the molecular level resulting from selective absorption of light at multiple wavelengths can now "hear color" including ICG dye kinetics in three dimensions from deep living tissues at high spatial resolution and in real time [133, 134].

Robotic-Assisted Surgery

The lowest threshold to introduce ICG based imaging is in areas of surgery that already routinely use optical devices such as endoscopes, laparoscopes, and microscopes. Integration of these devices with optics needed for ICG imaging is rather easy both technically and economically. Therefore, one of the most natural pioneering areas of ICG fluorescence imaging is robotic-assisted surgery. In addition, this area provides the unique potential opportunity for the development of applications of image data, i.e. machine vision for the automation of routine procedures.

A review of early works (2011–2012) on ICG imaging in robotic-assisted surgery is given by Marano et al. They cite six works using ICG imaging with robotic-assisted surgery. Since then over 40 papers have been published in about 1 year. Thus we can safely say that the interest of using ICG imaging in medical robotic is increasing quickly [135]. Here, we will briefly review some of the most recent and interesting contribution in this field.

Lee et al. have successfully used ICG for the localization of ureteral stenosis for seven patients during robot-assisted ureteroureterostomy [136]. They conclude that ICG administration aids in discerning healthy ureter from diseased tissue.

Manny et al. used ICG fluorescent imaging for robotic partial adrelectomy with good results [137]. They also used ICG to mark prostate tissue and sentinel lymphatic drainage by percutaneous, robot-guided injection. Their conclusion based on a population of 50 patients for radical prostatectomy is that ICG is safe, feasible, and reliable dye for prostate tissue and sentinel lymphatic drainage marking in majority of patients [138].

Harke et al., Borofsky et al., and Bjurlin et al. have used ICGA for aiding the robot-assisted partial nephrectomy, where ICGA helps to clamp only the tumor feeding vascular branches, while preserving the vasculature feeding the healthy kidney tissue [139–141].

Jafari et al. and Hellan et al. have used ICGA to assess optimal spatial perfusion for robot-assisted rectal surgery to reduce the risk of postoperative anastomotic leak. They conclude that ICGA may play a role in anastomotic tissue perfusion assessment and affect the anastomotic leak rate. However, larger prospective studies are needed to further validate this ICGA use [142, 143].

Angell et al. proposed an ICG fluorescence calibration method based on two doses. The first dose is small and used to estimate the optimal dose for tumor location in robot-assisted partial nephrectomy. This type of calibration is worth considering also in many other as well roboticed as manual operations [144].

Wagner et al. have used ICG to identify the contralateral pericardiophrenic neurovascular bundle and further the contralateral phrenic nerve in unilateral robotic thymectomy in order to achieve maximal thymic tissue resection. No injuries to the phrenic nerve occurred in the population of ten patients. They conclude that ICG has the potential to maximize thymic tissue resection, while reducing operative time and nerve injury [145].

Spinoglio et al. have used ICG for cholangiography in robotic cholecystectomy to prevent accidental intraoperational bile duct injury [146].

Rossi et al. have studied the optimal injection site (cervical or hysteroscopic) for detection of sentinel nodes for endometrical cancer. They found cervical injection to be better [64].

Imaging Development

Projective Display

The use of monitors for viewing ICG images and videos in intraoperative work is problematic in practice, because doctors need to mentally compare and fuse the patient and the displayed image, in addition to the obvious ergonomical problems. These can be solved by projector-based system that overlay the ICG image onto a patient. By using this projecting system, medical doctors can observe the ICG fluorescent area directly on the target tissue.

The existing special ICG cameras tend to be rather expensive. In order to make the system's cost lower, the signal is captured by using a normal camera, which is sensitive to infrared light while the signal is enhanced by controlling the excitation light and some image processing.

The structure of the ICG image projecting system is shown in Fig. 4.2. The target surface is illuminated by excitation light using infrared LEDs, and then the fluorescent image is captured by a conventional digital camera which is sensitive to infrared. The system has a demountable filter in front of the camera lens, which cuts the visible light. The excitation light can be switched from a PC through USB interface and the successive images captured by the camera are transferred to the PC. In order to display the captured fluorescent image overlaying onto the target surface, a small size projector is employed. After the camera and projector are geometrically calibrated, the captured image can be modified to registered projection onto the target surface (Fig. 4.2).

Generally, when the ICG fluorescent area is captured by a normal sensitivity camera, the signal is too weak to be detected clearly. Therefore, a method based on light switching and image processing is used. The data flow of the ICG image projecting system is shown in Fig. 4.3. First, a set of the excited and unexcited images is taken serially. The infrared LED as the excitation light is switched on by the PC, and the camera takes two images, P_t with infrared excitation illumination on and Q_{t-1} after turning it off. The switching of LED and the shutter of the camera are synchronized. The frame rate of the camera is 15 frames per second. Here t denotes the time index corresponding to shutter timing. Next, by subtracting the unexcited image Q_{t-1} from the excited image P_t, a differential image D_t is

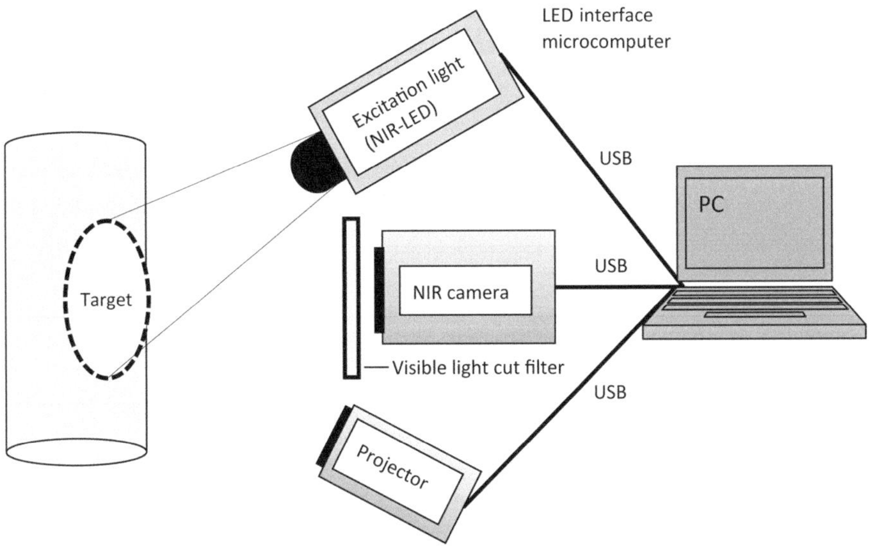

Fig. 4.2 The structure of the ICG image projecting system. The target surface is illuminated by infrared LEDs which can be switched on from the PC. The fluorescent image is captured by a personal-use camera sensitive to infrared. A near-infrared (NIR) filter is set in front of the camera lens

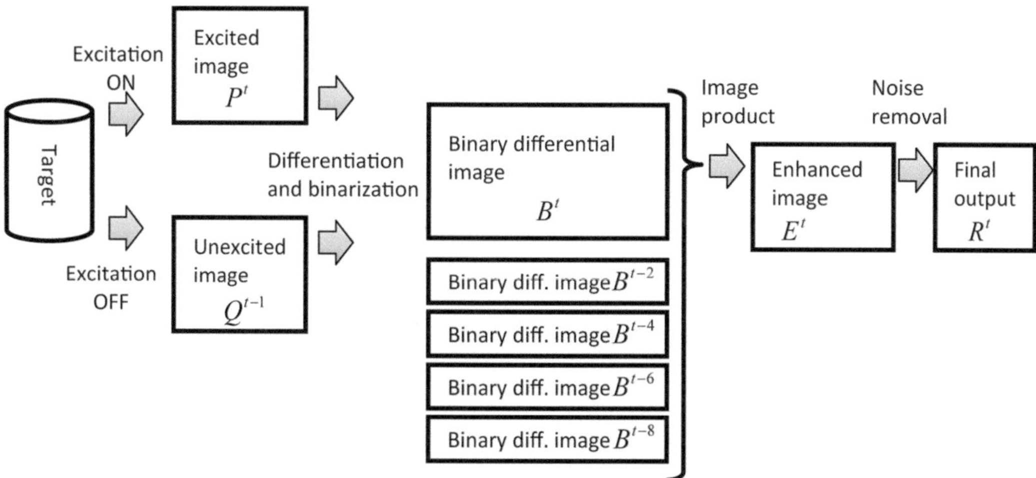

Fig. 4.3 Image data flow of the ICG image projecting system. First, a set of the excited and unexcited images is taken serially, then a binary differential image is calculated. An enhanced image is obtained by taking an image product, then the final output is got by applying noise removal

calculated. Then by thresholding the differential image D_t, a binary differential image B_t is derived. Since B_t is got by subtracting two temporally successive images, some artifacts may also appear. In order to remove these artifacts, five temporally successive images from B_{t-n} to B_t are used. By taking the product of these five images, an enhanced image E_t is obtained. By using this method, most of the random noise can also be removed. To further reduce noise, a 3 × 3 median filter is applied to the enhanced image E_t and then the final image R_t is obtained.

By conducting a porcine experiment, the feasibility of the ICG image projecting system was confirmed (Fig. 4.4). Since this system assumes that the geometry between the system and the target is static, an automatic update of the calibration as a background task need to be developed in order to make this system suitable for practical clinical work.

Indocyanine Green with Nanostructures

There have been several attempts to chemically modify ICG molecules to further improve the imaging quality and specificity. Unfortunately, that approach has not yet offered clinically potential dyes. An alternative way to modify ICG as a dye is to attach it non-covalently to other molecules and nano structures. For example, Kraft and Ho have very recently combined ICG with liposomes with good results with respect to both quantum efficiency and stability [147]. In addition, liposomes offer a further option for better targeting [148]. Toyota et al. have used liposomally formulated ICG derivatives for lymph node imaging in a mice model [149].

ICG can be combined with phospholipid-polyethylene glycol (PL-PEG) (ICG-PL-PEG) by utilizing non-covalent self-assembly chemistry [150, 151]. The dual functionality of ICG-PL-PEG for targeted optical imaging and photothermal therapy of cancer cells was successfully demonstrated. At conjugation with other therapeutic and imaging agents, this technology is prospective for multi-functional probing at cancer diagnosis and treatment. These nanoparticles were also shown to be good contrast agents for photoacoustic (PA) imaging [152] and as agents providing high selectivity of the PA destruction of cancer cells [153].

The feasibility of mapping a SLN and urinary bladder by using ICG conjugated with single-walled carbon nanotubes (SWNTs-ICG) as an effective PA contrasting agent was demonstrated in vivo [154]. Due to the accumulation of the conjugates in tissues, the SLNs and bladder were

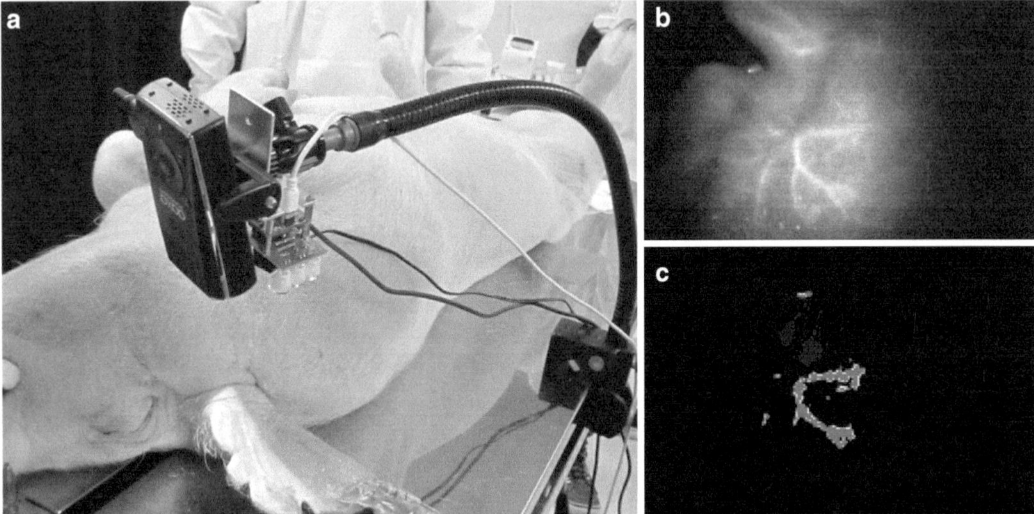

Fig. 4.4 (**a**) Geometrical setup of animal experiment. The feasibility of the ICG image projecting system was validated by conducting porcine experiments. The detection capacity and the feasibility of the system were evaluated in a practical clinical environment. ICG solution was injected into the blood vessel of the porcine ear, then images were captured by the ICG image projecting system. The primary vessels running around porcine ear can be seen clearly. (**b**) Original image taken by the near-infrared camera. (**c**) Resultant image of the ICG fluorescent area detection

clearly seen. The SWNTs–ICG conjugates in combination with PA imaging have good perspectives to be used for identifying SLNs in many types of the above-mentioned cancers and to track vesicoureteral reflux.

The liposome-encapsulated ICG (LP-ICG) has improved stability and fluorescence signal in comparison with free ICG. A newly designed mannosylated liposome-encapsulated ICG (M-LP-ICG) tested for a mouse model after subcutaneous injection showed that the signal from M-LP-ICG in SLN and other organs appeared early and disappeared quickly in comparison with signals from LP-ICG [155]. In addition, M-LP-ICG appears to increase the specificity of uptake and retention in macrophages.

Application of encapsulated ICG in conjunction with bacterial lectin was found to be effective in acceleration of cell lipolysis [156].

Conclusions and Future

During the last few years, the interest on using indocyanine green in intraoperative imaging has considerably increased due to the totally new applications in various branches of surgery. ICG provides a set of unique imaging methods ranging from angiography to lymphography and tumor imaging and marking. In addition, ICG can be used for therapy ranging from photodynamic therapy to tissue soldering. The devices needed in ICG imaging are so simple that the integration of ICG imaging devices to optical tools used in surgery like operational microscopes and endoscopes has been straightforward, which has greatly facilitated the fast introduction of these new applications in the routine clinical setting.

In ICGA there is plenty of room for technical developments, for example, in form of rapid and reliable flow dynamics analyses and easily processable and repeatable video playback loops, since rapid ICG re-injections generally suffer from lower contrast due to residual ICG inside the vessels. Also, one has to be aware of all the limiting factors of the technique such as, thick or atherosclerotic arterial wall which hinders the ICGA visualization.

Future developments in ICGA aim at better and faster information analysis from the collected ICG data. The most important research topics at the moment are flow distribution and flow velocity analysis. It seems that flow analysis provides additional information about the hemodynamic changes but these do not necessarily correlate with perfusion or quantitative flow. Apart from microscope integration, the ICG video angiography module has been also mounted on an endoscope camera allowing for visualization of structures which would otherwise stay hidden in the microscopes field of view [157].

There is an interest to the light-induced bactericidal efficiency of ICG at NIR laser irradiation of different types of bacterium strains, including antibiotic-resistant ones [158, 159] that can provide serious post-surgery complications. Therefore, bactericidal treatment in the framework of ICG-guided surgery may have sense for future ICG technologies.

A totally new area for ICG is in various oncological surgery imaging tasks including but not limited to intraoperational SLN harvesting, liver tumors, and tumor marking. Probably, many important oncological applications of ICG imaging are still uncovered providing plenty of clinical research topics.

Acknowledgements The work was carried out under the partial support from grants: 14.Z50.31.0004 to support scientific research projects implemented under the supervision of leading scientists at Russian institutions and Russian institutions of higher education, Russian Presidential grant NSh-703.2014.2, and FiDiPro, TEKES Program (40111/11), Finland (for V.T.), and Finnish Foundation for Cardiovascular Research (for O.V.).

References

1. Guyer DR, Puliafito CA, Mones JM, Friedman E, Chang W, Verdooner SR. Digital indocyanine-green angiography in chorioretinal disorders. Ophthalmology. 1992;99(2):287–91.
2. Alander JT, Kaartinen I, Laakso A, Patila T, Spillmann T, Tuchin VV, et al. A review of indocyanine green fluorescent imaging in surgery. Int J Biomed Imag. 2012;2012:940585.
3. Hoekstra LT, de Graaf W, Nibourg GA, Heger M, Bennink RJ, Stieger B, et al. Physiological and biochemical basis of clinical liver function tests: a review. Ann Surg. 2013;257(1):27–36.
4. Mullin EJ, Metcalfe MS, Maddern GJ. How much liver resection is too much? Am J Surg. 2005;190(1):87–97.
5. Fazakas J, Mandli T, Ther G, Arkossy M, Pap S, Fule B, et al. Evaluation of liver function for hepatic resection. Transplant Proc. 2006;38(3):798–800.
6. Cherrick GR, Stein SW, Leevy CM, Davidson CS. Indocyanine green: observations on its physical properties, plasma decay, and hepatic extraction. J Clin Invest. 1960;39:592–600.
7. Kamath PS, Wiesner RH, Malinchoc M, Kremers W, Therneau TM, Kosberg CL, et al. A model to predict survival in patients with end-stage liver disease. Hepatology. 2001;33(2):464–70.
8. Poon RT, Fan ST. Hepatectomy for hepatocellular carcinoma: patient selection and postoperative outcome. Liver Transpl. 2004;10(2 Suppl 1):S39–45.
9. Lee SG, Hwang S. How I do it: assessment of hepatic functional reserve for indication of hepatic resection. J Hepatobiliary Pancreat Surg. 2005;12(1):38–43.
10. Minagawa M, Makuuchi M, Takayama T, Ohtomo K. Selection criteria for hepatectomy in patients with hepatocellular carcinoma and portal vein tumor thrombus. Ann Surg. 2001;233(3):379–84.
11. Makuuchi M, Sano K. The surgical approach to HCC: our progress and results in Japan. Liver Transpl. 2004;10(2 Suppl 1):S46–52.
12. Stockmann M, Malinowski M, Lock JF, Seehofer D, Neuhaus P. Factors influencing the indocyanine green (ICG) test: additional impact of acute cholestasis. Hepatogastroenterology. 2009;56(91–92):734–8.
13. Miyamoto Y, Oho K, Kumamoto M, Toyonaga A, Sata M. Balloon-occluded retrograde transvenous obliteration improves liver function in patients with cirrhosis and portal hypertension. J Gastroenterol Hepatol. 2003;18(8):934–42.
14. Okuda K, Ohkubo H, Musha H, Kotoda K, Abe H, Tanikawa K. Marked delay in indocyanine green plasma clearance with a near-normal bromosulphophthalein retention test: a constitutional abnormality? Gut. 1976;17(8):588–94.
15. Vos JJ, Scheeren TW, Lukes DJ, de Boer MT, Hendriks HG, Wietasch JK. Intraoperative ICG plasma disappearance rate helps to predict absence of early postoperative complications after orthotopic liver transplantation. J Clin Monit Comput. 2013;27(5):591–8.
16. Du Z, Wei Y, Chen K, Chen X, Zhang Z, Li H, et al. Risk factors and criteria predicting early graft loss after adult-to-adult living donor liver transplantation. J Surg Res. 2013;187:673–82.
17. Escorsell A, Mas A, Fernandez J, Garcia-Valdecasas JC. Limitations of use of the noninvasive clearance of indocyanine green as a prognostic indicator of graft function in liver transplantation. Transplant Proc. 2012;44(6):1539–41.
18. Kang Y, Lee J, An Y, Jeon J, Choi C. Segmental analysis of indocyanine green pharmacokinetics for the reliable diagnosis of functional vascular insufficiency. J Biomed Opt. 2011;16(3):030504.
19. An Y, Jeon JW, Kwon K, Choi C. Application of dynamic indocyanine green perfusion imaging for evaluation of vasoactive effect of acupuncture: a

preliminary follow-up study on normal healthy volunteers. Med Devices (Auckl). 2014;7:17–21.
20. Kamolz LP, Andel H, Haslik W, Donner A, Winter W, Meissl G, et al. Indocyanine green video angiographies help to identify burns requiring operation. Burns. 2003;29(8):785–91.
21. Kamolz LP, Andel H, Auer T, Meissl G, Frey M. Evaluation of skin perfusion by use of indocyanine green video angiography: rational design and planning of trauma surgery. J Trauma. 2006;61(3): 635–41.
22. Bovill ES, Jansen L, Macadam S, Lennox P. Reduction-pattern mastectomy: vascularity of the inferior dermal flap. J Plast Reconstr Aesthet Surg. 2013;66(4):587–8.
23. Moyer HR, Losken A. Predicting mastectomy skin flap necrosis with indocyanine green angiography: the gray area defined. Plast Reconstr Surg. 2012;129(5):1043–8.
24. Newman MI, Jack MC, Samson MC. SPY-Q analysis toolkit values potentially predict mastectomy flap necrosis. Ann Plast Surg. 2013;70(5):595–8.
25. Phillips BT, Lanier ST, Conkling N, Wang ED, Dagum AB, Ganz JC, et al. Intraoperative perfusion techniques can accurately predict mastectomy skin flap necrosis in breast reconstruction: results of a prospective trial. Plast Reconstr Surg. 2012; 129(5):778e–88e.
26. Holm C, Mayr M, Hofter E, Becker A, Pfeiffer UJ, Muhlbauer W. Intraoperative evaluation of skin-flap viability using laser-induced fluorescence of indocyanine green. Br J Plast Surg. 2002;55(8):635–44.
27. Ayestaray B, Andreoletti JB. Microscope-integrated indocyanine green videoangiography for deep inferior epigastric perforator flap. J Plast Reconstr Aesthet Surg. 2013;66(4):583–5.
28. Fourman MS, Gersch RP, Phillips BT, Nasser A, Rivara A, Verma R, et al. Comparison of laser doppler and laser-assisted indocyanine green angiography prediction of flap survival in a novel modification of the McFarlane flap. Ann Plast Surg; 2014 (in press).
29. Green 3rd JM, Thomas S, Sabino J, Howard R, Basile P, Dryden S, et al. Use of intraoperative fluorescent angiography to assess and optimize free tissue transfer in head and neck reconstruction. J Oral Maxillofac Surg. 2013;71(8):1439–49.
30. Krishnan KG, Schackert G, Steinmeier R. The role of near-infrared angiography in the assessment of post-operative venous congestion in random pattern, pedicled island and free flaps. Br J Plast Surg. 2005; 58(3):330–8.
31. Losken A, Zenn MR, Hammel JA, Walsh MW, Carlson GW. Assessment of zonal perfusion using intraoperative angiography during abdominal flap breast reconstruction. Plast Reconstr Surg. 2012;129(4):618e–24e.
32. Mothes H, Donicke T, Friedel R, Simon M, Markgraf E, Bach O. Indocyanine-green fluorescence video angiography used clinically to evaluate tissue perfusion in microsurgery. J Trauma. 2004;57(5): 1018–24.
33. Piwkowski C, Gabryel P, Gasiorowskia L, Zielinski P, Murawa D, Roszak M, et al. Indocyanine green fluorescence in the assessment of the quality of the pedicled intercostal muscle flap: a pilot study. Eur J Cardiothorac Surg. 2013;44(1):e77–81.
34. Suzuki A, Fujiwara M, Mizukami T, Fukamizu H. Delayed distally-based super sural flap: evaluation by indocyanine green fluorescence angiography. J Plast Reconstr Aesthet Surg. 2008;61(4):467–9.
35. Tanaka K, Okazaki M. Visualization of blood supply to the 'vascularized nerve' with anterolateral thigh flap using indocyanine green fluorescence angiography. J Plast Reconstr Aesthet Surg. 2013;66(1): 146–7.
36. Holm C, Mayr M, Hofter E, Dornseifer U, Ninkovic M. Assessment of the patency of microvascular anastomoses using microscope-integrated near-infrared angiography: a preliminary study. Microsurgery. 2009;29(7):509–14.
37. Holm C, Dornseifer U, Sturtz G, Basso G, Schuster T, Ninkovic M. The intrinsic transit time of free microvascular flaps: clinical and prognostic implications. Microsurgery. 2010;30(2):91–6.
38. Betz CS, Zhorzel S, Schachenmayr H, Stepp H, Havel M, Siedek V, et al. Endoscopic measurements of free-flap perfusion in the head and neck region using red-excited Indocyanine Green: preliminary results. J Plast Reconstr Aesthet Surg. 2009;62(12): 1602–8.
39. Holm C, Tegeler J, Mayr M, Becker A, Pfeiffer UJ, Muhlbauer W. Monitoring free flaps using laser-induced fluorescence of indocyanine green: a preliminary experience. Microsurgery. 2002;22(7): 278–87.
40. Ballardini B, Santoro L, Sangalli C, Gentilini O, Renne G, Lissidini G, et al. The indocyanine green method is equivalent to the (9)(9)mTc-labeled radiotracer method for identifying the sentinel node in breast cancer: a concordance and validation study. Eur J Surg Oncol. 2013;39(12):1332–6.
41. Hirano A, Kamimura M, Ogura K, Kim N, Hattori A, Setoguchi Y, et al. A comparison of indocyanine green fluorescence imaging plus blue dye and blue dye alone for sentinel node navigation surgery in breast cancer patients. Ann Surg Oncol. 2012;19(13):4112–6.
42. Hirche C, Engel H, Hirche Z, Doniga S, Herold T, Kneser U, et al. Real-time lymphography by indocyanine green fluorescence: improved navigation for regional lymph node staging. Ann Plast Surg. 2013; 73:701–5.
43. Hirche C, Mohr Z, Kneif S, Murawa D, Hunerbein M. High rate of solitary sentinel node metastases identification by fluorescence-guided lymphatic imaging in breast cancer. J Surg Oncol. 2012; 105(2):162–6.

44. Jung SY, Kim SK, Kim SW, Kwon Y, Lee ES, Kang HS, et al. Comparison of sentinel lymph node biopsy guided by the multimodal method of indocyanine green fluorescence, radioisotope, and blue dye versus the radioisotope method in breast cancer: a randomized controlled trial. Ann Surg Oncol. 2014; 21(4):1254–9.
45. Kitai T, Kawashima M. Transcutaneous detection and direct approach to the sentinel node using axillary compression technique in ICG fluorescence-navigated sentinel node biopsy for breast cancer. Breast Cancer. 2012;19(4):343–8.
46. Polom K, Murawa D, Nowaczyk P, Rho YS, Murawa P. Breast cancer sentinel lymph node mapping using near infrared guided indocyanine green and indocyanine green: human serum albumin in comparison with gamma emitting radioactive colloid tracer. Eur J Surg Oncol. 2012;38(2):137–42.
47. Schaafsma BE, Verbeek FP, Rietbergen DD, van der Hiel B, van der Vorst JR, Liefers GJ, et al. Clinical trial of combined radio- and fluorescence-guided sentinel lymph node biopsy in breast cancer. Br J Surg. 2013;100(8):1037–44.
48. Sugie T, Sawada T, Tagaya N, Kinoshita T, Yamagami K, Suwa H, et al. Comparison of the indocyanine green fluorescence and blue dye methods in detection of sentinel lymph nodes in early-stage breast cancer. Ann Surg Oncol. 2013; 20(7):2213–8.
49. van der Vorst JR, Schaafsma BE, Hutteman M, Verbeek FP, Liefers GJ, Hartgrink HH, et al. Near-infrared fluorescence-guided resection of colorectal liver metastases. Cancer. 2013;119(18):3411–8.
50. Wishart GC, Loh SW, Jones L, Benson JR. A feasibility study (ICG-10) of indocyanine green (ICG) fluorescence mapping for sentinel lymph node detection in early breast cancer. Eur J Surg Oncol. 2012;38(8):651–6.
51. Yamamoto S, Maeda N, Yoshimura K, Oka M. Intraoperative detection of sentinel lymph nodes in breast cancer patients using ultrasonography-guided direct indocyanine green dye-marking by real-time virtual sonography constructed with three-dimensional computed tomography-lymphography. Breast. 2013;22(5):933–7.
52. Buckle T, van Leeuwen AC, Chin PT, Janssen H, Muller SH, Jonkers J, et al. A self-assembled multimodal complex for combined pre- and intraoperative imaging of the sentinel lymph node. Nanotechnology. 2010;21(35):355101. 355101-4484/21/35/355101. Epub 2010 Aug 6.
53. Gilmore DM, Khullar OV, Gioux S, Stockdale A, Frangioni JV, Colson YL, et al. Effective low-dose escalation of indocyanine green for near-infrared fluorescent sentinel lymph node mapping in melanoma. Ann Surg Oncol. 2013;20(7):2357–63.
54. Hayashi T, Furukawa H, Oyama A, Funayama E, Saito A, Yamao T, et al. Sentinel lymph node biopsy using real-time fluorescence navigation with indocyanine green in cutaneous head and neck/lip mucosa melanomas. Head Neck. 2012; 34(5):758–61.
55. Jain V, Phillips BT, Conkling N, Pameijer C. Sentinel lymph node detection using laser-assisted indocyanine green dye lymphangiography in patients with melanoma. Int J Surg Oncol. 2013;2013:904214.
56. Namikawa K, Tsutsumida A, Tanaka R, Kato J, Yamazaki N. Limitation of indocyanine green fluorescence in identifying sentinel lymph node prior to skin incision in cutaneous melanoma. Int J Clin Oncol. 2014;19(1):198–203.
57. Polom K, Murawa D, Rho YS, Spychala A, Murawa P. Skin melanoma sentinel lymph node biopsy using real-time fluorescence navigation with indocyanine green and indocyanine green with human serum albumin. Br J Dermatol. 2012;166(3):682–3.
58. Uhara H, Yamazaki N, Takata M, Inoue Y, Sakakibara A, Nakamura Y, et al. Applicability of radiocolloids, blue dyes and fluorescent indocyanine green to sentinel node biopsy in melanoma. J Dermatol. 2012;39(4):336–8.
59. Hutteman M, van der Vorst JR, Gaarenstroom KN, Peters AA, Mieog JS, Schaafsma BE, et al. Optimization of near-infrared fluorescent sentinel lymph node mapping for vulvar cancer. Am J Obstet Gynecol. 2012;206(1):89.e1–5.
60. Schaafsma BE, Verbeek FP, Peters AA, van der Vorst JR, de Kroon CD, van Poelgeest MI, et al. Near-infrared fluorescence sentinel lymph node biopsy in vulvar cancer: a randomised comparison of lymphatic tracers. BJOG. 2013;120(6):758–64.
61. Schaafsma BE, van der Vorst JR, Gaarenstroom KN, Peters AA, Verbeek FP, de Kroon CD, et al. Randomized comparison of near-infrared fluorescence lymphatic tracers for sentinel lymph node mapping of cervical cancer. Gynecol Oncol. 2012; 127(1):126–30.
62. Matheron HM, van den Berg NS, Brouwer OR, Kleinjan GH, van Driel WJ, Trum JW, et al. Multimodal surgical guidance towards the sentinel node in vulvar cancer. Gynecol Oncol. 2013; 131(3):720–5.
63. Nakamura Y, Fujisawa Y, Nakamura Y, Maruyama H, Furuta J, Kawachi Y, et al. Improvement of the sentinel lymph node detection rate of cervical sentinel lymph node biopsy using real-time fluorescence navigation with indocyanine green in head and neck skin cancer. J Dermatol. 2013;40(6): 453–7.
64. Rossi EC, Jackson A, Ivanova A, Boggess JF. Detection of sentinel nodes for endometrial cancer with robotic assisted fluorescence imaging: cervical versus hysteroscopic injection. Int J Gynecol Cancer. 2013;23(9):1704–11.
65. Yokoyama J, Fujimaki M, Ohba S, Anzai T, Yoshii R, Ito S, et al. A feasibility study of NIR fluorescent image-guided surgery in head and neck cancer based on the assessment of optimum surgical time as revealed through dynamic imaging. Onco Targets Ther. 2013;6:325–30.

66. Yokoyama J, Ooba S, Fujimaki M, Anzai T, Yoshii R, Kojima M, et al. Impact of indocyanine green fluorescent image-guided surgery for parapharyngeal space tumours. J Craniomaxillofac Surg. 2013; 42:835–8.
67. Akita S, Mitsukawa N, Kazama T, Kuriyama M, Kubota Y, Omori N, et al. Comparison of lymphoscintigraphy and indocyanine green lymphography for the diagnosis of extremity lymphoedema. J Plast Reconstr Aesthet Surg. 2013;66(6):792–8.
68. Mihara M, Hara H, Araki J, Narushima M, Iida T, Koshima I. Treatment of hand lymphedema with free flap transfer and lymphangiogenesis analysis after hand replantation using indocyanine green (ICG) lymphography and histological analysis. J Plast Reconstr Aesthet Surg. 2013;66(11):e338–40.
69. Yamamoto T, Yamamoto N, Yoshimatsu H, Narushima M, Koshima I. LEC score: a judgment tool for indication of indocyanine green lymphography. Ann Plast Surg. 2013;70(2):227–30.
70. Yamamoto T, Narushima M, Yoshimatsu H, Yamamoto N, Oka A, Seki Y, et al. Indocyanine green velocity: lymph transportation capacity deterioration with progression of lymphedema. Ann Plast Surg. 2013;71(5):591–4.
71. Yamamoto T, Narushima M, Yoshimatsu H, Yamamoto N, Kikuchi K, Todokoro T, et al. Dynamic indocyanine green (ICG) lymphography for breast cancer-related arm lymphedema. Ann Plast Surg. 2013;73:706–9.
72. Hara H, Mihara M, Seki Y, Todokoro T, Iida T, Koshima I. Comparison of indocyanine green lymphographic findings with the conditions of collecting lymphatic vessels of limbs in patients with lymphedema. Plast Reconstr Surg. 2013;132(6):1612–8.
73. Mihara M, Hayashi Y, Hara H, Iida T, Narushima M, Yamamoto T, et al. High-accuracy diagnosis and regional classification of lymphedema using indocyanine green fluorescent lymphography after gynecologic cancer treatment. Ann Plast Surg. 2014; 72(2):204–8.
74. Azuma S, Yamamoto T, Koshima I. Donor-site lymphatic function after microvascular lymph node transfer should be followed using indocyanine green lymphography. Plast Reconstr Surg. 2013;131(3): 443e–4e.
75. Liu HL, Pang SY, Chan YW. The use of a microscope with near-infrared imaging function in indocyanine green lymphography and lymphaticovenous anastomosis. J Plast Reconstr Aesthet Surg. 2014; 67(2):231–6.
76. Mihara M, Hara H, Hayashi Y, Iida T, Araki J, Yamamoto T, et al. Upper-limb lymphedema treated aesthetically with lymphaticovenous anastomosis using indocyanine green lymphography and noncontact vein visualization. J Reconstr Microsurg. 2012;28(5):327–32.
77. Mihara M, Murai N, Hayashi Y, Hara H, Iida T, Narushima M, et al. Using indocyanine green fluorescent lymphography and lymphatic-venous anastomosis for cancer-related lymphedema. Ann Vasc Surg. 2012;26(2):278.e1–6.
78. Yamamoto T, Narushima M, Yoshimatsu H, Seki Y, Yamamoto N, Oka A, et al. Minimally invasive lymphatic supermicrosurgery (MILS): indocyanine green lymphography-guided simultaneous multisite lymphaticovenular anastomoses via millimeter skin incisions. Ann Plast Surg. 2014;72(1):67–70.
79. Wrobel CJ, Meltzer H, Lamond R, Alksne JF. Intraoperative assessment of aneurysm clip placement by intravenous fluorescein angiography. Neurosurgery. 1994;35(5):970–3. Discussion 973.
80. Raabe A, Beck J, Gerlach R, Zimmermann M, Seifert V. Near-infrared indocyanine green video angiography: a new method for intraoperative assessment of vascular flow. Neurosurgery. 2003;52(1):132–9. Discussion 139.
81. Raabe A, Nakaji P, Beck J, Kim LJ, Hsu FP, Kamerman JD, et al. Prospective evaluation of surgical microscope-integrated intraoperative near-infrared indocyanine green videoangiography during aneurysm surgery. J Neurosurg. 2005;103(6):982–9.
82. Dashti R, Laakso A, Niemela M, Porras M, Hernesniemi J. Microscope-integrated near-infrared indocyanine green videoangiography during surgery of intracranial aneurysms: the Helsinki experience. Surg Neurol. 2009;71(5):543–50. Discussion 550.
83. Killory BD, Nakaji P, Maughan PH, Wait SD, Spetzler RF. Evaluation of angiographically occult spinal dural arteriovenous fistulae with surgical microscope-integrated intraoperative near-infrared indocyanine green angiography: report of 3 cases. Neurosurgery. 2011;68(3):781–7. Discussion 787.
84. Colby GP, Coon AL, Sciubba DM, Bydon A, Gailloud P, Tamargo RJ. Intraoperative indocyanine green angiography for obliteration of a spinal dural arteriovenous fistula. J Neurosurg Spine. 2009;11(6): 705–9.
85. Woitzik J, Horn P, Vajkoczy P, Schmiedek P. Intraoperative control of extracranial-intracranial bypass patency by near-infrared indocyanine green videoangiography. J Neurosurg. 2005;102(4): 692–8.
86. Schebesch KM, Proescholdt M, Hohne J, Hohenberger C, Hansen E, Riemenschneider MJ, et al. Sodium fluorescein-guided resection under the YELLOW 560 nm surgical microscope filter in malignant brain tumor surgery: a feasibility study. Acta Neurochir (Wien). 2013;155(4):693–9.
87. Nashef SA, Roques F, Sharples LD, Nilsson J, Smith C, Goldstone AR, et al. EuroSCORE II. Eur J Cardiothorac Surg. 2012;41(4):734–44. discussion 744-5.
88. Balacumaraswami L, Taggart DP. Intraoperative imaging techniques to assess coronary artery bypass graft patency. Ann Thorac Surg. 2007;83(6): 2251–7.

89. Rubens FD, Ruel M, Fremes SE. A new and simplified method for coronary and graft imaging during CABG. Heart Surg Forum. 2002;5(2):141–4.
90. Taggart DP, Choudhary B, Anastasiadis K, Abu-Omar Y, Balacumaraswami L, Pigott DW. Preliminary experience with a novel intraoperative fluorescence imaging technique to evaluate the patency of bypass grafts in total arterial revascularization. Ann Thorac Surg. 2003;75(3):870–3.
91. Reuthebuch O, Haussler A, Genoni M, Tavakoli R, Odavic D, Kadner A, et al. Novadaq SPY: intraoperative quality assessment in off-pump coronary artery bypass grafting. Chest. 2004;125(2):418–24.
92. Balacumaraswami L, Abu-Omar Y, Choudhary B, Pigott D, Taggart DP. A comparison of transit-time flowmetry and intraoperative fluorescence imaging for assessing coronary artery bypass graft patency. J Thorac Cardiovasc Surg. 2005;130(2):315–20.
93. Takahashi M, Ishikawa T, Higashidani K, Katoh H. SPY: an innovative intra-operative imaging system to evaluate graft patency during off-pump coronary artery bypass grafting. Interact Cardiovasc Thorac Surg. 2004;3(3):479–83.
94. Desai ND, Miwa S, Kodama D, Cohen G, Christakis GT, Goldman BS, et al. Improving the quality of coronary bypass surgery with intraoperative angiography: validation of a new technique. J Am Coll Cardiol. 2005;46(8):1521–5.
95. Detter C, Russ D, Iffland A, Wipper S, Schurr MO, Reichenspurner H, et al. Near-infrared fluorescence coronary angiography: a new noninvasive technology for intraoperative graft patency control. Heart Surg Forum. 2002;5(4):364–9.
96. Detter C, Wipper S, Russ D, Iffland A, Burdorf L, Thein E, et al. Fluorescent cardiac imaging: a novel intraoperative method for quantitative assessment of myocardial perfusion during graded coronary artery stenosis. Circulation. 2007;116(9):1007–14.
97. Ferguson Jr TB, Chen C, Babb JD, Efird JT, Daggubati R, Cahill JM. Fractional flow reserve-guided coronary artery bypass grafting: can intraoperative physiologic imaging guide decision making? J Thorac Cardiovasc Surg. 2013;146(4):824–835.e1.
98. Kogon B, Fernandez J, Kanter K, Kirshbom P, Vincent B, Maher K, et al. The role of intraoperative indocyanine green fluorescence angiography in pediatric cardiac surgery. Ann Thorac Surg. 2009; 88(2):632–6.
99. Igari K, Kudo T, Toyofuku T, Jibiki M, Inoue Y, Kawano T. Quantitative evaluation of the outcomes of revascularization procedures for peripheral arterial disease using indocyanine green angiography. Eur J Vasc Endovasc Surg. 2013;46(4):460–5.
100. Igari K, Kudo T, Uchiyama H, Toyofuku T, Inoue Y. Intraarterial injection of indocyanine green for evaluation of peripheral blood circulation in patients with peripheral arterial disease. Ann Vasc Surg. 2014;28:1280–5.
101. Terasaki H, Inoue Y, Sugano N, Jibiki M, Kudo T, Lepantalo M, et al. A quantitative method for evaluating local perfusion using indocyanine green fluorescence imaging. Ann Vasc Surg. 2013;27(8): 1154–61.
102. Klein A, Buschmann M, Babilas P, Landthaler M, Baumler W. Indocyanine green-augmented diode laser therapy vs. long-pulsed Nd:YAG (1064 nm) laser treatment of telangiectatic leg veins: a randomized controlled trial. Br J Dermatol. 2013;169(2): 365–73.
103. Ishizawa T, Bandai Y, Ijichi M, Kaneko J, Hasegawa K, Kokudo N. Fluorescent cholangiography illuminating the biliary tree during laparoscopic cholecystectomy. Br J Surg. 2010;97(9):1369–77.
104. Ishizawa T, Tamura S, Masuda K, Aoki T, Hasegawa K, Imamura H, et al. Intraoperative fluorescent cholangiography using indocyanine green: a biliary road map for safe surgery. J Am Coll Surg. 2009;208(1): e1–4.
105. Ishizawa T, Bandai Y, Hasegawa K, Kokudo N. Fluorescent cholangiography during laparoscopic cholecystectomy: indocyanine green or new fluorescent agents? World J Surg. 2010;34(10):2505–6.
106. Ishizawa T, Fukushima N, Shibahara J, Masuda K, Tamura S, Aoki T, et al. Real-time identification of liver cancers by using indocyanine green fluorescent imaging. Cancer. 2009;115(11):2491–504.
107. Gotoh K, Yamada T, Ishikawa O, Takahashi H, Eguchi H, Yano M, et al. A novel image-guided surgery of hepatocellular carcinoma by indocyanine green fluorescence imaging navigation. J Surg Oncol. 2009;100(1):75–9.
108. Lim C, Vibert E, Azoulay D, Salloum C, Ishizawa T, Yoshioka R, et al. Indocyanine green fluorescence imaging in the surgical management of liver cancers: current facts and future implications. J Visc Surg. 2014;151:117–24.
109. Tanaka T, Takatsuki M, Hidaka M, Hara T, Muraoka I, Soyama A, et al. Is a fluorescence navigation system with indocyanine green effective enough to detect liver malignancies? J Hepatobiliary Pancreat Sci. 2013;21:199–204.
110. Aoki T, Yasuda D, Shimizu Y, Odaira M, Niiya T, Kusano T, et al. Image-guided liver mapping using fluorescence navigation system with indocyanine green for anatomical hepatic resection. World J Surg. 2008;32(8):1763–7.
111. Yokoyama N, Otani T, Hashidate H, Maeda C, Katada T, Sudo N, et al. Real-time detection of hepatic micrometastases from pancreatic cancer by intraoperative fluorescence imaging: preliminary results of a prospective study. Cancer. 2012; 118(11):2813–9.
112. Ishizuka M, Kubota K, Kita J, Shimoda M, Kato M, Sawada T. Intraoperative observation using a fluorescence imaging instrument during hepatic resection for liver metastasis from colorectal cancer. Hepatogastroenterology. 2012;59(113):90–2.
113. Okusanya OT, Madajewski B, Segal E, Judy BF, Venegas OG, Judy RP, et al. Small portable interchangeable imager of fluorescence for fluorescence

guided surgery and research. Technol Cancer Res Treat; 2013 (in press).
114. Liu Y, Zhao YM, Akers W, Tang ZY, Fan J, Sun HC, et al. First in-human intraoperative imaging of HCC using the fluorescence goggle system and transarterial delivery of near-infrared fluorescent imaging agent: a pilot study. Transl Res. 2013;162(5):324–31.
115. Liu Y, Njuguna R, Matthews T, Akers WJ, Sudlow GP, Mondal S, et al. Near-infrared fluorescence goggle system with complementary metal-oxide-semiconductor imaging sensor and see-through display. J Biomed Opt. 2013;18(10):101303.
116. Sarder P, Gullicksrud K, Mondal S, Sudlow GP, Achilefu S, Akers WJ. Dynamic optical projection of acquired luminescence for aiding oncologic surgery. J Biomed Opt. 2013;18(12):120501.
117. Venugopal V, Park M, Ashitate Y, Neacsu F, Kettenring F, Frangioni JV, et al. Design and characterization of an optimized simultaneous color and near-infrared fluorescence rigid endoscopic imaging system. J Biomed Opt. 2013;18(12):126018.
118. Brouwer OR, Buckle T, Bunschoten A, Kuil J, Vahrmeijer AL, Wendler T, et al. Image navigation as a means to expand the boundaries of fluorescence-guided surgery. Phys Med Biol. 2012;57(10):3123–36.
119. Towle EL, Richards LM, Kazmi SM, Fox DJ, Dunn AK. Comparison of indocyanine green angiography and laser speckle contrast imaging for the assessment of vasculature perfusion. Neurosurgery. 2012;71(5):1023–30. Discussion 1030-1.
120. Ohba S, Yokoyama J, Fujimaki M, Ito S, Kojima M, Shimoji K, et al. Significant improvement in superselective intra-arterial chemotherapy for oral cancer by using indocyanine green fluorescence. Oral Oncol. 2012;48(11):1101–5.
121. Yokoyama J, Ohba S, Fujimaki M, Kojima M, Suzuki M, Ikeda K. Significant improvement in superselective intra-arterial chemotherapy for advanced paranasal sinus cancer by using indocyanine green fluorescence. Eur Arch Otorhinolaryngol. 2013;271:2795–801.
122. Piyawattanametha W, Ra H, Qiu Z, Friedland S, Liu JT, Loewke K, et al. In vivo near-infrared dual-axis confocal microendoscopy in the human lower gastrointestinal tract. J Biomed Opt. 2012;17(2):021102.
123. Foersch S, Heimann A, Ayyad A, Spoden GA, Florin L, Mpoukouvalas K, et al. Confocal laser endomicroscopy for diagnosis and histomorphologic imaging of brain tumors in vivo. PLoS One. 2012;7(7):e41760.
124. Huang B, Bates M, Zhuang X. Super-resolution fluorescence microscopy. Annu Rev Biochem. 2009;78:993–1016.
125. Piper SK, Habermehl C, Schmitz CH, Kuebler WM, Obrig H, Steinbrink J, et al. Towards whole-body fluorescence imaging in humans. PLoS One. 2013;8(12):e83749.
126. Esposito G, Rossi F, Matteini P, Scerrati A, Puca A, Albanese A, et al. In vivo laser assisted microvascular repair and end-to-end anastomosis by means of indocyanine green-infused chitosan patches: a pilot study. Lasers Surg Med. 2013;45(5):318–25.
127. Yanina IY, Tuchin VV, Navolokin NA, Matveeva OV, Bucharskaya AB, Maslyakova GN, et al. Fat tissue histological study at indocyanine green-mediated photothermal/photodynamic treatment of the skin in vivo. J Biomed Opt. 2012;17(5):058002.
128. Yanina IY, Trunina NA, Tuchin VV. Photoinduced cell morphology alterations quantified within adipose tissues by spectral optical coherence tomography. J Biomed Opt. 2013;18(11):111407.
129. Yanina IY, Bochko VA, Alander JT, Tuchin VV. Optical image analysis of fat cells for indocyanine green mediated near-infrared laser treatment. Laser Phys Lett. 2011;8(9):684–90.
130. Yanina IY, Trunina NA, Tuchin VV. Optical coherence tomographyof adipose tissue at photodynamic/photothermal treatment in vitro. J Innov Opt Health Sci 2013;6(1350010-1-7) http://www.worldscientific.com/doi/pdf/10.1142/S1793545813500107.
131. Klein A, Baumler W, Buschmann M, Landthaler M, Babilas P. A randomized controlled trial to optimize indocyanine green-augmented diode laser therapy of capillary malformations. Lasers Surg Med. 2013;45(4):216–24.
132. Liebert A, Sawosz P, Milej D, Kacprzak M, Weigl W, Botwicz M, et al. Assessment of inflow and washout of indocyanine green in the adult human brain by monitoring of diffuse reflectance at large source-detector separation. J Biomed Opt. 2011;16(4):046011.
133. Lutzweiler C, Razansky D. Optoacoustic imaging and tomography: reconstruction approaches and outstanding challenges in image performance and quantification. Sensors (Basel). 2013;13(6):7345–84.
134. Ntziachristos V, Razansky D. Optical and optoacoustic imaging. Recent Results Cancer Res. 2013;187:133–50.
135. Marano A, Priora F, Lenti LM, Ravazzoni F, Quarati R, Spinoglio G. Application of fluorescence in robotic general surgery: review of the literature and state of the art. World J Surg. 2013;37(12):2800–11.
136. Lee Z, Simhan J, Parker DC, Reilly C, Llukani E, Lee DI, et al. Novel use of indocyanine green for intraoperative, real-time localization of ureteral stenosis during robot-assisted ureteroureterostomy. Urology. 2013;82(3):729–33.
137. Manny TB, Pompeo AS, Hemal AK. Robotic partial adrenalectomy using indocyanine green dye with near-infrared imaging: the initial clinical experience. Urology. 2013;82(3):738–42.
138. Manny TB, Patel M, Hemal AK. Fluorescence-enhanced robotic radical prostatectomy using real-time lymphangiography and tissue marking with percutaneous injection of unconjugated indocyanine green: the initial clinical experience in 50 patients. Eur Urol. 2013;66:e15–6.
139. Harke N, Schoen G, Schiefelbein F, Heinrich E. Selective clamping under the usage of near-infrared fluorescence imaging with indocyanine green in

robot-assisted partial nephrectomy: a single-surgeon matched-pair study. World J Urol. 2013;32:1259–65.

140. Borofsky MS, Gill IS, Hemal AK, Marien TP, Jayaratna I, Krane LS, et al. Near-infrared fluorescence imaging to facilitate super-selective arterial clamping during zero-ischaemia robotic partial nephrectomy. BJU Int. 2013;111(4):604–10.
141. Bjurlin MA, Gan M, McClintock TR, Volpe A, Borofsky MS, Mottrie A, et al. Near-infrared fluorescence imaging: emerging applications in robotic upper urinary tract surgery. Eur Urol. 2014;65(4): 793–801.
142. Hellan M, Spinoglio G, Pigazzi A, Lagares-Garcia JA. The influence of fluorescence imaging on the location of bowel transection during robotic left-sided colorectal surgery. Surg Endosc. 2014;28: 1695–702.
143. Jafari MD, Lee KH, Halabi WJ, Mills SD, Carmichael JC, Stamos MJ, et al. The use of indocyanine green fluorescence to assess anastomotic perfusion during robotic assisted laparoscopic rectal surgery. Surg Endosc. 2013;27(8):3003–8.
144. Angell JE, Khemees TA, Abaza R. Optimization of near infrared fluorescence tumor localization during robotic partial nephrectomy. J Urol. 2013;190(5): 1668–73.
145. Wagner OJ, Louie BE, Vallieres E, Aye RW, Farivar AS. Near-infrared fluorescence imaging can help identify the contralateral phrenic nerve during robotic thymectomy. Ann Thorac Surg. 2012;94(2):622–5.
146. Spinoglio G, Priora F, Bianchi PP, Lucido FS, Licciardello A, Maglione V, et al. Real-time near-infrared (NIR) fluorescent cholangiography in single-site robotic cholecystectomy (SSRC): a single-institutional prospective study. Surg Endosc. 2013;27(6):2156–62.
147. Kraft JC, Ho RJ. Interactions of indocyanine green and lipid in enhancing near-infrared fluorescence properties: the basis for near-infrared imaging in vivo. Biochemistry. 2014;53(8):1275–83.
148. Zanganeh S, Xu Y, Hamby CV, Backer MV, Backer JM, Zhu Q. Enhanced fluorescence diffuse optical tomography with indocyanine green-encapsulating liposomes targeted to receptors for vascular endothelial growth factor in tumor vasculature. J Biomed Opt. 2013;18(12):126014.
149. Toyota T, Fujito H, Suganami A, Ouchi T, Ooishi A, Aoki A, et al. Near-infrared-fluorescence imaging of lymph nodes by using liposomally formulated indocyanine green derivatives. Bioorg Med Chem. 2014;22(2):721–7.
150. Zheng X, Xing D, Zhou F, Wu B, Chen WR. Indocyanine green-containing nanostructure as near infrared dual-functional targeting probes for optical imaging and photothermal therapy. Mol Pharm. 2011;8(2):447–56.
151. Zheng X, Zhou F, Wu B, Chen WR, Xing D. Enhanced tumor treatment using biofunctional indocyanine green-containing nanostructure by intratumoral or intravenous injection. Mol Pharm. 2012;9(3):514–22.
152. Zhong J, Yang S. Contrast-enhanced photoacoustic imaging using indocyanine green-containing nanoparticles. J Innov Opt Health Sci. 2014; 7(1350029-1-7) http://www.worldscientific.com/doi/abs/10.1142/S1793545813500296.
153. Zhong J, Yang S, Zheng X, Zhou T, Xing D. In vivo photoacoustic therapy with cancer-targeted indocyanine green-containing nanoparticles. Nanomedicine (Lond). 2013;8(6):903–19.
154. Koo J, Jeon M, Oh Y, Kang HW, Kim J, Kim C, et al. In vivo non-ionizing photoacoustic mapping of sentinel lymph nodes and bladders with ICG-enhanced carbon nanotubes. Phys Med Biol. 2012;57(23): 7853–62.
155. Jeong HS, Lee CM, Cheong SJ, Kim EM, Hwang H, Na KS, et al. The effect of mannosylation of liposome-encapsulated indocyanine green on imaging of sentinel lymph node. J Liposome Res. 2013;23(4):291–7.
156. Yanina IY, Kochubey VI, Tuchin VV, Portnov SA, Svenskaya YI, Gorin DA, et al. Effect of bacterial lectin on acceleration of fat cell lipolysis at in vitro diode laser treatment using encapsulated ICG. Proc SPIE 2012;8337:83370F-1-7 http://proceedings.spiedigitallibrary.org/proceeding.aspx?articleid=1312228.
157. Nishiyama Y, Kinouchi H, Senbokuya N, Kato T, Kanemaru K, Yoshioka H, et al. Endoscopic indocyanine green video angiography in aneurysm surgery: an innovative method for intraoperative assessment of blood flow in vasculature hidden from microscopic view. J Neurosurg. 2012;117(2): 302–8.
158. Topaloglu N, Gulsoy M, Yuksel S. Antimicrobial photodynamic therapy of resistant bacterial strains by indocyanine green and 809-nm diode laser. Photomed Laser Surg. 2013;31(4):155–62.
159. Tuchina ES, Tuchin VV, Khlebtsov BN, Khlebtsov NG. Phototoxic effect of conjugates of plasmonresonance nanoparticles with indocyanine green dye on *Staphylococcus aureus* induced by IR laser radiation. Quant Electron. 2011;41(4):354–9.

Section II

Imaging Systems

D-Light P System (Karl Storz)

5

Rutger M. Schols, Nicole D. Bouvy, Ronald M. van Dam, and Laurents P.S. Stassen

Abbreviations

CA	Cystic artery
CBD	Common bile duct
CD	Cystic duct
CVS	Critical view of safety
ICG	Indocyanine green
NIRF	Near-infrared fluorescence
TBR	Target-to-background ratio
WL	White light

Introduction

Our experiences with the laparoscopic fluorescence imaging system, after which Chap. 5 is named, are merely based on clinical feasibility studies, as previously reported [1, 2], performed during laparoscopic cholecystectomy. The latter concerns the most performed minimally invasive surgical procedure worldwide. Bile duct injury during this type of surgery is a rare but severe complication, with a reported incidence of 0.3–0.7 % [3–6]. Misidentification of the extrahepatic bile duct anatomy is the main cause of such injury [7]. Furthermore, vasculobiliary injury has negative impact on the outcome of bile duct injury repair [8].

To reduce the risk of bile duct injury, the Critical View of Safety (CVS) technique was introduced by Strasberg in 1995 [9] and recommended by SAGES in 2005 [10]. CVS is especially aimed at mobilizing the gallbladder neck from the liver, in order to obtain a circumferential identification of the cystic duct. CVS is established when two anatomical windows have been created during dissection of Calot's triangle: one window between the cystic artery, cystic duct and gallbladder, another window between the cystic artery, gallbladder and liver.

Near-infrared fluorescence (NIRF) laparoscopy with preoperative intravenous injection of indocyanine green (ICG) is a promising, innovative, and noninvasive imaging method for enhanced intraoperative visualization of biliary anatomy. It has the potential to improve the outcome of the laparoscopic cholecystectomy procedure [11, 12]. ICG is cleared quickly and exclusively by the liver after intravenous administration and has few, if any, side effects, making it an ideal fluorophore for this application.

Electronic supplementary material: The online version of this chapter (doi:10.1007/978-3-319-15678-1_5) contains supplementary material, which is available to authorized users. Videos can also be accessed at http://link.springer.com/chapter/10.1007/978-3-319-15678-1_5.

R.M. Schols, M.D., Ph.D. (✉)
N.D. Bouvy, M.D. Ph.D. • R.M. van Dam, M.D., Ph.D.
L.P.S. Stassen, M.D., Ph.D.
Department of Surgery, Maastricht University Medical Center, P. Debyelaan 25, Maastricht, Limburg 6229 HX, The Netherlands
e-mail: rutgermschols@hotmail.com; n.bouvy@mumc.nl; r.van.dam@mumc.nl; lps.stassen@mumc.nl

F.D. Dip et al. (eds.), *Fluorescence Imaging for Surgeons: Concepts and Applications*, DOI 10.1007/978-3-319-15678-1_5, © Springer International Publishing Switzerland 2015

The main goal of our clinical experiments was to test the practical applicability of the D-Light P System for NIRF imaging during elective laparoscopic cholecystectomy for obtaining real-time intraoperative early biliary and confirmative arterial imaging [1, 2].

Methods

As previously described [1, 2], laparoscopic cholecystectomies using the D-Light P System were performed at the Department of Surgery of the Maastricht University Medical Center (MUMC, Maastricht, The Netherlands). The institutional review board (IRB) of the MUMC approved of this study. Consecutive male and female patients, aged 18 years and above, scheduled for elective laparoscopic cholecystectomy and with written informed consent, were eligible for inclusion in this study. Indications for surgery were: cholecystolithiasis, cholecystitis, and cholecystectomy after biliary pancreatitis. Patients were excluded in case of liver or renal insufficiency, known iodine or ICG hypersensitivity and pregnancy (all are contraindications for the use of intravenous ICG).

Laparoscopic Fluorescence Imaging System

The experiences with fluorescence laparoscopy, described in this chapter, have been obtained using the D-Light P System (Tricam SLII; Karl Storz GmbH & CO. KG, Tuttlingen, Germany). This laparoscopic fluorescence imaging system includes a plasma light guide and 30° 10-mm rigid laparoscope, containing an optical filter system for white light, autofluorescence and ICG imaging. In the described clinical experiments the system was used for intraoperative conventional imaging (WL-mode) and near-infrared fluorescence imaging (ICG-mode). The system is equipped with a foot pedal, allowing the surgeon to easily switch from WL-mode to ICG-mode, and back. Because of the instantaneous changing of images and the stable position of the laparoscope, anatomical orientation can be maintained. However, direct fluorescence image-overlay on the conventional anatomical image is not yet possible with this system.

ICG Administration

All patients received an intravenous injection of 2.5 mg of ICG (Infracyanine®; SERB, France) directly after induction of anesthesia, to obtain intraoperative NIRF of the extra-hepatic bile ducts. In part of the study population, a repeat intravenous injection of 2.5 mg of ICG was administered at establishment of CVS for concomitant arterial and biliary fluorescence laparoscopy.

Surgical and Study Procedure

Laparoscopic cholecystectomy was performed according to the Dutch Guidelines and Best Practice [13], based on the CVS-technique. Initial NIRF imaging was at the first view of the liver hilum. Then, NIRF laparoscopy was conducted every 5–10 min until establishment of CVS. Simultaneous fluorescence cholangiography and angiography was obtained from 10 s up to approximately 1 min after repeat intravenous ICG injection at establishment of CVS. See Fig. 5.1 for a flowchart of the study procedures.

Intraoperatively a researcher systematically registered whether the localization of the common bile duct, cystic duct, and cystic artery could be identified at set time points, comparing the WL-camera mode to NIRF-mode. For agreement on the identification of the aforementioned structures, the attending surgeon was consulted. A structure was scored as "identified" if its localization was confirmed with great certainty by the experienced surgeon. In case of the common bile duct this does explicitly not mean that it was surgically exposed, as this is contradictory to the CVS-technique.

During all laparoscopic cholecystectomies, the full surgical procedure was recorded on DVD. After completion of surgery, length of time

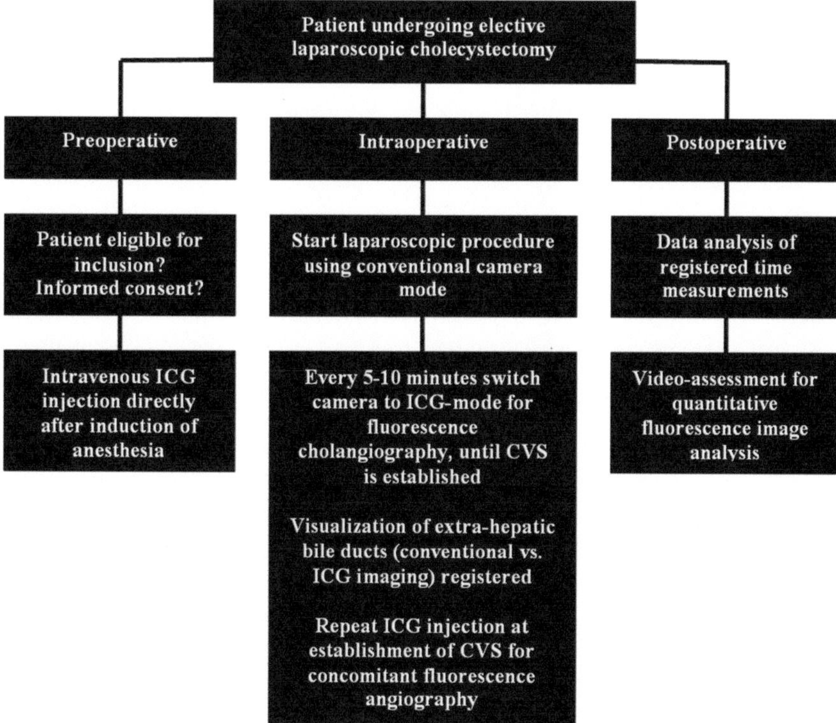

Fig. 5.1 Flowchart of the study procedures regarding the clinical experiments described in this chapter. With permission from: Schols RM, Bouvy ND, van Dam RM, Masclee AA, Dejong CH, Stassen LP; Combined vascular and biliary fluorescence imaging in laparoscopic cholecystectomy. Surgical Endoscopy 2013;27(12), page 4514 © in 2013 Springer Science and Business Media, New York [2]

between the "introduction of the laparoscope" until "the first recognition of the following components" was calculated, based on the intraoperative registration: common bile duct (CBD), cystic duct (CD), cystic artery (CA), and Critical View of Safety (CVS). Using conventional imaging the CBD is regularly not displayed; this is partly a result of the CVS technique.

Quantitative Analysis of Fluorescence Images

For objective assessment of the degree of fluorescence illumination in the extra-hepatic bile ducts and artery, OsiriX 5.5.1 Imaging Software was used. The fluorescence images were analyzed by determining target-to-background ratio (TBR). TBR was defined as the mean fluorescence intensity (FI) of two point regions of interest (ROIs) in the target (i.e. CBD, CD or CA) minus the mean fluorescence intensity of two background (BG) ROIs in the liver hilum, divided by the mean fluorescence intensity of the two background ROIs in the liver hilum; in formula: TBR = (FI of target − FI of BG)/FI of BG.

Statistical Analysis

Regarding the primary endpoint (i.e. moment of clear visualization of the extra-hepatic bile duct anatomy), a paired *T*-test was applied for determination of possible significant differences between the time measurements from "introduction of laparoscope" until "identification of CD/CBD"; comparing fluorescence imaging with conventional imaging.

Results

Thirty patients, undergoing an elective laparoscopic cholecystectomy, were included in this study [1, 2]. Patient characteristics are summarized in Table 5.1.

The median time interval from ICG injection until first fluorescence laparoscopic recordings with the D-Light P System was 34 (19–67) min. Time until first fluorescence visualization depended mainly on whether adhesiolysis had to be conducted before initial exposure of the liver hilum could be obtained. After start of surgery, the common bile duct could be identified significantly earlier using NIRF mode compared with WL-mode (median 23 min and 33 min respectively; *p-value<0.001*). Using fluorescence laparoscopy the cystic duct was delineated after an average of 25 min, whereas using WL-mode this took on average of 36 min (*p-value<0.001*). Intraoperative observations are summarized in Table 5.2.

Using fluorescence cholangiography, the CBD and the CD could be clearly visualized and delineated before dissection of Calot's triangle respectively in 25/30 patients (83 %) and 29/30 patients (97 %). In four of five cases (17 %), in which the CBD could not be visualized using fluorescence imaging, a chronic inflammation of the gallbladder was present. One patient appeared to have a chronic, focally active cholecystitis. BMI of these patients ranged from 26.35 to 31.14 kg/m^2. However, in the 25 patients in whom the CBD could be delineated before dissection, there were also cases of chronic inflamed gallbladders and BMIs within the same range as these 5 patients.

Conventional laparoscopy provided certainty on the course of the CBD and the CD in respectively 22/30 patients and 29/30 patients. See Figs. 5.2, 5.3 and 5.4 for an example of the peroperative WL-images and corresponding NIRF-images during the dissection of CVS. A short video fragment illustrating the application of near-infrared fluorescence cholangiography

Table 5.1 Patient characteristics

No. of patients	30
Gender	11 male
	19 female
Age (y)	53 (26–81)
Body Mass Index (kg/m^2)	26.7 (19.7–36.8)
Indication for surgery	20 cholecystolithiasis
	8 cholecystitis
	2 cholecystectomy after biliary pancreatitis
ICG administration	2.5 mg directly after induction of anesthesia ($n=30$)
	2.5 mg at establishment of CVS ($n=15$)

ICG = Indocyanine green; CVS = Critical View of Safety
With permission from: Schols RM, Bouvy ND, van Dam RM, Masclee AA, Dejong CH, Stassen LP; Combined vascular and biliary fluorescence imaging in laparoscopic cholecystectomy. Surgical Endoscopy 2013;27(12), page 4514 © in 2013 Springer Science and Business Media, New York [2]

Table 5.2 Intraoperative observations

	Fluorescence cholangiography ($n=30$)				Fluorescence angiography ($n=15$)	
Clear identification of…	Cystic duct		Common bile duct		Cystic artery	
	NIRF	WL	NIRF	WL	NIRF	WL
Patients	29/30 (97 %)	29/30 (97 %)	25/30 (83 %)	22/30 (73 %)	13/15[a] (87 %)	13/15[a] (87 %)
Median time in minutes [range]	25 [5–49]	36 [9–69]	23 [5–65]	33 [9–65]		
	*p<0.001**		*p<0.001**			

NIRF = near-infrared fluorescence mode of the laparoscope; WL = white light camera mode of the laparoscope
[a]In 2 cases the cystic artery was already dissected before repeat ICG injection
*Significant difference ($p<0.05$)
With permission from: Schols RM, Bouvy ND, van Dam RM, Masclee AA, Dejong CH, Stassen LP; Combined vascular and biliary fluorescence imaging in laparoscopic cholecystectomy. Surgical Endoscopy 2013;27(12), page 4514 © in 2013 Springer Science and Business Media, New York [2]

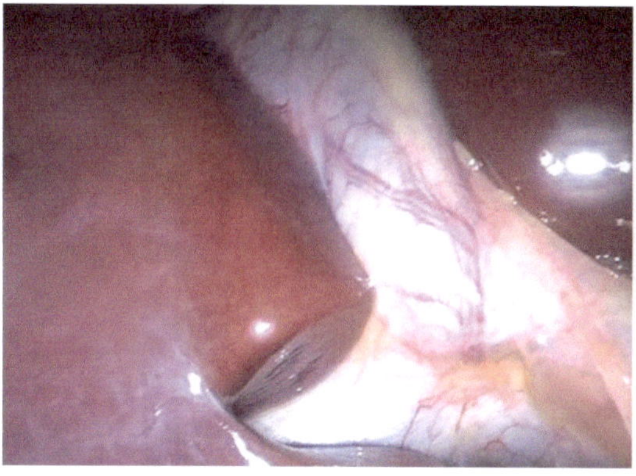

Fig. 5.2 WL image (*left*) and corresponding fluorescence image (*right*) at first view of the liver hilum; images were recorded directly after initial exposure of the liver hilum. The liver and extra-hepatic bile ducts are illuminated *bright blue*. With permission from Schols RM, Bouvy ND, Masclee AA, van Dam RM, Dejong CH, Stassen LP. Fluorescence cholangiography during laparoscopic cholecystectomy: a feasibility study on early biliary tract delineation. Surgical Endoscopy 2013;27(5):1534 © in 2013 Springer Science and Business Media, New York [1]

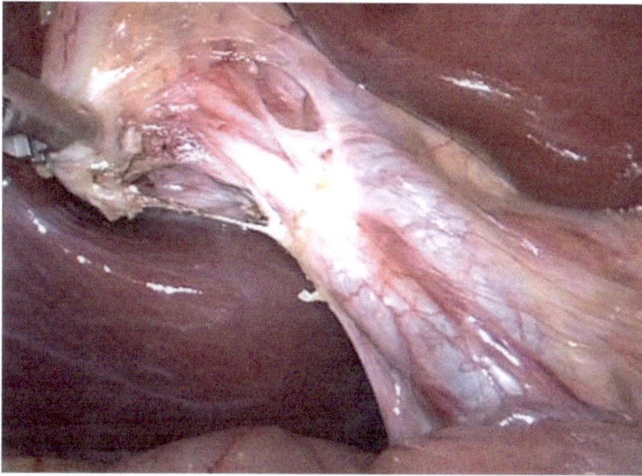

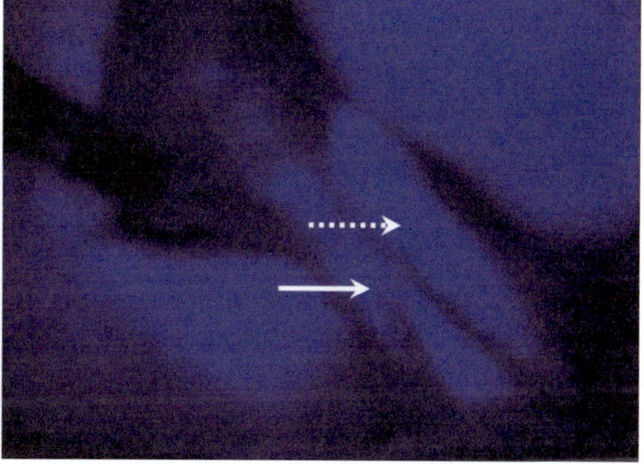

Fig. 5.3 WL image (*left*) and corresponding fluorescence image (*right*) during dissection of Calot's triangle; note the parallel junction between the cystic duct (*continuous arrow*) and the common hepatic duct (*interrupted arrow*) in the fluorescence image. The liver remains illuminated as well. TBR of CD was 3.6 and TBR of CBD was 3.9 (28 min after ICG injection). With permission from Schols RM, Bouvy ND, Masclee AA, van Dam RM, Dejong CH, Stassen LP. Fluorescence cholangiography during laparoscopic cholecystectomy: a feasibility study on early biliary tract delineation. Surgical Endoscopy 2013;27(5):1534 © in 2013 Springer Science and Business Media, New York [1]

during one of the surgical procedures is included as online supplement (see Video 5.1).

In one patient a parallel course [14] of the cystic duct and common hepatic duct was identified (see Fig. 5.3).

In one patient conversion to open cholecystectomy followed, due to insufficient sight on the liver hilum after an early perforation of a very thin-walled gallbladder, causing persistent nuisance due to bile leakage.

Critical view of safety was obtained in a median of 45 min after incision. Concomitant fluorescence angiography of the cystic artery (see Fig. 5.5 and Video 5.2) was successful in 13 of 15 patients (87 %). In two patients, the cystic artery was already ligated in an earlier stage of dissection.

In the total operation time (median 90 min) 1 up to 2 min was spent for the intraoperative use of the fluorescence technique. As a preventive measure, in 3/30 patients the gall bladder was punctured and drained during surgery to facilitate manipulation; this did neither compromise fluorescence imaging of the biliary tract nor the postoperative course.

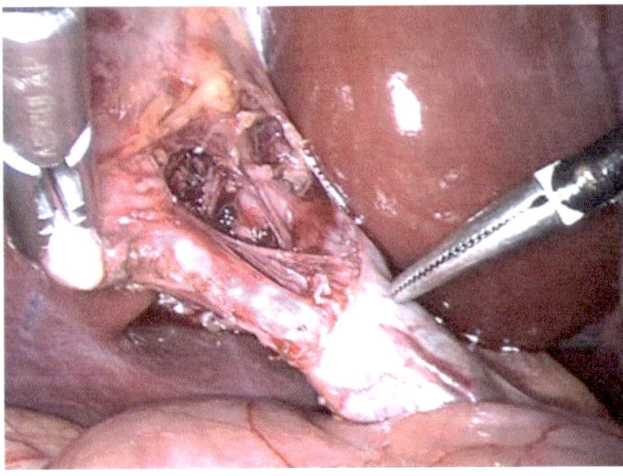

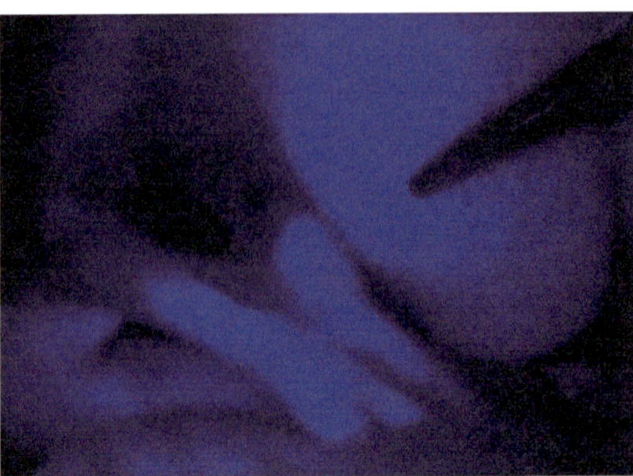

Fig. 5.4 WL image (*left*) and corresponding fluorescence image (*right*) in a later stage of dissection of Calot's triangle; the fluorescence image confirms the divergent course of the two separate bile ducts as already identified in an earlier stage (see Fig. 5.2). The liver remains illuminated. TBR of CD of 7.9 and TBR of CBD of 6.3 (36 min post-ICG injection). With permission from Schols RM, Bouvy ND, Masclee AA, van Dam RM, Dejong CH, Stassen LP. Fluorescence cholangiography during laparoscopic cholecystectomy: a feasibility study on early biliary tract delineation. Surgical Endoscopy 2013;27(5):1534 © in 2013 Springer Science and Business Media, New York [1]

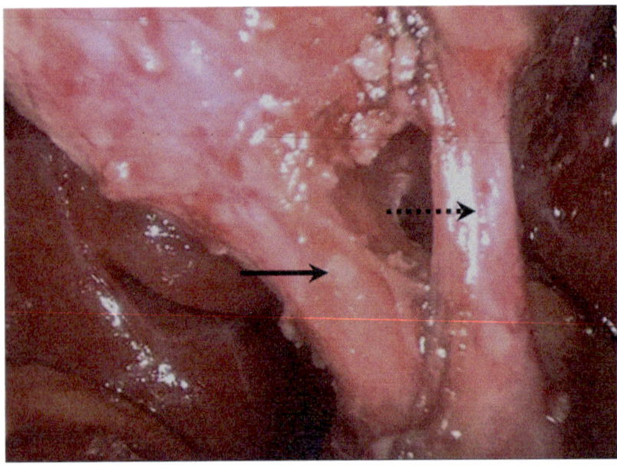

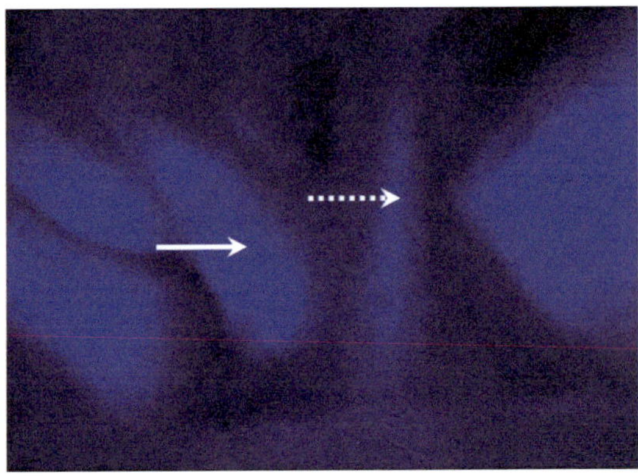

Fig. 5.5 WL image (*left*) and corresponding fluorescence image (*right*) at establishment of Critical View of Safety; snap-shots were taken 10 s after repeat ICG injection. The cystic duct (*continuous arrow*) and cystic artery (*interrupted arrow*) are clearly exposed. With permission from: Schols RM, Bouvy ND, van Dam RM, Masclee AA, Dejong CH, Stassen LP; Combined vascular and biliary fluorescence imaging in laparoscopic cholecystectomy. Surgical Endoscopy 2013;27(12), page 4514 © in 2013 Springer Science and Business Media, New York [2]

Target-to-Background Ratio

Mean TBR at establishment of CVS (median 63 min after ICG administration) amounted to 5.6 for the CD, 6.8 for the CBD, and 4.8 for the CA.

Complications

No per- or postoperative complications regarding the extra-hepatic bile ducts or arteries occurred. There was one conversion from a laparoscopic to an open procedure. None of the patients developed any adverse reactions to the injected ICG.

Discussion

We report our first clinical experiences with the D-Light P System during elective laparoscopic cholecystectomy, as previously described [1, 2]. The addition of the NIRF imaging technique was useful for earlier bile duct delineation during dissection of CVS [1]. Next, it was found that repeat

ICG injection at establishment of CVS was helpful in the confirmation of the course of the cystic artery [2]. The fluorescence technique provided a significantly earlier identification and a clear delineation (high TBR) of the common bile duct, the cystic duct and the cystic artery, therewith assisting in safe and efficient dissection of the gallbladder according to the CVS technique.

The NIRF technique after ICG injection has been evaluated in open, laparoscopic, single-incision laparoscopic and robotic single-site cholecystectomy [15–20]. Promising results were presented for successful intraoperative identification of the common bile duct and the cystic duct, compared to conventional laparoscopic imaging. Our observations showed moreover that fluorescence laparoscopy provided significantly earlier identification of the extra-hepatic bile ducts during dissection.

Real-time simultaneous imaging of the bile ducts and the arterial anatomy (i.e. hepatic and cystic arteries) can also be obtained [15, 21–23]. Our experiments confirmed the possibility of concomitant vascular (i.e. cystic artery) and biliary (i.e. cystic duct and common bile duct) imaging at establishment of Critical View of Safety. Implementation of conducting both fluorescence cholangiography and angiography can be helpful in cases in which the establishment of CVS is challenging. Furthermore, it may facilitate the intraoperative detection of the bile ducts and possible concomitant vasculobiliary injury (e.g. hepatic artery). The latter is an increasingly recognized complication of laparoscopic cholecystectomy [24]. Such combined injury negatively affects the outcome of bile duct injury repair [8, 25].

Compared with other available laparoscopic fluorescence imaging systems [26], the D-Light P System offers the advantage that it is equipped with a foot pedal allowing the surgeon to easily switch between conventional and fluorescence imaging using the same laparoscope.

A disadvantage of the current and the other available devices is that it does not yet possess the ability to real-time overlay the NIRF images with the conventional WL images. Another improvement could be contrast enhancement of Calot's triangle during NIRF imaging for even better delineation of the biliary tract. The present plasma light guide appeared quite rigid and is, in combination with the 30° laparoscope, sometimes inconvenient to handle during surgery.

With respect to the image quality of the D-Light P System itself, first clinical experiences with an updated version (Image 1 High Definition) have just been reported [27]. Improved sensitivity of the CCD chip not only increased fluorescence brightness but also allowed for better anatomical orientation in the fluorescence mode.

The negligible extension of operation duration in this study, related to the use of the fluorescence technique, will probably further decrease with growing experience. Looking at the results of the quantitative fluorescence image analysis, the moment of ICG administration (immediately after induction of anesthesia) seems appropriate for the laparoscopic cholecystectomy procedure: a timely and continuous intraoperative detection of biliary structures can be obtained. Also from a practical and logistical point of view the applied timing of administration is appropriate.

In our feasibility study not only uncomplicated cholecystolithiasis patients, but also more complicated cases, such as cholecystitis or cholecystectomy after biliary pancreatitis, were included. Fluorescence cholangiography and angiography were successful in almost all cases, and no difference was found in image quality between the uncomplicated and complicated procedures. However, presence of factors as (either chronic or active) cholecystitis or obesity might compromise the efficacy of the new imaging modality. The actual impact of such conditions should be further investigated by conducting a univariate and multivariate analysis in a study population (large enough for this kind of analysis) to determine the risk factors. This was not possible in the current study population.

It could be argued that the fluorescence laparoscopy technique is especially useful for the less experienced surgeon to perform safe laparoscopic gallbladder removal, or in more complicated cholecystectomies. However, it may be most effective when performed routinely in all cases. Increased costs are involved in terms of the light source, camera, and fluorescent dye. However, it is anticipated that worldwide, standard application of this technique will be accompanied by cheaper devices and shorter operation duration.

Fluorescence laparoscopy could also be implemented as a final check of the liver hilum before closing the pneumoperitoneum, in order to rule out or detect bile duct or vasculobiliary injury. This could be checked through the presence or absence of fluorescence contrast leakage into the abdomen.

The NIRF laparoscopy technique possesses potential to enter the clinical practice in the near future. However, optimization of the current laparoscopic imaging systems would make them more easily applicable in the operating room. Examples of such optimization are: the ability to display the fluorescence image in anatomical context; increased penetration depth at which the imaging system can detect a fluorophore in tissues (currently up to maximum of 1 cm); increased fluorescence capabilities of contrast agents [28].

In light of the last comment, we recently conducted animal experiments to assess the performance of a pre-clinical near-infrared dye CW800-CA. It concerns a carboxylate of IRDye® 800CW (LI-COR, Lincoln, Nebraska), a tetrasulfonated heptamethine indocyanine. Emission of this dye is quite similar to ICG (also at 800 nm). However, after intravenous injection it is not only more rapidly cleared by the liver and excreted into bile, but also by the kidneys and excreted into the urine. As a result this dye is eligible for imaging of the biliary anatomy and of the ureters [29].

Most importantly, we found that the D-Light P System, primarily developed and commercialized for ICG imaging, seems also suitable for CW800-CA fluorescence laparoscopy. Promising results for enhanced fluorescence visualization of biliary anatomy and ureters were achieved [30, 31].

References

1. Schols RM, Bouvy ND, Masclee AA, van Dam RM, Dejong CH, Stassen LP. Fluorescence cholangiography during laparoscopic cholecystectomy: a feasibility study on early biliary tract delineation. Surg Endosc. 2013;27:1530–6.
2. Schols RM, Bouvy ND, van Dam RM, Masclee AA, Dejong CH, Stassen LP. Combined vascular and biliary fluorescence imaging in laparoscopic cholecystectomy. Surg Endosc. 2013;27:4511–7.
3. Flum DR, Dellinger EP, Cheadle A, Chan L, Koepsell T. Intraoperative cholangiography and risk of common bile duct injury during cholecystectomy. JAMA. 2003;289:1639–44.
4. Fletcher DR, Hobbs MS, Tan P, Valinsky LJ, Hockey RL, Pikora TJ, Knuiman MW, Sheiner HJ, Edis A. Complications of cholecystectomy: risks of the laparoscopic approach and protective effects of operative cholangiography: a population-based study. Ann Surg. 1999;229:449–57.
5. Nuzzo G, Giuliante F, Giovannini I, Ardito F, D'Acapito F, Vellone M, Murazio M, Capelli G. Bile duct injury during laparoscopic cholecystectomy: results of an Italian national survey on 56 591 cholecystectomies. Arch Surg. 2005;140:986–92.
6. Waage A, Nilsson M. Iatrogenic bile duct injury: a population-based study of 152 776 cholecystectomies in the Swedish Inpatient Registry. Arch Surg. 2006;141:1207–13.
7. Way LW, Stewart L, Gantert W, Liu K, Lee CM, Whang K, Hunter JG. Causes and prevention of laparoscopic bile duct injuries: analysis of 252 cases from a human factors and cognitive psychology perspective. Ann Surg. 2003;237:460–9.
8. Sarno G, Al-Sarira AA, Ghaneh P, Fenwick SW, Malik HZ, Poston GJ. Cholecystectomy-related bile duct and vasculobiliary injuries. Br J Surg. 2012;99: 1129–36.
9. Strasberg SM, Hertl M, Soper NJ. An analysis of the problem of biliary injury during laparoscopic cholecystectomy. J Am Coll Surg. 1995;180:101–25.
10. Kuwada T. Highlights of the Society of American Gastrointestinal and Endoscopic Surgeons 2005, Annual Meeting April 13–16, 2005; Ft Lauderdale, FL. Medscape General Surgery: Conference reports, http://www.medscape.com/viewarticle/506432
11. Agarwal BB. Patient safety in laparoscopic cholecystectomy. Arch Surg. 2009;144:979. author reply 979.
12. Buddingh KT, Nieuwenhuijs VB, van Buuren L, Hulscher JB, de Jong JS, van Dam GM. Intraoperative assessment of biliary anatomy for prevention of bile duct injury: a review of current and future patient safety interventions. Surg Endosc. 2011;25:2449–61.
13. Lange JF, Stassen LPS. Best practice in verband met de techniek van laparoscopische cholecystectomie (Critical View of Safety [CVS] in 7 stappen).

Werkgroep Endoscopische Chirurgie van de Nederlandse vereniging voor Heelkunde. 2006; http://www.heelkunde.nl/uploads/_6/re/_6reZZkgrYUAuCG6uvcN-A/richtlijn_galsteen.pdf
14. Johnston EV, Anson BJ. Variations in the formation and vascular relationships of the bile ducts. Surg Gynecol Obstet. 1952;94:669–86.
15. Tagaya N, Shimoda M, Kato M, Nakagawa A, Abe A, Iwasaki Y, Oishi H, Shirotani N, Kubota K. Intraoperative exploration of biliary anatomy using fluorescence imaging of indocyanine green in experimental and clinical cholecystectomies. J Hepatobiliary Pancreat Sci. 2010;17:595–600.
16. Ishizawa T, Bandai Y, Ijichi M, Kaneko J, Hasegawa K, Kokudo N. Fluorescent cholangiography illuminating the biliary tree during laparoscopic cholecystectomy. Br J Surg. 2010;97:1369–77.
17. Ishizawa T, Kaneko J, Inoue Y, Takemura N, Seyama Y, Aoki T, Beck Y, Sugawara Y, Hasegawa K, Harada N, Ijichi M, Kusaka K, Shibasaki M, Bandai Y, Kokudo N. Application of fluorescent cholangiography to single-incision laparoscopic cholecystectomy. Surg Endosc. 2011;25:2631–6.
18. Aoki T, Murakami M, Yasuda D, Shimizu Y, Kusano T, Matsuda K, Niiya T, Kato H, Murai N, Otsuka K, Kusano M, Kato T. Intraoperative fluorescent imaging using indocyanine green for liver mapping and cholangiography. J Hepatobiliary Pancreat Sci. 2010; 17:590–4.
19. Spinoglio G, Priora F, Bianchi PP, Lucido FS, Licciardello A, Maglione V, Grosso F, Quarati R, Ravazzoni F, Lenti LM. Real-time near-infrared (NIR) fluorescent cholangiography in single-site robotic cholecystectomy (SSRC): a single-institutional prospective study. Surg Endosc. 2013;27:2156–62.
20. Buchs NC, Hagen ME, Pugin F, Volonte F, Bucher P, Schiffer E, Morel P. Intra-operative fluorescent cholangiography using indocyanine green during robotic single site cholecystectomy. Int J Med Robot. 2012;8: 436–40.
21. Ashitate Y, Stockdale A, Choi HS, Laurence RG, Frangioni JV. Real-time simultaneous near-infrared fluorescence imaging of bile duct and arterial anatomy. J Surg Res. 2012;176:7–13.
22. Mitsuhashi N, Kimura F, Shimizu H, Imamaki M, Yoshidome H, Ohtsuka M, Kato A, Yoshitomi H, Nozawa S, Furukawa K, Takeuchi D, Takayashiki T, Suda K, Igarashi T, Miyazaki M. Usefulness of intraoperative fluorescence imaging to evaluate local anatomy in hepatobiliary surgery. J Hepatobiliary Pancreat Surg. 2008;15:508–14.
23. Kaneko J, Ishizawa T, Masuda K, Kawaguchi Y, Aoki T, Sakamoto Y, Hasegawa K, Sugawara Y, Kokudo N. Indocyanine green reinjection technique for use in fluorescent angiography concomitant with cholangiography during laparoscopic cholecystectomy. Surg Laparosc Endosc Percutan Tech. 2012;22:341–4.
24. Pulitano C, Parks RW, Ireland H, Wigmore SJ, Garden OJ. Impact of concomitant arterial injury on the outcome of laparoscopic bile duct injury. Am J Surg. 2011;201:238–44.
25. Schmidt SC, Langrehr JM, Raakow R, Klupp J, Steinmuller T, Neuhaus P. Right hepatic lobectomy for recurrent cholangitis after combined bile duct and right hepatic artery injury during laparoscopic cholecystectomy: a report of two cases. Langenbecks Arch Surg. 2002;387:183–7.
26. Ankersmit M, van der Pas MH, van Dam DA, Meijerink WJ. Near infrared fluorescence lymphatic laparoscopy of the colon and mesocolon. Colorectal Dis. 2011;13 Suppl 7:70–3.
27. Verbeek FP, Schaafsma BE, Tummers QR, van der Vorst JR, van der Made WJ, Baeten CI, Bonsing BA, Frangioni JV, van de Velde CJ, Vahrmeijer AL, Swijnenburg RJ. Optimization of near-infrared fluorescence cholangiography for open and laparoscopic surgery. Surg Endosc. 2014;28:1076.
28. Schols RM, Bouvy ND, van Dam RM, Stassen LP. Advanced intraoperative imaging methods for laparoscopic anatomy navigation: an overview. Surg Endosc. 2013;27:1851–9.
29. Frangioni JV, Tanaka E, Borud LJ. Intraoperative imaging methods. Patent Cooperation Treaty WO/2008/076467, http://patentscope.wipo.int/search/en/WO2008076467
30. Schols RM, Lodewick TM, Bouvy ND, van Dam DA, Meijerink WJHJ, van Dam GM, Dejong CHC, Stassen LPS. Near-infrared fluorescence laparoscopy of the cystic duct and artery in pigs: performance of a preclinical dye. J Laparoendosc Adv Surg Tech. 2014;24(5):318–22.
31. Schols RM, Lodewick TM, Bouvy ND, van Dam GM, Dejong CH, Stassen LP. Application of a new dye for near-infrared fluorescence laparoscopy of the ureters: demonstration in a pig model. Dis Colon Rectum. 2014;57(3):407–11

Robotic Surgery Using Firefly System

6

Giuseppe Spinoglio, Alessandra Marano, and Giampaolo Formisano

Introduction

The imaging technique based on indocyanine green (ICG) fluorescence has been widely used for more than 40 years, especially to study blood flow and microcirculation [1]. In general surgery, this method was first applied to perform sentinel lymph node (SLN) biopsies in patients affected by breast and colorectal cancer [2, 3]. Since then, ICG fluorescence has been widely employed in many surgical procedures at the beginning with open approach and later on with laparoscopic technique by means of different imaging systems.

In 2010, a near infrared laser light system has been integrated in the da Vinci® Si™ HD robotic system (da Vinci® Surgical System; Intuitive Surgical Inc., Sunnyvale, California) allowing a fluorescence-guided surgery; it is capable of emitting laser light that is closer to infrared light with the ability to switch between white light and near infrared (NIR) light view in real time. The NIR-fluorescence imaging system of the da Vinci® Si™ HD robot comprises: an endoscope that is able to provide visible light and near infrared light images, a 3D HD stereoscopic camera connected to the endoscope, and an endoscopic illuminator that through a flexible cable provides the endoscope surgeon with both near-infrared and visible lighting (Fig. 6.1). The surgeon can quickly switch between normal viewing mode (white light) to fluorescence (NIR) with a pedal on the surgical console.

There are many fields of application of ICG fluorescence in robotic general surgery, some experimental and still evolving, that include the intraoperative (IO) fluorescent cholangiography to assess biliary anatomy, the evaluation of bowel stump perfusion, the lymph nodes (LN) mapping, and the SLN biopsy in cancer surgery. In addition, ICG fluorescence can be used to endoscopically tattooing the colonic, rectal, and gastric lesions. We describe our experience with current applications of fluorescence in robotic general surgery.

G. Spinoglio, M.D. (✉)
Center of Robotic Surgery, Department of General Surgery, Humanitas Clinical and Research Hospital IRCCS, Rozzano, MI, Italy
e-mail: gspinoglio@icloud.com

A. Marano, M.D.
Department of General Surgery, Humanitas Clinical and Research Hospital IRCCS, Rozzano, MI, Italy
e-mail: alessandra.marano@hotmail.com

G. Formisano, M.D.
Department of General and Minimally Invasive Surgery, Misericordia Hospital, Grosseto, Italy
e-mail: giampaoloformisano@hotmail.com

Fluorescent Cholangiography in Single-Site™ Robotic Cholecystectomy

Bile duct injury (BDI) is a serious and feared complication of cholecystectomy: the rate of these lesions passed from 0.2 % at the time of open cholecystectomy [4] to 0.2–0.5 % in the

F.D. Dip et al. (eds.), *Fluorescence Imaging for Surgeons: Concepts and Applications*,
DOI 10.1007/978-3-319-15678-1_6, © Springer International Publishing Switzerland 2015

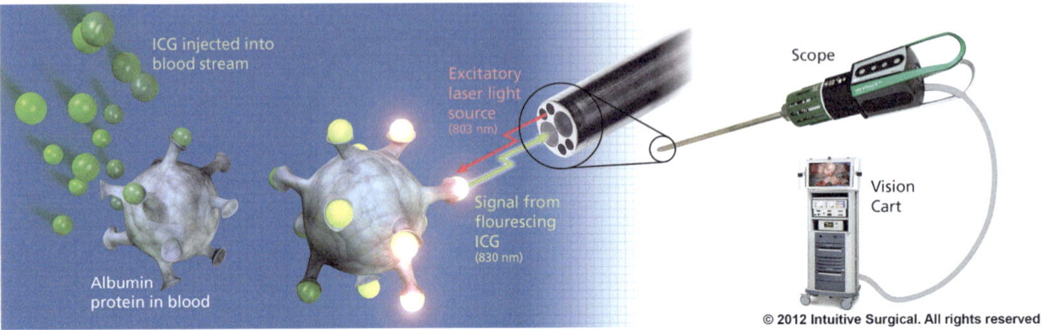

Fig. 6.1 da Vinci® indocyanine green (ICG) near infrared (NIR) fluorescence Imaging Vision System. With permission from Intuitive Surgical, Inc © 2012

current era of laparoscopic cholecystectomy [5, 6]. In order to better evaluate biliary anatomy and prevent BDI, the adoption of some crucial dissection strategies and the use of a routinary intraoperative cholangiography (IOC) have been recently emphasized [7].

While recognizing the importance of IOC, it has several disadvantages such as a longer operative time, requirement for a multidisciplinary team, staff and patient exposure to radiation and interruption of the regular workflow. Moreover, it is not always easy to interpret an IOC because it provides static images and this can often cause a "misperception" of the biliary anatomy that is considered the primary cause of BDI rather than the lack of skill or knowledge [8]. Thus, as already mentioned by some authors [7], there is a need for a "simpler method of locating the course of the ductal system during the operation, something simpler than cholangiography or ultrasonography."

The use of the ICG fluorescence imaging system could be a valid solution, as already reported during laparoscopic cholecystectomy [9–11]. ICG NIR-cholangiography is a non-invasive method and it does not require X-rays or bulky equipment such as the C-bow. Moreover, it permits direct real-time IO visualization of biliary anatomy, alternating between white light view and NIR-imaging with a simple switch system.

The abovementioned advantages of the use of ICG fluorescent cholangiography have been recently extended to robotic surgery with promising outcomes, especially during a single site approach [12–14]. The use of the NIR fluorescent vision system, integrated into the Single-Site™ da Vinci® platform has demonstrated to increase safety during single site surgery. This method allows for a safe, real time view of the anatomy of the extra-hepatic biliary tract and a time-efficient dissection of Calot's triangle.

Technique

A dose of 2.5 mg of ICG is administered intravenously during patient preparation by anesthesia, about 30–45 min before surgery. If fluorescence is not detected in the liver after 60 min, an additional dose of 2.5 mg ICG is again injected intravenously. SSRC starts in the usual fashion [15]. Once the exposure of Calot's triangle is achieved, the camera view is switched to fluorescence mode for an initial identification of the biliary anatomy (Fig. 6.2). Then, the dissection of Calot's triangle begins with the incision of the peritoneum and continues as usual, alternatively switching from white to NIR light allowing views of the fluorescent bile ducts in real time. In this way, the surgeon can follow a road map for a safe skeletonization of the cystic duct and the cystic artery, especially in presence of accessory structures (Fig. 6.3). The cystic duct may be clipped under fluorescence before sectioning, especially if it is very short and if there are problems in the biliary confluence. If there are problems with the vascular anatomy during the cystic artery skeletonization, it is possible to proceed with a further injection of 2.5 mg of ICG and, after 10–20 s, obtain a view of the hepatic and cystic arteries

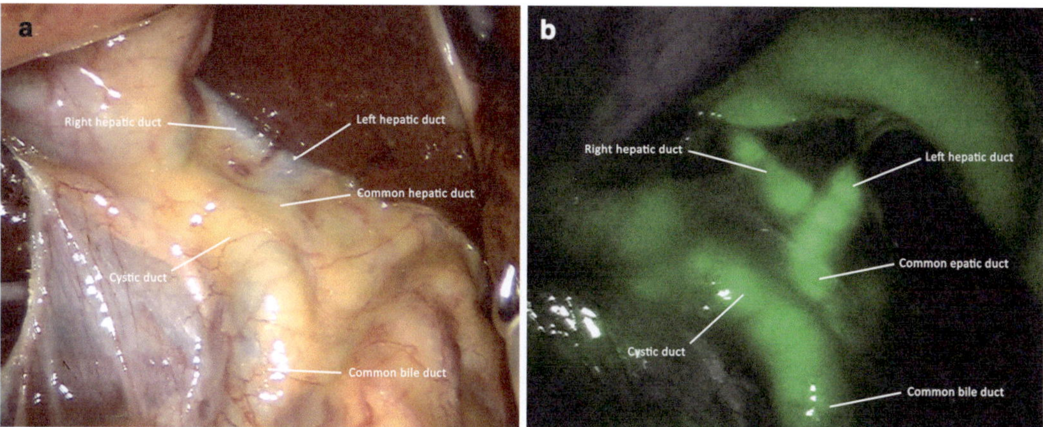

Fig. 6.2 White light (**a**) and near infrared (NIR) fluorescence (**b**) view of the biliary anatomy before Calot's dissection

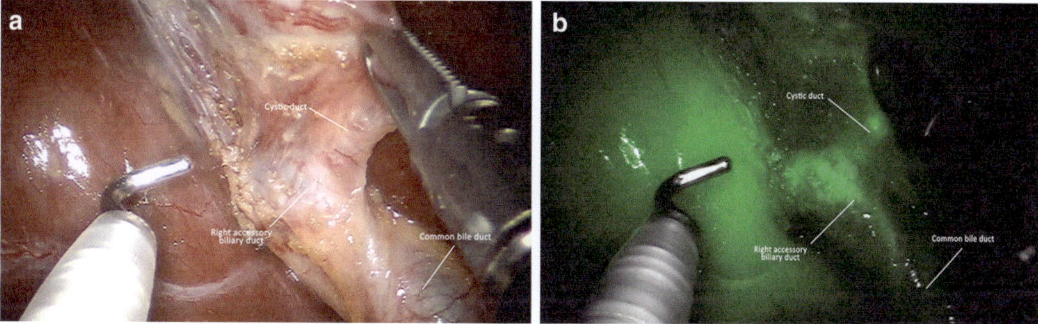

Fig. 6.3 White light (**a**) and near infrared (NIR) fluorescence (**b**) view of a right accessory biliary duct before sectioning the cyst duct

and their divisions, and avoid any damage to anomalous branches, especially the branch to the sixth hepatic segment.

During the detachment of the gallbladder from the liver bed, the use of fluorescence to define the boundary between the gallbladder and liver bed is useful, especially in cases of a thin or an intrahepatic gallbladder, and to visualize any aberrant Luschka ducts (Fig. 6.4). After the procedure, a final fluorescence view of the operative field may be prudent.

Personal Experience and Considerations

From July 2011 to December 2013, 104 patients underwent SSRC with ICG NIR-cholangiography for symptomatic cholelithiasis and gallbladder polyposis. Five out of 104 patients suffered from acute cholecystitis (4.8 %). Mean BMI was 24.7 kg/m^2.

Mean operative time and mean console time were 71 min and 24.1 min, respectively. No statistically significant differences in operative times between standard SSRC (our previous experience) and NIR-fluorescence imaging system SSRC were observed. There were no conversions, BDIs, intra- or postoperative complications or adverse events; mean hospital stay was 1.6 days.

The rates of visualization of the cystic duct, the common hepatic duct, and the common bile duct were 94 %, 71 %, 74 % prior to Calot's dissection, respectively, and 99 %, 93 %, 98 % after Calot's dissection, respectively. At least one biliary structure was visualized in 100 out of 104 patients (96 %) before Calot's dissection, and in 100 % of cases after Calot's dissection.

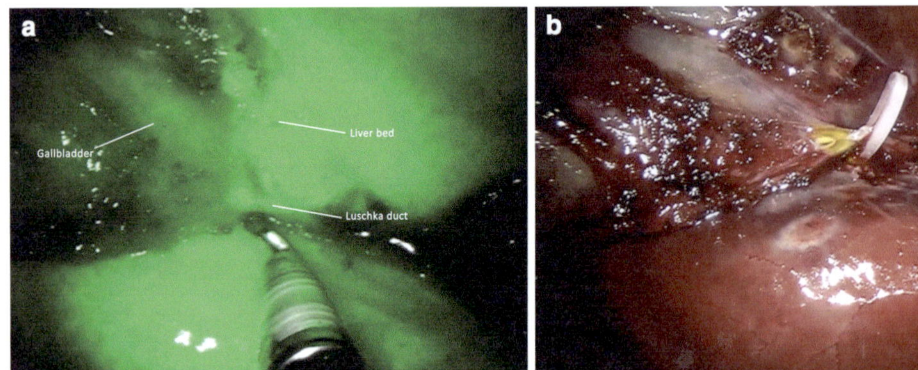

Fig. 6.4 Near infrared (NIR) fluorescence view of Luschka duct before sectioning (**a**) and white light view of Luschka duct after sectioning (**b**)

The advantages of this method compared to the traditional radiological intraoperative cholangiography are many:

- There is no interruption of the regular workflow since images are highlighted on the surgical field during the operation and the surgeon can operate both in white light and fluorescence.
- The interpretation of images is simpler because they appear in real time and can be checked with surgical maneuvers of moving structures whilst they are in view. This is in contrast to the traditional IOC images that are fixed on the screen of the radiology equipment with the surgeon working with static information.
- It is possible to control the section of the cystic duct with a clear distinction of the bile ducts.
- During the detachment of the gallbladder from the liver bed, its wall can be better highlighted and any aberrant Luscka ducts easily found.
- The fluorescent bile that usually comes out of the stump of the sectioned cystic duct or of the gallbladder in case of perforation is always clearly visible, therefore the system could potentially highlight any bile leaks. However, this event is not reported in available studies.
- If necessary, the vascular anatomy of the hepatic artery and cystic artery can be verified.
- C-bow or other equipment is not needed, thus avoiding the un-docking and re-docking of the robotic system.
- The procedure does not require any additional time compared to normal SSRC.

The main limitations of ICG-NIR fluorescent cholangiography are obesity and inflammations since fluorescence has the weakness of limited tissue penetration under NIR light [14]. Furthermore, the capability of this method to recognize biliary gallstones or other obstructions is low and of course IOC remains the best method to assess choledocolithiasis.

We can conclude that fluorescent cholangiography is a simple, fast, safe, and effective procedure. It allows clear real-time identification of extra-hepatic biliary anatomy in almost all patients, thus implementing the well-known advantages of SSRC over the traditional single-incision laparoscopic approach. It requires neither interaction of multidisciplinary teams nor additional time, it does not expose patients and staff to radiation and is not burdened by adverse reactions.

Fluorescent Intraoperative (IO) Near Infra-Red Imaging During Colorectal Surgery

The use of fluorescence in robotic colorectal surgery can be useful for some applications currently under development:

- real-time identification of vascular anatomy
- evaluation of colorectal anastomotic perfusion
- lymph node (LN) mapping and sentinel lymph node (SLN) biopsy
- tattooing for the localization of colorectal tumors.

Real-Time Identification of Vascular Anatomy

ICG can be easily injected (about 5–7 mg in a bolus) into the blood circulation during colorectal surgery, when blood vessels are exposed, allowing a direct visual observation until its elimination by hepatic clearance in 3–4 min [16]. Because of these characteristics, IO fluorescence imaging is a simple and effective method that helps surgeons to identify vascular structures and so the point section of the bowel, especially in not standardized colic resections, such as resection of the transverse and left colonic angle. Moreover, in case of anatomical abnormalities that can impair blood supply, it is ostensible that this new system could reduce the conversion rate to the open approach. Recently, Bae et al. have underlined the utility of this method in three case reports [17].

Evaluation of Colorectal Anastomotic Perfusion

The perfusion of the intestinal stumps is one of the most important factors to safely perform colorectal anastomosis [18]. The evaluation of the adequacy of stumps' perfusion is mainly based on some parameters such as presence of an active bleeding from the section line, pulsatility of mesenteric vessels, peristalsis, and the lack of discoloration of bowel segments. These findings are extremely subjective and could represent an inaccurate method to evaluate the risk of anastomotic leak [19, 20].

Many different solutions have been proposed and applied in cases of minimally invasive approaches: laser Doppler flowmetry, visible light spectroscopy, fluorescence laser angiography, narrow band laser imaging techniques, and near infrared reflection spectroscopy [20, 21]. Even if some shortcomings are present, these techniques seem to be feasible methods to evaluate colorectal anastomotic perfusion. More recently, the NIR-fluorescence imaging system of the da Vinci® Si™ HD surgical platform has been introduced and may represent an additional innovative tool.

This device can be used to assess in real time both macroscopic vascular anatomy and perfusion of intestinal stumps. Subsequently, it can support and reassure surgeons in choosing the section point of the bowel during left and right hemicolectomy and anterior resection of the rectum. It can also be useful in non-standardized colonic resections (i.e. transverse colon, splenic flexure), since vascular abnormalities can impair blood supply.

Technique

Surgery is progressively carried out with the da Vinci® Si™ HD surgical system following the usual technique of vascular control and preparation of intestinal segments. Then, the optimal point of transection is evaluated and marked by the surgeon with (visible) light. A dose of 5.0–10 mg of ICG is now administered intravenously and after approximately 30–45 s, the surgeon switches to fluorescent vision (Fig. 6.5).

If the site chosen for the section does not appear to be sufficiently perfused, the section line may be revised and the stapler can be moved to a more proximal/distal location according to the best fluorescence reflected point. In case of doubt, the test can be repeated after waiting for a few minutes to allow the dye to washout. However, depending on the tissue, the green fluorescent intensity appears different and at different times.

Regarding the colonic stumps, the vessels of the epiploic appendices and mesentery turn green first, then the green spreads across the intestinal wall. The antimesenteric border of the descending and transverse colon is always a little paler, because the vascularization of the tenia is less intense due to the thickness of muscle tissue. The perfused segments gradually become green until they assume a bright green color, in contrast with the gray segments that are not well vascularized. Further checks may be carried out before and after performing the anastomosis.

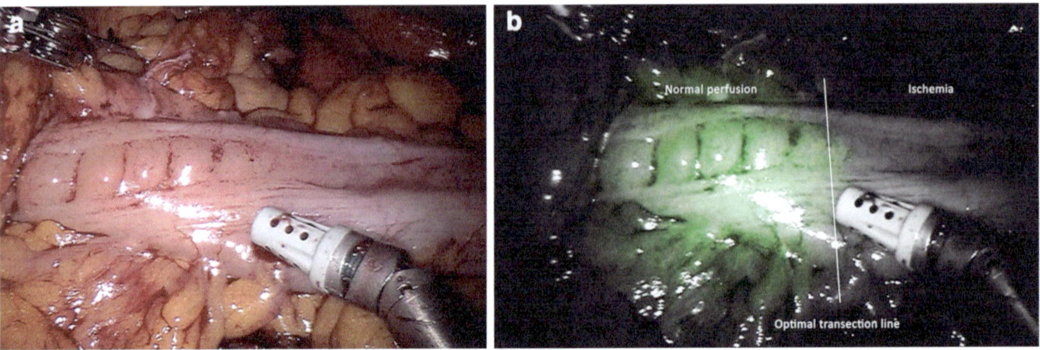

Fig. 6.5 Transection line assessment during left colectomy with white light (**a**) and near infrared (NIR) fluorescence (**b**) view

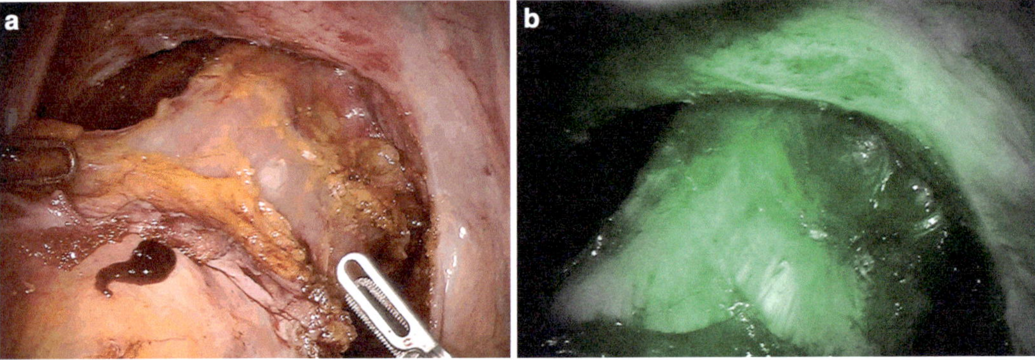

Fig. 6.6 White light (**a**) and near infrared (NIR) fluorescence (**b**) view of rectal transaction line during robotic anterior resection. Fluorescence imaging system confirms the good perfusion of the rectal stump

With regard to the rectal stump, the pelvic wall turns green first (as does the uterus in women), because it is highly vascularized. After few seconds, the rectal stump colors up allowing the assessment of perfusion at the selected point (Fig. 6.6). The ends of the section lines of the rectum in Knight-Griffen anastomosis or of the colon in latero-lateral anastomosis are often referred to as critical points of leakage for their potential minor irroration: particular attention should be paid to their perfusion.

Personal Experience and Considerations

The potential advantages of bowel perfusion assessment using fluorescence system integrated into the da Vinci® system appear interesting but so far only limited clinical data can be found in literature.

Jafari et al. [22] have recently analyzed the effectiveness of NIR-ICG in reducing the rate of anastomotic leak after robotic-assisted low anterior resection (LAR) for rectal cancer in a retrospective case–control study. They compared LAR with and without fluorescence imaging and reported a change in the proximal transection point in 3 out of 16 patients (19 %) and a reduced leak rate of 6 % when compared to 18 % for the control group. Moreover, they described that the overall risk of leak decreased of 12 % in the ICG fluorescence group.

Promising outcomes have been also reported by Bae et al. [17] in two case reports. The authors applied the ICG fluorescence during robotic LAR for cancer to better demark the ischemic area in the distal rectum so that the surgeon was helped to define the distal resection margin. Patients had an uneventful postoperative course.

Hellan et al. [23] have recently published the outcomes of a prospective multicenter study evaluating the impact of fluorescence imaging on visualization of perfusion and subsequent change of transection point during left-sided robotic colorectal surgery. Fluorescence imaging was applied on 40 patients resulted in a change of the proximal transection location in 40 % (16/40) of the candidates: two patients (5 %) with a change in transection line developed an anastomotic leak at postoperative days 15 and 40. Authors conclude that fluorescence imaging provides additional information during determination of transection line in left-sided colorectal procedures and this results in a significant change of transection location, particularly at the proximal transection site.

From September 2011 to December 2013, we have performed 100 full-robotic multiport colorectal resections for benign and malignant diseases (unpublished data) using fluorescence for bowel stump evaluation: 40 right colectomies, 30 low and ultralow anterior rectal resections, 25 left colectomies, and 5 splenic flexure resections. We routinely chose the transection line, after mesenteric section and bowel preparation, according to intestinal perfusion as shown by fluorescence imaging system. There were no intra-operative or anesthetic complications associated with the injection of ICG dye and fluorescence bowel assessment was easily accomplished in real-time in 100 % of the cases with no additional operative time.

We registered one case of anastomotic leak (1 %) in a 78-year-old obese male patient affected by sigmoid cancer. He presented a Grade C anastomotic leakage on PO day IV and therefore he underwent emergency laparoscopic surgery with toilette and endoscopic clips placement.

In most cases, visual inspection and caution are sufficient to perform a well vascularized anastomosis, especially for experienced surgeons. However there are difficult cases, both for patient conditions (such as obesity, diabetes, inflammatory disease, etc.) and for the type of anastomosis (i.e. ultralow anterior rectal resection, splenic flexure resection), in which the evaluation of the perfusion is important, even if only as confirmation, especially in presence of thick mesocolon and short mesentery. Moreover, particularly during minimally invasive surgery where there is a loss of tactile feedback, it appears that this application might be clearly useful.

A limitation of this method can be represented by the fluorescence of the peritoneum due to an accidental spillage in the course of injection for tattoo or lymph node mapping. In this case it may be difficult to appreciate the fluorescence of the stump tied to the perfusion, as distinct from the one linked to impregnation of the tissue. Nevertheless, considering both promising published outcomes and our large experience, we conclude that to date, ICG fluorescence imaging is the more objective method to assess the bowel stump perfusion.

Lymph Node (LN) Mapping and Sentinel Lymph Node (SLN) Biopsy

The prognosis and quality of life of patients with colorectal cancer depends on the extent of the tumor, the characteristics of onset, and the quality of surgical care. In particular, a correct locoregional lymphadenectomy is mandatory for tumor staging and treatment. Following identical pathophysiological principles, the complete mesocolic excision (CME) and the total mesorectal excision (TME) allow to maximize lymph node yield with an en-bloc resection of the tumor and its relative nodes.

However, concerns about lymphatic spread in colorectal cancer still exist and examples of abnormal extension are represented by lateral pelvic LNs in rectal cancer and periaortic LNs in left colon cancers. Therefore, over the years, more aggressive surgical treatments have been advocated, such as lateral pelvic lymphadenectomy for rectal cancer. On the other hand, surgical oncology is evolving toward less aggressive approaches and, in this scenario, LN mapping with ICG can be considered a valid tool in performing a "tailored" surgery in patients with advanced stages of the disease.

The SLN biopsy is nowadays considered the standard of care in the treatment of breast cancer

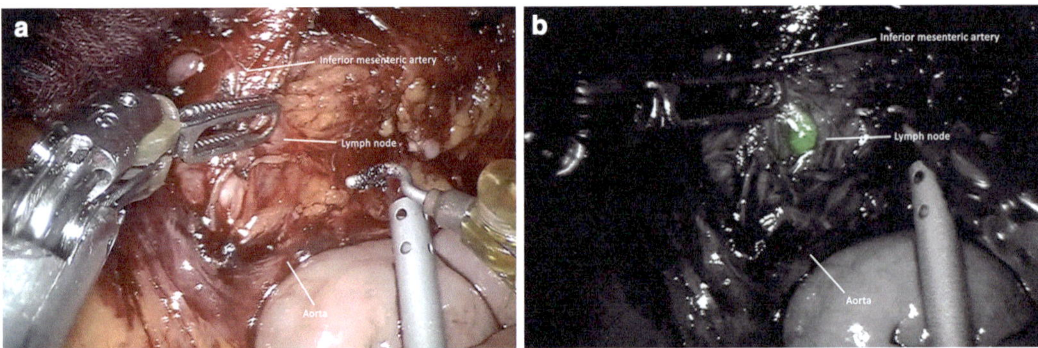

Fig. 6.7 Lymph node mapping during robotic anterior rectal resection. White light (**a**) and near infrared (NIR) fluorescence (**b**) view of lymph nodes at the distal portion of inferior mesenteric artery

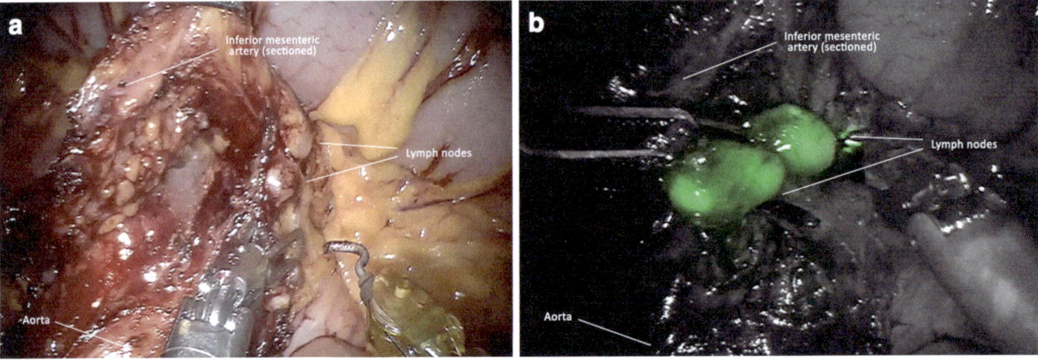

Fig. 6.8 Lymph node mapping during robotic anterior rectal resection. White light (**a**) and near infrared (NIR) fluorescence (**b**) view of lymph nodes at the origin of inferior mesenteric artery

and melanoma, with the aim of avoiding unnecessary locoregional lymphadenectomies and thus limiting postoperative morbidity. Although its value in colon cancer has not yet been established, its primary purpose is to upstage tumors (stage I and II) by providing the pathologist with one or two lymph nodes for detailed evaluation. Therefore, a reliable sentinel node harvesting technique may alter the management in colon cancer treatment whose micrometastases would remain undetected by conventional pathological examination.

Technique

When injected subserosally or submucosally, the ICG is drained through the network of lymphatic vessels. Within 10–15 min it reaches the first LN, 1 or 2 h later it reaches the regional LNs where it remains for about 24–48 h. The technique differs in the timing of injection of the ICG to assess the LN mapping rather than to identify the SLN.

Regarding LN mapping, 1 or 2 cm^3 of ICG solution (2.5–5 mg) are injected endoscopically around the tumor in the submucosa from 3 to 24 h before surgery. The injection should be done by introducing the needle tangentially to the wall and not perpendicularly in order to avoid any risk of perforation of the intestinal wall with consequent diffusion of ICG in the peritoneum or in the surrounding tissues. Operating in NIR light shows the first LN before the peritoneal dissection and, following the dissection, all LNs are highlighted from where the ICG has been drained (Figs 6.7 and 6.8).

The LNs are removed en bloc if present in typical sites by a standard lymphadenectomy, they are removed with a "berry picking" technique when present in unusual locations (periperiaortic

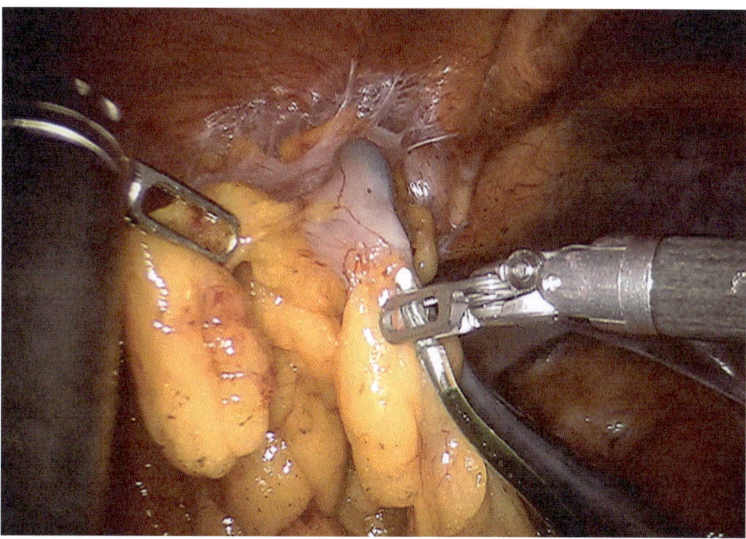

Fig. 6.9 Intraoperative technique of subserosal indocyanine green (ICG) injection for sentinel lymph node (SLN) biopsy during colonic resection

and pericaval in the left colon, pelvic side walls in the rectum).

To locate the SLN, the dye is injected intraoperatively into the subserosa or into the submucosa by the endoscopist if the identification of a small tumor is also needed. The intraoperative technique involves the insertion, by the assistant, through the trocar of a Butterfly infusion set connected to a syringe containing the ICG solution. The surgeon at the console guides and inserts the needle tangentially into the subserosa around the tumor and the infusion of a small amount of ICG (0.75–1.25 mg) is made by the assistant (Fig. 6.9).

The needle is removed whilst maintaining suction to prevent leakage of ICG into the peritoneum; the dose of ICG is identical to that injected endoscopically. A few minutes after the injection, one or two fluorescent lymphatic channels are displayed which run beneath the serosa and then drain into the SLN (Fig. 6.10).

Personal Experience and Considerations

To date and to the best of our knowledge, no published reports exist about LN mapping and SLN biopsy in robotic colorectal surgery.

The advantages of the definition of a LN map, that sometimes permits a more extensive but guided lymphadenectomy, could be the compromise solution, based on objective data, between extended and standard lymphectomy with the aim of a "tailored surgery." The removal of LNs outside of the typical sites shown by the fluorescence has an important significance for correct staging rather than for surgical treatment and, therefore, may be performed with a "berry picking" technique that, guided by the fluorescence view, can be more precise, meticulous and certainly less traumatic than a regional extended lymphadenectomy. The robotic platform allows to easily perform these surgical steps, thanks to improved dexterity, 3D imaging and real-time switching between visible light and NIR-fluorescence.

The SLN biopsy guided by ICG fluorescence imaging has been shown to increase the detection rate with a low rate of false negatives in breast cancer, melanoma, cancer of the anus and in the stomach [2, 3, 22, 24, 25]. Some studies have been published about SLN in colorectal surgery, especially with the radioguided technique, but the role of a biopsy for the detection of micrometastases by immunohistochemistry is currently under investigation and its clinical significance has not yet been defined [26–28]. Wang et al. [29] have shown that ex-vivo localization and immunohistochemical detection of SLN micrometastasis in patients with colorectal cancer can upgrade tumor staging.

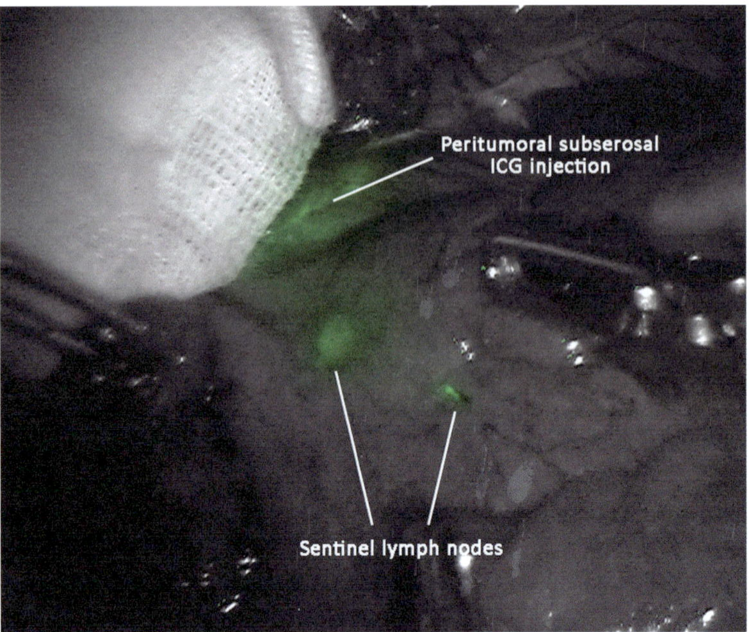

Fig. 6.10 Near infrared (NIR) fluorescence view of paracolic sentinel lymph nodes (SLNs) during colonic resection

An ongoing multicenter, randomized clinical trial is currently recruiting participants (NCT01097265) to evaluate the effect of SLN micrometastases and adjuvant chemotherapy on survival.

The application of fluorescence in colorectal surgery for SLN is still limited to experimental studies, but the feasibility of the method during left colonic resections has already been demonstrated by Hutteman et al. [30]. Recently, a study protocol for SLN identification using the robotic NIR fluorescent imaging system has been submitted for approval to the Ethical Committee of our institution. The objectives of the study are to identify in "real time" SLN in patients affected by stage I–II colon cancer with ICG fluorescence technique integrated into the da Vinci® Surgical System and to histologically evaluate by immunohistochemistry the SLN harvested: by processing all collected data, the aim of this prospective study is to establish the prognostic significance of upstaging and consequences for adjuvant therapy in this subgroup of patients.

Nevertheless, our preliminary experience showed some disadvantages. In cases of SLN detection and LN mapping, the peritumoral injection of ICG prior to surgery has sometimes represented a negative factor because of excessive or inadequate diffusion of the fluorescent dye around the tumor, spreading of the ICG inside the abdominal cavity and subsequent failure in SLN detection. Moreover, in patients affected by rectal cancer who underwent neoadjuvant radiotherapy, ICG injection did not make visible any LN: we suppose that this was due to lymphatic damage induced by the radiation.

The possibility of viewing the SLN in extraperitoneal rectal neoplasms is limited by the capacity of penetration of the NIR light that decreases drastically with tissue thicknesses greater than 0.5–1 mm. Likewise, the intraoperative Butterfly needle injection can be difficult and often remains too superficial and spreads the dye into the peritoneum. An important caveat in the removal of fluorescent LN with the "berry picking" technique is to gently manipulate them since, even without fractures, the dye can easily spread staining the robotic forceps and, consequently, the surrounding tissues.

Several technical details need to be optimized, including the dose of the dye that was progressively reduced over time in our experience. Moreover, great attention to injection technique is needed to avoid gross contamination. This, in addition to the right selection criteria of patients, can help surgeons to improve the accuracy of detection.

The ICG fluorescence imaging system for LN mapping and SLN biopsy during robotic colorectal surgery is a novel and challenging procedure: it may offer new opportunities to perform a "tailored" lymphadenectomy and to extent the boundaries of SLN biopsy.

Tattooing for the Localization of Colorectal Tumors

To date, Indian ink is the most common agent used for perioperative colorectal tumor localization in case of small or flat lesions or those that were removed endoscopically and are difficult to identify during minimally invasive surgery. However, the use of this tattooing agent has been associated with complications such as perforation of the gut, spreading ink in the peritoneal cavity, paracolic abscesses and adhesions [31, 32]. Therefore, some authors have proposed the use of ICG for endoscopic colonic tattooing before laparoscopic-assisted colectomy and have demonstrated the feasibility of the technique without adverse events [33].

In our preliminary experience, we used the preoperative tattooing with ICG in order to identify in real time the tumor during fully robotic colorectal surgery by means of da Vinci® fluorescence vision system. Regarding rectal cancer, before starting the surgery, a minimal dose of 2.5 mg of ICG is injected via a rectoscope with a 25-G butterfly needle built on a laparoscopic grasper: the jab has to be peritumoral and it must not pass in the peritoneum (Fig. 6.11). For colonic lesions, an alternative solution is the intraoperative endoscopy, closing the colon upstream with enterostasis, so as to perform a tattoo visually guided by the surgeon.

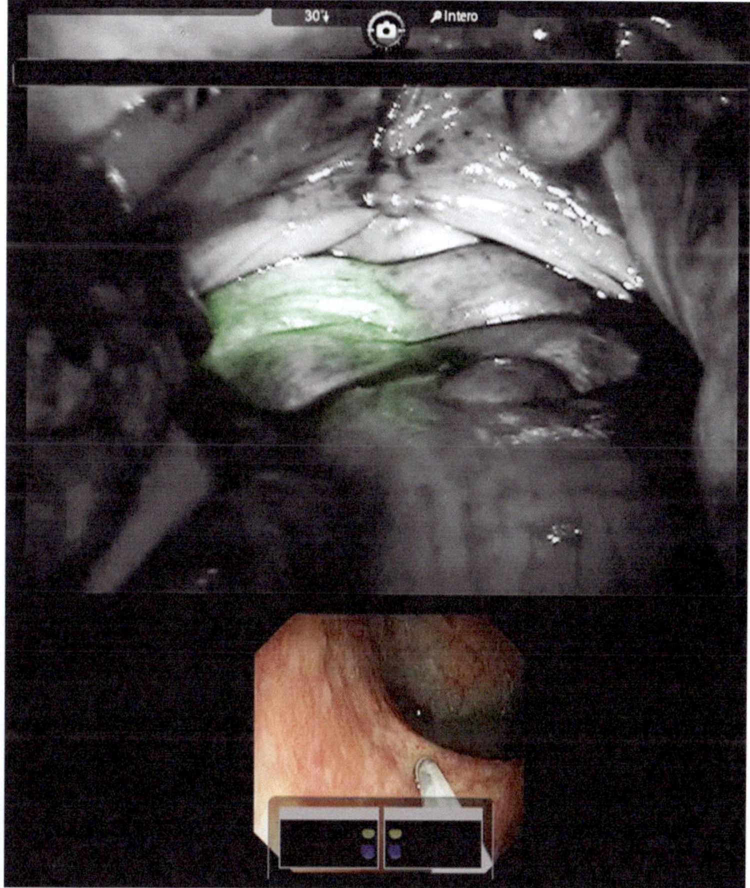

Fig. 6.11 Intraperitoneal near infrared (NIR) fluorescence and endoscopic view during rectal tumor tattooing

The use of ICG fluorescent imaging appeared safe and useful for perioperative colorectal tumor localization in almost all cases; nevertheless, in a few cases we noticed some disadvantages related to the accuracy of detection. Indeed, the peritumoral injection of the ICG before the surgery caused an excessive or inadequate diffusion of the fluorescent dye around the tumor, spreading the ICG inside the abdominal cavity. Moreover, the timing between ICG injection and the surgery is another important limitation, especially in a proximal colonic cancer, because an endoscopy can over-inflate the colon, making the operation difficult or impossible.

Further experiences need to be focused on optimizing the dose of the dye, the time of the injection and the way to do it, in addition to the right selection criteria of the patients, in order to improve the accuracy of the detection.

Fluorescent IO Near Infra-Red Imaging During Gastric LN Dissection for Cancer

ICG fluorescence may be applied during robotic gastrectomy for cancer to better understand the lymphatic drainage. Indeed, LN dissection in gastric cancer surgery is a very important factor not only for exact acquisition of stage but also for proper treatment. Since the identification of complete removal of LN in dissected nodal stations by naked eye is almost impossible to achieve, ICG fluorescence may play a crucial role in identifying any residual LNs in dissected area.

Currently, a prospective trial is in progress to compare the number of retrieved LN in each nodal station after additional application of NIR fluorescence imaging during laparoscopic and robotic gastrectomy (NCT01926743). This method might help surgeons to perform a more complete D1+ or D2 LN dissection.

References

1. Landsman ML, Kwant G, Mook GA, Zijlstra WG. Light-absorbing properties, stability, and spectral stabilization of indocyanine green. J Appl Physiol. 1976;40(4):575–83.
2. Kitai T, Inomoto T, Miwa M, Shikayama T. Fluorescence navigation with indocyanine green for detecting sentinel lymph nodes in breast cancer. Breast Cancer. 2005;12(3):211–5.
3. Kusano M, Tajima Y, Yamazaki K, Kato M, Watanabe M, Miwa M. Sentinel node mapping guided by indocyanine green fluorescence imaging: a new method for sentinel node navigation surgery in gastrointestinal cancer. Dig Surg. 2008;25(2):103–8.
4. Morgenstern L, Wong L, Berci G. Twelve hundred open cholecystectomies before the laparoscopic era. A standard for comparison. Arch Surg. 1992;127(4):400–3.
5. Sinha S, Hofman D, Stoker DL, Friend PJ, Poloniecki JD, Thompson MM, et al. Epidemiological study of provision of cholecystectomy in England from 2000 to 2009: retrospective analysis of Hospital Episode Statistics. Surg Endosc. 2013;27(1):162–75.
6. Harboe KM, Bardram L. The quality of cholecystectomy in Denmark: outcome and risk factors for 20,307 patients from the national database. Surg Endosc. 2011;25(5):1630–41.
7. Berci G, Hunter J, Morgenstern L, Arregui M, Brunt M, Carroll B, et al. Laparoscopic cholecystectomy: first, do no harm; second, take care of bile duct stones. Surg Endosc. 2013;27(4):1051–4.
8. Way LW, Stewart L, Gantert W, Liu K, Lee CM, Whang K, et al. Causes and prevention of laparoscopic bile duct injuries: analysis of 252 cases from a human factors and cognitive psychology perspective. Ann Surg. 2003;237(4):460–9.
9. Ishizawa T, Tamura S, Masuda K, Aoki T, Hasegawa K, Imamura H, et al. Intraoperative fluorescent cholangiography using indocyanine green: a biliary road map for safe surgery. J Am Coll Surg. 2009;208(1):e1–4.
10. Ishizawa T, Bandai Y, Kokudo N. Fluorescent cholangiography using indocyanine green for laparoscopic cholecystectomy: an initial experience. Arch Surg. 2009;144(4):381–2.
11. Ishizawa T, Bandai Y, Hasegawa K, Kokudo N. Fluorescent cholangiography during laparoscopic cholecystectomy: indocyanine green or new fluorescent agents? World J Surg. 2010;34(10):2505–6.
12. Buchs NC, Hagen ME, Pugin F, Volonte F, Bucher P, Schiffer E, et al. Intra-operative fluorescent cholangiography using indocyanine green during robotic single site cholecystectomy. Int J Med Rob. 2012;8(4):436–40.
13. Spinoglio G, Priora F, Bianchi PP, Lucido FS, Licciardello A, Maglione V, et al. Real-time near-infrared (NIR) fluorescent cholangiography in single-site robotic cholecystectomy (SSRC): a single-institutional prospective study. Surg Endosc. 2013;27(6):2156–62.
14. Buchs NC, Pugin F, Azagury DE, Jung M, Volonte F, Hagen ME, et al. Real-time near-infrared fluorescent cholangiography could shorten operative time during robotic single-site cholecystectomy. Surg Endosc. 2013;27(10):3897–901.

15. Spinoglio G, Lenti LM, Maglione V, Lucido FS, Priora F, Bianchi PP, et al. Single-site robotic cholecystectomy (SSRC) versus single-incision laparoscopic cholecystectomy (SILC): comparison of learning curves. First European experience. Surg Endosc. 2012;26(6):1648–55.
16. Tobis S, Knopf J, Silvers C, Yao J, Rashid H, Wu G, et al. Near infrared fluorescence imaging with robotic assisted laparoscopic partial nephrectomy: initial clinical experience for renal cortical tumors. J Urol. 2011;186(1):47–52.
17. Bae SU, Baek SJ, Hur H, Baik SH, Kim NK, Min BS. Intraoperative near infrared fluorescence imaging in robotic low anterior resection: three case reports. Yonsei Med J. 2013;54(4):1066–9.
18. Vignali A, Gianotti L, Braga M, Radaelli G, Malvezzi L, Di Carlo V. Altered microperfusion at the rectal stump is predictive for rectal anastomotic leak. Dis Colon Rectum. 2000;43(1):76–82.
19. Karliczek A, Harlaar NJ, Zeebregts CJ, Wiggers T, Baas PC, van Dam GM. Surgeons lack predictive accuracy for anastomotic leakage in gastrointestinal surgery. Int J Colorectal Dis. 2009;24(5):569–76.
20. Kudszus S, Roesel C, Schachtrupp A, Hoer JJ. Intraoperative laser fluorescence angiography in colorectal surgery: a noninvasive analysis to reduce the rate of anastomotic leakage. Langenbecks Arch Surg. 2010;395(8):1025–30.
21. Sherwinter DA, Gallagher J, Donkar T. Intraoperative transanal near infrared imaging of colorectal anastomotic perfusion: a feasibility study. Colorectal Dis. 2013;15:91.
22. Jafari MD, Lee KH, Halabi WJ, Mills SD, Carmichael JC, Stamos MJ, et al. The use of indocyanine green fluorescence to assess anastomotic perfusion during robotic assisted laparoscopic rectal surgery. Surg Endosc. 2013;27(8):3003–8.
23. Hellan M, Spinoglio G, Pigazzi A, Lagares-Garcia JA. The influence of fluorescence imaging on the location of bowel transection during robotic left-sided colorectal surgery. Surg Endosc. 2014;28:1695.
24. Morton DL, Thompson JF, Essner R, Elashoff R, Stern SL, Nieweg OE, et al. Validation of the accuracy of intraoperative lymphatic mapping and sentinel lymphadenectomy for early-stage melanoma: a multicenter trial. Multicenter Selective Lymphadenectomy Trial Group. Ann Surg. 1999;230(4):453–63. discussion 63–5.
25. Miyashiro I, Miyoshi N, Hiratsuka M, Kishi K, Yamada T, Ohue M, et al. Detection of sentinel node in gastric cancer surgery by indocyanine green fluorescence imaging: comparison with infrared imaging. Ann Surg Oncol. 2008;15(6):1640–3.
26. Adell G, Boeryd B, Franlund B, Sjodahl R, Hakansson L. Occurrence and prognostic importance of micrometastases in regional lymph nodes in Dukes' B colorectal carcinoma: an immunohistochemical study. Eur J Surg. 1996;162(8):637–42.
27. Noura S, Yamamoto H, Ohnishi T, Masuda N, Matsumoto T, Takayama O, et al. Comparative detection of lymph node micrometastases of stage II colorectal cancer by reverse transcriptase polymerase chain reaction and immunohistochemistry. J Clin Oncol. 2002;20(20):4232–41.
28. Palma RT, Waisberg J, Bromberg SH, Simao AB, Godoy AC. Micrometastasis in regional lymph nodes of extirpated colorectal carcinoma: immunohistochemical study using anti-cytokeratin antibodies AE1/AE3. Colorectal Dis. 2003;5(2):164–8.
29. Wang FL, Shen F, Wan DS, Lu ZH, Li LR, Chen G, et al. Ex vivo localization and immunohistochemical detection of sentinel lymph node micrometastasis in patients with colorectal cancer can upgrade tumor staging. Diagn Pathol. 2012;7:71.
30. Hutteman M, Choi HS, Mieog JS, van der Vorst JR, Ashitate Y, Kuppen PJ, et al. Clinical translation of ex vivo sentinel lymph node mapping for colorectal cancer using invisible near-infrared fluorescence light. Ann Surg Oncol. 2011;18(4):1006–14.
31. Gianom D, Hollinger A, Wirth HP. Intestinal perforation after preoperative colonic tattooing with India ink. Swiss Surg. 2003;9(6):307–10. Darmperforation nach praoperativer kolonoskopischer Tuschemarkierung.
32. Yano H, Okada K, Monden T. Adhesion ileus caused by tattoo-marking: unusual complication after laparoscopic surgery for early colorectal cancer. Dis Colon Rectum. 2003;46(7):987.
33. Miyoshi N, Ohue M, Noura S, Yano M, Sasaki Y, Kishi K, et al. Surgical usefulness of indocyanine green as an alternative to India ink for endoscopic marking. Surg Endosc. 2009;23(2):347–51.

7 Fluorescence Imaging Systems (PDE, HyperEye Medical System, and Prototypes in Japan)

Takeaki Ishizawa and Norihiro Kokudo

Introduction

Recently, fluorescence imaging for visualizing biological structures during surgery has become one of the most active fields of basic and translational research. A MEDLINE search using the keywords "fluorescence imaging" and "surgery" revealed a dramatic increase in the number of publications, reaching more than 1,200 articles during the year 2012 [1]. The driving forces of the development of this research field are the advances in imaging techniques and the distribution of commercially available fluorescence imaging systems using indocyanine green (ICG) as a source of fluorescence. We describe the history of the development and characteristics of fluorescence imaging systems for open and laparoscopic surgery that are currently available in Japan.

PDE®

PDE® (Hamamatsu Photonics, Hamamatsu, Japan) is composed of a camera unit (0.5 kg) and a small control unit (2.8 kg). The camera unit consists of a charge-coupled device (CCD) camera that filters out light with wavelengths less than 820 nm and 36 light-emitting diodes with a wavelength of 760 nm (Fig. 7.1a). Originally, the PDE® system was used to detect sentinel lymph nodes during breast surgery. Since Aoki and colleagues reported the clinical application of ICG-fluorescence imaging for identifying hepatic segments to be removed in 2008 [2], this system has become widely recognized among surgeons. It was rapidly applied as a new technique for intraoperative identification of biological structures, such as bile ducts (fluorescence cholangiography [3, 4]), primary and secondary liver cancer [5, 6], and lymph drainage [7, 8].

The major advantage of PDE® lies in its size and portability, which enables the surgeon to hold the camera unit (covered with a sterilized drape) and easily obtain fluorescence images from various angles during surgery. With this

Electronic supplementary material: The online version of this chapter (doi:10.1007/978-3-319-15678-1_7) contains supplementary material, which is available to authorized users. Videos can also be accessed at http://link.springer.com/chapter/10.1007/978-3-319-15678-1_7.

T. Ishizawa, M.D., Ph.D., F.A.C.S
Department of Gastroenterological Surgery, Cancer Institute Hospital, Japanese Foundation for Cancer Research, Koto-ku, Tokyo, Japan

Hepato-Biliary-Pancreatic Surgery Division, Department of Surgery, Graduate School of Medicine, University of Tokyo, 7-3-1 Hongo, Bunkyo-ku, Tokyo 113-8655, Japan
e-mail: tish-tky@umin.ac.jp

N. Kokudo, M.D., Ph.D. (✉)
Hepato-Biliary-Pancreatic Surgery Division, Department of Surgery, Graduate School of Medicine, University of Tokyo, 7-3-1 Hongo, Bunkyo-ku, Tokyo 113-8655, Japan
e-mail: KOKUDO-2SU@h.u-tokyo.ac.jp

F.D. Dip et al. (eds.), *Fluorescence Imaging for Surgeons: Concepts and Applications*, DOI 10.1007/978-3-319-15678-1_7, © Springer International Publishing Switzerland 2015

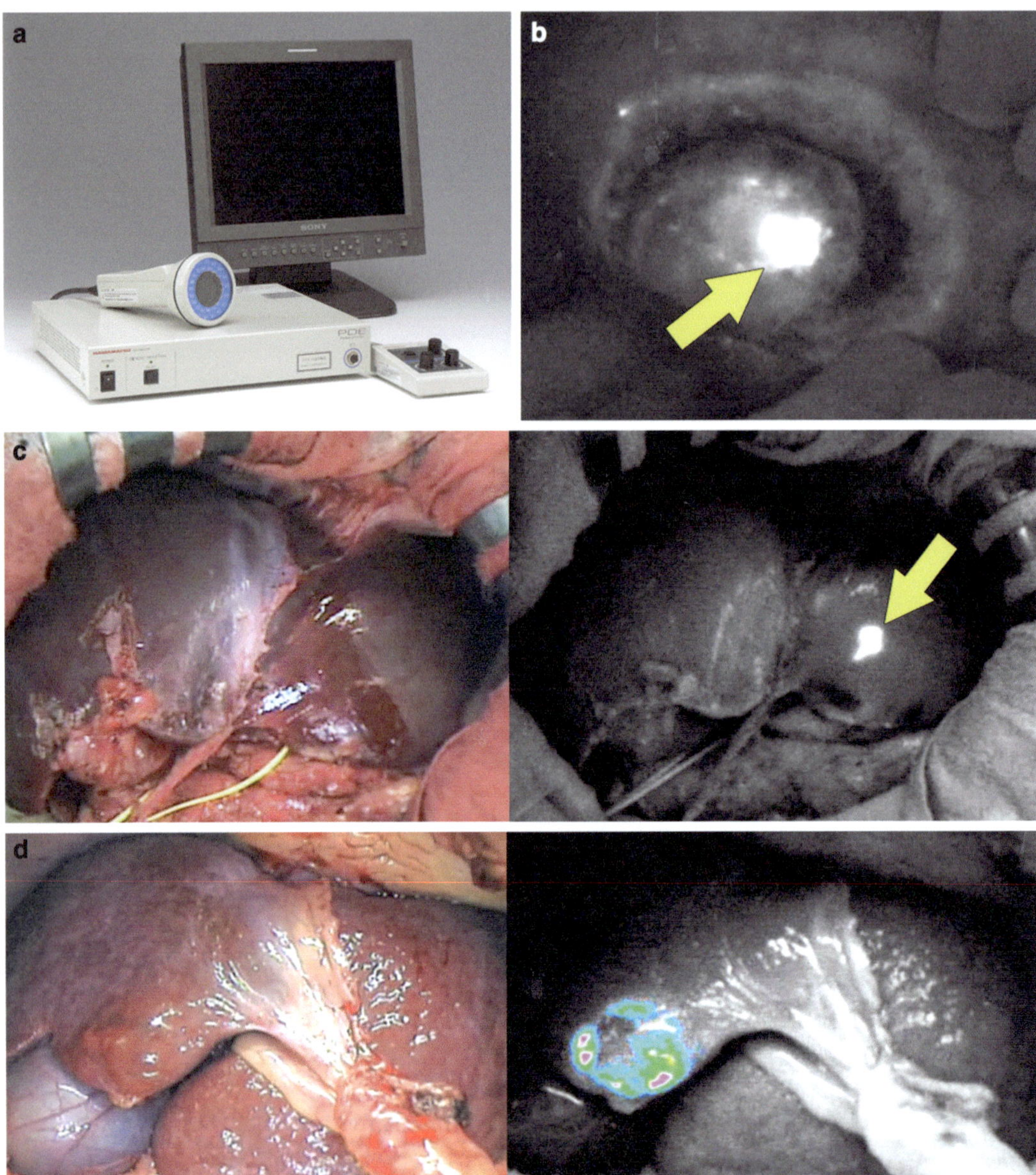

Fig. 7.1 PDE® system and its application to liver resection. (**a**) Camera and control units of PDE®. (**b**) Hepatocellular carcinoma (HCC) (*arrow*) is visualized on the liver surface by PDE®. (**c**) Colorectal liver metastasis (*arrow*) is identified on the liver surface prior to liver resection by pde-neo®, which can switch between color images and fluorescence images. (**d**) Prototype imaging system enables visualization of indocyanine green (ICG)-specific fluorescence emitted by an HCC as pseudo-color images (see Video 7.1)

system, the ICG fluorescence emitted from target organs is visualized on a background of monochromatic structures using autofluorescence of other organs within near-infrared range (Fig. 7.1b). The second clinically available model (pde-neo®) consists of a CCD camera for obtaining color images and enables switching to fluorescence images from color images (Fig. 7.1c). Currently, the next-generation model is under development to extract ICG fluorescence from autofluorescence, thereby showing ICG-specific fluorescence as pseudo-color images on a monochromatic background (Fig. 7.1d).

HyperEye Medical System

The HyperEye Medical System (Mizuho Ikakogyo, Tokyo, Japan) (Fig. 7.2a) was originally devised to identify coronary arteries during cardiovascular surgery [9]. It was first marketed in Japan in 2010. Since then, application of this system has been extended to sentinel node biopsy [10] and fluorescence cholangiography/angiography during digestive surgery and liver transplantation [11].

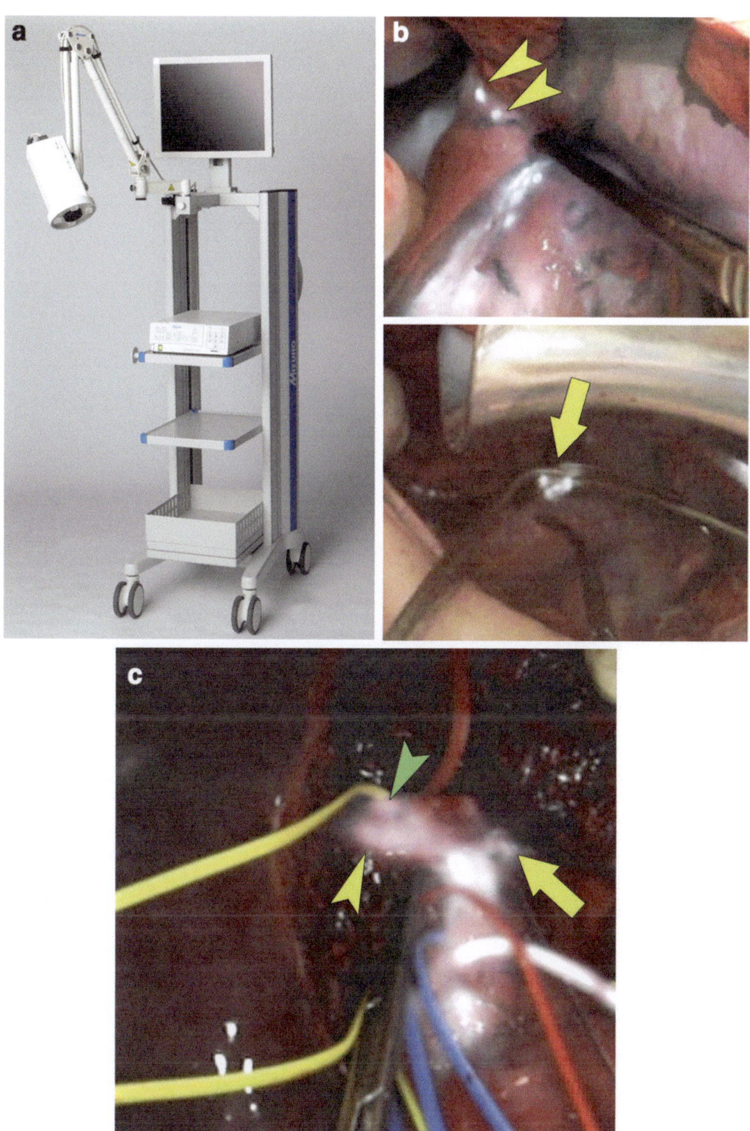

Fig. 7.2 HyperEye Medical System and its applications to hepatobiliary surgery. (**a**) Appearance of the HyperEye Medical System. (**b**) HyperEye Medical System enables visualization of fluorescing tiny nodules of HCC invading the diaphragm (*arrowheads*) on color images (*top*) during resection of disseminated nodules caused by percutaneous radiofrequency ablation. Simultaneous visualization of fluorescence images and color images helps surgeons detect disseminated nodules to be removed (*arrow*, *bottom*) (see Video 7.2). (**c**) Fluorescence cholangiography following intrabiliary injection of ICG solution (0.025 mg/mL) identifies the confluence between the left hepatic duct (*arrow*) and branches of the right hepatic ducts draining the right lateral sector (*yellow arrowhead*) and right paramedian sector (*green arrowhead*) during donor right hepatectomy (see Video 7.3)

This system enables ICG-fluorescence images to be superimposed on color images, which helps surgeons understand spatial relations between the target structures containing ICG and surrounding organs (Fig. 7.2b, c). In the original model, the camera imaging head with a 10× optical zoom was mounted on the imaging system to obtain ICG-fluorescence images from outside the surgical field without blurring of images caused by hand movement. A newer model, composed of a portable camera unit such as PDE®, is being developed for marketing in countries outside of Japan.

Prototypes of Laparoscopic Fluorescence Imaging System

In addition to commercially available laparoscopic fluorescence imaging systems described in the following chapters, several prototypes have been used in the clinical setting in Japan. In 2004, Nimura and colleagues reported clinical application of infrared ray electronic endoscopy (Olympus Optical, Tokyo, Japan) for detecting sentinel lymph nodes during gastric cancer surgery through absorption of infrared

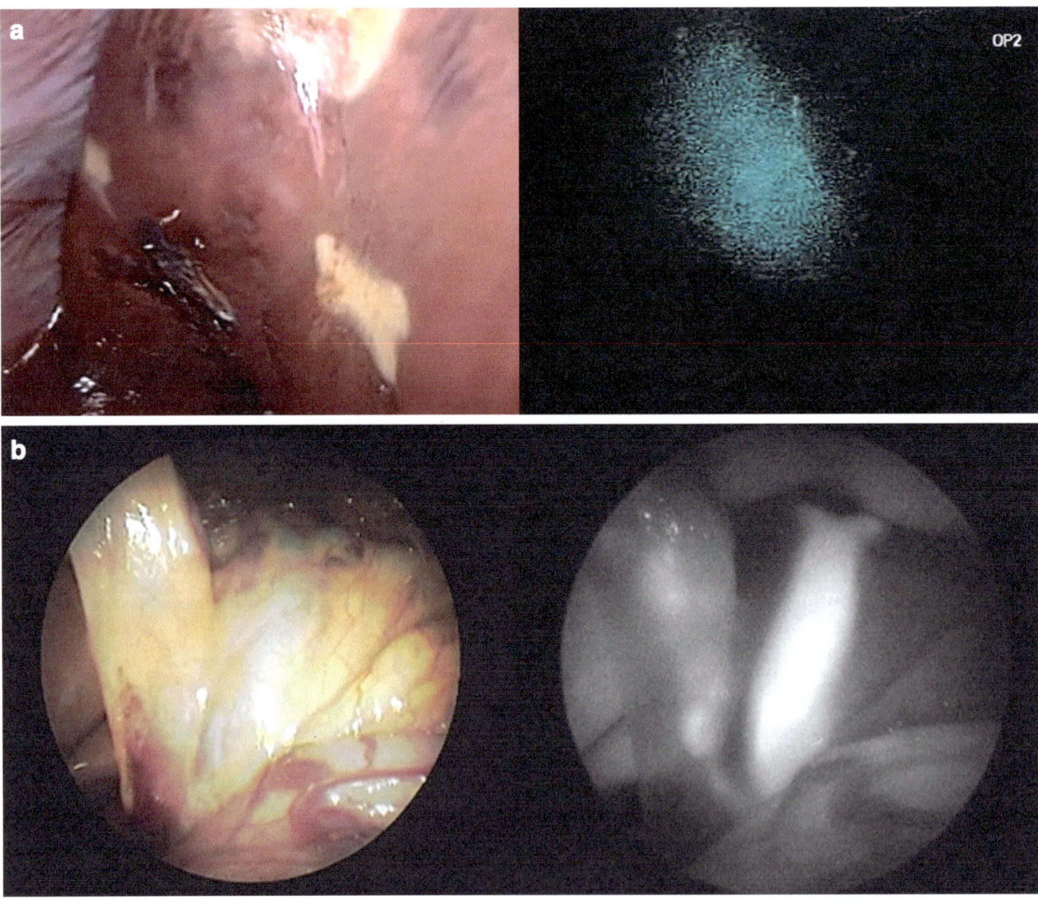

Fig. 7.3 Prototype fluorescence imaging systems for laparoscopic surgery. (**a**) Colorectal liver metastasis is visualized on the liver surface during laparoscopic hepatectomy using the prototype fluorescence imaging system (Olympus). (**b**) Prototype fluorescence imaging system (Hamamatsu Photonics and Shinko Optical) delineates the anatomy of the extrahepatic bile ducts during laparoscopic cholecystectomy

rays by ICG [12]. This technique led to the development of a laparoscopic imaging system visualizing ICG fluorescence under illumination by near-infrared light (Olympus Medical Systems, Tokyo, Japan) (Fig. 7.3a). Another prototype laparoscopic imaging system (Hamamatsu Photonics and Shinko Optical, Tokyo, Japan) (Fig. 7.3b) has been developed for sentinel node navigation surgery in gastrointestinal cancer [13] and evaluation of the placental vascular network in the treatment of twin–twin transfusion syndrome [14]. Although these standard-definition models already provide high-contrast, sensitive fluorescence images sufficient for real-time identification of the bile ducts [15], liver cancer [16], and hepatic segments [17] during laparoscopic surgery, the quality of both the color and fluorescence images should be further improved to meet the standard of high-definition imaging and display before marketing.

Conclusion

Although the development of imaging systems has promoted clinical application of ICG-fluorescence imaging as a novel intraoperative diagnostic tool, further technical advances are essential if fluorescence imaging is to play a greater role in enhancing the safety and efficacy of surgery.

Financial Disclosure: The authors have no relevant affiliations or financial involvement with any organization or entity with a financial interest in or financial conflict with the subject matter or materials discussed in this manuscript. This includes employment, consultancies, honoraria, stock ownership or options, expert testimony, grants or patents received or pending, and royalties. No writing assistance was used in the production of this manuscript.

Funding/Support: This work was supported by grants from the Takeda Science Foundation, the Ministry of Education, Culture, Sports, Science, Technology of Japan, and the Ministry of Health, Labour, and Welfare of Japan, and Global Leader Program for Social Design and Management.

References

1. Ishizawa T, Kokudo N. The beginning of a new era of digestive surgery guided by fluorescence imaging. Liver Cancer. 2014;3:6.
2. Aoki T, Yasuda D, Shimizu Y, Odaira M, Niiya T, Kusano T, et al. Image-guided liver mapping using fluorescence navigation system with indocyanine green for anatomical hepatic resection. World J Surg. 2008;32:1763–7.
3. Mitsuhashi N, Kimura F, Shimizu H, Imamaki M, Yoshidome H, Ohtsuka M, et al. Usefulness of intraoperative fluorescence imaging to evaluate local anatomy in hepatobiliary surgery. J Hepatobiliary Pancreat Surg. 2008;15:508–14.
4. Ishizawa T, Tamura S, Masuda K, Aoki T, Hasegawa K, Imamura H, et al. Intraoperative fluorescent cholangiography using indocyanine green: a biliary road map for safe surgery. J Am Coll Surg. 2008;208:e1–4.
5. Ishizawa T, Fukushima N, Shibahara J, Masuda K, Tamura S, Aoki T, et al. Real-time identification of liver cancers by using indocyanine green fluorescent imaging. Cancer. 2009;115:2491–504.
6. Gotoh K, Yamada T, Ishikawa O, Takahashi H, Eguchi H, Yano M, et al. A novel image-guided surgery of hepatocellular carcinoma by indocyanine green fluorescence imaging navigation. J Surg Oncol. 2009; 100:75–9.
7. Hirono S, Tani M, Kawai M, Okada K, Miyazawa M, Shimizu A, et al. Identification of the lymphatic drainage pathways from the pancreatic head guided by indocyanine green fluorescence imaging during pancreaticoduodenectomy. Dig Surg. 2012;29:132–9.
8. Mihara M, Murai N, Hayashi Y, Hara H, Iida T, Narushima M, et al. Using indocyanine green fluorescent lymphography and lymphatic-venous anastomosis for cancer-related lymphedema. Ann Vasc Surg. 2012;278:e1–6.
9. Handa T, Katare RG, Nishimori H, Wariishi S, Fukutomi T, Yamamoto M, et al. New device for intraoperative graft assessment: HyperEye charge-coupled device camera system. Gen Thorac Cardiovasc Surg. 2010;58:68–77.
10. Yamauchi K, Nagafuji H, Nakamura T, Sato T, Kohno N. Feasibility of ICG fluorescence-guided sentinel node biopsy in animal models using the HyperEye Medical System. Ann Surg Oncol. 2011;18:2042–7.
11. Kawaguchi Y, Ishizawa T, Masuda K, Sato S, Kaneko J, Aoki T, et al. Hepatobiliary surgery guided by a novel fluorescent imaging technique for visualizing hepatic arteries, bile ducts, and liver cancers on color images. J Am Coll Surg. 2011;212:e33–9.
12. Nimura H, Narimiya N, Mitsumori N, Yamazaki Y, Yanaga K, Urashima M. Infrared ray electronic endoscopy combined with indocyanine green injection for detection of sentinel nodes of patients with gastric cancer. Br J Surg. 2004;91:575–9.

13. Kusano M, Tajima Y, Yamazaki K, Kato M, Watanabe M, Miwa M. Sentinel node mapping guided by indocyanine green fluorescence imaging: a new method for sentinel node navigation surgery in gastrointestinal cancer. Dig Surg. 2008;25:103–8.
14. Harada K, Miwa M, Fukuyo T, Watanabe S, Enosawa S, Chiba T. ICG fluorescence endoscope for visualization of the placental vascular network. Minim Invasive Ther Allied Technol. 2009;18:1–5.
15. Ishizawa T, Bandai Y, Ijichi M, Kaneko J, Hasegawa K, Kokudo N. Fluorescent cholangiography illuminating the biliary tree during laparoscopic cholecystectomy. Br J Surg. 2010;97:1369–77.
16. Kudo H, Ishizawa T, Tani K, Harada N, Ichida A, Shimizu A, et al. Application of indocyanine-green fluorescence imaging for visualizing hepatic malignancy during laparoscopic surgery. Surg Endosc. 2014;28:2504.
17. Ishizawa T, Zuker NB, Kokudo N, Gayet B. Positive and negative staining of hepatic segments by use of fluorescent imaging techniques during laparoscopic hepatectomy. Arch Surg. 2012;147:393–4.

The Pinpoint System

8

Jana L. Lewis and Danny A. Sherwinter

Theoretical Basis

In 2005, Novadaq (Mississauga, Ontario) obtained 510(k) approval for the Spy Imaging System. Combining a laser and a specialized camera head, the Spy system became the first commercially viable real-time fluorescence imaging system for intraoperative assessment of tissue perfusion.

The Spy system uses a laser to generate a tight band of light in the near infrared spectrum (803 nm). This wavelength excites the compound, indocyanine green (ICG), causing it to emit light at a different, higher, wavelength. ICG is a cyanine dye, which fluoresces at 830 nm when exposed to light in the near infrared spectrum. Filters in the Spy system camera block out light with wavelengths below 821 nm (including the excitation wavelength) allowing it to isolate the fluorescence of the ICG.

Since ICG remains bound to plasma proteins and is thus confined to the intravascular space, angiographic images can be obtained following bolus injection. Since ICG also does not appreciably leak into the interstitium, ICG angiography can be repeated multiple times. In this way it differs from other fluorophores, such as fluorescein, which historically have been used to evaluate intraoperative bowel viability. Other important data such as rate of clearance can be calculated and may correlate with venous outflow. Although affected by cardiac output, ICG can usually be visualized within 30 s of injection and vascularized tissue remains hyper-fluorescent for a minimum of 2–3 min.

After a first pass in the vascular compartment, ICG is taken up almost exclusively by the liver and excreted in bile. This characteristic gives ICG the added benefit of making it useful for intraoperative cholangiography. ICG administration is not affected by renal function but should be avoided in patients with cirrhosis. ICG has been used clinically for over 30 years with a high degree of safety. Reactions are usually mild but can, in rare cases, present with anaphylaxis.

The Spy system is optimized for open surgery and is designed with a large camera head on a boom, which is arrayed above the surgical field for maximizing surgical field viewing, often imaging an entire surgical bed at one time. An additional advantage of this system is software, which quantifies tissue perfusion. The surgeon can assign normally perfused tissue as a baseline and then the system calculates a ratio (shown as a percentage) of perfusion in the area of question in relation to the normal tissue.

Electronic supplementary material: The online version of this chapter (doi:10.1007/978-3-319-15678-1_8) contains supplementary material, which is available to authorized users. Videos can also be accessed at http://link.springer.com/chapter/10.1007/978-3-319-15678-1_8.

J.L. Lewis, M.D. • D.A. Sherwinter, M.D. (✉)
Department of Surgery, Maimonides Medical Center, 4802 10th Ave, Brooklyn, NY 11219, USA
e-mail: DSherwinter@MaimonidesMed.org

F.D. Dip et al. (eds.), *Fluorescence Imaging for Surgeons: Concepts and Applications*, DOI 10.1007/978-3-319-15678-1_8, © Springer International Publishing Switzerland 2015

Perfusion quantification gives the surgeon an objective measure of tissue blood flow.

The Spy system projects the fluorescence imaging as a black and white image, with poorly perfused tissue appearing dark and well perfused tissue white. The Spy system has progressively gaining traction for the evaluation of tissue perfusion in open surgical procedures. Plastic reconstructive surgery has been the primary beneficiary of this technology with the Spy system being used extensively in free flap and microvascular procedures [1].

With the widespread adoption of minimally invasive techniques in virtually all fields of surgery, Novadaq began developing an endoscopic version of their Spy system in 2006. What grew out of this effort was an endoscopic system initially called the Spy Scope and subsequently rebranded The Pinpoint system. The Pinpoint is fundamentally similar to the open Spy system and similarly uses NIR for excitation and ICG as its fluorophore.

The Pinpoint is not the first endoscopic NIR system to be introduced. NIR systems capable of being used in laparoscopic surgery have been extensively reported and used experimentally [2–4]. These previously developed systems were able to assess fluorescence endoscopically but have one significant shortcoming in common. These previously developed endoscopic NIR imaging systems were capable of either imaging fluorescence or white light BUT *not both*. To use these systems the surgeon has to continually halt dissection and switch modes from white light to fluorescence mode, or use a separate dedicated NIR camera capable of only fluorescence imaging, to obtain the fluorescent views. The reason for this is that in all previous systems the white light needed for the standard laparoscopic image interferes with the NIR making it impossible to obtain both views simultaneously. Because of this, using these systems is very cumbersome. Surgeons found themselves with the added benefits provided by NIR but were unable to operate in real-time with this information. Furthermore, the context necessary to accurately interpret the NIR images was lost and the operative flow interrupted.

The Pinpoint solved this problem by incorporating both a high definition (HD) white light and near infra-red (NIR) fluorescence excitation and acquisition systems into a single camera head. Both images can then viewed simultaneously by using proprietary software which displays the NIR image as an overlay over the standard white light image in real-time as neon green highlighting. Since the standard white light view and NIR are present throughout the procedure, the NIR information retains its context, allowing the user to continue the operation and more readily interpret the data. NIR thus becomes not just abstract information but rather true augmented vision. For the first time, using a Pinpoint system, a surgeon can truly operate with real-time fluorescence guidance.

Pinpoint Development Timeline

The earliest working prototype, dubbed “Gen-1” and referred to as the Spy-Scope (Fig. 8.1), was developed and shown at the SAGES conference in 2009. Although still not commercially available nor FDA approved the belief was that by putting a powerful tool such as this in the hands of surgeons the indications would follow, and this is precisely what has happened. Since 2009, multiple studies have been completed and many more are ongoing looking at a myriad of indications for this groundbreaking technology. The system name was later changed to Pinpoint and was commercially launched in January 2012 as the Pinpoint “Gen-2” (Fig. 8.2). This was followed a few months later by the “Gen-3” which is the model currently available on the market (Fig. 8.3). Each subsequent version came with improvements in form factor, optics and other upgrades but fundamentally the system has remained unchanged.

Starting in July 2011, Novadaq began integrating spy-imaging technology into the 3-D HD imaging system of the daVinci robot (Intuitive Surgical, Sunnyvale, CA). The robotic version of this technology is referred to as Firefly. The Firefly is identical to the Pinpoint, except for minor differences in the white light portion of the

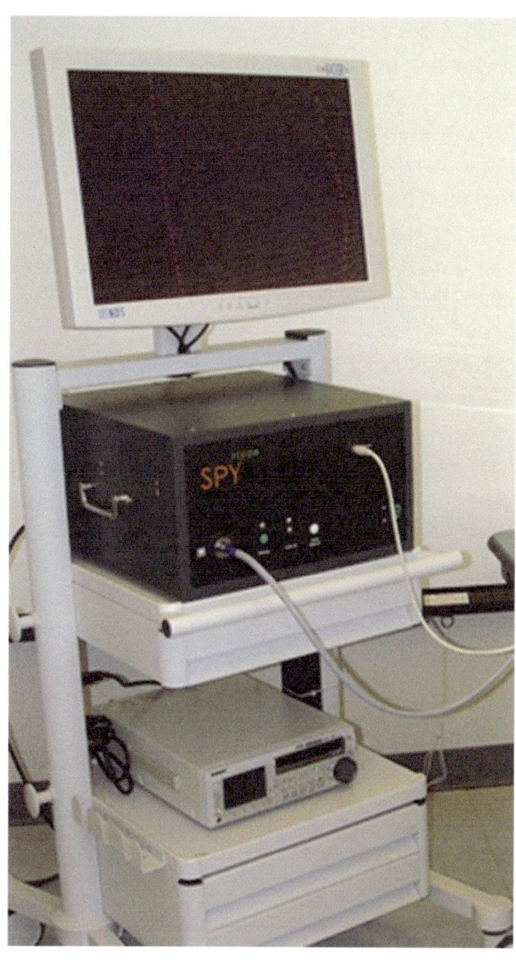

Fig. 8.1 The earliest working prototype, dubbed "Gen-1" and referred to as the Spy-Scope

Fig. 8.3 "Gen-3" (the model currently available on the market)

Fig. 8.2 The Pinpoint System was commercially launched in January 2012 as the Pinpoint "Gen-2"

image. In the daVinci Firefly, the green fluorescent data is projected on a background of a black and white laparoscopic image rather than the full colored image of the Pinpoint. There is significant overlap between work done with the Pinpoint and Firefly since both are fundamentally endoscopic NIR systems but we will be limiting our discussion in this chapter to the Pinpoint.

Our group and others have mainly concentrated on the use of Pinpoint for the following four major indications and we will be dividing the following sections of this chapter into these four fields of interest. Firstly, will be a discussion regarding our work with the Pinpoint in colorectal perfusion and prevention of anastomotic leak. Secondly, we will discuss our experience using the Pinpoint for intraoperative cholangiography primarily during laparoscopic cholecystectomy. Thirdly, we will discuss our early experience using the pinpoint for sentinel lymph node dissection for colorectal and gastric malignancies. Lastly, we will discuss other uses and potential future directions of this technology.

Colorectal Evaluation of Perfusion

The influence of perfusion on colorectal anastomotic healing, and conversely the role of ischemia on anastomotic dehiscence has been thoroughly reported [5–7]. Anastomotic leak is a devastating complication in colorectal surgery and is especially prevalent in low rectal anastomoses with some reports describing a leak rate as high as 20 % [8]. Besides for the acute effects of sepsis, there is long-term morbidity including: stricture formation, bowel dysfunction, and an increased risk of locoregional cancer recurrence [9, 10]. Ischemia is the likely predisposing factor in the majority of cases. Anatomic variations of the vascular supply to the colon (especially the descending and sigmoid colon), atherosclerotic disease of the mesocolic vasculature, injury to the vasa recta by over-dissection, tension on the anastomosis, and mis-oriented bowel all affect the final blood supply to the anastomotic ring and hence impact healing and by extension can lead to leak.

In order to be useful, information regarding perfusion must be available intraoperatively when intervention is possible. However, the clinical incorporation of technologies capable of intraoperatively evaluating perfusion has been limited. This is primarily because these technologies are highly user-dependent, difficult to use and involve a steep learning curve and thus have never become commercially available [11]. We hypothesized that by using a visually based tool capable of being advanced under direct vision to the point of interest and then able to assess the entire area in question would resolve the problems experienced with these previous perfusion assessment tools. Furthermore, since mucosa is more susceptible than serosa to minor alterations in perfusion, transanal assessment of flow should be an especially sensitive test for bowel ischemia [12–14].

We evaluated the use of Pinpoint to *transanally* evaluate anastomotic perfusion in patients undergoing colorectal anastomoses [15]. The study was approved by the ethics review board of our institution. We enrolled seven patients undergoing LAR for benign or malignant disease of the sigmoid or rectum with anastomoses expected to lie within 25 cm of the anal verge. Prospectively, demographic data, indications for surgery, distance of the anastomosis from the anal verge, and other factors known to affect anastomotic healing were collected. After completion of the resection and the creation of an anastomosis using a double stapled, end-to-end technique, the doughnuts were checked for completeness and an air leak test was performed.

Although the Pinpoint was designed for laparoscopic use, visual inspection of the anastomosis and evaluation of the proximal and distal segments for viability was made possible by using a rigid procto-sigmoidoscope (Welch Allyn, Skaneateles Falls, NY) mated with a laparoscopic trocar to allow the use of the Pinpoint system transanally (Fig. 8.4) [16].

The Pinpoint scope was inserted into the rectum to the level of the anastomosis and the staple line and proximal and distal margins of the anastomosis were visualized. 1 ml (2.5 mg reconstituted) of ICG was injected via a peripheral IV catheter as a rapid bolus just prior to imaging. Fluorescence was evaluated subjectively by the surgeon as being hypo-, iso-, or hyperfluorescent as compared to surrounding bowel using the distal rectum as the benchmark of maximal fluorescence. An ICGA was considered normal if the distal colon, proximal rectum, and distal rectum were all iso-fluorescent (Fig. 8.5). A decision was then made regarding the need for revision and/or proximal diversion. All patients were followed up at least twice postoperatively, at 1 week and 1 month post discharge.

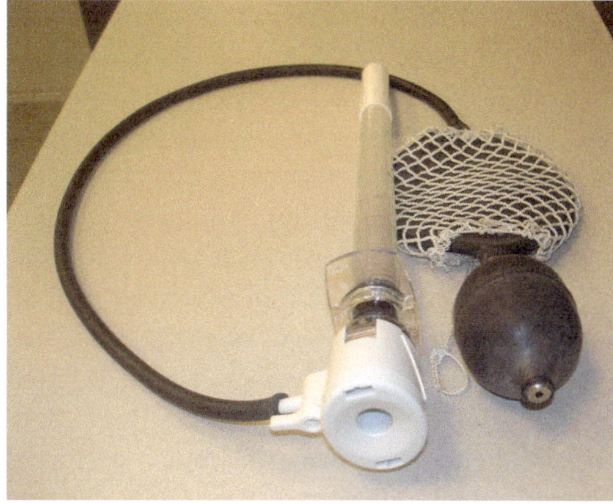

Fig. 8.4 Although the Pinpoint was designed for laparoscopic use, visual inspection of the anastomosis and evaluation of the proximal and distal segments for viability was made possible by using a rigid procto-sigmoidoscope (Welch Allyn, Skaneateles Falls, NY) mated with a laparoscopic trocar to allow the use of the Pinpoint system transanally

Five of seven patients were women. Mean age was 70.29 (+11.44, range 51–85) and operative time ranges from 173 to 296 min (mean 218.00 min, +43.11). Mean length of hospital stay was 4.71 days (+1.38, range 3–6). All anastomotic doughnuts were complete and none of the anastomoses leaked during air testing.

Normal, pink and healthy appearing mucosa and a complete anastomotic staple line were noted in all cases. The sequence of uptake (colon → rectum vs. rectum → colon) varied, but in all subjects both the proximal and distal margins of the anastomosis appeared iso-fluorescent.

None of the patients in this group experienced a clinically significant anastomotic leak and no adverse events were identified due to the use of the ICG or Pinpoint system. At the 1 month follow up, all were doing well.

We concluded that due to the devastating consequences of an anastomotic leak, there is a need for a modality that can accurately and consistently assess bowel perfusion in real time. The pinpoint system is the first system that allowed a

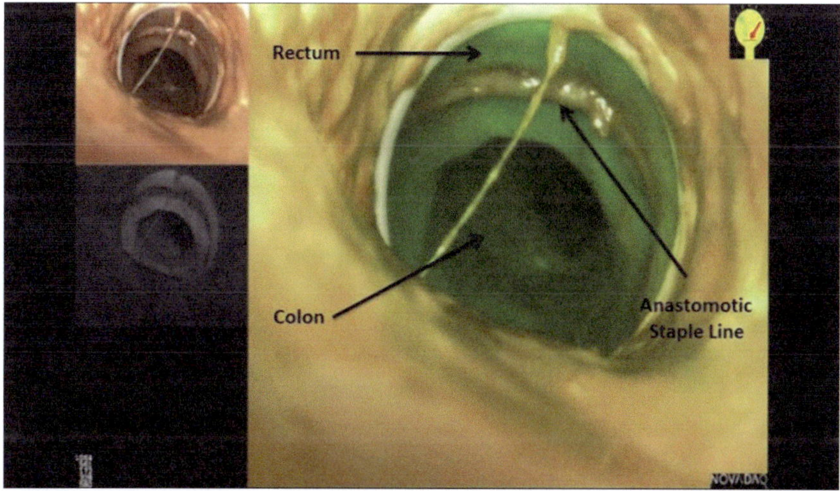

Fig. 8.5 An ICGA (indocyanine green) was considered normal if the distal colon, proximal rectum, and distal rectum were all iso-fluorescent

transanal view of mucosal viability and evaluation of perfusion to the anastomotic ring.

Success in imaging these first seven patients and the outstanding images the Pinpoint generated encouraged us to open this study up to an additional 100 patients undergoing colorectal resections for benign and malignant disease.

This study was approved by the Ethics Review Board at our institution. Twenty consecutive patients undergoing LAR for benign or malignant disease were reviewed [17]. Patient demographics, indications for surgery, comorbid conditions, and other factors known to affect healing (steroid use, previous pelvic irradiation, smoking history) were entered prospectively into a database. All operations were performed laparoscopically using a standard double-stapled technique. Following completion of the anastomosis, the mucosa of the colon, rectum and anastomotic staple line were visually inspected with a standard rigid procto-sigmoidoscope. An air leak test was then performed and the decision to revise the anastomosis and/or divert the fecal stream with a loop ileostomy was based on clinical grounds.

Using a similar protocol, we injected 1 ml of ICG via a peripheral IV catheter as a single rapid bolus, NIR angiographic images were obtained transanally using the Pinpoint system. Uptake of ICG was evaluated subjectively by the operating surgeon and video recorded. Fluorescence was evaluated by employing a scoring system. A fluorescence score (FS) of 1–5 (1—no uptake, 5—maximal uptake) for each segment of the anastomosis and the staple line was assigned to each patient. Additional parameters including sequence of uptake and time to maximal excitation of all bowel segments were recorded. All patients were followed up postoperatively at 1 week and 1 month.

Of the 20 patients, 14 were women. The mean age was 67.2 (SD + 14.9). The indications for surgery included malignant disease of the colon or rectum and diverticular disease. The mean distance of the anastomosis from the anal verge was 11 (+3) cm, operative time ranged from 121 to 234 min (mean 170 + 33.4). Complete anastomotic doughnuts and a negative air leak test were achieved in all patients. Out of 20 patients, four exhibited abnormality of flow on ICG evaluation. Three exhibited a significant hypofluorescence (FS 3) and one had patchy fluorescence (FS 2). Uptake of colonic mucosa preceded the rectum in the majority of cases (17/22, 85 %) and the staple line fluoresced in only a minority of cases (1/20, 5 %). In two of the patients with hypofluorescence, the surgeon had already decided to perform a diverting loop ileostomy based on visual evaluation of the perfusion of the proximal bowel in one patient and the very low level of the anastomosis in the second patient. Both diverted patients did well postoperatively. The other two patients with low FS had adequate perfusion by clinical parameters, yet appeared to have inadequate uptake of ICG in the proximal colon on imaging. As an early feasibility study and as per protocol, no change in operative management was made based on the ICGA findings. Postoperatively, both patients experienced delayed return of bowel function and febrile episodes, and CT scans performed revealed peri-anastomotic collections consistent with minor anastomotic leaks. They were treated conservatively with antibiotics and subsequently showed clinical improvement and were discharged home on postoperative days 10 and 13. Follow-up imaging confirmed resolution of the collection in both cases. The remaining patients were discharged home with a mean hospital stay of 4.75 (+1.0) days. At 1-month follow-up, all 20 patients were doing well.

In light of the findings in these 20 patients plus an additional 19 patients accrued to that point, the IRB was petitioned to amend the study protocol to allow for intraoperative decision making based on ICGA findings (Video 8.1). The following 41 consecutive patients undergoing colectomy for benign or malignant disease were managed based on clinical *and* ICGA findings (Group B) and then compared retrospectively to the initial 39 patients managed by clinical parameters alone (Group A) (unpublished data). Patient demographics, indications for surgery, comorbid conditions, and other factors known to affect healing did not differ significantly between groups. In Group B, prior to ICG imaging clinical assessment of both the proximal and distal margins of resection were made and clips placed at the planned transection sites. In Group B, four changes in management were made based on ICGA findings (Video 8.2). All of these revisions manifested as more extensive resections of the proximal or distal colon due to hypo-fluorescence of these segments on ICG angiography. In each of these cases clinical assessment had failed to identify areas of hypo-perfusion and would have led to an anastomosis in these regions. In one of the cases requiring more extensive resection, an additional 12 in. of left colon was found to have been devascularized but had appeared normal by examination alone. No patients had diversions based on ICGA findings. Patients in each group were followed up at 1 week and 1 month postoperatively.

In Group A, two patients manifested minor leaks, requiring prolonged hospitalizations and antibiotic treatment (see above) while in Group B no leaks were identified.

These encouraging results have led to an expanded interest internationally in fluorescence for the assessment of colorectal anastomotic perfusion. Although the cause of anastomotic leak is most certainly multifactorial, our data indicate that eliminating hypo-perfusion can significantly impact outcomes. Further study in a multicenter and ultimately in a randomized control study termed the "perfusion assessment in laparoscopic left anterior resection" or PILLAR II and PILLAR III trials respectively will hopefully bear out these early findings.

Intraoperative Cholangiography

Intraoperative cholangiography (IOC) has been shown to be helpful in delineating biliary anatomy and may prevent main duct injury during laparoscopic cholecystectomy. A number of large population-based reports support this contention and indicate that routine IOC does appear to further reduce the incidence of BDI or at minimum allows for earlier recognition and repair of injuries [18–20]. Despite these benefits, the need for ionizing radiation and injection of a contrast agent directly into the biliary system has prevented IOC from becoming routine. Fluorescent cholangiography using an intravenously injected fluorophore and near infrared (NIR) imaging provides similar anatomical detail without the drawbacks of radiographic cholangiography (Video 8.3).

The Pinpoint system is ideal for this application since it provides NIR cholangiography mated with the standard high definition (HD) white light laparoscopic view producing an uninterrupted, augmented view of the anatomy (Fig. 8.6). The benefits of NIR cholangiography include:

- Does not require puncture of the biliary system
- Quick
- No ionizing radiation
- Provides a real-time continuous or on-demand roadmap
- Allows for visualization and dissection of the tissue planes while simultaneously providing real-time cholangiography
- Retains orienting landmarks
- Does not affect the flow of procedure

But also has a number of drawbacks:

- Only 3 mm penetration and therefore:
- Cannot visualize structures covered by fat or inflammation (e.g. morbidly obese patients, acute cholecystitis)
- Cannot be used to evaluate the common bile duct for stones
- Liver uptake can overwhelm fluorescence of the individual structures
- Cost

In a randomized control study approved by the ethics review board at our institution and registered with clinicaltrials.gov Identifier # NCT01424215, 327 patients undergoing laparoscopic cholecystectomy were assessed for inclusion into the study but only 100 agreed to be randomized, 50 acted as the control group while the other 50 received 1 cm^3 of indocyanine green (ICG) intravenously prior to the procedure. Operative time, time to identification of the major structures, and the surgeon's assessment of visualization, safety, and resident autonomy were evaluated. Time to recognition of structures was defined as the time from intra-abdominal

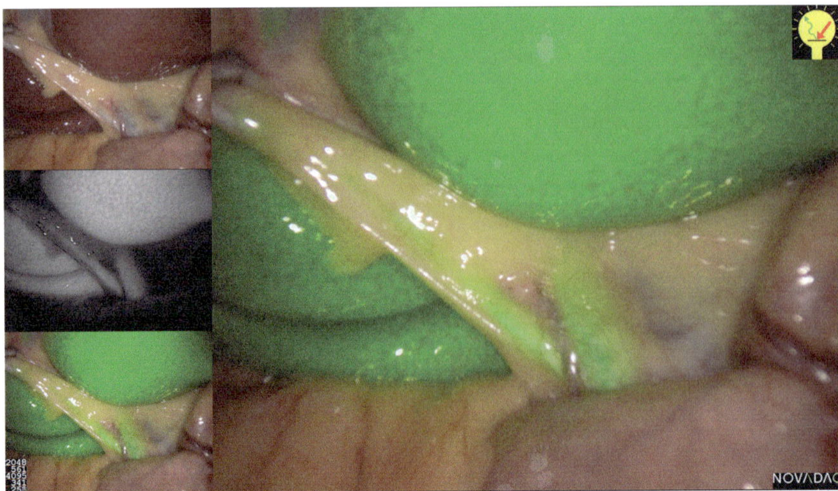

Fig. 8.6 The Pinpoint system provides near infra-red (NIR) cholangiography mated with the standard high definition (HD) white light laparoscopic view producing an uninterrupted, augmented view of the anatomy

camera entry to identification of either of the following:

1. The Cystic duct and either the CBD or CHD OR
2. CBD and CHD.

We found no difference in mean operative times between the control and ICG groups (46.1 min and 52.4 min respectively) but mean time to identification of the key anatomical structures (19.7 min and 14.2 min respectively) was significantly reduced in the ICG group. Based on a subjective questionnaire administered to the surgeons, Pinpoint did not significantly improve the safety of the procedure or the ability of the surgeon to identify important structures but did allow the surgeon to give trainees greater operative autonomy.

In five cases the NIRC changed management:
One case of anomolous anatomy (Fig. 8.7)
One slipped cystic duct clip
Three cystic duct stones (Video 8.4)

The results of this study are encouraging and indicate that although the Pinpoint system does not supplant the critical view technique it speeds recognition of key structures by providing real-time contextualized cholangiographic images without being obtrusive or interrupting the operative flow. It may be especially helpful for surgeons early in the learning curve to help in identifying anatomy and preventing injury.

The one patient with an anomalous bile duct seen on NIR cholangiography (NIRC) was a 28-year-old female with a history of biliary colic and ultrasonographic evidence of cholelithiasis undergoing a laparoscopic cholecystectomy. An anomalous duct was identified during dissection and development of the critical view of safety [21]. Recognition of this anomalous anatomy was made possible solely due to the Pinpoint. Without NIRC this structure would have gone unnoticed and assuredly divided under the mistaken conclusion that it was a vascular structure ("posterior branch of the cystic artery") or part of the areolar peri-cystic tissue. Because it was identified as ductal (due to the presence of fluorescence), the possibility that this represented an aberrant right hepatic duct was raised leading us to control the cystic duct by divided distal to the anomalous duct. The excellent detail provided by the NIR cholangiography helped identify this structure as part of the biliary tree and prevented inadvertent injury.

As many as 23 % of patients undergoing laparoscopic cholecystectomies have anomalous ductal anatomy [22, 23]. The use of NIRC may be helpful in preventing injuries to these structures that would likely go unrecognized.

Three patients in this study had obstructing stones in their cystic duct. This was visualized by abrupt cutoff of fluorescence in the mid cystic

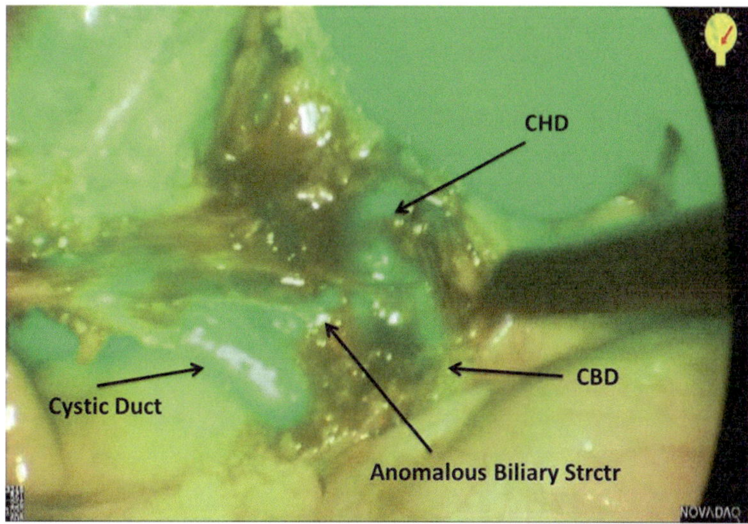

Fig. 8.7 Anomolous anatomy

duct and failure of the gall bladder (GB) to fill with dye. We do not routinely palpate the cystic duct prior to clipping and dividing the duct. Furthermore, we tend to divide high on the CD, i.e. close to the GB to avoid CBD injury. These stones would therefore likely have been missed and left behind if not for the NIR cholangiography. The clinical significance of a retained cystic duct stone is unknown but may contribute to the "post-cholecystectomy syndrome" [24–26]. In each of these cases, the stone was "milked" into the GB prior to controlling and dividing the cystic duct.

One other important consideration when discussing cholecystectomy is cost. As a high volume procedure with over one million of these procedures performed annually in the US, cost becomes an extremely important factor when assessing new technologies. Due to the rare nature of main duct injuries, between 0.2% and 0.8 % [27–33], a study adequately powered to evaluate the clinical value of this expensive technology will be difficult.

Sentinel Lymph Node Dissection

Sentinel lymph node (SLN) dissection has been described for a number of primary malignancies and has become standard of care for breast and melanoma. SLN dissection for other tumors such as colon gastric and gynecologic has been extensively reported with conflicting results and thus remains non-standard. This is not the forum for an extensive review of this topic but as minimally invasive surgery has begun to dominate these procedures, SLN dissection has become virtually impossible. Standard blue dyes are difficult to visualize laparoscopically and although available, laparoscopic gamma probes for the detection of radiotracers have a very steep learning curve [34].

The Pinpoint is a perfect fit for this application. It allows visualization of nodes even in the retropertioneum or covered by fat. These can then be dissected laparoscopically under direct visualization with real-time lymphatic mapping. Also ICG is a small molecule and rapidly flows from the site of injection into the lymphatic channels and then to the SLN.

Some of the applications currently being aggressively researched include NIR-SLN for gastric and colon cancer. Our group is currently prospectively collecting data on patients undergoing laparoscopic gastric resections for cancer.

Although in our institution we routinely perform D2 lymphadenectomy (LD) this is not the standard in most western centers. The reason for this is no improved survival and an increase in morbidity and mortality seen in western studies comparing D1 with D2 LD [35]. This is in contrast to Asian centers which have reported excellent results and improved survival with increasing LD.

We hypothesize that instead of a complete D2 LD, SLN may allow even less experienced centers to perform a "directed D2" LD. For example, a gastric lesion with a SLN and lymphatic channels leading to the right gastric basin could allow the surgeon to avoid an extensive dissection of the nodes at the pancreatic and splenic stations (10,11p,11d) thus improving outcomes. This technique might be especially applicable in western centers since it would allow for avoidance of the complications associated with extensive LD but still produce equivalent oncologic results.

The technique for gastric SLN dissection involves the submucosal injection of ICG endoscopically in a peri-tumoral location. Within approx. 60 s the lymphatic channels and the SLN are identified and marked (Video 8.5). In our series to date this technique has been highly successful in identifying the SLN but whether this node impacts the extent of LD remains to be seen and is pending further data collection and study.

Future Directions

ICG can readily be bound to a number of other molecules including antibodies [36] and nanoparticles [37]. A number of groups are currently working on these combo molecules potentially opening a whole new world of image-guided laparoscopic surgery. Antibodies to tumors, the ureters, or nerve bundles are only a few of the possible designer molecules which may someday be part of the toolbox available to surgeons for image-guided surgery using real-time fluorescence imaging with the Pinpoint.

References

1. Liu DZ, Mathes DW, Zenn MR, Neligan PC. The application of indocyanine green fluorescence angiography in plastic surgery. J Reconstr Microsurg. 2011;27(6):355–64.
2. Mitsuhashi N, Kimura F, Shimizu H, Imamaki M, Yoshidome H, Ohtsuka M, et al. Usefulness of intraoperative fluorescence imaging to evaluate local anatomy in hepatobiliary surgery. J Hepatobiliary Pancreat Surg. 2008;15:508–14.
3. Tagaya N, Shimoda M, Kato M, Nakagawa A, Abe A, Iwasaki Y, Oishi H, Shirotani N, Kubota KJ. Intraoperative exploration of biliary anatomy using fluorescence imaging of indocyanine green in experimental and clinical cholecystectomies. J Hepatobiliary Pancreat Sci. 2010;17(5):595–600.
4. Stiles BM, Adusumilli PS, Bhargava A, Fong Y. Fluorescence cholangiography in a mouse model; an innovative method for improved laparoscopic identification of the biliary anatomy. Surg Endosc. 2006;20:1291–5.
5. Vignali A, Fazio VW, Lavery IC, Milsom JW, Church JM, Hull TL, Strong SA, Oakley JR. Factors associated with the occurrence of leaks in stapled rectal anastomoses: a review of 1,014 patients. J Am Coll Surg. 1997;185(2):105–13.
6. Vignali A, Gianotti L, Braga M, et al. Altered microperfusion at the rectal stump is predictive for rectal anastomotic leak. Dis Colon Rectum. 2000;43:76–82.
7. Novell JR, Lewis AA. Peroperative observation of marginal artery bleeding: a predictor of anastomotic leakage. Br J Surg. 1990;77(2):137–8.
8. Matthiessen P, Hallböök O, Andersson M, Rutegård J, Sjödahl R. Risk factors for anastomotic leakage after anterior resection of the rectum. Colorectal Dis. 2004;6(6):462–9.
9. Kudszus S, Roesel C, Schachtrupp A, Höer JJ. Intraoperative laser fluorescence angiography in colorectal surgery: a noninvasive analysis to reduce the rate of anastomotic leakage. Langenbecks Arch Surg. 2010;395(8):1025–30.
10. Choi HK, Law WL, Ho JW. Leakage after resection and intraperitoneal anastomosis for colorectal malignancy: analysis of risk factors. Dis Colon Rectum. 2006;49(11):1719–25.
11. Urbanavičius L, Pattyn P, Van de Putte D, Venskutonis D How to assess intestinal viability during surgery: a review of techniques. World J Gastrointest Surg. 2011;3(5):59–69.
12. Ahn H, Lindhagen J, Nilsson GE, Salerud EG, Jodal M, Lundgren O. Evaluation of laser Doppler flowmetry in the assessment of intestinal blood flow in cat. Gastroenterology. 1985;88:951–7.
13. Gore RW, Bohlen HG. Microvascular pressures in rat intestinal muscle and mucosal villi. Am J Physiol. 1977;80:H685–93.
14. Brolin RE, Bibbo C, Petschenik A, Reddell MT, Semmlow JL. Comparison of ischemic and reperfusion injury in canine bowel viability assessment. J Gastrointest Surg. 1997;1:511–6.
15. Sherwinter DA. Transanal near-infrared imaging of colorectal anastomotic perfusion. Surg Laparosc Endosc Percutan Tech. 2012;22(5):433–6.
16. Sherwinter DA. A novel adaptor converts a laparoscope into a high-definition rigid sigmoidoscope. Surg Innov. 2013;20(4):411–3.
17. Sherwinter DA, Gallagher J, Donkar T. Intraoperative transanal near infrared imaging of colorectal

anastomotic perfusion: a feasibility study. Colorectal Dis. 2013;15(1):91–6.
18. Buddingh KT, Morks AN, Ten Cate Hoedemaker HO, Blaauw CB, van Dam GM, Ploeg RJ, Hofker HS, Nieuwenhuijs VB. Documenting correct assessment of biliary anatomy during laparoscopic cholecystectomy. Surg Endosc. 2012;26(1):79–85.
19. Flum DR, Dellinger EP, Cheadle A, Chan L, Koepsell T. Intraoperative cholangiography and risk of common bile duct injury during cholecystectomy. JAMA. 2003;289(13):1639–44.
20. Buddingh KT, Nieuwenhuijs VB, van Buuren L, Hulscher JB, de Jong JS, van Dam GM. Intraoperative assessment of biliary anatomy for prevention of bile duct injury: a review of current and future patient safety interventions. Surg Endosc. 2011;25(8): 2449–61.
21. Sherwinter DA. Identification of anomolous biliary anatomy using near-infrared cholangiography. J Gastrointest Surg. 2012;16(9):1814–5.
22. Corbitt Jr JD, Leonetti LA. One thousand and six consecutive laparoscopic intraoperative cholangiograms. JSLS. 1997;1(1):13–6.
23. Puente SG, Bannura GC. Radiological anatomy of the biliary tract: variations and congenital abnormalities. World J Surg. 1983;7(2):271–6.
24. Tantia O, Jain M, Khanna S, Sen B. Post cholecystectomy syndrome: role of cystic duct stump and reintervention by laparoscopic surgery. J Minim Access Surg. 2008;4:71–5.
25. Flörcken H. Gallenblasenregeneration mit Steinrecidiv nach Cholecystectomie. Deutsch Z Chir. 1912; 113:604.
26. Ponsky JL, Dumot J. Retained gallbladder/cystic duct remnant calculi as a cause of postcholecystectomy pain. Surg Endosc. 2002;16:981–4.
27. de Santibañes E, Palavecino M, Ardiles V, et al. Bile duct injuries: management of late complications. Surg Endosc. 2006;20:1648–53.
28. Fischer JE. Is damage to the common bile duct during laparoscopic cholecystectomy an inherent risk of the operation? Am J Surg. 2009;197:829–32.
29. Fiore NF, Ledniczky G, Wiebke EA, et al. An analysis of perioperative cholangiography in one thousand laparoscopic cholecystectomies. Surgery. 1997;122: 817–23.
30. Shea JA, Healey MJ, Berlin JA, et al. Mortality and complications associated with laparoscopic cholecystectomy: a meta-analysis. Ann Surg. 1996;224:609–20.
31. Fletcher DR, Hobbs MST, Tan P, et al. Complications of cholecystectomy: risks of the laparoscopic approach and protective effects of operative cholangiography. A population-based study. Ann Surg. 1999;229:449–57.
32. Vecchio R, MacFadyen BV, Latteri S. Laparoscopic cholecystectomy: an analysis of 114,005 cases of United States series. Int Surg. 1998;83:215–9.
33. Russell JC, Walsh SJ, Mattie AS, et al. Bile duct injuries, 1989-1993: a statewide experience. Arch Surg. 1996;131:383–8.
34. de Haas RJ, Wicherts DA, Hobbelink MG, van Diest PJ, Vleggaar FP, Borel Rinkes IH, van Hillegersberg R. Sentinel lymph node mapping in colon cancer using radiocolloid as a single tracer: a feasibility study. Nucl Med Commun. 2012;33(8):832–7.
35. McCulloch P, Nita ME, Kazi H, Gama-Rodrigues J. Extended versus limited lymph nodes dissection technique for adenocarcinoma of the stomach. Cochrane Database Syst Rev 2004;(4):CD001964.
36. Ogawa M, Regino CA, Seidel J, Green MV, Xi W, Williams M, Kosaka N, Choyke PL, Kobayashi H. Dual-modality molecular imaging using antibodies labeled with activatable fluorescence and a radionuclide for specific and quantitative targeted cancer detection. Bioconjug Chem. 2009;20(11):2177–84.
37. Bunschoten A, Buckle T, Kuil J, Luker GD, Luker KE, Nieweg OE, van Leeuwen FW. Targeted non-covalent self-assembled nanoparticles based on human serum albumin. Biomaterials. 2012;33(3):867–75.

Economic Impact of Fluorescent Cholangiography

9

Emanuele Lo Menzo, Fernando D. Dip, Samuel Szomstein, and Raul J. Rosenthal

Introduction

Since its introduction in 1988 in France, laparoscopic cholecystectomy (LC) has gained exponentially in popularity and it is now considered the standard of care. Because of the prevalence of gallbladder pathology, LC remains one of the most common operations in general surgery, accounting for approximately 750,000 procedure per year in the USA [1]. In spite of the widespread popularity of the procedure and the improvement of the surgical techniques, the incidence of serious complications, in particular extrahepatic bile duct injury (BDI), remains stable. In fact, the incidence of BDI during LC is estimated between 0.2 and 0.7 %, which is significantly higher than the open counterpart (0.16–0.2 %) [2, 3]. Among the most common cause of BDI are surgeon's inexperience and misinterpretation of the images (laparoscopic or fluoroscopic in the case of intraoperative cholangiogram). Main risk factors for BDI remain the presence of inflammation at the infundibulum (acute cholecystitis), impacted stones with distorted anatomy, biliary pancreatitis, bleeding in the surgical field, a contracted gallbladder that prevents proper retraction and better display of the triangle of Calot, aberrant biliary and vascular anatomy (they often coexist).

Besides the negative consequences for the quality of life of post-BDI patients, often young and sentenced to a lifelong need of percutaneous and operative interventions, an increased mortality has been reported [4]. Furthermore, BDI are almost invariably associated with malpractice litigation with often negative judgments for the surgeons and hefty settlements. This in addition to the added expense of the medical care of these patients determines an overall enormous increase in expenses. Depending on the need for several revisional procedures or not, the cost of BDI has been calculated between $587,000 and $669,000 per patient. If we extrapolate these costs for the known incidence of BDI, the resultant added expense would be between $1,870,000,000 and $7,500,000,000 per year [5].

E. Lo Menzo, M.D., Ph.D. (✉)
Digestive Disease Institute, Cleveland Clinic Florida, 2950 Cleveland Clinic Boulevard, Weston, FL 33331, USA
e-mail: lomenze@ccf.org

F.D. Dip, M.D., M.A.A.C.
Section of Minimally Invasive Surgery, Department of General and Vascular Surgery, The Bariatric and Metabolic Institute, Cleveland Clinic Florida, 2950 Cleveland Clinic Boulevard, Weston, FL 33331, USA

Oncological Surgical Division, Division of Surgical Research, Department of Surgery, Hospital de Clinicas Buenos Aires, University of Buenos Aires, Buenos Aires, Argentina

S. Szomstein, M.D., F.A.C.S., F.A.S.M.B.S.
R.J. Rosenthal, M.D., F.A.C.S.
Section of Minimally Invasive Surgery, Department of General and Vascular Surgery, The Bariatric and Metabolic Institute, Cleveland Clinic Florida, 2950 Cleveland Clinic Boulevard, Weston, FL 33331, USA
e-mail: szomsts@ccf.org

F.D. Dip et al. (eds.), *Fluorescence Imaging for Surgeons: Concepts and Applications*,
DOI 10.1007/978-3-319-15678-1_9, © Springer International Publishing Switzerland 2015

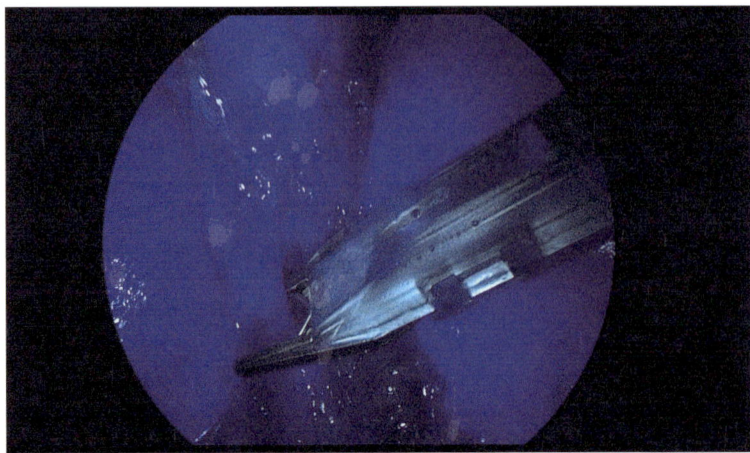

Fig. 9.1 Fluoroscopic view of the clip applier on the cystic duct. The clip applier is visible thanks to the intense reflected light from the liver

Fluorescent imaging is a methodology that, although developed several years ago, has found new applications relatively recently. The original concept of utilizing the property of certain substances to fluoresce after being illuminated by light at a particular wavelength spectrum has been used for clinical purposes for approximately 60 years.

The recent expansion of the clinical application of fluorescence in medicine derives also from the technological advancement in the manufacturing of video processing, and the better understanding of the biochemical properties of the fluorescent substances. Recently, this technology has been utilized to visualize the biliary tree anatomy during cholecystectomies.

We describe the potential advantages of fluorescent cholangiography (FC) over the currently used methodologies implemented to enhance the intraoperative recognition of the extrahepatic biliary tree anatomy.

Conventional Methods of Intraoperative Recognition of Biliary Tree Anatomy

Much debate exists on the preferred method to avoid or reduce BDI during cholecystectomy. Several techniques have been advocated to minimize the risk of BDI. The ideal technique should have the following characteristics: high specificity and sensitivity, simple, available real time, able to perform expeditiously without elevated added costs, repeatable during the procedure as necessary, safe for the patient and the healthcare providers without the administration of radiation, easy to learn and to teach. As expected based on the multitude of options described, no perfect one has been recognized.

Critical of View of Safety (CVS)

Among the most popular options the critical of view of safety (CVS) is often considered as the gold standard, even if level 1 data are lacking [6]. This methodology was originally described by Strasberg in 1995 and consists of the complete mobilization of the proximal gallbladder from the liver bed, clearance of the triangle of Calot (Fig. 9.1) until segment V of the liver is visible through the window, presence of only the cystic artery and duct between the gallbladder and the hepatoduodenal ligament [7]. As a consequence, the inability to achieve the CVS is considered an indication for conversion to open. Although several authors have reported a decrease in the incidence of BDI after the routine adoption of the CVS technique, the inability to completely eliminate such occurrence has lead to pursuing additional options [8].

Intraoperative Cholangiogram (IOC)

The most commonly utilized technique to assess the intraoperative biliary anatomy is the intraoperative cholangiography (IOC). This technique consists in the direct injection of contrast media in the biliary tree (the gallbladder, cystic duct, or more rarely common bile duct) and visualization via real-time fluoroscopic imaging equipment. Extensive literature exists for both the routine and selective use of IOC. The proponents of routine IOC report a simpler utilization and learning curve for both the operator and the staff because of the consistency of its application. These factors contribute to the overall efficiency with which IOC can be performed in the reporting medical centers (average added time 7–10 min) [9]. Also the routine use allows for more confident interpretation of the images by the surgeon. In fact some of the criticisms of IOC is the possibility of misinterpreting the images obtained, obviating the benefits of the added cost and time. Although conflicting evidence exists, at least two cohort studies have shown a reduction in BDI when routine IOC is implemented. In fact, Flum et al. found a reduction of almost 50 % in the rate of BDI when IOC was routinely used [10]. Similar rates of reduction in BDI injury with the use of routine IOC were demonstrated by Fletcher et al. [11].

The cost of IOC remains a very controversial topic. In fact evidence of both the cost effectiveness of routine IOC and selective IOC exists. The proponents of routine IOC report costs of performing IOC of around $675 per case [5]. This number multiplied by the number of LC performed yearly results in an added cost to the operations of over $1,000,000,000. When the potential cost of BDI injury and its incidence are considered, however, routine IOC would result in savings estimated between $164,000,000 and $1,290,000,000 per year [5]. Obviously the cost advantage is only beneficial if calculated in large scale, whereas for the individual surgeon the time and money saved would outweigh the low probability of incurring in BDI. In addition, BDI is linked to high probability of litigation often resulting in large settlements and payouts (between $5,000,000 and over $6 million) [5]. Unfortunately, the calculations of costs are flowed by approximations and extrapolations. In fact, often the calculation of the cost of each IOC only includes the tangible expenses of the supplies utilized, not taking into consideration costs related to equipment's depreciation, maintenance, OR time (estimated about $1,000/h), radiologist fees. Under this perspective it will require 714 IOC to prevent one BDI, which translates in a projected cost of $371,000 [12, 13].

One of the major criticisms to IOC is the basic concept of the technique itself. In fact, in order to perform an IOC the cystic duct has to be incised. This critical step is by definition the lowest degree of BDI (Bismuth 1) if the main bile duct is mistaken for the cystic. So even if the IOC is fundamental in order to recognize the anatomy, the chance of BDI is not diminished, but only recognized early. Undoubtedly, early recognition and repair of BDI has been linked with significantly better outcomes and cost saving [14]. Finally, the use of routine IOC is certainly superior in the diagnosis of CBD stones and results in decreased re-admission rates [14]. The latter factor should be accounted for in the cost analysis for routine IOC.

Cholecystocholangiography (CCC)

In order to obviate the opening of the biliary tree to insert the cholangiogram catheter, the technique of cholecystocholangiography (CCC) has been described. In the latter, the contrast is injected directly in the gallbladder and so higher pressure and quantity of contrast are required. In spite of the advantages of this technique in terms of avoidance of cystic duct cannulation and faster execution compared to the standard IOC, in randomized trial the image quality was inferior to the IOC and there was more than doubling of the radiation exposure [15]. An alternative to this technique is the injection of contrast directly at the infundibulum of the gallbladder using a particular occluding clamp (the "Kumar clamp"). The success rate of this technique has been reported in 80–90 % of the cases, but its superiority to the standard IOC still needs to be proven [16].

Laparoscopic Ultrasound (LUS)

The popularity of laparoscopic ultrasound (LUS) led to the implementation of such technique for the assessment of the biliary anatomy. In fact, LUS can visualize the CBD, the CBD–cystic duct junction and the vascular anatomy. Among the advantages of the technique are the lack of radiation exposure, the repeatability, and the short procedural time. The success rate in the visualization of the structures has been reported in over 90 % of the cases [17]. Also the cost is minimal since it only include the capital expense of laparoscopic probes that are already likely available as they serve multiple other applications in laparoscopy (liver, pancreatic surgery, etc.). Unfortunately, the procedure has never become popular among the general surgeons because of its steep learning curve in the interpretation of the images.

Fluorescent Cholangiography (FC)

The advantages of the use of fluorescence in surgery are determined by the possibility to obtain a high contrast image between the target structure and the background thanks to the different light wavelengths utilized to illuminate and record [18]. Additionally, the simplicity of the technique along with its safety and low cost makes it for an ideal technology for the expanding field of image-guided surgery.

Indocyanine green (ICG) was the first fluorescent dye described. This was initially utilized for near-infrared (NIR) photography in 1955 [18]. The first clinical application was in retinal angiography in the 1970s [19]. In order to visualize the blood vessels in a dark background, a proper light source is necessary to illuminate the ICG with wavelengths between 750 and 800 nm. The resulting fluorescence with wavelengths over 800 nm is captured using a camera with filters able to avoid the interference of the two light wavelengths, in particular the brighter excitation light. The ability to work in the NIR spectrum allows for more penetration into the tissues. Other substances like fluorescein, instead, are visible without the need of special cameras, but give information of more superficial layers.

The clinical applications of ICG imaging span from diagnostic modalities (such as tomography, retinal, cerebral, coronary angiographies) to therapeutic applications in lymphatic mapping, assessment of tissue perfusion, cancer identification and targeting. More recently ICG imaging has found a specific application in laparoscopy, by its property of visualize particular anatomic structure, allowing for more accurate and safe procedures.

The property of ICG to bind with plasma proteins allows for confinement into the vascular system. The binding to lipoproteins in particular is so rapid that the ICG can be safely re-administered in a short time interval [18]. The ICG is then rapidly, and almost exclusively, eliminated by the liver into the bile. The specific property of ICG to fluoresce at different wavelengths makes it for a potential specific molecular carrier. Unfortunately, once the ICG is conjugated to antibodies loses its fluorescence until is dissociated from them [18].

The ability of rapidly binding to plasma proteins after an intravenous bolus makes ICG an optimal substance to utilize in angiography. In fact, ICG angiography has been described in neurosurgery to verify complete obliteration of AV malformations and in general to evaluate the intracerebral circulation in real time at the time of surgery and in a reproducible manner. Also ICG angiography is utilized to assess patency of revascularization procedure such as coronary bypass grafting. Both the peak fluorescence intensity and the temporal slope of fluorescence in the tissue can help quantify tissue perfusion in a more objective manner. Similarly, the use of near-infrared spectroscopy (NIRS) can assess and quantify the degree of liver perfusion after injection of ICG [19, 20].

The overall safety profile of ICG is excellent. In fact, due to its minimal production of singlet oxygen, the potential for phototoxicity, of the retina in particular, is very low. Although in small amounts, ICG contains iodine and could generate allergic reaction in predisposed individuals. In alternative, a compound without iodine, infracyanine green (IfCG) has been developed. This is less than ten times retina toxic than ICG [18]. Because of its lack of metabolites ICG is readily excreted into the bile, which makes it the ideal substance for cholangiography. This property also limits the dose necessary to visualize the extrahepatic biliary tree to 0.5 mg/kg of body weight, well below the threshold of side effects described with doses above 5 mg/kg.

One of the main limitations of the use of ICG in cholangiography is related to the limited penetration of the tissues. In fact, the tissue optical window depends on the ability of the NIR light to reach the target tissue. The variable presence of connective tissue, worse in the presence of more intra-abdominal fat or acute inflammatory changes (edema), limits the utility of this methodology. In our experience of over 100 laparoscopic cholecystectomies with intraoperative fluorescent cholangiography, we were able to visualize at least one of the extrahepatic biliary structures in 100 % of the cases even in the presence of inflammation (unpublished data).

In order to obtain the best definition, the excitation light should be filtered so it does not mix with the fluorescent one. In order to do this the excitation light, especially if it is a broad spectrum one, should be filtered to block the longer wavelengths. In fact, the fluorescent light has only a small portion of the spectrum of the exciting light and it would not be able to be visualized without filtration. This problem is obviated if a monochromic light such as the laser is utilized. Since the normally used charge coupled device (CCD) sensor cameras are normally equipped with a filter that cuts the NIR wavelengths in order to reduce interference, special filters are then necessary to obtain the visualize the fluorescent image.

More recently, the utilization of fluorescence in surgery has led to the development of several techniques to visualize the bile ducts. As previously mentioned, ICG has the unique feature of completely binding to serum protein and exclusive excretion into the biliary tree. Initially, this property was utilized by injecting high ICG doses intravenously and the bile ducts could be identified without any special filters because of the dark blue discoloration [21]. It was soon obvious that the discoloration when observed at the visible light spectrum (380–600 nm) had a penetration of only few mm. This was impractical in many cases in which thick connective tissue or inflammation was present at the level of the infundibulum. In addition, the leak of such concentration of dye would obscure the anatomy and potentially impair the surgeon's ability to perform a safe dissection. On the other hand, the significant advantage of lack of radiation, real time visualization, and, more importantly, the possibility of identify the extrahepatic biliary anatomy prior to any dissection/injury fomented the interest and research in this field [22].

At the same time, infrared cameras were developed and Liu et al. were able to delineate the extrahepatic anatomy by infusing saline into the biliary tract of pigs at different temperature than core and contrasting the image with the body temperature [23]. Obviously, this was an impractical technique because of the quick equilibration on the saline temperature and the need for repetitive injections.

It was in 2009 that Ishizawa et al. describe for the first time the use of ICG excited with NIR light and its fluorescence captured with a filtered camera for the purpose of obtaining visualization of the biliary tree anatomy [24]. In their series of 52 patients, Ishizawa et al. reported a 100 % visualization of the cystic duct and 96 % visualization of the cystic duct-common hepatic (CHD) junction prior to any dissection of Calot's triangle [24]. Although they found that FC was helpful in determining the level of division of the cystic duct in order to avoid leaving cystic duct stones behind, the methodology is not helpful in identifying CBD stones [24].

The quality of the images obtained during the fluorescent cholangiography (FC), although improved, still has the limitation of being projected over a dark background (Fig. 9.1). This limits the ability of the surgeon to visualize his/her instruments at the same time of the visualization of the biliary structures. More recently, however, optical (multispectral NIRF imaging systems) and software advancements resulted in the possibility of simultaneously acquiring or superimposing xenon light and fluorescent images in the same operative field, allowing for real-time dissection of the structures while visualizing the biliary tree [25].

In spite of the multiple publications on the feasibility and results of FC, up until very recently no data existed on the cost implication of such technique.

In 2014, our group published a cost analysis comparing FC to IOC. In the 43 patients studied FC was successfully carried out in 100 % of the patients as compared to 93 % in the IOC group

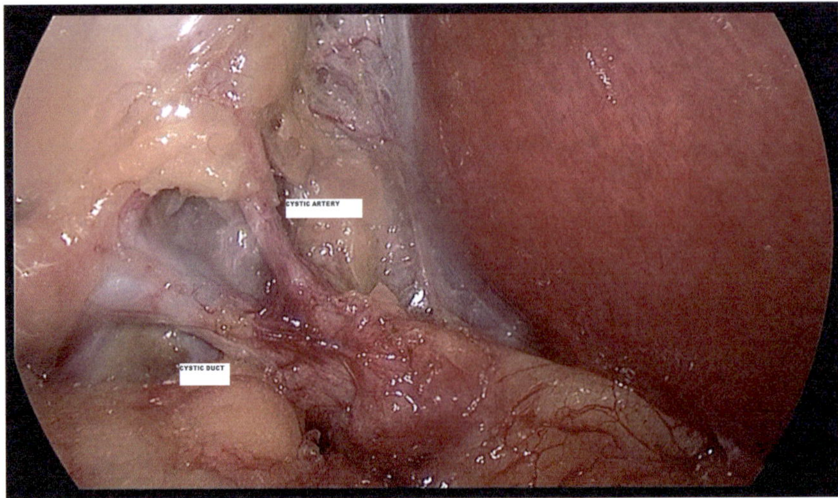

Fig. 9.2 Laparoscopic view of the triangle of Calot dissected

[9]. The cystic duct was identified in 98 % of the patients with FC, the CBD in 79 %, and the CHD in 58 % of the cases prior to any dissection [9]. This is one of the primary procedural differences between FC and IOC. In fact, injuries during the dissection of Calot's triangle or the cannulation of the cystic duct occur before obtaining the IOC. Conversely, the ability to identify these structures before any dissection takes place constitutes one of the main advantages of FC (Figs. 9.2 and 9.3). Furthermore, the availability of the information in real time and on the same screen facilitates the steps of the dissection. The FC can also be repeated at the surgeon's discretion by just switching the view from xenon light to filtered light by activating a pedal switch.

As far as the costs involved in the procedure, FC requires injection of ICG that has a calculated price of an average of $14 ($0.16/kg) [9]. The operating room time for the FC, calculated on the basis of $1,000/h, would be equivalent to $18, considering that in our experience it takes on an average 0.71 min to complete [9, 12]. In fact the ability to obtain the fluoroscopic picture by simply activating a pedal allows for quick and repetitive alternation between xenon and filtered light during the key portions of the triangle of Calot dissection. Finally the cost of the FC system, which includes the light box, light cord, camera and laparoscope is estimated 20 % more than the standard high definition laparoscopic system [9]. These figures are in net contrast with the cost of IOC. In fact, according to our experience, the overall cost of IOC is on average $778 per case as compared to $32 per case of the FC ($p<0.0001$) [9]. The expenses per case do not include the capital costs for the fluoroscopic C-arm (approximately $75,000) and for the FC equipment previously mentioned. The significantly superior cost of the IOC derives from the different components necessary. The minor expenses are the dye ($6.50), the cholangiogram catheter ($6.34), and radiology report ($20). The more significant are the costs of the use of the fluoroscopic equipment and technician ($618), and operative time cost ($128). In fact, even in the hands of routine users of IOC, the time of execution of IOC remains significantly longer than FC (7.1 min vs. 0.71 min $p<0.0001$) [9].

Based on all the previously mentioned qualities, FC is emerging as the preferred method of identification of the extrahepatic bile duct anatomy. Further advances in the quality of the images will contribute to the wider acceptance of this methodology. Furthermore, FC appears to be easily reproducible and requires a minimum learning curve.

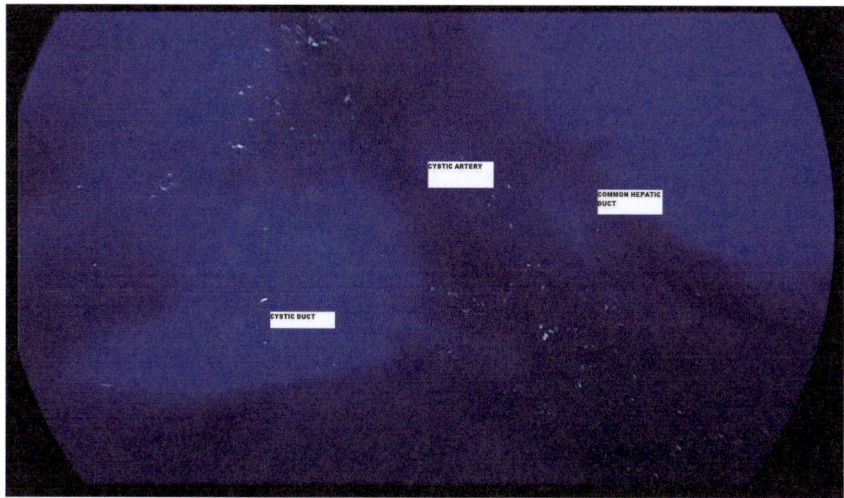

Fig. 9.3 Fluoroscopic image of the triangle of Calot dissected shown in Fig. 9.1. Note the intense blue background of the liver. The cystic artery projects in black

Conclusions

BDI remains a worrisome and potentially devastating complication of cholecystectomy. Despite the vast application of the laparoscopic technique in gallbladder surgery, the rates of BDI have not changed and continue to be higher than the open counterpart. No ideal method of prevention of BDI has been described. IOC offers the ability of early recognition of BDI and CBD stones, but at a significant cost and also exposes healthcare professional to significant doses of radiation. Because of the technique of execution of the IOC, often the injury has already occurred before its completion. FC has recently emerged as an alternative to other techniques in the definition of the biliary tract anatomy during LC. The main advantage remains the ability to recognize the biliary anatomy prior to any dissection. The low cost and simplicity of the technique make FC a potential better alternative to be applied routinely.

References

1. Flum DR, Dellinger EP, Cheadle A, Chan L, Koepsell T. Intraoperative cholangiography and risk of common bile duct injury during cholecystectomy. JAMA. 2003;289(13):1639–44. Available from: http://www.ncbi.nlm.nih.gov/pubmed/12672731.
2. Gigot JF. Bile duct injury during laparoscopic cholecystectomy: risk factors, mechanisms, type, severity and immediate detection. Acta Chir Belg. 2003;103(2):154–60. Available from: http://www.ncbi.nlm.nih.gov/pubmed/12768857.
3. Nuzzo G, Giuliante F, Giovannini I, Ardito F, D'Acapito F, Vellone M, et al. Bile duct injury during laparoscopic cholecystectomy: results of an Italian national survey on 56 591 cholecystectomies. Arch Surg. 2005;140(10):986–92. Available from: http://www.ncbi.nlm.nih.gov/pubmed/16230550.
4. Kern KA. Malpractice litigation involving laparoscopic cholecystectomy. Cost, cause, and consequences. Arch Surg. 1997;132(4):392–7. discussion 397–8. Available from: http://www.ncbi.nlm.nih.gov/pubmed/9108760.
5. Podnos YD, Gelfand DV, Dulkanchainun TS, Wilson SE, Cao S, Ji P, et al. Is intraoperative cholangiography during laparoscopic cholecystectomy cost effective? Am J Surg. 2001;182(6):663–9. Available from: http://www.ncbi.nlm.nih.gov/pubmed/11839335.
6. Strasberg SM, Brunt LM. Rationale and use of the critical view of safety in laparoscopic cholecystectomy. J Am Coll Surg. 2010;211(1):132–8. Available from: http://www.ncbi.nlm.nih.gov/pubmed/20610259.
7. Strasberg SM, Hertl M, Soper NJ. An analysis of the problem of biliary injury during laparoscopic cholecystectomy. J Am Coll Surg. 1995;180(1):101–25. Available from: http://www.ncbi.nlm.nih.gov/pubmed/8000648.
8. Yamashita Y, Kimura T, Matsumoto S. A safe laparoscopic cholecystectomy depends upon the establishment of a critical view of safety. Surg Today. 2010;40(6):507–13. Available from: http://www.ncbi.nlm.nih.gov/pubmed/20496131.

9. Dip FD, Asbun D, Rosales-Velderrain A, Lo Menzo E, Simpfendorfer CH, Szomstein S, et al. Cost analysis and effectiveness comparing the routine use of intraoperative fluorescent cholangiography with fluoroscopic cholangiogram in patients undergoing laparoscopic cholecystectomy. Surg Endosc. 2014;28:1838. Available from: http://www.ncbi.nlm.nih.gov/pubmed/24414461.
10. Flum DR, Koepsell T, Heagerty P, Sinanan M, Dellinger EP. Common bile duct injury during laparoscopic cholecystectomy and the use of intraoperative cholangiography: adverse outcome or preventable error? Arch Surg. 2001;136(11):1287–92. Available from: http://www.ncbi.nlm.nih.gov/pubmed/11695975.
11. Fletcher DR, Hobbs MS, Tan P, Valinsky LJ, Hockey RL, Pikora TJ, et al. Complications of cholecystectomy: risks of the laparoscopic approach and protective effects of operative cholangiography: a population-based study. Ann Surg. 1999;229(4):449–57. Available from: http://www.pubmedcentral.nih.gov/articlerender.fcgi?artid=1191728&tool=pmcentrez&rendertype=abstract.
12. Livingston EH, Miller JAG, Coan B, Rege RV. Costs and utilization of intraoperative cholangiography. J Gastrointest Surg. 2007;11(9):1162–7. Available from: http://www.ncbi.nlm.nih.gov/pubmed/17602271.
13. Flum DR, Flowers C, Veenstra DL. A cost-effectiveness analysis of intraoperative cholangiography in the prevention of bile duct injury during laparoscopic cholecystectomy. J Am Coll Surg. 2003;196(3):385–93. Available from: http://www.ncbi.nlm.nih.gov/pubmed/12648690.
14. Sajid MS, Leaver C, Haider Z, Worthington T, Karanjia N, Singh KK. Routine on-table cholangiography during cholecystectomy: a systematic review. Ann R Coll Surg Engl. 2012;94(6):375–80. Available from: http://www.pubmedcentral.nih.gov/articlerender.fcgi?artid=3954316&tool=pmcentrez&rendertype=abstract.
15. Wills VL, Jorgensen JO, Hunt DR. A randomized controlled trial comparing cholecystocholangiography with cystic duct cholangiography during laparoscopic cholecystectomy. Aust N Z J Surg. 2000;70(8):573–7. Available from: http://www.ncbi.nlm.nih.gov/pubmed/10945550.
16. Kumar SS. Laparoscopic cholangiography: a new method and device. J Laparoendosc Surg. 1992;2(5):247–54. Available from: http://www.ncbi.nlm.nih.gov/pubmed/1421544.
17. Machi J, Tateishi T, Oishi AJ, Furumoto NL, Oishi RH, Uchida S, et al. Laparoscopic ultrasonography versus operative cholangiography during laparoscopic cholecystectomy: review of the literature and a comparison with open intraoperative ultrasonography. J Am Coll Surg. 1999;188(4):360–7. Available from: http://www.ncbi.nlm.nih.gov/pubmed/10195719.
18. Alander JT, Kaartinen I, Laakso A, Pätilä T, Spillmann T, Tuchin VV, et al. A review of indocyanine green fluorescent imaging in surgery. Int J Biomed Imaging. 2012;2012:940585. Available from: http://www.pubmedcentral.nih.gov/articlerender.fcgi?artid=3346977&tool=pmcentrez&rendertype=abstract.
19. Flower RW. Injection technique for indocyanine green and sodium fluorescein dye angiography of the eye. Invest Ophthalmol. 1973;12(12):881–95. Available from: http://www.ncbi.nlm.nih.gov/pubmed/4203466.
20. Jiao LR, El-Desoky AA, Seifalian AM, Habib N, Davidson BR. Effect of liver blood flow and function on hepatic indocyanine green clearance measured directly in a cirrhotic animal model. Br J Surg. 2000;87(5):568–74. Available from: http://www.ncbi.nlm.nih.gov/pubmed/10792311.
21. Pertsemlidis D. Fluorescent indocyanine green for imaging of bile ducts during laparoscopic cholecystectomy. Arch Surg. 2009;144(10):978. Available from: http://www.ncbi.nlm.nih.gov/pubmed/19841371.
22. Buddingh KT, Nieuwenhuijs VB, van Buuren L, Hulscher JBF, de Jong JS, van Dam GM. Intraoperative assessment of biliary anatomy for prevention of bile duct injury: a review of current and future patient safety interventions. Surg Endosc. 2011;25(8):2449–61. Available from: http://www.pubmedcentral.nih.gov/articlerender.fcgi?artid=3142332&tool=pmcentrez&rendertype=abstract.
23. Liu JJ, Alemozaffar M, McHone B, Dhanani N, Gage F, Pinto PA, et al. Evaluation of real-time infrared intraoperative cholangiography in a porcine model. Surg Endosc. 2008;22(12):2659–64. Available from: http://www.pubmedcentral.nih.gov/articlerender.fcgi?artid=2737323&tool=pmcentrez&rendertype=abstract.
24. Ishizawa T, Bandai Y, Ijichi M, Kaneko J, Hasegawa K, Kokudo N. Fluorescent cholangiography illuminating the biliary tree during laparoscopic cholecystectomy. Br J Surg. 2010;97(9):1369–77. Available from: http://www.ncbi.nlm.nih.gov/pubmed/20623766.
25. Themelis G, Yoo JS, Soh K-S, Schulz R, Ntziachristos V. Real-time intraoperative fluorescence imaging system using light-absorption correction. J Biomed Opt. 2009;14(6):064012. Available from: http://www.ncbi.nlm.nih.gov/pubmed/20059250.

Real-Time Near-Infrared Fluorescent Cholangiography During Robotic Single-Site Cholecystectomy

10

Nicolas C. Buchs

Introduction

With the introduction of minimally invasive cholecystectomy a couple of decades ago, an increased rate of biliary injury was observed [1, 2]. In fact, the risk of biliary injury at least doubled and rose up to 0.7 % with a laparoscopic approach [1, 3–6]. While the technique has evolved and the safety has increased, several authors and surgical societies have advocated the liberal use of intra-operative cholangiography in order to reduce this risk [7–9]. Indeed, the main reason of biliary injury remains a misunderstanding or a misinterpretation of the biliary anatomy [2, 4, 10, 11].

On the other hand, the development of fluorescent cholangiography is not new. Pioneers have reported encouraging data using indocyanine green (ICG) for many years [12]. Indeed, ICG has several favorable characteristics: it remains confined in the vascular compartment by binding to plasma proteins; it is excreted rapidly and fairly exclusively by the bile; and it has a very low toxicity [13]. Thus, there has been interest in the use of ICG to investigate the extrahepatic biliary anatomy, as shown by several groups [14–16]. Among them, Ishizawa and colleagues have probably the largest experience with this technology. Their wide knowledge has been reported in several publications [15, 17–19] showing not only the feasibility and the safety of this approach but also and especially the possibility to recognize the biliary anatomy. Obviously, this could lead to a reduction of biliary injury.

Recently, single-site surgery has emerged as safe and feasible for various procedures [20–22]. Among them, cholecystectomy is probably the best reported indication. Several randomized controlled studies have found good outcomes in comparison to standard laparoscopy [23, 24]. These results were confirmed in a meta-analysis [25]. On the other hand, performing an intra-operative cholangiography becomes more difficult by a single-site approach even if feasible [26]. In this context, the concept of a critical view of safety becomes all important [27, 28], i.e. a clear hepatocystic triangle, lower gallbladder dissected off the cystic plate, and only two structures entering the gallbladder [29].

In 2011, a new robotic single-site platform was launched by Intuitive Surgical Inc (Sunnyvale, CA). The concept behind this new platform was

Electronic supplementary material: The online version of this chapter (doi:10.1007/978-3-319-15678-1_10) contains supplementary material, which is available to authorized users. Videos can also be accessed at http://link.springer.com/chapter/10.1007/978-3-319-15678-1_10.

N.C. Buchs, M.D. (✉)
Department of Surgery, Clinic for Visceral and Transplantation Surgery, University Hospital of Geneva, Faculty of Medicine, University of Geneva, Rue Gabriel-Perret-Gentil, 4, Geneva 1211, Switzerland
e-mail: Nicolas.c.buchs@hcuge.ch

F.D. Dip et al. (eds.), *Fluorescence Imaging for Surgeons: Concepts and Applications*, DOI 10.1007/978-3-319-15678-1_10, © Springer International Publishing Switzerland 2015

to improve the ergonomics, since this remains the major drawback of the single-site approach [30]. The initial experience with robotic cholecystectomy was encouraging, with a low complications and conversion rate [31–33].

However, few centers have reported their experience of robotic single-site cholecystectomy (RSSC) using an intra-operative cholangiography [32]. Clearly, performing a cholangiography takes time and often requires dedocking the robotic system (removing the robot from the patient).

As for laparoscopy, fluorescent technology has been introduced for single-site surgery; firstly, of course, for standard laparoscopic single-site cholecystectomy [18], but also more recently for RSSC [14, 16]. While the aim of an intra-operative cholangiography is not limited to exclude a biliary injury but also to rule out a common bile duct stone, fluorescent cholangiography might help the robotic surgeon during RSSC to quickly recognize the anatomy and to safely perform the procedure. In addition, this technique is real-time, non-ionizing, and can be used as often as needed.

Surgical Technique

For a standard RSSC, the technique has been largely described previously [32]. To summarize, the patient is positioned in a slightly reverse Trendelenburg. For the procedure, a peri-umbilical skin incision of about 2.5 cm is done. An open access to the abdominal cavity is formed and a special silicon port with four access points is installed (Fig. 10.1). The da Vinci Si Surgical System (Intuitive Surgical Inc., Sunnyvale, CA) comes over the patient's head.

An 8.5-mm straight port for a camera, two curved 5-mm cannulas—crossing at the level of the abdominal wall—and a 5-mm straight laparoscopic trocar are introduced through this port. The robotic platform automatically switches the arm control to facilitate instrument control.

The procedure starts by retracting the gallbladder upward. The dissection of Calot's triangle is performed using the monopolar hook (Fig. 10.2). Once the cystic artery and the cystic duct are dissected (critical view of safety) (Fig. 10.3), a Hemolock (Teleflex, Medical,

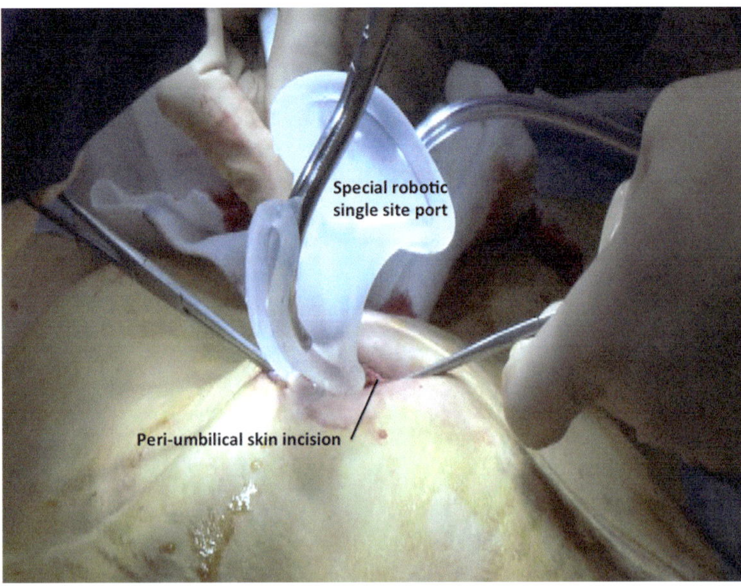

Fig. 10.1 Port placement with a peri-umbilical skin incision (around 2.5 cm)

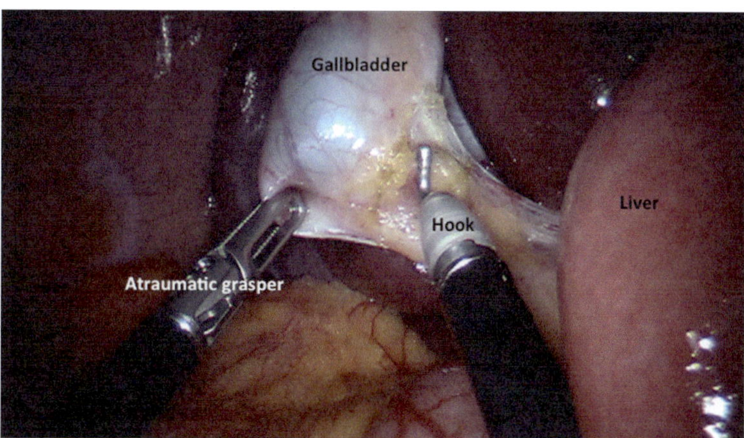

Fig. 10.2 Dissection of Calot's triangle using a monopolar hook and an atraumatic grasper

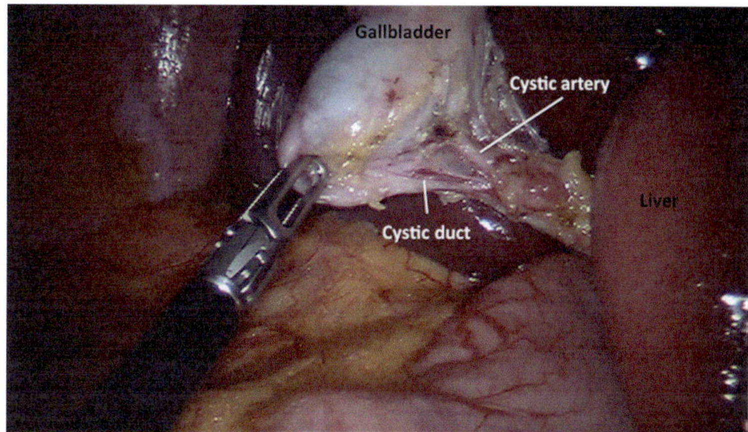

Fig. 10.3 View of the cystic duct and the cystic artery

Ireland) is placed on each side of the cystic artery (Fig. 10.4), and then on each side of the cystic duct (Fig. 10.5). The structures are divided using a robotic scissor. A retrograde dissection is performed (Fig. 10.6 and Video 10.1). The gallbladder is placed in a special endoscopic bag. The robot is dedocked and the specimen is retrieved (Fig. 10.7). The abdominal wall is carefully closed using an absorbable suture.

The technique remains the same during a RSSC with fluorescent cholangiography, as reported previously [14]. To summarize, this fluorescence-capable da Vinci Si HD vision system is an endoscopic imaging system for high definition (HD) visible light and near-infrared fluorescence imaging. The system components include a surgical endoscope capable of visible light and near infrared imaging, a 3DHD stereoscopic camera head that couples to the endoscope and an endoscopic illuminator that provides visible light and near infrared illumination through the surgical endoscope via a flexible light-guide. The fluorescence-capable illuminator provides lighting for the surgical field. The system displays the live video image on the 3DHD stereo viewer and the touchscreen. If switched to fluorescence imaging, the Camera Control Unit processes and displays the resulting images as a fluorescent overlay on a black and white image. The surgeon can quickly switch between normal mode and fluorescence (NIR) by either making adjustments in the control menu at the console or initiation at the finger clutch [14].

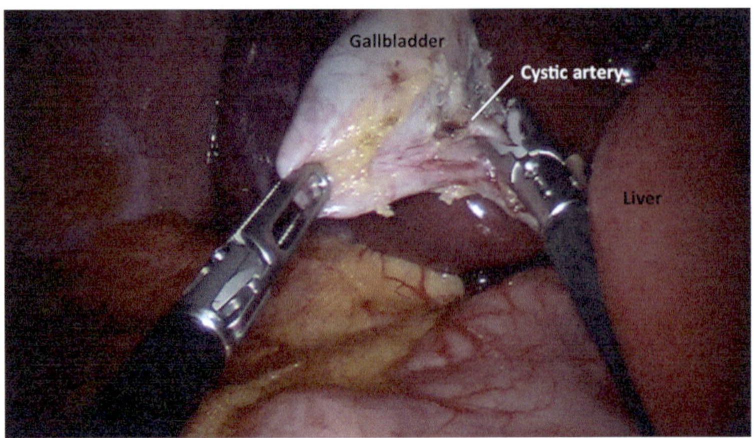

Fig. 10.4 Clipping of the cystic artery

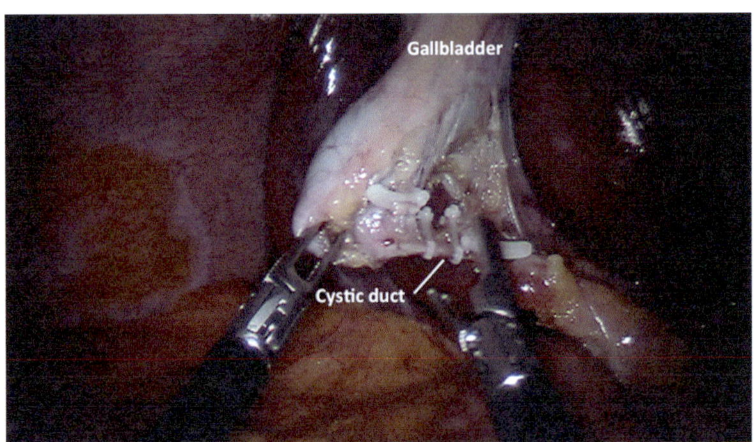

Fig. 10.5 Clipping of the cystic duct

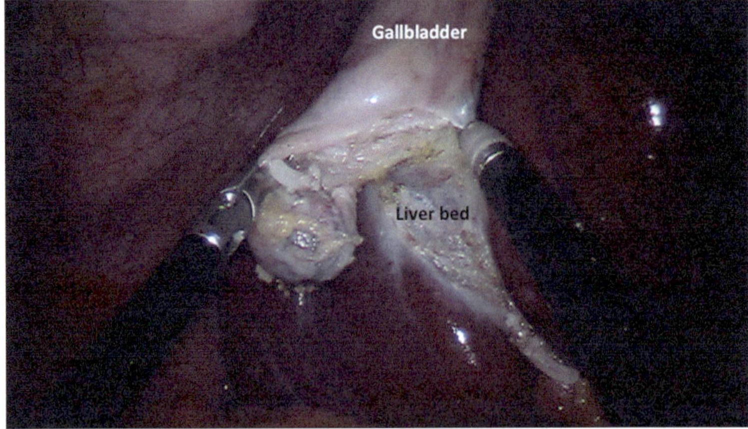

Fig. 10.6 Retrograde dissection of the gallbladder

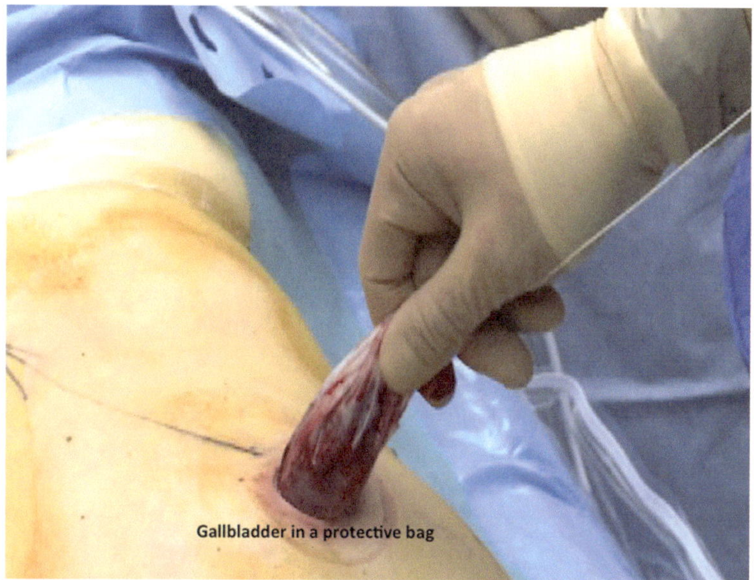

Fig. 10.7 Retrieval of the specimen in a wound protector bag

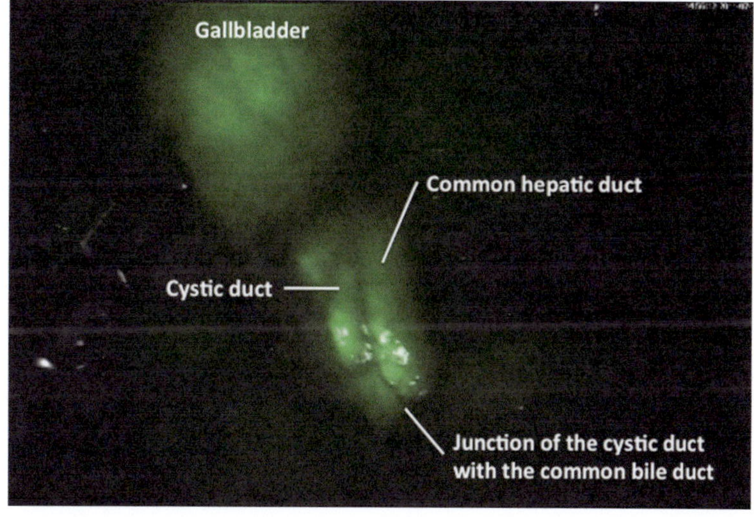

Fig. 10.8 Near-infrared fluorescent cholangiography before the dissection of the Calot's triangle

The ICG imaging agent is a sterile, water-soluble tricarbocyananine dye with a peak spectral absorption at 800–810 nm in blood plasma or blood. ICG is administered intravenously. ICG is nearly exclusively eliminated by the liver into the bile and does not undergo enterohepatic recirculation [34]. Physiologically, ICG appears unconjugated in the bile about 8 min after injection. ICG removal from the blood depends on liver blood flow, parenchymal cellular function, and biliary excretion. 2.5 mg of ICG is administered intravenously during patient preparation by anesthesia, approximately 30–45 min before the start of the case. A second dose of 2.5 mg ICG can be administered intravenously if fluorescence is not detected in the liver approximately 45 min after injection of the first dose or if additional questions regarding perfusion arise during surgery. The RSSC starts as described above. Once the usual retraction of the gallbladder cranially and laterally is achieved exposing the triangle of Calot, the camera view is switched to the fluorescence image (Fig. 10.8 and Video 10.2) and a first attempt to visualize the biliary anatomy can be

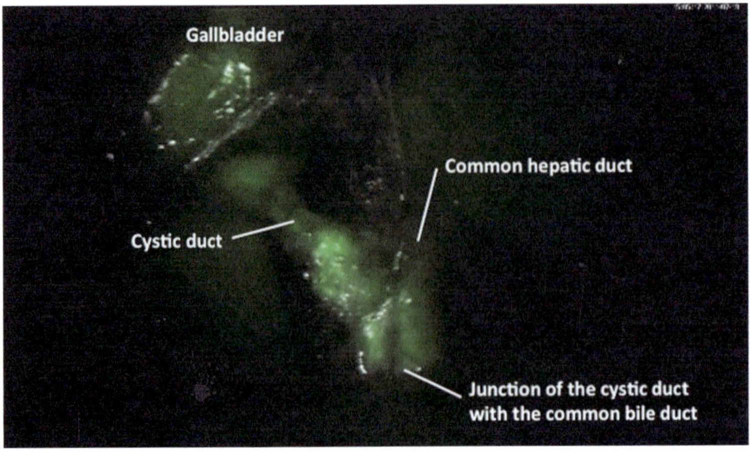

Fig. 10.9 Near-infrared fluorescent cholangiography after the dissection of Calot's triangle

performed. Then, the triangle of Calot is dissected in the usual fashion until the cystic duct and artery are skeletonized for at least 1 cm (Fig. 10.9 and Video 10.3). The camera view can be switched to the fluorescence image as desired to view the biliary anatomy during this surgery [14].

Table 10.1 Visualization of biliary extrahepatic anatomy using near-infrared fluorescent cholangiography

	Before dissection	After dissection
Cystic duct (%)	91	100
Common hepatic duct (%)	50	66.7
Common bile duct (%)	33.3	83.3
Junction between cystic and common bile duct (%)	25	58.3

Exploration of the Biliary Anatomy

Several groups have reported their experience with near-infrared fluorescent cholangiography using ICG. Although the feasibility and the safety seem well reported for a standard laparoscopic cholecystectomy [15, 17–19], the results for RSSC are scarcer. Indeed, only a couple of centers, including ours, have reported interesting and encouraging outcomes [14, 16].

As a potential advantage, the ICG administration is done before surgery and the biliary structures can be analyzed before any dissection, which is not the case for cholangiography requiring extensive dissection of the cystic duct. Moreover, the surgeons can obtain a road map of the biliary tree in real-time and at any time during dissection of Calot's triangle [18]. Table 10.1 summarizes our current data regarding the use of ICG for biliary extrahepatic anatomy assessment.

Spinoglio et al. [16] found relatively similar results. Indeed, in a selected population of 45 patients, they reported different rates of visualization prior to Calot's triangle dissection: 93 % for the cystic duct, 88 % for the common hepatic duct, 91 % for the common bile duct, and 88 % for the junction. After dissection, this rate rose up to 97 % for all the structures. As in our experience [14], at least one biliary structure was visualized in all patients [16].

The interest of a real-time biliary assessment remains to exclude a biliary anatomical variation as it was reported for standard laparoscopic cholecystectomy [35]. Regarding the robotic experience, Calatayud et al. [36] reported an interesting case of aberrant biliary canaliculus, identified by ICG. Obviously, this anatomical malformation could have been missed by a standard approach with potentially dramatic consequences.

Overall, using ICG during RSSC can help to assess the extrahepatic biliary anatomy and thus could reduce the risk of biliary injury by misinterpretation of the biliary anatomy. However, there are no available data to verify this hypothesis. On the other hand, it seems obvious that a clear interpretation of anatomy, especially in real-time and before any dissection, might prevent a biliary injury. Further studies are still required to define the exact role of real-time near-infrared fluorescent cholangiography during RSSC, even if the preliminary data are encouraging.

Other Potential Advantages

Considering the capability to analyze and recognize quickly the extrahepatic biliary anatomy, ICG fluorescent cholangiography might help to shorten the operative time. Indeed, a quick recognition of the cystic and common bile duct is a key step during a cholecystectomy.

Recently, we have shown that ICG fluorescent cholangiography was able to significantly shorten significantly the operative time in normal Body Mass Index (BMI) patients [37]. When considering patients with a BMI less than 25 undergoing RSSC with ICG in comparison to standard RSSC, the operative time was reduced by 24 min ($p=0.06$). In addition, considering only patients undergoing RSSC with ICG, the operative time was shortened by 25 min in low BMI patients compared to patients with BMI >25 ($p=0.004$) [37].

This interesting finding can be explained by the tissue penetration of near-infrared light. Indeed, ICG can be visualized only in a couple of millimeters (typically 5–10 mm) [18]. In obese patients, the Calot's triangle is often covered by a thick layer of fat, which might limit a quick recognition of the main biliary structures and thus slow down the dissection phase. The same is true in the case of severe inflammation. Ishizawa et al. [18] reported that even in these challenging situations, fluorescent cholangiography can be helpful by frequently using fluorescent imaging during cystic dissection.

Inclusion and Exclusion Criteria

The current experience reported in the literature was undergone under IRB approval. While the near-infrared fluorescent cholangiography can help to assess the extrahepatic biliary anatomy, it is not validated to rule out a common bile duct stone and, thus strict selection criteria should be followed [14, 37].

Initially, the inclusion criteria included patients between 18 and 80 years old with symptoms consistent with gallbladder disease and gallstones confirmed by ultrasound, and with ASA score 1–3. Currently, age and ASA score are not strict but relative criteria. Of course, a preoperative proof of gallbladder disease remains mandatory.

On the other hand, initially the exclusion criteria were acute cholecystitis, biliary pancreatitis or suspicion of common bile duct stones, and/or perturbation of a liver test. Currently, acute disease does not represent a strict contraindication. In case of suspicion of common bile duct stone, a preoperative imaging ruling out this pathology should be performed. Of note, Ishizawa et al. [17] performed fluorescent cholangiography in 52 patients during standard laparoscopic cholecystectomy and showed that fluorescent images might identify gallstones in the cystic duct. However, they failed to diagnose common bile duct stones.

Moreover, previous upper abdominal open surgery was also considered as a relative contraindication. In addition, a history of adverse reaction to ICG or any contraindication to ICG label was also considered as reasons not to proceed with this approach. Even if the risk of allergic reaction remains low, caution should be undertaken before injection of ICG. In fact, ICG administration is reported to be safe and well tolerated, without any side effects in various robotic studies [14, 16, 37]. Previous reports [15, 18, 38] have confirmed the safety of this drug for several other indications. The risk of anaphylactic reactions is approximately 0.003 % at doses exceeding 0.5 mg/kg [39] and thus the risk remains quite small considering the small amount of ICG (2.5 mg) injected intravenously for cholangiography [18].

Finally, pregnancy should also be considered as a contraindication.

Comments and Future Developments

The use of ICG fluorescent cholangiography opens interesting avenues. As was shown for standard laparoscopy, the interest of this technology for RSSC is obvious today. In addition, ICG helps to recognize the anatomy and thus adds not only safety but also speed during the dissection phase.

Recently, the development of augmented reality has been reported in various medical fields notably for robotic surgery [40, 41]. Liver surgery is one example [42], and could be considered as a good model. On the other hand, several technical problems still exist such as plastic deformation during dissection and mobilization, fixed landmarks for calibration, and real-time resynchronization [42].

While the possibilities of augmented reality seem endless, ICG fluorescent cholangiography might help to overcome part of the above limits. Indeed, a real-time cholangiography directly projected into the surgical console might be a good option to compare this information to the preoperative biliary images [43]. In addition, the integration of these images directly coupled to real intra-operative images would be the obvious next step for image-guided biliary surgery.

Conclusions

The introduction of real-time near-infrared fluorescent cholangiography during RSSC can help in assessing the biliary anatomy, and might reduce the risk of biliary injury by a better anatomical identification. In addition, ICG might reduce the operative time in normal BMI patients, by fast identification of main biliary structures and by excluding any anatomical variation. Finally, fluorescent images might help in the future to integrate a virtually enhanced world directly inside the robotic console.

References

1. Adamsen S, Hansen OH, Funch-Jensen P, Schulze S, Stage JG, Wara P. Bile duct injury during laparoscopic cholecystectomy: a prospective nationwide series. J Am Coll Surg. 1997;184(6):571–8.
2. Connor S, Garden OJ. Bile duct injury in the era of laparoscopic cholecystectomy. Br J Surg. 2006;93(2):158–68.
3. Caputo L, Aitken DR, Mackett MC, Robles AE. Iatrogenic bile duct injuries. The real incidence and contributing factors–implications for laparoscopic cholecystectomy. Am Surg. 1992;58(12):766–71.
4. Jones DB, Soper NJ. Complications of laparoscopic cholecystectomy. Annu Rev Med. 1996;47:31–44.
5. Orlando 3rd R, Russell JC, Lynch J, Mattie A. Laparoscopic cholecystectomy. A statewide experience. The Connecticut Laparoscopic Cholecystectomy Registry. Arch Surg. 1993;128(5):494–8.
6. Smith EB. Complications of laparoscopic cholecystectomy. J Natl Med Assoc. 1992;84(10):880–2.
7. Buddingh KT, Nieuwenhuijs VB, van Buuren L, Hulscher JB, de Jong JS, van Dam GM. Intraoperative assessment of biliary anatomy for prevention of bile duct injury: a review of current and future patient safety interventions. Surg Endosc. 2011;25(8):2449–61.
8. Buddingh KT, Weersma RK, Savenije RA, van Dam GM, Nieuwenhuijs VB. Lower rate of major bile duct injury and increased intraoperative management of common bile duct stones after implementation of routine intraoperative cholangiography. J Am Coll Surg. 2011;213(2):267–74.
9. Flum DR, Dellinger EP, Cheadle A, Chan L, Koepsell T. Intraoperative cholangiography and risk of common bile duct injury during cholecystectomy. JAMA. 2003;289(13):1639–44.
10. Soper NJ, Flye MW, Brunt LM, Stockmann PT, Sicard GA, Picus D, et al. Diagnosis and management of biliary complications of laparoscopic cholecystectomy. Am J Surg. 1993;165(6):663–9.
11. Way LW, Stewart L, Gantert W, Liu K, Lee CM, Whang K, et al. Causes and prevention of laparoscopic bile duct injuries: analysis of 252 cases from a human factors and cognitive psychology perspective. Ann Surg. 2003;237(4):460–9.
12. Stiles BM, Adusumilli PS, Bhargava A, Fong Y. Fluorescent cholangiography in a mouse model: an innovative method for improved laparoscopic identification of the biliary anatomy. Surg Endosc. 2006;20(8):1291–5.
13. Marano A, Priora F, Lenti LM, Ravazzoni F, Quarati R, Spinoglio G. Application of fluorescence in robotic general surgery: review of the literature and state of the art. World J Surg. 2013;37(12):2800–11.
14. Buchs NC, Hagen ME, Pugin F, Volonte F, Bucher P, Schiffer E, et al. Intra-operative fluorescent cholangiography using indocyanine green during robotic single site cholecystectomy. Int J Med Robot. 2012;8(4):436–40.
15. Ishizawa T, Bandai Y, Kokudo N. Fluorescent cholangiography using indocyanine green for laparoscopic cholecystectomy: an initial experience. Arch Surg. 2009;144(4):381–2.
16. Spinoglio G, Priora F, Bianchi PP, Lucido FS, Licciardello A, Maglione V, et al. Real-time near-infrared (NIR) fluorescent cholangiography in

single-site robotic cholecystectomy (SSRC): a single-institutional prospective study. Surg Endosc. 2013;27(6):2156–62.
17. Ishizawa T, Bandai Y, Ijichi M, Kaneko J, Hasegawa K, Kokudo N. Fluorescent cholangiography illuminating the biliary tree during laparoscopic cholecystectomy. Br J Surg. 2010;97(9):1369–77.
18. Ishizawa T, Kaneko J, Inoue Y, Takemura N, Seyama Y, Aoki T, et al. Application of fluorescent cholangiography to single-incision laparoscopic cholecystectomy. Surg Endosc. 2011;25(8):2631–6.
19. Ishizawa T, Tamura S, Masuda K, Aoki T, Hasegawa K, Imamura H, et al. Intraoperative fluorescent cholangiography using indocyanine green: a biliary road map for safe surgery. J Am Coll Surg. 2009; 208(1):e1–4.
20. Bucher P, Pugin F, Morel P. Single-port access prosthetic repair for primary and incisional ventral hernia: toward less parietal trauma. Surg Endosc. 2011; 25(6):1921–5.
21. Bucher P, Pugin F, Morel P. Single-port access laparoscopic radical left colectomy in humans. Dis Colon Rectum. 2009;52(10):1797–801.
22. Bucher P, Pugin F, Buchs N, Ostermann S, Charara F, Morel P. Single port access laparoscopic cholecystectomy (with video). World J Surg. 2009;33(5):1015–9.
23. Bucher P, Pugin F, Buchs NC, Ostermann S, Morel P. Randomized clinical trial of laparoendoscopic single-site versus conventional laparoscopic cholecystectomy. Br J Surg. 2011;98(12):1695–702.
24. Madureira FA, Manso JE, Madureira Fo D, Iglesias AC. Randomized clinical study for assessment of incision characteristics and pain associated with LESS versus laparoscopic cholecystectomy. Surg Endosc. 2013;27(3):1009–15.
25. Zhong X, Rui YY, Zhou ZG. Laparoendoscopic single-site versus traditional laparoscopic surgery in patients with cholecystectomy: a meta-analysis. J Laparoendosc Adv Surg Tech A. 2012;22(5):449–55.
26. Yeo D, Mackay S, Martin D. Single-incision laparoscopic cholecystectomy with routine intraoperative cholangiography and common bile duct exploration via the umbilical port. Surg Endosc. 2012;26(4): 1122–7.
27. Buddingh KT, Hofker HS, ten Cate Hoedemaker HO, van Dam GM, Ploeg RJ, Nieuwenhuijs VB. Safety measures during cholecystectomy: results of a nationwide survey. World J Surg. 2011;35(6):1235–41.
28. Buddingh KT, Nieuwenhuijs VB. The critical view of safety and routine intraoperative cholangiography complement each other as safety measures during cholecystectomy. J Gastrointest Surg. 2011;15(6): 1069–70.
29. Rawlings A, Hodgett SE, Matthews BD, Strasberg SM, Quasebarth M, Brunt LM. Single-incision laparoscopic cholecystectomy: initial experience with critical view of safety dissection and routine intraoperative cholangiography. J Am Coll Surg. 2010; 211(1):1–7.
30. Ayloo SM, Buchs NC, Addeo P, Bianco FM, Giulianotti PC. Traditional versus single-site placement of adjustable gastric banding: a comparative study and cost analysis. Obes Surg. 2011;21(7): 815–9.
31. Morel P, Buchs NC, Iranmanesh P, Pugin F, Buehler L, Azagury DE, et al. Robotic single-site cholecystectomy. J Hepatobiliary Pancreat Sci. 2014;21:18.
32. Morel P, Hagen ME, Bucher P, Buchs NC, Pugin F. Robotic single-port cholecystectomy using a new platform: initial clinical experience. J Gastrointest Surg. 2011;15(12):2182–6.
33. Vidovszky TJ, Carr AD, Farinholt GN, Ho HS, Smith WH, Ali MR. Single-site robotic cholecystectomy in a broadly inclusive patient population: a prospective study. Ann Surg. 2014;260:134.
34. Faybik P, Hetz H. Plasma disappearance rate of indocyanine green in liver dysfunction. Transplant Proc. 2006;38(3):801–2.
35. Sherwinter DA. Identification of anomolous biliary anatomy using near-infrared cholangiography. J Gastrointest Surg. 2012;16(9):1814–5.
36. Calatayud D, Milone L, Elli EF, Giulianotti PC. ICG-fluorescence identification of a small aberrant biliary canaliculus during robotic cholecystectomy. Liver Int. 2012;32(4):602.
37. Buchs NC, Pugin F, Azagury DE, Jung M, Volonte F, Hagen ME, et al. Real-time near-infrared fluorescent cholangiography could shorten operative time during robotic single-site cholecystectomy. Surg Endosc. 2013;27(10):3897–901.
38. Schaafsma BE, Mieog JS, Hutteman M, van der Vorst JR, Kuppen PJ, Lowik CW, et al. The clinical use of indocyanine green as a near-infrared fluorescent contrast agent for image-guided oncologic surgery. J Surg Oncol. 2011;104(3):323–32.
39. Speich R, Saesseli B, Hoffmann U, Neftel KA, Reichen J. Anaphylactoid reactions after indocyanine-green administration. Ann Intern Med. 1988;109(4): 345–6.
40. Volonte F, Buchs NC, Pugin F, Spaltenstein J, Schiltz B, Jung M, et al. Augmented reality to the rescue of the minimally invasive surgeon. The usefulness of the interposition of stereoscopic images in the Da Vinci robotic console. Int J Med Robot. 2013;9(3):e34–8.
41. Volonte F, Pugin F, Buchs NC, Spaltenstein J, Hagen M, Ratib O, et al. Console-integrated stereoscopic OsiriX 3D volume-rendered images for da Vinci colorectal robotic surgery. Surg Innov. 2013;20(2):158–63.
42. Buchs NC, Volonte F, Pugin F, Toso C, Fusaglia M, Gavaghan K, et al. Augmented environments for the targeting of hepatic lesions during image-guided robotic liver surgery. J Surg Res. 2013;184(2):825–31.
43. Volonte F, Buchs NC, Pugin F, Spaltenstein J, Jung M, Ratib O, et al. Stereoscopic augmented reality for da Vincii robotic biliary surgery. Int J Surg Case Rep. 2013;4(4):365–7.

Fluorescent-Guided Liver Surgery: Paul Brousse Experiences and Perspective

11

Mohamed Bekheit and Eric Vibert

Introduction

Dr Irwin Fox reported the clinical use of indocyanine green (ICG) for the first time in 1960 at Mayo Clinic [1]. Its high safety profile and sensitivity makes it appealing for many clinical applications. Since it is exclusively removed from the body via the biliary system, the study of liver function using the indocyanine green became widely adopted [2]. The indocyanine green 15 min retention test is commonly used for preoperative assessment of liver reserve [3]. Furthermore, its plasma clearance rate was a significant predictor of liver regeneration after portal vein embolization (PVE) [4].

Recently, with the development of easily installed near-infrared camera systems, the live intraoperative evaluation of liver fluorescence became available. In general, the evaluation of tissues uptake and excretion of ICG requires prior to surgery administration. The time window between injection and evaluation ranges from 24 to 72 h. On the other hand, evaluation of vascular structures and perfusion requires intraoperative administration of ICG. In most instances the required preoperative dose is between 0.25 and 0.5 mg/kg. Intraoperatively, we inject 1 cm of a diluted ICG solution once or twice for a total dose of 0.25 mg.

In Paul Brousse, we first used the infrared camera systems PDE developed by Hamamatsu Photonics (Hamamatsu, Japan) in 28 patients operated for colorectal liver metastases or hepatocellular carcinoma by laparotomy. All patients received intravenously ICG at a dose of 0.5 mg/kg between 12 and 24 h before the surgical procedure. Four (14.3 %) patients presented new lesions that were neither detected by preoperative imaging nor by intraoperative ultrasound. Three (75 %) of those four patients had lesions confirmed to be CRLM. This sensitivity was supposed to improve the surgical radicality in 11 % of our series [5].

Because of these encouraging results of PDE and our leading French experience in this field, since our group performs around 230 liver resections and 130 liver transplantations by year, we tested a new device (Fluobeam®) developed by a French company (Fluoptic, Grenoble) (Fig. 11.1).

M. Bekheit, M.B.Ch.B., M.Sc., M.R.C.S.
Hepatobiliary and Liver Transplant, Paul Brousse Hospital, 12 Avenue Paul Vaillant Couturier, Villejuif 94800, France
e-mail: dr_mohamedbekheit@hotmail.com

E. Vibert, M.D., Ph.D. (✉)
Hepatobiliary and Liver Transplant, Paul Brousse Hospital, 12 Avenue Paul Vaillant Couturier, Villejuif 94800, France

Centre Hépato-Biliaire, Hôpital Paul Brousse, 14 Avenue Paul Vaillant Couturier, 94800 Villejuif, France
e-mail: eric.vibert.pbr@gmail.com

F.D. Dip et al. (eds.), *Fluorescence Imaging for Surgeons: Concepts and Applications*, DOI 10.1007/978-3-319-15678-1_11,

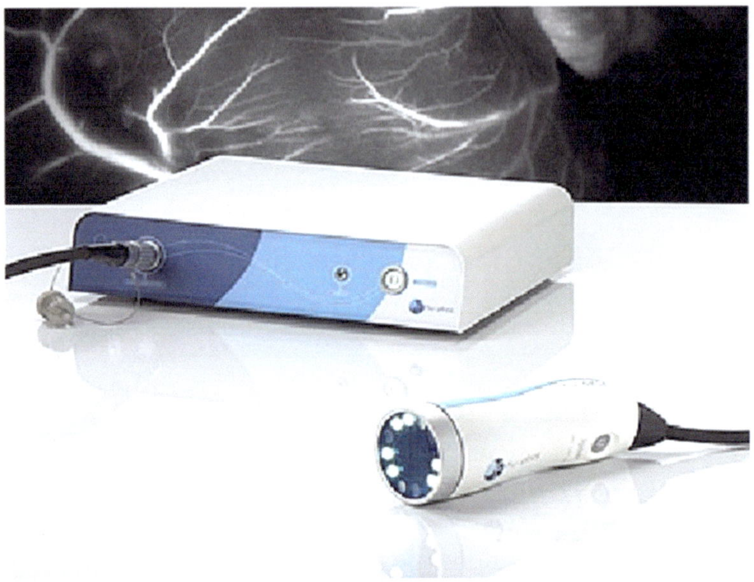

Fig. 11.1 Fluobeam® medical device; central processing unit and the camera head

Fluobeam® Medical Device

The Fluobeam® medical device allows visualizing, on a computer screen, the flow, the distribution and/or the accumulation of near-infrared fluorescent agents such as ICG, injected to patient before and/or during surgery for various indications such as:

Intraoperative visualization and detection of primary liver tumors and/or hepatic metastases located less than 0.5 cm from liver surface.

Intraoperative better visualization of tumor margin on the raw liver surface during parenchyma transaction.

Intraoperative identification of bile ducts.

Intraoperative assessment of liver vascularization.

Fluobeam® is composed of (Fig. 11.1):

1. *Optical head*, which is composed of the camera and the optical elements such as fibers, filters, lens, and diffusers. The optical head is also the place where light emission occurs: laser and LEDs (Light Emitting Diodes).
2. *Control box* including all the elements to control, monitor, and supply power to the optical head.
3. *Cable* including electrical wires and optical fibers. Cable is mechanically linked to the optical head, and can be plugged/unplugged to/from the electrical case through a dedicated connector.
4. *Power supply* and RJ 45 cable.
5. *Imaging software (Fluosoft™) (Fig. 11.2)*, which allows real-time visualization of the fluorescent images acquired by the optical head. It also contains the control commands of the Fluobeam® and allows the recording of images, sequence of images and videos.

Fluobeam® is a class IIa CE marked medical device. It allows the visualization of fluorescent moiety with a maximum emission between 760 and 850 nm for an excitation wavelength of 750 nm. Fluobeam® includes a class 1 laser source which is safe under all conditions of normal use. The laser light is conveyed from the control box to the optical head via an optical fiber. Two laser beams are emitted from the optical head shedding light on an elliptic area of 15 cm × 10 cm at a distance of 20 cm from the optical head.

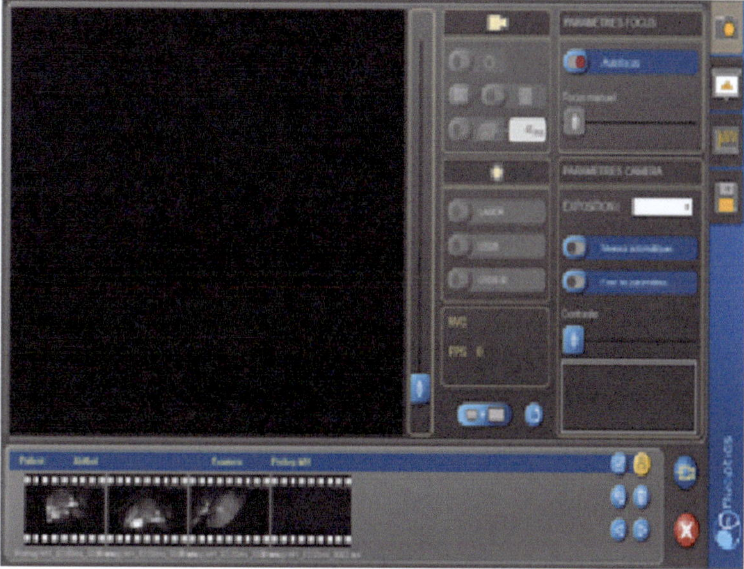

Fig. 11.2 Fluosoft™ screen shot demonstrating the various functional commands

Fluobeam® includes an illumination ring based on white and near-infrared LEDs in order to provide a natural white illumination during the surgery. A zoom of 10× can be directly driven by the user via two push-buttons located on the front side of the optical head, which allow changing of the field of view from a maximum of 20 cm × 15 cm to a minimum of 2 cm × 1.5 cm. For distance ranging from 20 to 50 cm between the optical head and the examined area, the camera autofocus ensures a precise focusing and leads to sharp image. Depending on the intensity of the fluorescence signal, the exposure time of the camera can be adjusted from 1 ms to 1 s which allows the acquisition of videos with a frequency ranging from 25 images per second to 1 image per second.

One of the advantages of Fluobeam® is the software accompanying the devices that allow to optimize the exposure according to the intensity of fluorescence. These settings were automatic or manual and allowed to improve largely the quality of images that could be recorded on the software as images and movies. A subsequent analysis of the intensity of fluorescence in specific area of the image according to time could be made on the same software.

As for the other devices, the main drawback of Fluobeam® device was the relatively large size of the camera head that renders the exposure of the upper part of the liver surface difficult.

Our Experience with Hepatic Malignant Tumors

The small number of patients that we had analyzed to date with Fluobeam® is not sufficient to conclude concerning the accuracy of this device to detect underestimated superficial lesion. We presented in this chapter some images that we had obtained with Fluobeam® in some specific situation to demonstrate the quality of images obtainable by this device.

Currently, we aim at increasing the specificity of this technique especially in patients with abnormal liver function and/or cirrhosis where we observed more lesions were falsely interpreted as potentially malignant compared to patients without cirrhosis (Fig. 11.3).

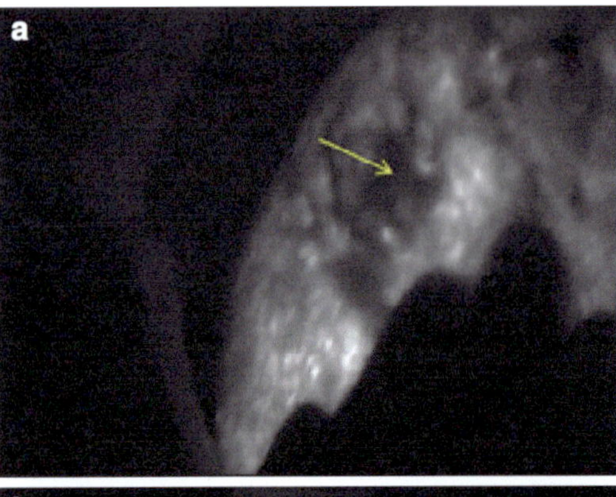

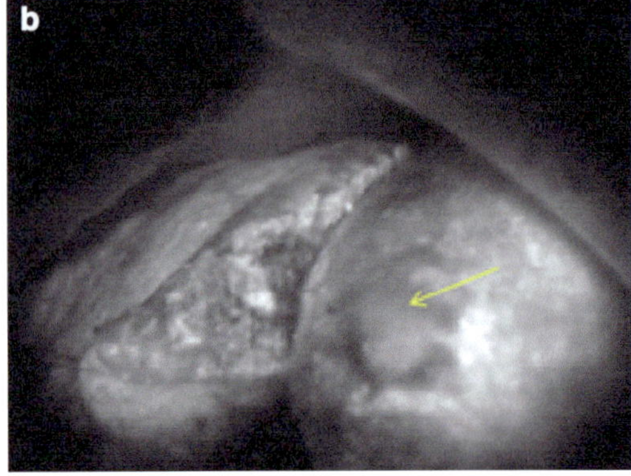

Fig. 11.3 (**a**) Undifferentiated HCC (*arrow*) on a cirrhotic liver background. (**b**) Moderately differentiated HCC (*arrow*) over cirrhotic liver

Hepatocellular Cancers

Fluorescence intensity correlated well with the differentiation of the HCC when analyzed by fluorescence microscopy [6]. In live fluorescence-guided surgery, we were able to notice the difference in the HCC differentiation based on the brightness (Fig. 11.3). Although this was done in retrospect, after the histopathologic examination of the specimens, it demonstrates that real-time fluorescence could be useful in tumor grading. We are currently working on stratification and quantification of this relation.

Due to its relatively uncommon incidence, there is no specific report on the use of indocyanine green during surgery for fibrolamellar carcinoma. In our experience, fibrolamellar carcinoma captures well the indocyanine green and subsequently emits strong fluorescent signal (Fig. 11.4).

Cholangiocarcinoma

The same concept is applicable for cholangiocarcinoma. The tumor retains the indocyanine green

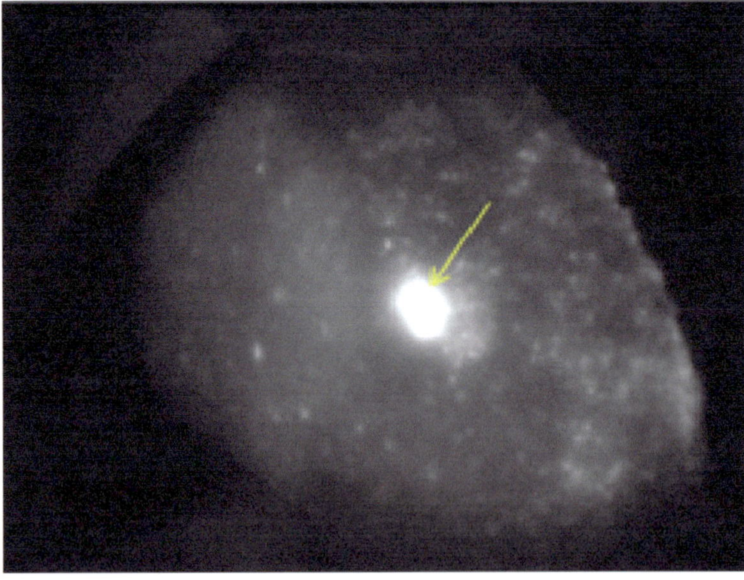

Fig. 11.4 Fibrolamellar HCC (*arrow*) on a non-cirrhotic background

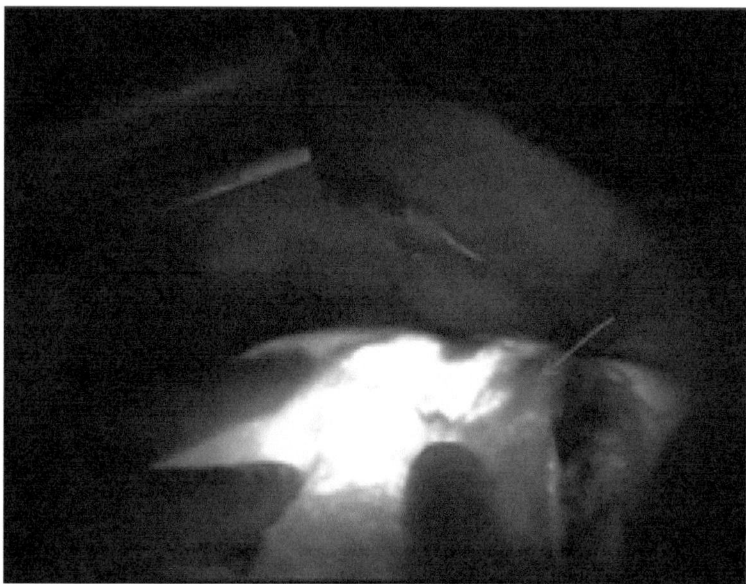

Fig. 11.5 Bright cholangiocarcinoma with surrounding safety margin (*arrow*) during anatomical resection

and emits strong bright fluorescence on exposure to the near infrared light. Real-time determination of the free resection margin is of paramount importance to minimize the recurrence. As shown in Fig. 11.5, fluorescence imaging was a useful tool in ascertaining the tumor safety margin.

Colorectal Liver Metastasis

Metastases to the liver from the colorectal cancers did not capture indocyanine green therefore they appear as dark masses or spots; contrary to the liver tissue that captures and emits fluorescence; giving them a bright image. Since metastasis are surrounded by a rim of compressed hepatic cells, these hepatic cells are responsible for the hyperfluorescent rim surrounding the dark tumor mass (Fig. 11.6) [7].

Achieving an R0 resection is an oncologic target during surgery for liver tumors. Theoretically, the smaller the lesion the less compression it exerts on the surrounding tissue and the lower difference between tumor and non-tumor brightness. Nonetheless, the use of fluorescence has been reported to improve the detection rate of tumors smaller than 3 mm compared to the combined use of CT and IOUS [8]. This would lead to improving the radicality of surgery. In Fig. 11.7, small contralateral tumor was detected during right hepatectomy for colorectal liver metastasis following PVE.

Visualization of Bile Duct

Real-time fluorescence could be useful in delineation of the bile duct anatomy. The direct visualization might allow to safely ligate and cut sacrifiable ducts (Fig. 11.8). Although we did not have the opportunity to test it, bile leak could be also detected as demonstrated in other studies.

Potential limitations of the current real-time fluorescence in surgery for hepatic tumors:

Beside the limitation in detecting tumors deeper than 0.5 cm, there is, nonetheless, some inaccuracies related to the real-time fluorescence assessment using the endocyanine green techniques. Not every illuminating lesion is cancer. For instance, biliary proliferative lesions and benign regenerative nodules could appear, as well, bright (Fig. 11.9) [7].

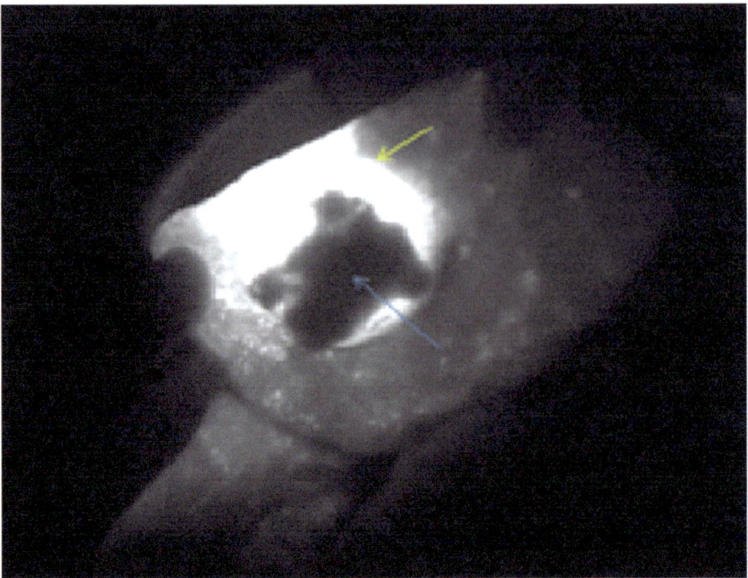

Fig. 11.6 Dim colorectal liver metastasis (*blue long arrow*) with surrounding bright rim of compressed liver tissue (*yellow shorter arrow*)

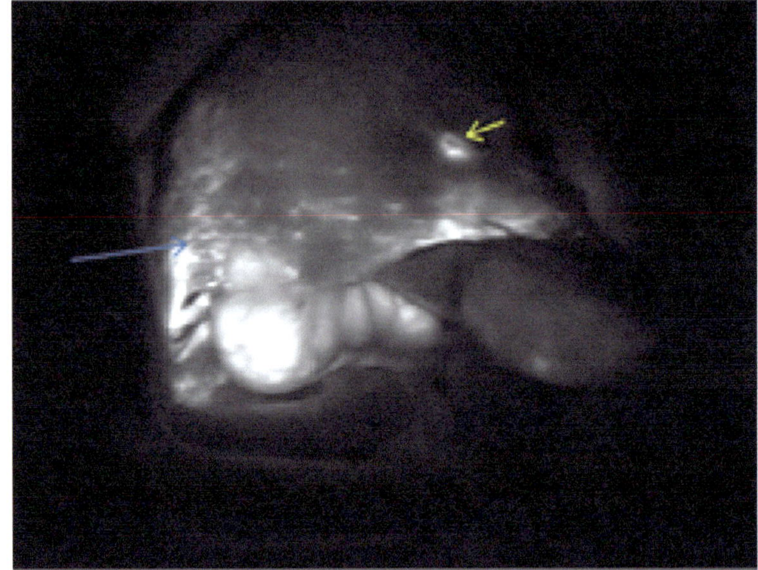

Fig. 11.7 Parenchymal liver appearance after portal vein embolization (*long blue arrow*) with ICG injection 24 h prior to surgery for right hepatectomy for colorectal liver metastasis and the appearance of small lesion in the left lobe (*short yellow arrow*)

Our Experience with Liver Transplantation

Anticipation of early graft function could facilitate preventive measures against graft loss. The integrity of hepatic microcirculation has been directly linked to the early post-implant graft function [9]. Reduction of sinusoidal blood flow, due to reduced portal venous flow, along with cellular death are the end results of the different implicated pathophysiologic pathways [10]. Disorders in sinusoidal perfusion could be reflected through fluorescence uptake and emission [11].

Our group previously reported the value of the ICG elimination rate as a predictor for post-liver transplant graft function and in the evaluation of

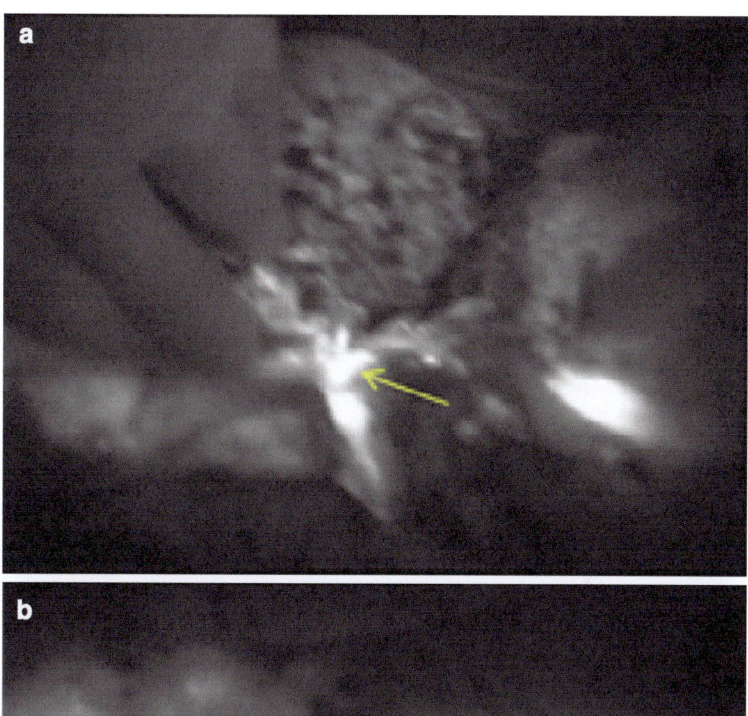

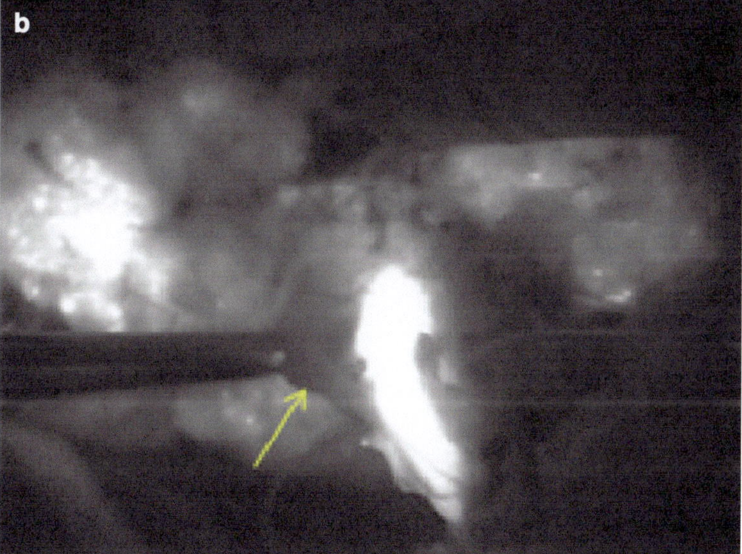

Fig. 11.8 (**a**) Visualization of the bile ducts during liver transection demonstrating preserved integrity of the main ducts. The *arrow* marks the biliary confluence. (**b**) forceps grasping the cystic duct (*arrow*) which does not show fluorescence emission. Contrary to the bright main biliary system

hepatic artery thrombosis [12, 13]. It was possible to quantify the ICG uptake and excretion and correlate these rates with the degree of ischemia–reperfusion injury. Recently, we started to use the Fluobeam technology for intraoperative evaluation of the liver perfusion at the microcirculation level.

In theory, the real-time assessment of the hepatic microcirculation should be feasible through analysis of the homogeneity of fluorescence emission. Homogeneous fluorescence could reflect the homogeneity in microcircular perfusion and cellular capturing of fluorescent material and vice versa. We currently aim at testing this hypothesis on micro- and macroscale levels in a pig model.

The initial crude data highlight the presence of two main patterns of perfusion disorders that could be described as systematic and unsystematic. The pattern is unsystematic when there is patchy or mottled surface fluorescence, and this pattern is highly probable in conditions related to disturbances in the hepatic microcirculation

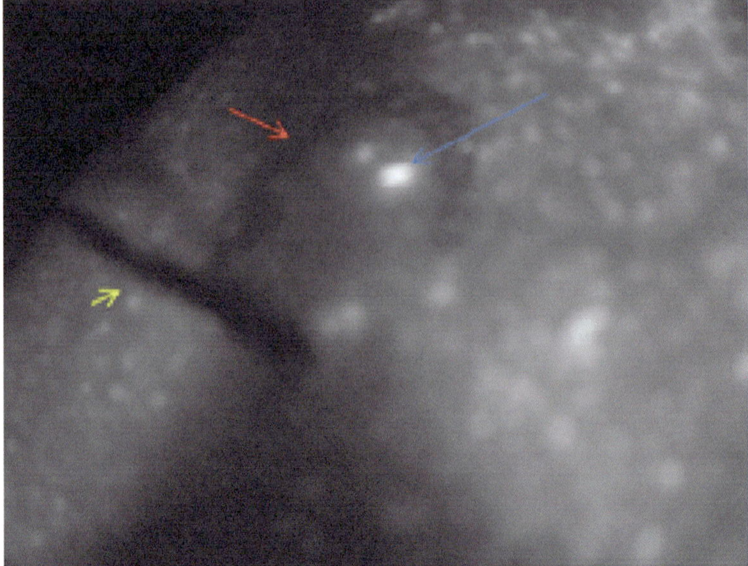

Fig. 11.9 Small false-positive lesions (*blue arrow*), surrounded by burned ring (*red arrow*) during electrocautery dissection (*yellow arrow*)

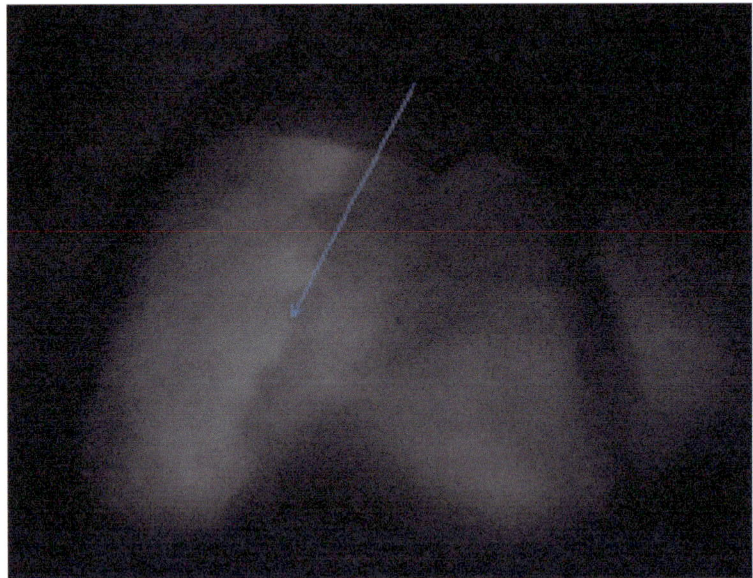

Fig. 11.10 Perfusion difference between the right and left hepatic hemilivers suggesting left inflow problem. The *arrow* marks the line of demarcation between the perfused right and less perfused left hemilivers

(i.e. ischemia/reperfusion injury). Unlike the patchy pattern, the systematic pattern is described when the fluorescence difference is related to a major anatomical region. In the last case the perfusion disturbance is related to a major vascular structure (i.e. arterial or portal branch(s)). Figure 11.10 demonstrates a case with left hepatic hypofluorescence that was consistent with the Doppler finding of left arterial spasm, which recovered 10 min later.

The presence of a systematic perfusion disorder warrants further examination of the extrahepatic inflow and outflow systems in real time with an extra-bolus of ICG. Vascular patency could be confirmed by the presence of indocyanine flow particularly in small vessels for which the sensation of a pulse is sometime difficult (Fig. 11.11).

Bile excretion is a good indicator of regain of function after liver transplantation [14]. The immediate secretion of bile after hepatic reperfusion

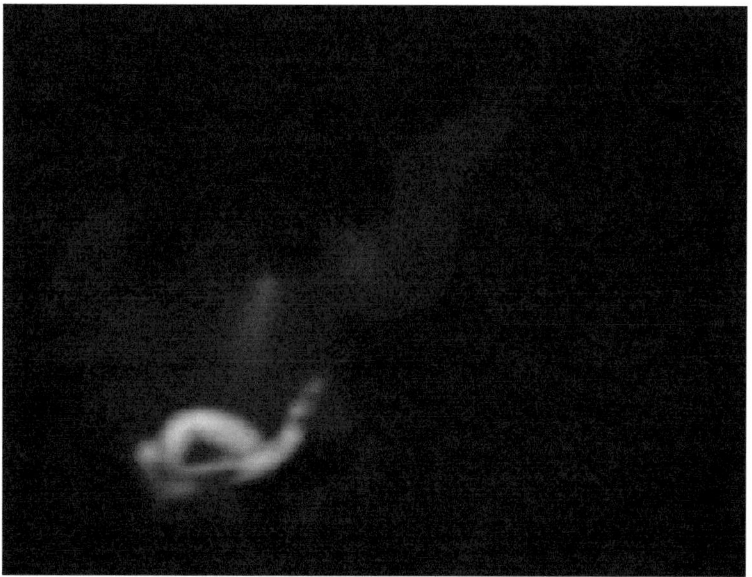

Fig. 11.11 Arterial illumination confirming the patency through the anastomosis

is a predictor of the early good graft function [15]. Besides, real time evaluation of the integrity of the biliary anastomosis could be possible.

Summary

Real-time fluorescence evaluation using Fluobeam® is an interesting tool of evaluation of the liver condition at reperfusion during transplantation and of the oncologic radicality during hepatic surgery for cancer that has many perspectives.

The authors declare no conflict of interest

References

1. Fox IJ, Wood EH. Indocyanine green: physical and physiologic properties. Proc Staff Meet Mayo Clin. 1960;35:732–44.
2. Landsman ML, Kwant G, Mook GA, Zijlstra WG. Light-absorbing properties, stability, and spectral stabilization of indocyanine green. J Appl Physiol. 1976;40(4):575–83.
3. Kubota K, Makuuchi M, Kusaka K, Kobayashi T, Miki K, Hasegawa K, Harihara Y, Takayama T. Measurement of liver volume and hepatic functional reserve as a guide to decision-making in resectional surgery for hepatic tumors. Hepatology. 1997;26(5):1176–81.
4. Mihara K, Sugiura T, Okamura Y, Kanemoto H, Mizuno T, Moriguchi M, Aramaki T, Uesaka K. A predictive factor of insufficient liver regeneration after preoperative portal vein embolization. Eur Surg Res. 2013;51(3–4):118–28.
5. Lim C, Vibert E, Azoulay D, Salloum C, Ishizawa T, Yoshioka R, Mise Y, Sakamoto Y, Aoki T, Sugawara Y, Hasegawa K, Kokudo N. Indocyanine green fluorescence imaging in the surgical management of liver cancers: current facts and future implications. J Visc Surg. 2014;151:117. http://dx.doi.org/10.1016/j.jviscsurg.2013.11.003.
6. Ishizawa T, Masuda K, Urano Y, Kawaguchi Y, Satou S, Kaneko J, Hasegawa K, Shibahara J, Fukayama M, Tsuji S, Midorikawa Y, Aburatani H, Kokudo N. Mechanistic background and clinical applications of indocyanine green fluorescence imaging of hepatocellular carcinoma. Ann Surg Oncol. 2014;21(2):440–8.
7. Ishizawa T, Fukushima N, Shibahara J, Masuda K, Tamura S, Aoki T, Hasegawa K, Beck Y, Fukayama M, Kokudo N. Real-time identification of liver cancers by using indocyanine green fluorescent imaging. Cancer. 2009;115(11):2491–504.
8. Peloso A, Franchi E, Canepa MC, Barbieri L, Briani L, Ferrario J, Bianco C, Quaretti P, Brugnatelli S, Dionigi P, Maestri M. Combined use of intraoperative ultrasound and indocyanine green fluorescence imaging to detect liver metastases from colorectal cancer. HPB (Oxford). 2013;15(12):928–34.
9. Puhl G, Schaser K-D, Pust D, Köhler K, Vollmar B, Menger MD, Neuhaus P, Settmacher U. Initial hepatic microcirculation correlates with early graft function

in human orthotopic liver transplantation. Liver Transpl. 2005;11(5):555–63.
10. Shimizu H, Miyazaki M, Ito H, Nakagawa K, Ambiru S, Kato A, Nukui Y, Nozawa S, Nakajima N. Mechanism of cold ischemia-reperfusion-induced graft injury after orthotopic liver transplantation in rats. Hepatogastroenterology. 2000;48(37):216–9.
11. Vollmar B, Glasz J, Leiderer R, Post S, Menger MD. Hepatic microcirculatory perfusion failure is a determinant of liver dysfunction in warm ischemia-reperfusion. Am J Pathol. 1994;145(6):1421.
12. Levesque E, Saliba F, Benhamida S, Ichai P, Azoulay D, Adam R, Castaing D, Samuel D. Plasma disappearance rate of indocyanine green: a tool to evaluate early graft outcome after liver transplantation. Liver Transpl. 2009;15(10):1358–64.
13. Levesque E, Hoti E, Azoulay D, Adam R, Samuel D, Castaing D, Saliba F. Non-invasive ICG-clearance: a useful tool for the management of hepatic artery thrombosis following liver transplantation. Clin Transplant. 2011;25(2):297–301.
14. Fan YD, Praet M, Vanzieleghem B, Vanwynsberghe D, Stoop D, Leroux-Roels G, Delanghe J, de Hemptinne B. Effects of re-arterialization on early graft function and regeneration in the rat model of heterotopic auxiliary liver transplantation. Eur Surg Res. 2000;32(1):11–7.
15. Gontarczyk GW, Lagiewska B, Pacholczyk M, Trzebicki J, Jureczko L, Kolacz M, Kosieradzki M, Adadynski L, Wasiak D, Rowinski W. Intraoperative blood flow measurements and liver allograft function: preliminary results. Transplant Proc. 2006;38(1):234–6.

Fluorescence Imaging in Laparoscopic Hepatectomy

12

Yoshikuni Kawaguchi, Takeo Nomi, David Fuks, Norihiro Kokudo, and Brice Gayet

Abbreviations

ICG Indocyanine green
IOUS Intraoperative ultrasound

Introduction

In the last two decades, laparoscopic hepatectomy has been increasingly performed throughout the world [1–6]. A recent review of the world literature including 2,804 cases indicated that laparoscopic hepatectomy is associated with decreased blood loss, reduced postoperative morbidity, and shorter length of stay, with no significant difference in oncological outcomes as compared to open hepatectomy [7]. However, disadvantages of laparoscopic hepatectomy lacking in complete hepatic view and tactile feedback may cause anatomical segments to be resected unidentifiable during surgery, resulting in specific postoperative complications.

To compensate these disadvantages, we have applied fluorescence imaging technique utilizing indocyanine green (ICG) for identifying the biliary ducts or segments [8] during laparoscopic hepatectomy. In principle, ICG-fluorescence imaging technique is based on the fact that protein-bound ICG emits light with a peak wavelength of around 830 nm when illuminated with near-infrared light [9]. This technique has been clinically applied to hepatobiliary surgery [10], especially for intraoperative visualization of bile [11–14], liver cancers [15–17], and hepatic segments to be resected [8, 18, 19] as well as for evaluation of regional portal uptake in the liver [20, 21].

We herein describe the performance of laparoscopic hepatectomy under the navigation of ICG-fluorescence imaging for more accurate intraoperative detection of the biliary ducts and anatomical segments.

Electronic supplementary material: The online version of this chapter (doi:10.1007/978-3-319-15678-1_12) contains supplementary material, which is available to authorized users. Videos can also be accessed at http://link.springer.com/chapter/10.1007/978-3-319-15678-1_12.

Y. Kawaguchi, M.D. • N. Kokudo, M.D., Ph.D.
Hepato-Biliary-Pancreatic Surgery Division, Department of Surgery, Graduate School of Medicine, University of Tokyo, 7-3-1 Hongo, Bunkyo-ku, Tokyo 113-8655, Japan
e-mail: yokawaguchi-tky@umin.ac.jp; KOKUDO-2SU@h.u-tokyo.ac.jp

T. Nomi, M.D., Ph.D.
Department of Surgery, Nara Medical University, 1-24-6, Ayameike-kita, Nara 639-0126, Japan
e-mail: t.nomi45@gmail.com

D. Fuks, M.D., Ph.D. • B. Gayet, M.D., Ph.D. (✉)
Department of Digestive Diseases, Institut Mutualiste Montsouris, Université Paris Descartes, 42 boulevard Jourdan, Paris 75014, France
e-mail: davidfuks80@gmail.com; brice.gayet@imm.fr

F.D. Dip et al. (eds.), *Fluorescence Imaging for Surgeons: Concepts and Applications*,
DOI 10.1007/978-3-319-15678-1_12, © Springer International Publishing Switzerland 2015

Fluorescence Imaging Systems

Two fluorescence imaging systems are currently available for laparoscopic surgery: the preclinical system comprising a CCD camera (410,000 pixels in total) and a xenon light source (Olympus medical systems, Tokyo, Japan) and standard- or high-definition D-Light P System (Karl Storz, Tuttlingen, Germany). Since 2011, we have utilized the former fluorescence imaging system for laparoscopic hepatectomy.

Identification of Biliary Duct

Administration of ICG

Fluorescence cholangiography is performed by intrabiliary injection of ICG (0.025 mg/mL of Diagnogreen; Daiichi Sankyo Co., Tokyo, Japan) or by preoperative intravenous injection of ICG (2.5 mg in 1 mL of normal saline) [12–14].

Clinical Application for Identification of the Confluence of the Left and the Right Hepatic Duct

Figure 12.1 shows a case of a patient who underwent laparoscopic right hepatectomy for multiple colorectal liver metastases. One mL of ICG (2.5 mg) was injected intravenously after intubation in the operating room to decrease the consequences of anaphylaxis, even if the risk of adverse event following administration of ICG is small [22]. In addition, this timing of administration was chosen because fluorescence of ICG in the bile was visible in 10–15 min after intravenous injection and continued for more than 2 h [23]. After the fluorescence camera was placed above the hepatoduodenal ligament, fluorescent cholangiography was performed by changing the full-color images to fluorescent image using the switch in the equipment. The extrahepatic bile ducts in the hepatoduodenal ligament were visualized without dissection (Fig. 12.1a and Video 12.1). After ligation and division of the right hepatic artery and the right portal vein, fluorescence cholangiography was performed again. The confluence of the left hepatic duct and the right hepatic duct as well as the tiny bile duct from segment I were clearly delineated (Fig. 12.1b). Consequently, the right hepatic duct was ligated, referring to the fluorescence images of the bile ducts.

The major advantage of fluorescence cholangiography during laparoscopic hepatectomy is that it enables visualization of the biliary tract in relation to the surrounding structures from various angles in real-time. In addition, it can save time and avoid bile duct injury associated with insertion of a tube for the injection of contrast material when using radiographic cholangiography [24]. Fluorescence cholangiography is expected to complement conventional radiographic cholangiography, though it has limitations in tissue-penetration ability of ICG fluorescence up to 5–10 mm [15] and delineation of the intrahepatic bile ducts.

Clinical Application for Identification of the Left Hepatic Duct with Severe Adhesion

Figure 12.2 shows a case of a patient who underwent laparoscopic left hepatectomy for two colorectal liver metastases 3 years after open sigmoidectomy and lymphadenectomy of hepatoduodenal ligament. Due to the severe adhesions around hepatoduodenal ligament following lymphadenectomy, the anatomy of the hepatic vessels and the biliary ducts were not identified by gross appearance. To avoid their injury, we attempted to identify them using a fluorescence imaging technique. This showed clear fluorescence of the extrahepatic ducts in the hepatoduodenal ligament (Fig. 12.2a and Video 12.2). Based on the information, the left portal vein and the hilar plate containing the left hepatic duct were dissected and identified (Fig. 12.2b), after which the left portal vein was ligated (Video 12.2).

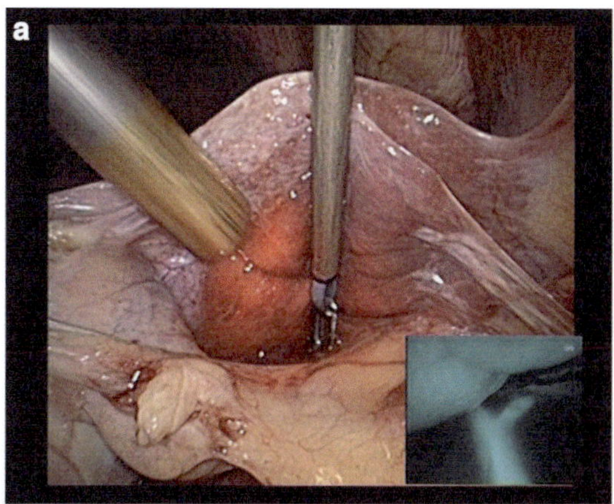

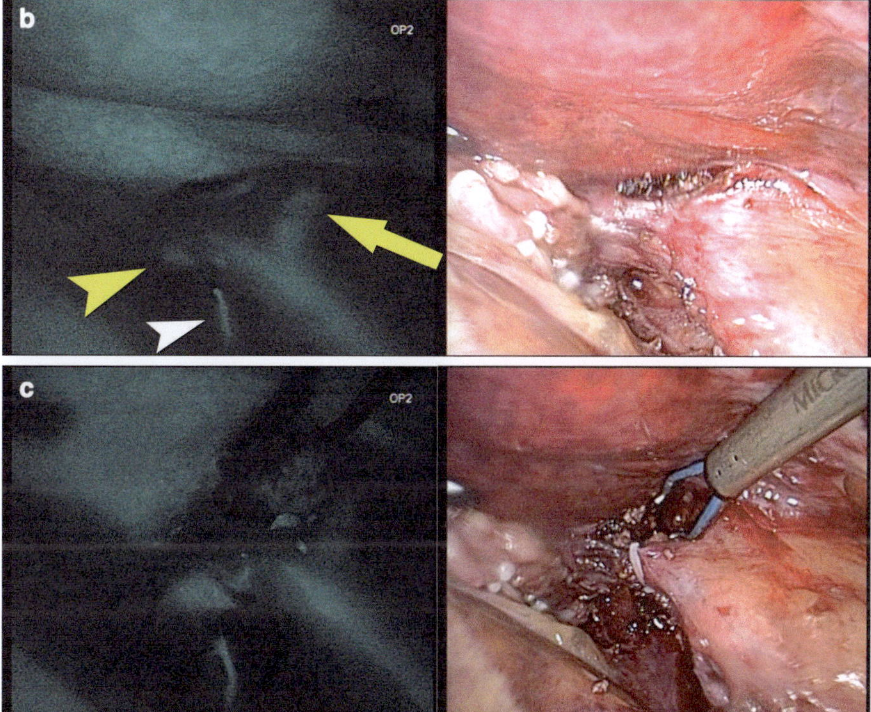

Fig. 12.1 (**a**) ICG-Fluorescence imaging visualized the extrahepatic bile ducts in the hepatoduodenal ligament. (**b**) After ligation and division of the right hepatic artery and the right portal vein, ICG-fluorescence cholangiography was performed again. The confluence of the left hepatic duct (*arrow*) and the right hepatic duct (*arrowhead*) as well as the tiny bile duct from segment I (*white arrowhead*) were clearly delineated. (**c**) The right hepatic duct was ligated, referring to the fluorescence images of the bile ducts

ICG-fluorescence cholangiography is useful for identifying the anatomy around the liver with severe adhesions following previous surgery. This technique may complement the theoretical disadvantage of laparoscopic hepatectomy lacking tactile feedback and serves as a guide for surgeons to identify the anatomy of the hepatic vessels during surgery.

Identification of Segments

Administration of ICG

For identification of segments as fluorescence, ICG (0.025 mg in 10 mL of normal saline) was administered into the portal branch through a needle

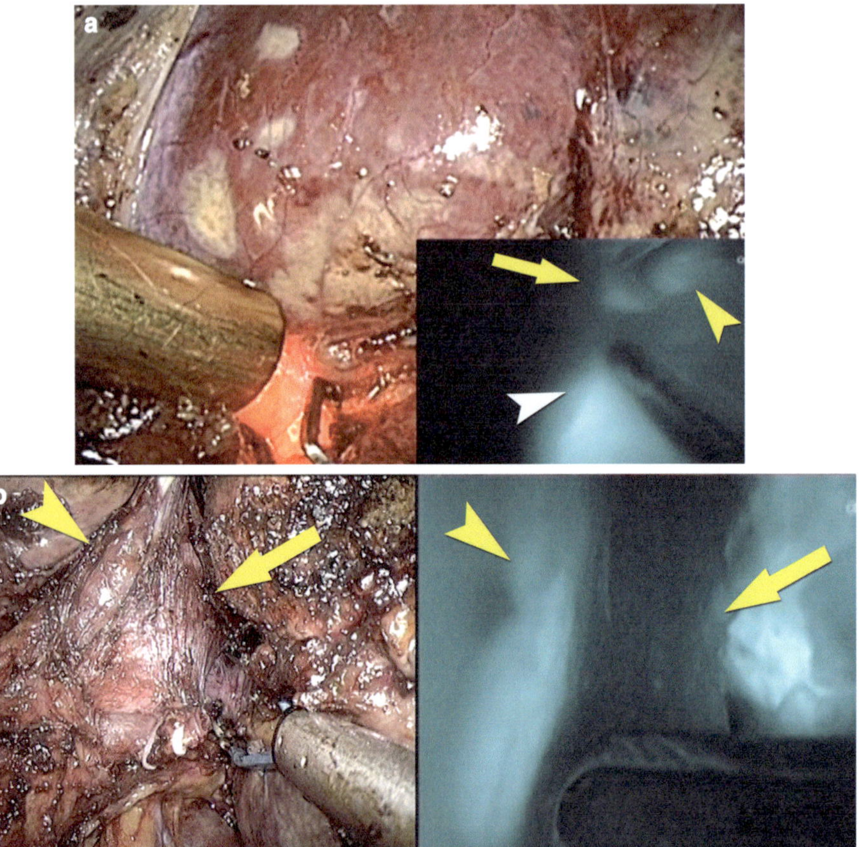

Fig. 12.2 (**a**) ICG-Fluorescence imaging visualized the extrahepatic bile ducts which were grossly unidentifiable because of the severe adhesions around the hepatoduodenal ligament. (**b**) After dissection of the left portal vein and the hilar plate, ICG-fluorescence cholangiography was performed again. The left hepatic duct in the hilar plate was visualized as fluorescence, while the left portal vein was identified as the defect of fluorescence

after its puncture. In contrast, for identification of segments as the defect of fluorescence, ICG (2.5 mg in 1 mL of normal saline) was injected intravenously after clamping of the glissonian pedicle with interest.

Clinical Application of Positive-Staining Technique

Figure 12.3 shows a case of a patient with a colorectal liver metastasis located in segment IV after right hepatectomy. Due to the severe adhesions around the liver and the regeneration of the remnant left liver after previous right hepatectomy, both vascularization in the liver and the tumor-bearing segment were unclear. In order to identify its segment, the suspected portal branch of the tumor-bearing segment was identified by IOUS and punctured with a 22-gauge spinal needle, which was introduced through the abdominal wall and a guide hole on the flexible laparoscopic ultrasonographic probe (BK medical, Herlev, Denmark). After ICG (0.025 mg in 10 mL of normal saline) was adminiştered into the portal branch through the needle, the liver surface corresponding to segment IVb started to fluoresce. The liver surface of the segment IVb provided clear fluorescence less than 5 min after the injection, after which the resection of the anatomical segment IVb was performed (Video 12.3).

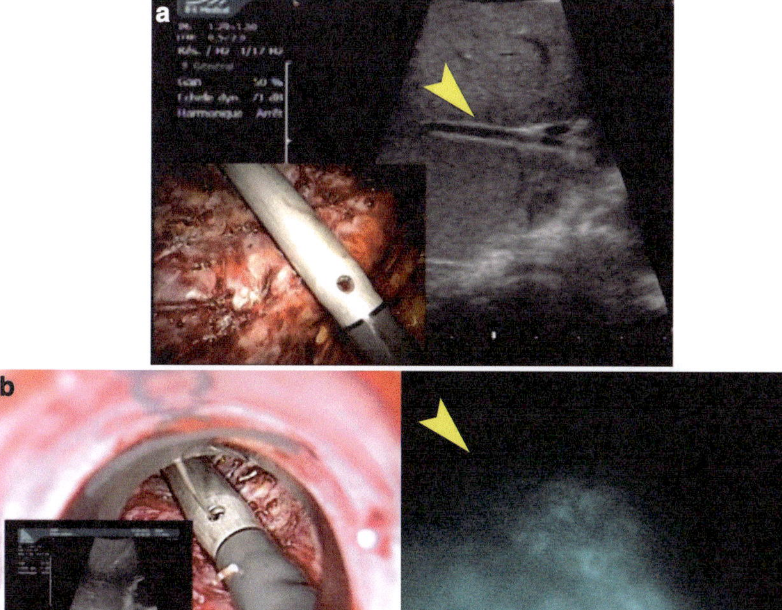

Fig. 12.3 (**a**) IOUS identified the portal vein flowing into the segment IVb (*arrowhead*) where the tumor was located. (**b**) After the puncture of the suspected portal vein of the tumor-bearing segment (*left*) and injection of ICG, the border between the segment IVb and the segment II/III was identified (*arrowheads, right*)

Clinical Application of Negative-Staining Technique

Figure 12.4 shows a case of a patient with a colorectal liver metastasis located in segment VI. Anatomical resection of segment VI was attempted to ensure the surgical margin, because intraoperative ultrasound (IOUS) revealed the tumor near from the root of the glissonian pedicle (Fig. 12.4a and Video 12.4). In order to identify the hepatic segment, the glissonian pedicle for segment VI was dissected while using IOUS. After temporarily clamping its root (Fig. 12.4b) and intravenous injection of ICG (2.5 mg), all hepatic segments except for segment VI provides clear fluorescence less than 5 min after injection (Fig. 12.4c). The resection line was marked with ultrasonic laparoscopic coagulation shears (Fig. 12.4d), after which anatomical resection of segment VI was performed. ICG-fluorescence imaging detected no bile leak on the raw surface of the liver after resection (Video 12.4).

Usefulness of Identification of Segments Using ICG-Fluorescence Imaging

These techniques which enable identification of hepatic segments as defect of fluorescence were reported as the positive- or negative-staining technique in authors' group [8]. The advantage of the negative-staining technique is not technically complicated as compared to the positive-staining technique [8] which requires to perform an image-guided puncture of a portal branch during surgery under laparoscopic control or through abdominal wall before surgery. Although laparoscopic anatomical liver resection is a technically demanding procedure [25], surgeons need to perform it to achieve equivalent oncological outcome for the treatment of hepatocellular carcinoma [26, 27] compared to open hepatectomy. These techniques guide surgeons to clearly identify the hepatic segments before transection of the liver parenchyma, ensuring accurate and efficacious laparoscopic anatomical

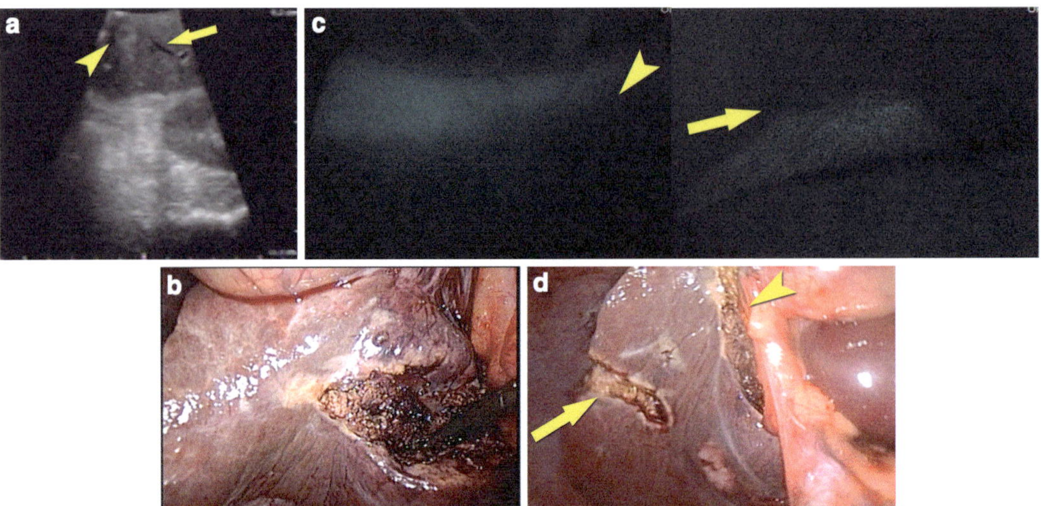

Fig. 12.4 (**a**) IOUS identified a tumor (*arrowhead*) and the glissonian pedicle flowing into the segment VI (*arrow*). (**b**) The gross appearance of the liver surface before injection of ICG with the glissonian pedicle clamped. (**c**) Fluorescence imaging on the liver surface of the segment V and VI (*left*) and that of the segment VI and VII (*right*). Segment VI was visualized as defect of fluorescence, and the borderline between segment V and VI (*arrowhead*) and that between segment VI and VII (*arrow*) were identified, respectively. (**d**) The gross appearance of a tumor and segment VI which was identified and marked with ultrasonic laparoscopic coagulation shears (*arrowhead* and *arrow*)

hepatectomy and probably less postoperative secondary bile leak or abscess due to a devascularized parenchyma in the remnant liver.

Conclusion

Fluorescence imaging techniques enable visualization of the biliary duct and the anatomical segments during laparoscopic hepatectomy in real-time. These techniques are expected to complement conventional intraoperative imaging technique, such as intraoperative ultrasonography and radiographic cholangiography, enhancing the safety and efficacy of laparoscopic hepatectomy.

Financial Disclosure: Dr Gayet is a consultant for Olympus. No writing assistance was utilized in the production of this manuscript.

Funding/Support: This work was supported by French State Funds managed by the ANR within the Investissements d'Avenir programme (Labex CAMI) under reference ANR-11-LABX-0004.

References

1. Vibert E, Perniceni T, Levard H, Denet C, Shahri NK, Gayet B. Laparoscopic liver resection. Br J Surg. 2006;93:67–72.
2. Otsuka Y, Tsuchiya M, Maeda T, Katagiri T, Isii J, Tamura A, et al. Laparoscopic hepatectomy for liver tumors: proposals for standardization. J Hepatobiliary Pancreat Surg. 2009;16:720–5.
3. Nitta H, Sasaki A, Fujita T, Itabashi H, Hoshikawa K, Takahara T, et al. Laparoscopy-assisted major liver resections employing a hanging technique: the original procedure. Ann Surg. 2010;251:450–3.
4. Nguyen KT, Marsh JW, Tsung A, Steel JJ, Gamblin TC, Geller DA. Comparative benefits of laparoscopic vs open hepatic resection: a critical appraisal. Arch Surg. 2011;146:348–56.
5. Cannon RM, Brock GN, Marvin MR, Buell JF. Laparoscopic liver resection: an examination of our first 300 patients. J Am Coll Surg. 2011;213:501–7.
6. Hwang DW, Han HS, Yoon YS, Cho JY, Kwon Y, Kim JH, et al. Laparoscopic major liver resection in Korea: a multicenter study. J Hepatobiliary Pancreat Sci. 2013;20:125–30.
7. Nguyen KT, Gamblin TC, Geller DA. World review of laparoscopic liver resection-2,804 patients. Ann Surg. 2009;250:831–41.
8. Ishizawa T, Zuker NB, Kokudo N, Gayet B. Positive and negative staining of hepatic segments by use of

fluorescent imaging techniques during laparoscopic hepatectomy. Arch Surg. 2012;147:393–4.
9. Landsman ML, Kwant G, Mook GA, Zijlstra WG. Light-absorbing properties, stability, and spectral stabilization of indocyanine green. J Appl Physiol. 1976;40:575–83.
10. Verbeek FP, van der Vorst JR, Schaafsma BE, Hutteman M, Bonsing BA, van Leeuwen FW, et al. Image-guided hepatopancreatobiliary surgery using near-infrared fluorescent light. J Hepatobiliary Pancreat Sci. 2012;19:626–37.
11. Mitsuhashi N, Kimura F, Shimizu H, Imamaki M, Yoshidome H, Ohtsuka M, et al. Usefulness of intraoperative fluorescence imaging to evaluate local anatomy in hepatobiliary surgery. J Hepatobiliary Pancreat Surg. 2008;15:508–14.
12. Ishizawa T, Tamura S, Masuda K, Aoki T, Hasegawa K, Imamura H, et al. Intraoperative fluorescent cholangiography using indocyanine green: a biliary road map for safe surgery. J Am Coll Surg. 2009; 208:e1–4.
13. Ishizawa T, Bandai Y, Ijichi M, Kaneko J, Hasegawa K, Kokudo N. Fluorescent cholangiography illuminating the biliary tree during laparoscopic cholecystectomy. Br J Surg. 2010;97:1369–77.
14. Kawaguchi Y, Ishizawa T, Masuda K, Sato S, Kaneko J, Aoki T, et al. Hepatobiliary surgery guided by a novel fluorescent imaging technique for visualizing hepatic arteries, bile ducts, and liver cancers on color images. J Am Coll Surg. 2011;212:e33–9.
15. Ishizawa T, Fukushima N, Shibahara J, Masuda K, Tamura S, Aoki T, et al. Real-time identification of liver cancers by using indocyanine green fluorescent imaging. Cancer. 2009;115:2491–504.
16. Gotoh K, Yamada T, Ishikawa O, Takahashi H, Eguchi H, Yano M, et al. A novel image-guided surgery of hepatocellular carcinoma by indocyanine green fluorescence imaging navigation. J Surg Oncol. 2009; 100:75–9.
17. Kawaguchi Y, Aoki T, Ishizawa T, Arita J, Satou S, Kaneko J, et al. Education and imaging: hepatobiliary and pancreatic: identification of recurrent hepatocellular carcinoma by intraoperative fluorescent imaging. J Gastroenterol Hepatol. 2013;28:587.
18. Aoki T, Murakami M, Yasuda D, Shimizu Y, Kusano T, Matsuda K, et al. Intraoperative fluorescent imaging using indocyanine green for liver mapping and cholangiography. J Hepatobiliary Pancreat Sci. 2010;17:590–4.
19. Harada N, Ishizawa T, Muraoka A, Ijichi M, Kusaka K, Shibasaki M, et al. Fluorescence navigation hepatectomy by visualization of localized cholestasis from bile duct tumor infiltration. J Am Coll Surg. 2010;210:e2–6.
20. Kawaguchi Y, Ishizawa T, Miyata Y, Yamashita S, Masuda K, Satou S, et al. Portal uptake function in veno-occlusive regions evaluated by real-time fluorescent imaging using indocyanine green. J Hepatol. 2013;58:247–53.
21. Kawaguchi Y, Sugawara Y, Ishizawa T, Satou S, Kaneko J, Tamura S, et al. Identification of veno-occlusive regions in a right liver graft after reconstruction of vein segments 5 and 8: application of indocyanine green fluorescence imaging. Liver Transpl. 2013;19:778–9.
22. Speich R, Saesseli B, Hoffmann U, Neftel KA, Reichen J. Anaphylactoid reactions after indocyanine-green administration. Ann Intern Med. 1988;109: 345–6.
23. Cherrick GR, Stein SW, Leevy CM, Davidson CS. Indocyanine green: observations on its physical properties, plasma decay, and hepatic extraction. J Clin Invest. 1960;39:592–600.
24. White TT, Hart MJ. Cholangiography and small duct injury. Am J Surg. 1985;149:640–3.
25. Ishizawa T, Gumbs AA, Kokudo N, Gayet B. Laparoscopic segmentectomy of the liver: from segment I to VIII. Ann Surg. 2012;256:959.
26. Eguchi S, Kanematsu T, Arii S, Okazaki M, Okita K, Omata M, et al. Comparison of the outcomes between an anatomical subsegmentectomy and a non-anatomical minor hepatectomy for single hepatocellular carcinomas based on a Japanese nationwide survey. Surgery. 2008;143:469–75.
27. Hasegawa K, Kokudo N, Imamura H, Matsuyama Y, Aoki T, Minagawa M, et al. Prognostic impact of anatomic resection for hepatocellular carcinoma. Ann Surg. 2005;242:252–9.

Section III

Clinical Applications of Intraoperative Fluorescence Imaging

Identification of Hepatic Segment 13

Masaki Ueno

Introduction

Liver surgery is complicated and requires understanding of three-dimensional vascular and biliary anatomies. In performing hepatic resection, there are two types of liver resection. One is anatomical resection and the other is non-anatomical resection. Anatomical resection means systematic removal of a hepatic segment including tumor-bearing portal branches and their associating liver parenchyma with tumor. As hepatocellular carcinoma (HCC) tends to invade into a portal vein and to metastasize the perfusion area of the invaded portal vein, this portal vein-oriented anatomical resection has been thought to be theoretically effective in eradicating intrahepatic metastasis of HCC and ideal procedure [1–3].

In addition, anatomical resection is also preferred because of reducing blood loss and rate of biliary complication via the preservation of the vascular and biliary structure of the remnant liver.

Therefore, in order to perform anatomical resection, identification of hepatic segment during surgery is quite important.

Conventional Methods for Hepatic Segmentation

Glissonian approach has been used previously for hepatic segmentation [4]. This technique consists of dissection of Glissonian sheath from liver parenchyma toward the targeted tumor-bearing Glissonian pedicle. By clamping the targeted Glissonian pedicle, the planned resection area becomes ischemic and borderline is depicted. Although this method could depict surface of the liver, it could not depict intraparenchymal cut plane.

A dye staining method under intraoperative ultrasonography (IOUS) was another conventional method [5]. According to this technique, the tumor-bearing portal pedicle was punctured under IOUS guidance and then 5 ml of indigo carmine dye was injected into that branch. The stained area became evident on the liver surface and guided appropriate anatomical resection plane. However, this procedure was also impossible to depict intraparenchymal cut plane and the staining was transient and disappeared soon.

Electronic supplementary material: The online version of this chapter (doi:10.1007/978-3-319-15678-1_13) contains supplementary material, which is available to authorized users. Videos can also be accessed at http://link.springer.com/chapter/10.1007/978-3-319-15678-1_13.

M. Ueno, M.D., Ph.D. (✉)
Second Department of Surgery, Wakayama Medical University, 811-1 Kimiidera, Wakayama 641-8510, Japan
e-mail: ma@wakayama-med.ac.jp

F.D. Dip et al. (eds.), *Fluorescence Imaging for Surgeons: Concepts and Applications*,
DOI 10.1007/978-3-319-15678-1_13, © Springer International Publishing Switzerland 2015

NIRF Imaging as a Novel Method for Liver Segmentation

Optical imaging using near-infrared fluorescence (NIRF) with indocyanine green (ICG) injection is recently recognized as a novel technique for intraoperative and real-time imaging diagnosis in various surgical fields such as sentinel node navigation of foregut and breast cancer [6–8], checking intraoperative vascular flow of neurovascular surgery [9], checking graft patency of cardiovascular surgery [10], and liver surgery [11–16].

In the field of liver surgery, there are three ways usage for intraoperative NIRF imaging with ICG. One is detection of small metastatic nodules on liver surface [11–13], another is liver mapping [14, 15], and the other is checking bile leakage [16].

In this chapter, we show a novel intraoperative technique for identifying hepatic segment using NIRF imaging under ICG injection and discuss about the usefulness of this technique.

Materials and Methods

Fluorescent Agent

ICG has been widely used in estimating liver functional reserve. ICG has attractive features of very low toxicity and is thought to shift about 90 % dose of injected ICG from vessel into hepatocyte, if the liver functional reserve is normally maintained.

Also ICG is widely used as a fluorescent agent and absorption and emission peaks are at the wavelength of 760 nm and 830 nm, respectively in the blood.

NIRF Imaging System

In this chapter, we used following two NIRF imaging systems: Photo Dynamic Eye (PDE; Hamamatsu photonics, Hamamatsu, Japan) (Fig. 13.1a) and HyperEye Medical System (HEMS; Mizuho medical, Tokyo, Japan) (Fig. 13.1b).

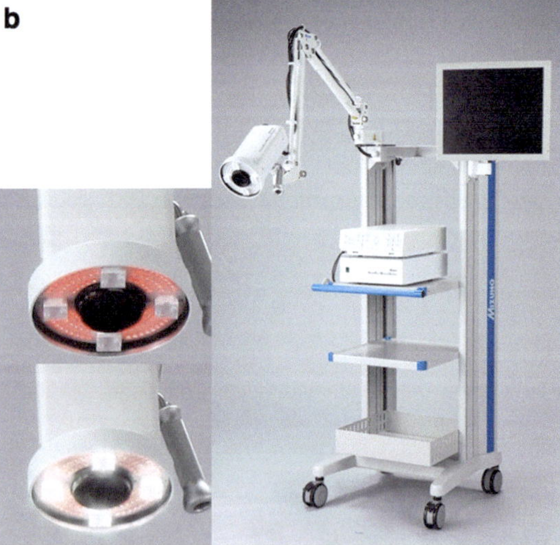

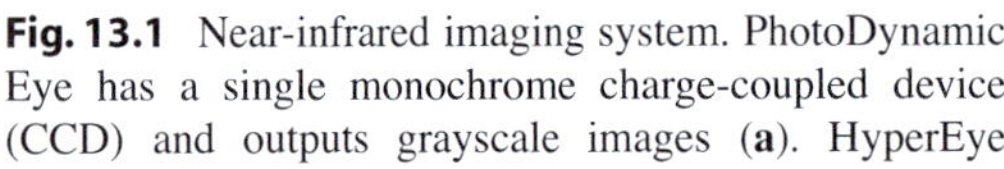

Fig. 13.1 Near-infrared imaging system. PhotoDynamic Eye has a single monochrome charge-coupled device (CCD) and outputs grayscale images (**a**). HyperEye Medical System has a single CCD which has red, green, blue, and infrared filter on it and could output grayscale and color + infrared images (**b**)

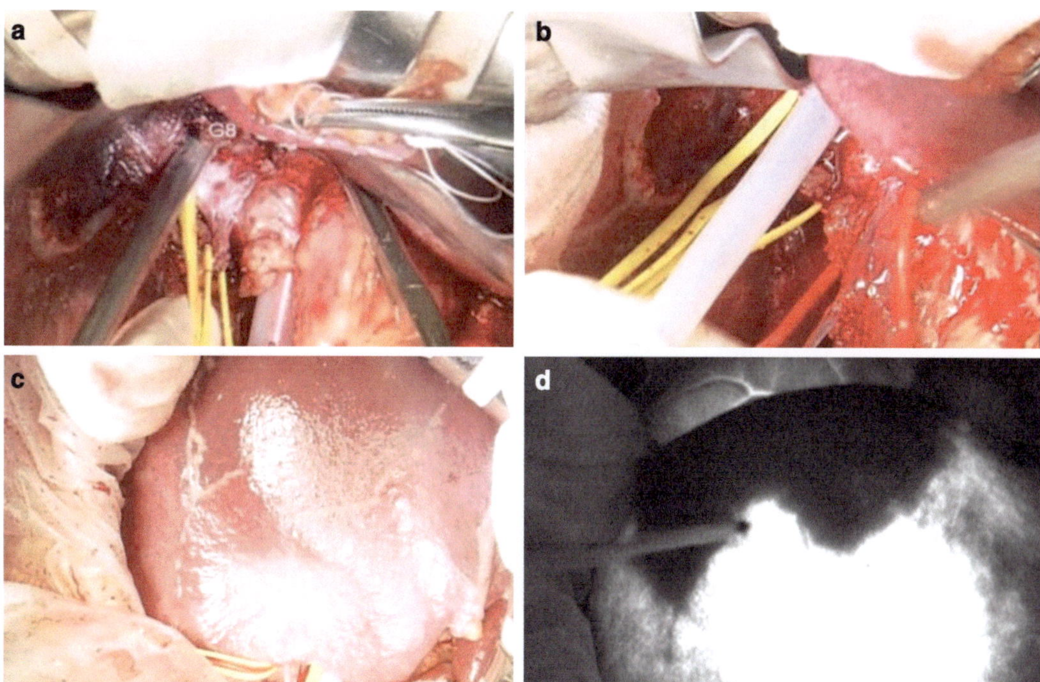

Fig. 13.2 A case of counter perfusion method for segment VIII resection is shown. Firstly, dissect of Glissonian branch (G8) from Hilar plate (**a**). Next, clamp G8 pedicle (**b**). Segment VIII become ischemic but not so clearly visible (**c**). After peripheral venous injection of 1 ml ICG water solution (2.5 mg/mL), segment VIII was clearly depicted by NIRF imaging (**d**)

PDE irradiate near-infrared (NIR) wavelength of 760 nm by using red light-emitting diode (LED) and band pass filter and observe the emission peak at the wavelength of 830 nm through long wave pass filter and a single monochrome charge-coupled device (CCD) with high sensitivity in NIR waves.

HEMS also irradiate NIR wavelength of 760 nm and observe the wavelength of 830 nm as well as the PDE. Different from the PDE, HEMS is installed single special CCD image sensor which has red, green, blue filter and also infrared filter on it and enables to detect both visible and NIR radiation simultaneously and outputs real-time color and NIR imaging.

Technique of ICG Injection

There are two ways to perform ICG injection: counter perfusion method and direct perfusion method.

Counter Perfusion Method

Counter perfusion method is based on the Glissonian approach. Procedure of counter perfusion method is as follows: Firstly, Glissonian sheath is dissected from liver parenchyma toward the tumor-bearing Glissonian pedicle from the Hilar plate and encircle the Glissonian pedicle (Fig. 13.2a). Next, ligation or tourniquet is placed and blood inflow of the planned resection area is clamped (Fig. 13.2b). Inflow occluded area becomes ischemic but unclear (Fig. 13.2c). Then ICG is intravenously injected. Observation of NIRF imaging is started about 20 s after ICG injection. Blood perfusion area, namely remnant hepatic segment, gradually generates fluorescence of ICG. On the other hand, blood nonperfusion area, namely planned resection segment, does not show the fluorescence at all (Fig. 13.2d).

Direct Perfusion Method

Direct perfusion method is the same way as the indigo carmine dye staining method. If a targeted

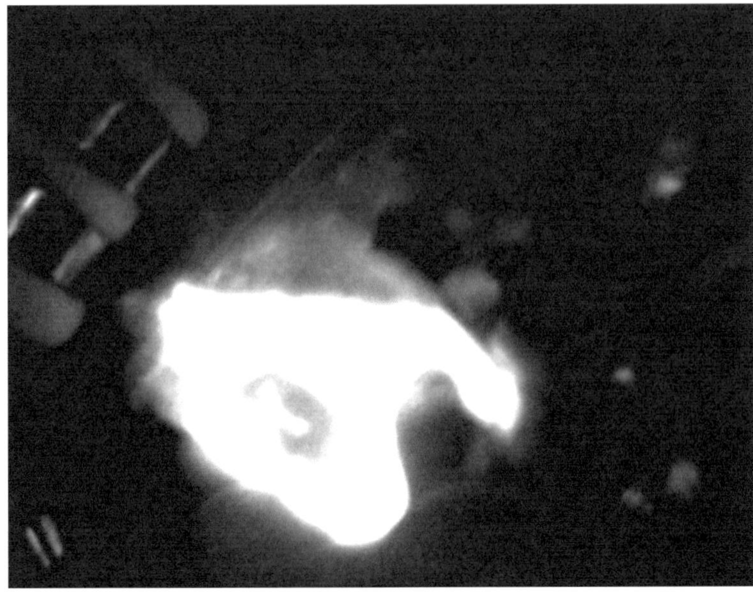

Fig. 13.3 A case of direct perfusion method for segment VI resection is shown. Immediately after direct injection of 1 ml of 5 mg/ml ICG water solution to the target portal vein, Pringle maneuver was placed to avoid passing ICG away from liver parenchyma. Fluorescence of planned resection area was maintained when Pringle maneuver was placed

Glissonian pedicle is far from the Hilar plate and technically the counter perfusion method is difficult, this procedure is applied. Procedure is as follows: Puncture with 23- or 21-gauge needle directly the target Glissonian pedicle (precisely, the target portal vein pedicle) under IOUS guidance. In practically, we punctured about 1 cm distal from the target Glissonian pedicle to avoid straying ICG to some other Glissonian branches. Next, inject 1 ml of 5 mg/ml ICG water solution slowly. Immediately after ICG injection, perform Pringle maneuver to avoid passing ICG away from liver parenchyma. After that, observe NIRF imaging (Fig. 13.3).

Imaging Results

Images of the Counter Perfusion Method During Liver Resection

About 60 s after injection, namely in portal phase, fluorescence had maximum intensity in the remnant liver area and clearly depicted the planned resection area. Compared with macroscopic viewing, the border line was obvious. Moreover, the intensity of fluorescence was maintained during surgery (Fig. 13.2c, d).

A movie of right anterior sectionectomy using this counter perfusion method is shown (Video 13.1). After preparation of the Pringle maneuver and the Glissonian approach, the Glissonian pedicle of segment V + VIII was encircled and clamped all the time during surgery. And then 1 ml of 2.5 mg/ml ICG water solution was injected from peripheral vein. After ICG injection, NIRF image was observed. In this case, HEMS camera was used for capturing NIRF image. About 60 s after injection, fluorescence had maximum intensity in the remnant liver area and clearly depicted the planned resection area. Based on this fluorescent borderline, liver resection was performed. During liver resection, NIRF image was observed repeatedly and resection line or transection plane could be visually checked. By observing uniform fluorescence on the cut surface, we could confirm the resection was carried out as planned (Fig. 13.4a). If the resection was not carried out as planned, the cut surface does not show the fluorescence (Fig. 13.4b).

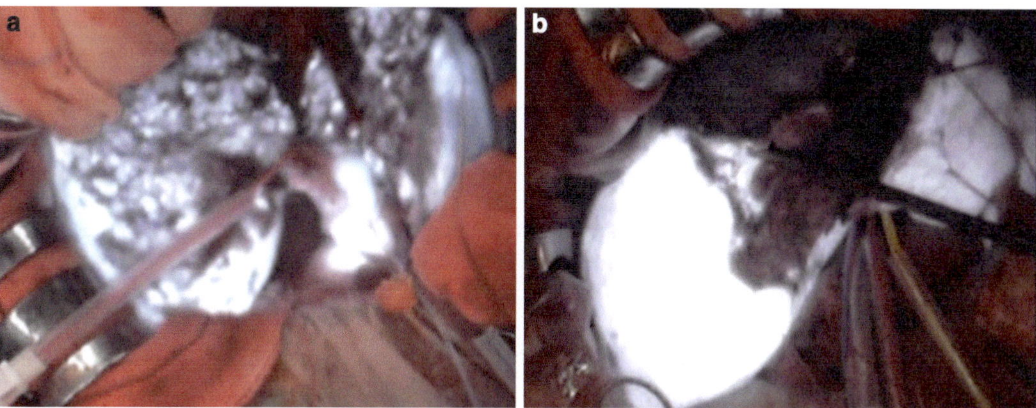

Fig. 13.4 If the dissection is carried out as planned, the fluorescence can be observed on the cut surface (**a**). If not, the cut surface will not show the fluorescence (**b**)

Images of the Direct Staining Method During Liver Resection

A movie of segmentectomy VI using direct perfusion method is shown (Video 13.2). Immediately after injection, segment VI had strong fluorescence. As blood inflow was occulted by Pringle maneuver, ICG remained at segment VI and was maintained fluorescence during the Pringle maneuver. Under this situation, the segment with fluorescence was resected.

Comments

In this chapter, the usefulness of NIRF imaging with ICG for visualizing hepatic segment during hepatectomy was presented. Liver anatomy that divides the hepatic parenchyma into Segment I to VIII based on the branching form of Glissonian pedicle has been familiar worldwide and which is recognized as the Couinaud's segmentation. Using this Couinaud's segmentation, current liver surgeons express the tumor-existent area and proposed anatomical resection.

In order to identify the hepatic segment during surgery, counter perfusion and direct perfusion methods were proposed as novel and useful techniques. In order to select which method is better, preoperative surgical simulation is important. Recently, three-dimensional liver anatomy reconstructed by contrast-enhanced multi-detector row CT images has been easily acquired using various applications softwares. Owing to these preoperative images, accurate intraoperative procedure could be achieved [17, 18].

In practice, hemihepatectomy or left lateral sectionectomy may not always need this NIRF imaging because of simple cut plane. Right anterior, right posterior, and left medial sectionectomy and Segmentectomy of V will have a benefit by counter perfusion method. Segmentectomy of VI, VII, and VIII, which are sometimes technically difficult for adapting counter perfusion method, will have a benefit by applying direct perfusion method.

Conventionally, Glissonian approach and indigo carmine dye staining have been useful for detecting segment of surface liver parenchyma [4, 5]. However, especially in repeat hepatectomy or in cirrhotic liver parenchyma, these methods were sometimes not so clear and difficult to depict resection line. Moreover, these methods were impossible to depict planned transection plane during surgery. Additionally, indigo carmine staining method required clamping hepatic artery for obtaining clear visualization and required some injection techniqques. That might be troublesome. However, our NIRF imaging procedures could clearly depict hepatic segment both surface and intraparenchymal of the liver. Additionally, the effect could be maintained during surgery.

Regarding the difference between the counter perfusion method and direct perfusion method, there are both advantages and disadvantages.

As for the counter perfusion method, the most noteworthy point is that the resection line and plane is represented by ICG fluorescence all the time during surgery. In other words, the counter perfusion method might become a navigation surgery. However, applying this method to segmentectomy or smaller anatomical resection is technically difficult, because this method is based on Glissonian approach. That would be a weak point of counter perfusion method.

As for the direct perfusion method, this is an easy way with accurate puncture technique and would back up the weak point of counter perfusion method. However, weak point of this method is that the ICG fluorescence is not maintained all the time. The fluorescence is limited when the Pringle maneuver is firstly placed. If the first Pringle maneuver is released, the ICG remaining in the blood will circulate and distribute to the remnant liver. As a result, whole the liver will exhibit the ICG fluorescence. Therefore, during the first Pringle maneuver, it is necessary to perform parenchyma resection until the transection direction is fixed [14].

Recently, laparoscopic procedure of liver resection has become popular [19]. Therefore, NIRF devices also must keep up with the progress of laparoscopic procedures. As several articles already reported clinical usage of laparoscopic NIRF imaging, our procedures also would be performed laparoscopically in future [20, 21].

In conclusion, intraoperative identification of hepatic segment using NIRF imaging with ICG is a novel procedure and will overcome the weak point of conventional methods. This method will be applied to navigation surgery of the liver and will be a help for laparoscopic anatomical resection in future.

References

1. Eguchi S, Kanematsu T, Arii S, et al. Comparison of the outcomes between an anatomical subsegmentectomy and a non-anatomical minor hepatectomy for single hepatocellular carcinomas based on a Japanese nationwide survey. Surgery. 2008;143:469–75.
2. Agrawal S, Belghiti J. Oncologic resection for malignant tumors of the liver. Ann Surg. 2011;253: 656–65.
3. Hasegawa K, Kokudo N, Imamura H, et al. Prognostic impact of anatomic resection for hepatocellular carcinoma. Ann Surg. 2005;242:252–9.
4. Takasaki K, Kobayashi S, Tanaka S, Saito A, Yamamoto M, Hanyu F. Highly anatomically systematized hepatic resection with glissonean sheath code transection at the hepatic hilus. Int Surg. 1990;75: 73–7.
5. Makuuchi M, Hasegawa H, Yamazaki S. Ultrasonically guided subsegmentectomy. Surg Gynecol Obstet. 1985;161:346–50.
6. Kusano M, Tajima Y, Yamazaki K, Kato M, Watanabe M, Miwa M. Sentinel node mapping guided by indocyanine green fluorescence imaging: a new method for sentinel node navigation surgery in gastrointestinal cancer. Dig Surg. 2008;25:103–8.
7. Troyan SL, Kianzad V, Gibbs-Strauss SL, et al. The flare intraoperative near-infrared fluorescence imaging system: a first-in-human clinical trial in breast cancer sentinel lymph node mapping. Ann Surg Oncol. 2009;16:2943–52.
8. Hirono S, Tani M, Kawai M, et al. Identification of the lymphatic drainage pathways from the pancreatic head guided by indocyanine green fluorescence imaging during pancreaticoduodenectomy. Dig Surg. 2012;29:132–9.
9. Raabe A, Beck J, Gerlach R, Zimmermann M, Seifert V. Near-infrared indocyanine green video angiography: a new method for intraoperative assessment of vascular flow. Neurosurgery. 2003;52:132–9. discussion 139.
10. Rubens FD, Ruel M, Fremes SE. A new and simplified method for coronary and graft imaging during cabg. Heart Surg Forum. 2002;5:141–4.
11. Gotoh K, Yamada T, Ishikawa O, et al. A novel image-guided surgery of hepatocellular carcinoma by indocyanine green fluorescence imaging navigation. J Surg Oncol. 2009;100:75–9.
12. Ishizawa T, Fukushima N, Shibahara J, et al. Real-time identification of liver cancers by using indocyanine green fluorescent imaging. Cancer. 2009;115: 2491–504.
13. Morita Y, Sakaguchi T, Unno N, et al. Detection of hepatocellular carcinomas with near-infrared fluorescence imaging using indocyanine green: its usefulness and limitation. Int J Clin Oncol. 2013;18:232–41.
14. Aoki T, Yasuda D, Shimizu Y, et al. Image-guided liver mapping using fluorescence navigation system with indocyanine green for anatomical hepatic resection. World J Surg. 2008;32:1763–7.
15. Uchiyama K, Ueno M, Ozawa S, et al. Combined intraoperative use of contrast-enhanced ultrasonography imaging using a sonazoid and fluorescence navigation system with indocyanine green during anatomical hepatectomy. Langenbecks Arch Surg. 2011;396:1101–7.
16. Kaibori M, Ishizaki M, Matsui K, Kwon AH. Intraoperative indocyanine green fluorescent imaging

for prevention of bile leakage after hepatic resection. Surgery. 2011;150:91–8.
17. Saito S, Yamanaka J, Miura K, et al. A novel 3d hepatectomy simulation based on liver circulation: application to liver resection and transplantation. Hepatology. 2005;41:1297–304.
18. Takamoto T, Hashimoto T, Ogata S, et al. Planning of anatomical liver segmentectomy and subsegmentectomy with 3-dimensional simulation software. Am J Surg. 2013;206:530–8.
19. Buell JF, Cherqui D, Geller DA, et al. The international position on laparoscopic liver surgery: the Louisville statement, 2008. Ann Surg. 2009;250:825–30.
20. Jafari MD, Lee KH, Halabi WJ, et al. The use of indocyanine green fluorescence to assess anastomotic perfusion during robotic assisted laparoscopic rectal surgery. Surg Endosc. 2013;27:3003–8.
21. Cahill RA, Anderson M, Wang LM, Lindsey I, Cunningham C, Mortensen NJ. Near-infrared (nir) laparoscopy for intraoperative lymphatic roadmapping and sentinel node identification during definitive surgical resection of early-stage colorectal neoplasia. Surg Endosc. 2012;26:197–204.

Intraoperative Evaluation of Regional Portal Uptake Function

14

Yoshikuni Kawaguchi, Takeaki Ishizawa, and Norihiro Kokudo

Abbreviations

CT	Computed tomography
FI	Fluorescence intensity
ICG	Indocyanine green
TLV	Total liver volume

Introduction

It is important to identify the distribution and the level of portal perfusion into hepatic regions during surgery for preventing postoperative hepatic dysfunction [1, 2]. Although resection of the regions without portal flow is accepted well, it has been unclear whether the regions without hepatic venous flow (so called veno-occlusive regions) are needed to be resected or not, because the level of portal perfusion which may represent part of liver function in such regions has been unclear. In order to prevent postoperative hepatic dysfunction in veno-occlusive regions, stumps of the major hepatic vein tributaries are sometimes reconstructed during liver resection and living donor liver transplantation [1, 3, 4]. Hashimoto et al. [5] first reported a technique for the evaluation of sinusoidal perfusion in veno-occlusive regions in the remnant liver after graft procurement, using intraoperative ICG injection and near-infrared spectroscopy. However, their technique is somewhat complex in clinical settings and has limitations in demonstrating the intersegmental difference in the portal uptake function at the same time. Recently, the author's group have applied fluorescence imaging technique with administration of indocyanine green (ICG) [6–11] for evaluation of regional portal uptake in the liver [2].

We herein describe ICG-fluorescence imaging technique evaluating regional portal uptake in the liver during surgery and clinical application for liver resection and liver transplantation to improve the safety of hepatobiliary surgery.

Electronic supplementary material: The online version of this chapter (doi:10.1007/978-3-319-15678-1_14) contains supplementary material, which is available to authorized users. Videos can also be accessed at http://link.springer.com/chapter/10.1007/978-3-319-15678-1_14.

Y. Kawaguchi, M.D. • N. Kokudo, M.D., Ph.D. (✉)
Hepato-Biliary-Pancreatic Surgery Division, Department of Surgery, Graduate School of Medicine, University of Tokyo, 7-3-1 Hongo, Bunkyo-ku, Tokyo 113-8655, Japan
e-mail: yokawaguchi-tky@umin.ac.jp; KOKUDO-2SU@h.u-tokyo.ac.jp

T. Ishizawa, M.D., Ph.D., F.A.C.S
Department of Gastroenterological Surgery, Cancer Institute Hospital, Japanese Foundation for Cancer Research, 7-3-1 Hongo, Koto-ku, Tokyo Japan

Hepato-Biliary-Pancreatic Surgery Division, Department of Surgery, Graduate School of Medicine, University of Tokyo, 7-3-1 Hongo, Bunkyo-ku, Tokyo 113-8655, Japan
e-mail: tish-tky@umin.ac.jp

F.D. Dip et al. (eds.), *Fluorescence Imaging for Surgeons: Concepts and Applications*, DOI 10.1007/978-3-319-15678-1_14, © Springer International Publishing Switzerland 2015

Evaluation of Regional Portal Uptake Using ICG-Fluorescence Imaging

Following liver resection or procedure of liver transplantation, the camera head of the florescence imaging system (PDE, Hamamatsu Photonics, Hamamatsu, Japan) was positioned 20 cm above the liver surface and surgical lights were turned off. The output of the light-emitting diodes (760 nm) was set at 0.21 mW/cm^2, after which fluorescence images of the liver surface were observed after intravenous injection of ICG (2.5 μg/1 mL of remnant liver volume calculated based on preoperative computed tomography (CT)). Figure 14.1 shows the remnant right liver after procurement of the left liver graft with the middle hepatic vein. Following intravenous administration of ICG, fluorescence imaging revealed the demarcation among non-veno-occlusive regions and possible veno-occlusive regions corresponding to segment V and VIII (Fig. 14.1a). In order to quantitate the regional portal uptake in veno-occlusive regions, the FI value can be calculated using a luminance-analyzing software (U11437, Hamamatsu Photonics) (Fig. 14.1b). The FI values on the liver surface increased linearly during first 200 s and then reached a plateau in an average, and could be converted to ICG concentration using the following formula obtained by the relation between the concentration of ICG and the FI value; ICG concentration = 0.167 × exp (0.014 × FI) [2]. We assume that the estimated ICG concentrations converted from the FI values at a plateau reflect potential function in the regions with interest.

In our previous study [2], the portal uptake function in veno-occlusive regions is usually 30–40 % of that in the non-veno-occlusive regions. These techniques enable visualization of the extent of regions with decreased portal uptake and serve as a guide for the need/lack for venous reconstruction. No matter how accurately volume-analyzing software based on preoperative three-dimensional CT calculate the volume of the regions with interest, its result is virtual and does not enable estimate the degree of portal uptake. Fluorescence images may reflect the actual extent and degree of the portal uptake in the regions with interest as a consequence of a complex system of uptake into the liver, although these techniques make it impossible to observe the deep regions in the liver. In the next part, the authors show clinical application of these techniques.

Clinical Application

Figure 14.2 shows a case of a patient who underwent extended right hepatectomy with excision of the middle hepatic vein for the tumor located in the segment VIII. Our previous criteria for reconstruction of the hepatic vein are described elsewhere [12]. Briefly, the stump of the hepatic vein in the veno-occlusive regions is reconstructed, when the remnant liver volume without venous congestion is less than 40 % of the total liver volume (TLV) for the patient with normal liver function. In this case, volume-analyzing software based on three-dimensional CT estimated the regional liver volume as non-veno-occlusive regions drained by the left hepatic vein (314 mL, 36.9 % of TLV) and veno-occlusive regions drained by the tributary of the middle hepatic vein (77 mL, 9.1 % of TLV) (Fig. 14.2a). In order to determine the need/lack for venous reconstruction, we applied fluorescence imaging technique for evaluation of veno-occlusive regions, after which the regions with decreased portal uptake were identified (Fig. 14.2b). The estimated concentration of ICG introduced in this region was 1.6 μg/mL and corresponded to 30.8 % of that in non-veno-occlusive regions (5.2 μg/mL). The functional remnant liver volume in this case was estimated as 337 mL (40 % of TLV) based on the formula: (functional remnant liver volume) = (liver volume without venous occlusion) + (liver volume with venous occlusion) × (the ratio of the concentration of ICG in the veno-occlusive regions to that in the non-veno-occlusive regions) [2]. Based on the information above, extended right hepatectomy was performed and the stump of the tributary of

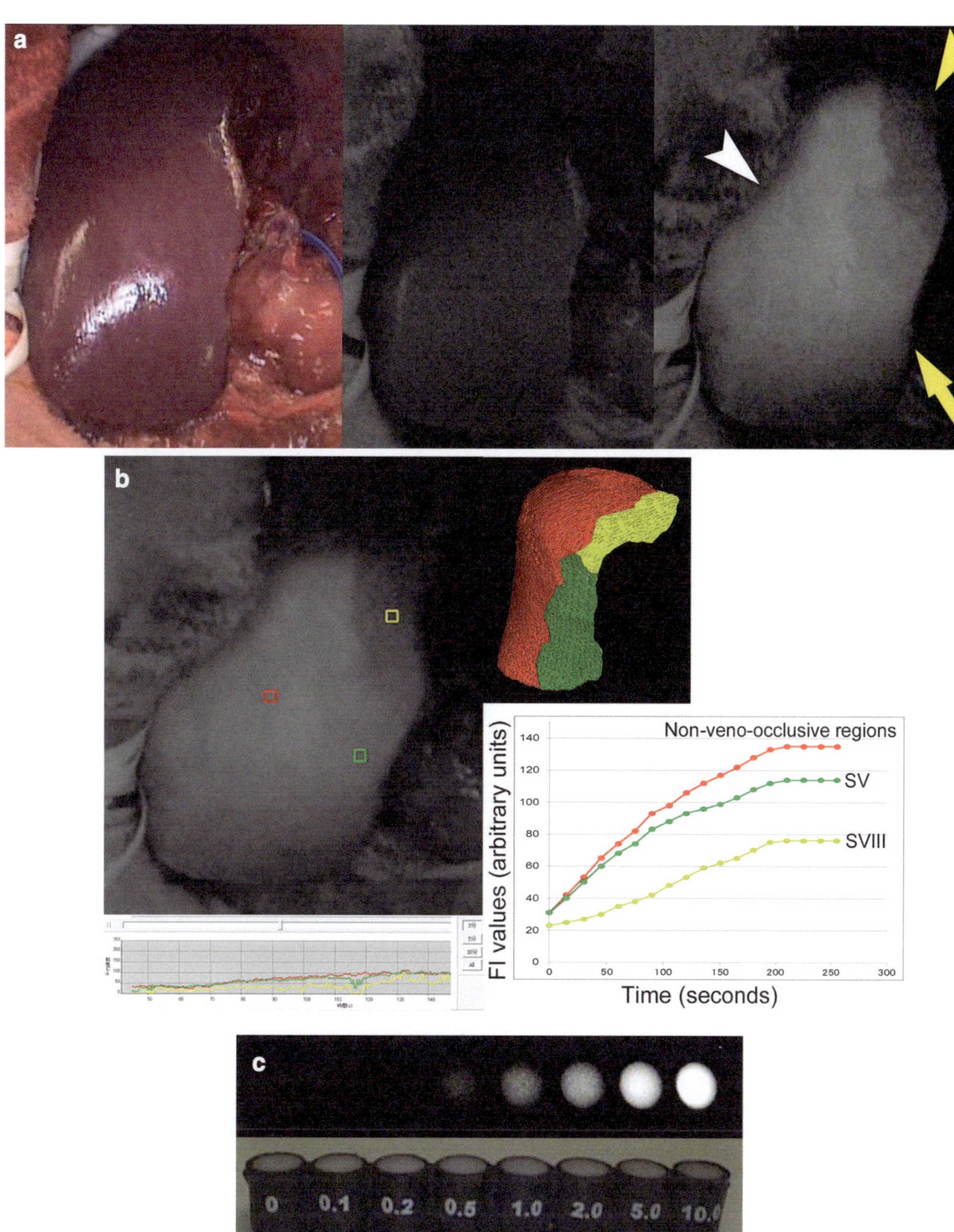

Fig. 14.1 Evaluation of the portal uptake function in veno-occlusive regions by intraoperative ICG-fluorescence imaging. (**a**) Fluorescence imaging following intravenous injection of ICG reveals the demarcation among possible veno-occlusive regions in S5 (*arrow*) and S8 (*arrowhead*) and non-veno-occlusive regions (*white arrowhead*); *left*, gross appearance, *middle*, fluorescence images just before ICG injection, and *right*, fluorescence images 240 s after the injection. (**b**) By using a luminance-analyzing software, intersegmental changes in the fluorescence intensity, which reflect the distribution of ICG after intravenous injection, can be evaluated quantitatively (*left*). FI in the non-veno-occlusive regions, segment V, and segment VIII were continuously recorded (*lower right*). Corresponding 3D reconstruction of CT images reveals the regions drained by the right hepatic vein (*red*) and middle hepatic vein tributaries draining segment VIII (*yellow*) and segment V (*green*) in the remnant right liver (*upper right*). (**c**) An artificial model showing the relationship between the ICG concentrations and the FI values. The FI values increase with elevation of the concentration of phantoms of ICG diluted with ethanol and epoxy resin; the unit of the concentration is expressed as μmol/L

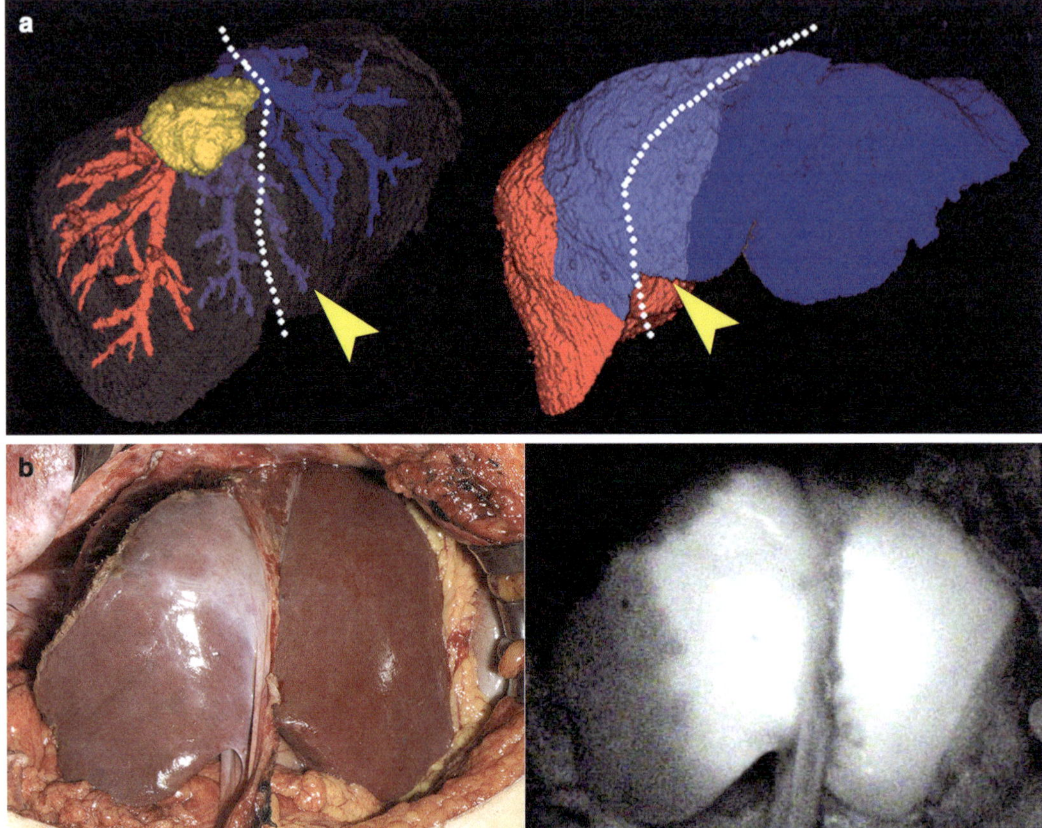

Fig. 14.2 Evaluation of the regional portal uptake in the postoperative remnant liver. (**a**) A tumor location and planned resection line (*white dotted line*) was demonstrated on preoperative 3D-CT. Due to the division of MHV tributaries (*left*, *arrowhead*), S4 (*right*, *arrowhead*) was estimated to be the veno-occlusive regions after liver resection. (**b**) Fluorescence imaging following intravenous injection of ICG reveals the demarcation between the regions with and without venous occlusion (*left*, gross appearance, and *right*, fluorescent images 240 s after the injection). Please see Video 14.1

the middle hepatic vein was not reconstructed. The patient was discharged postoperative day 17 without presenting hepatic dysfunction.

These evaluations were performed in approximately 10 min in total. One of the advantages of these techniques is that they enable visualization of the degree and the extent of the portal uptake in the regions as a result of surgical procedure. These techniques are expected to more accurately estimate the functional volume in the remnant liver, ensuring the safety of liver resection and liver transplantation.

Conclusion

ICG-fluorescence imaging enables estimation of regional portal uptake in veno-occlusive regions which usually decreases to approximately 30–40 % of that in non-veno-occlusive regions. This result and these techniques are expected to be applied for estimation of functional reserve in the postoperative remnant liver, enhancing the safety of liver resection and liver transplantation.

Financial Disclosure: The authors have no relevant affiliations or financial involvement with any organization or entity with a financial interest in or financial conflict with the subject matter or materials discussed in this manuscript. This includes employment, consultancies, honoraria, stock ownership or options, expert testimony, grants or patents received or pending, and royalties. No writing assistance was utilized in the production of this manuscript.

Funding/Support: This work was supported by grants from the Takeda Science Foundation, the Kanae Foundation for the Promotion of Medical Science, the Pancreas Research Foundation of Japan, the Canon Foundation in Europe, and the Ministry of Education, Culture, Sports, Science and Technology of Japan (No. 23689060).

References

1. Sano K, Makuuchi M, Miki K, Maema A, Sugawara Y, Imamura H, et al. Evaluation of hepatic venous congestion: proposed indication criteria for hepatic vein reconstruction. Ann Surg. 2002;236:241–7.
2. Kawaguchi Y, Ishizawa T, Miyata Y, Yamashita S, Masuda K, Satou S, et al. Portal uptake function in veno-occlusive regions evaluated by real-time fluorescent imaging using indocyanine green. J Hepatol. 2013;58:247–53.
3. Kakazu T, Makuuchi M, Kawasaki S, Miyagawa S, Nakazawa Y, Kubota T, et al. Reconstruction of the middle hepatic vein tributary during right anterior segmentectomy. Surgery. 1995;117:238–40.
4. Cattral MS, Greig PD, Muradali D, Grant D. Reconstruction of middle hepatic vein of a living-donor right lobe liver graft with recipient left portal vein. Transplantation. 2001;71:1864–6.
5. Hashimoto T, Miki K, Imamura H, Sano K, Satou S, Sugawara Y, et al. Sinusoidal perfusion in the veno-occlusive region of living liver donors evaluated by indocyanine green and near-infrared spectroscopy. Liver Transpl. 2008;14:872–80.
6. Kubota K, Kita J, Shimoda M, Rokkaku K, Kato M, Iso Y, et al. Intraoperative assessment of reconstructed vessels in living-donor liver transplantation, using a novel fluorescence imaging technique. J Hepatobiliary Pancreat Surg. 2006;13:100–4.
7. Mitsuhashi N, Kimura F, Shimizu H, Imamaki M, Yoshidome H, Ohtsuka M, et al. Usefulness of intraoperative fluorescence imaging to evaluate local anatomy in hepatobiliary surgery. J Hepatobiliary Pancreat Surg. 2008;15:508–14.
8. Ishizawa T, Fukushima N, Shibahara J, Masuda K, Tamura S, Aoki T, et al. Real-time identification of liver cancers by using indocyanine green fluorescent imaging. Cancer. 2009;115:2491–504.
9. Ishizawa T, Bandai Y, Ijichi M, Kaneko J, Hasegawa K, Kokudo N. Fluorescent cholangiography illuminating the biliary tree during laparoscopic cholecystectomy. Br J Surg. 2010;97:1369–77.
10. Gotoh K, Yamada T, Ishikawa O, Takahashi H, Eguchi H, Yano M, et al. A novel image-guided surgery of hepatocellular carcinoma by indocyanine green fluorescence imaging navigation. J Surg Oncol. 2009; 100:75–9.
11. Kawaguchi Y, Ishizawa T, Masuda K, Sato S, Kaneko J, Aoki T, et al. Hepatobiliary surgery guided by a novel fluorescent imaging technique for visualizing hepatic arteries, bile ducts, and liver cancers on color images. J Am Coll Surg. 2011; 212:e33–9.
12. Mise Y, Hasegawa K, Satou S, Aoki T, Beck Y, Sugawara Y, et al. Venous reconstruction based on virtual liver resection to avoid congestion in the liver remnant. Br J Surg. 2011;98:1742–51.

Identification of Hepatocellular Carcinoma

15

Yoshifumi Morita, Takanori Sakaguchi, Hirotoshi Kikuchi, Naoki Unno, and Hiroyuki Konno

Introduction

Hepatocellular carcinoma (HCC) is the fifth most common cancer and the third most common cause of cancer-related death worldwide [1]. Accurate identification of the number, size, and location of HCCs is required for determining the optimal therapeutic strategy. Formerly, computed tomography (CT) during arteriography (CTA) and arterial portography (CTAP) were the standard preoperative examinations. CTA and CTAP are particularly effective for identifying small hepatic lesions [2]. Recently, contrast-enhanced ultrasound (US) using microbubble agent has improved the reliability of US in the assessment of liver tumors [3]. Additionally, magnetic resonance imaging (MRI) using gadoxetic acid (Gd-EOB DTPA) is becoming the primary imaging technique in HCC, because it offers more accurate detection and diagnosis of liver tumors than contrast-enhanced CT [4]. Using these contrast agents, we can evaluate not only intratumoral blood flow but also cellular characteristics such as Kupffer cell-like or hepatocyte-like phenotype. Even with these advances in diagnostic modalities, preoperatively unidentified HCCs are occasionally found during hepatectomy [5]. Contrast-enhanced intraoperative ultrasound (IOUS) has been shown to be useful in identifying small HCCs. However, it is difficult to detect small, unpalpable liver tumors just beneath the liver surface by IOUS. It is clear that overlooked cancer can cause early recurrence and can worsen prognosis. Therefore, a new, sensitive technique is required to detect such potentially overlooked small HCCs.

Fluorescence imaging has emerged and developed as a powerful tool for visualizing target molecules, mainly in the field of basic research. In particular, near-infrared (NIR) fluorescence imaging has attracted great interest because of its broad usefulness. NIR (wave length of 700–1,000 nm) can deeply penetrate the optical window in biological tissue with reduced scattering and minimal interfering absorption [6].

However, the number of clinically applicable fluorescent agents is limited. Indocyanine green (ICG) has been approved by the US Food and Drug Administration (FDA) and is widely used in procedures such as ophthalmic angiography, measurement of cardiac output, and hepatic function [7–9]. ICG is a negatively charged, amphiphilic, water-soluble tricarbocyanine with a molecular

Electronic supplementary material: The online version of this chapter (doi:10.1007/978-3-319-15678-1_15) contains supplementary material, which is available to authorized users. Videos can also be accessed at http://link.springer.com/chapter/10.1007/978-3-319-15678-1_15.

Y. Morita, M.D., Ph.D. (✉) • T. Sakaguchi, M.D., Ph.D.
H. Kikuchi, M.D., Ph.D. • N. Unno, M.D., Ph.D.
H. Konno, M.D., Ph.D.
Second Department of Surgery, Hamamatsu University School of Medicine, 1-20-1 Handayama, Higashi-ku, Hamamatsu 431-3192, Japan
e-mail: yoshi-mo@hama-med.ac.jp; saka1119@hama-med.ac.jp; kikuchih@hama-med.ac.jp; unno@hama-med.ac.jp; kon_6416@hama-med.ac.jp

F.D. Dip et al. (eds.), *Fluorescence Imaging for Surgeons: Concepts and Applications*,
DOI 10.1007/978-3-319-15678-1_15, © Springer International Publishing Switzerland 2015

mass of 776 Da. ICG is one of the least toxic imaging agents; the only known adverse reaction is rare anaphylaxis. ICG has an absorption peak of 800 nm in vivo and is green in color. After intravenous administration, ICG has a short half-life of 150–180 s and is excreted exclusively by the liver. ICG becomes fluorescent under NIR light after binding to serum proteins. The excitation and emission wavelengths in serum are 765 nm and 840 nm, respectively [10]. Thus, the fluorescence wavelength is longer than that of excitation and is specific to ICG, and a high signal-to-noise ratio can be obtained by filtering out light at wavelengths other than that of the fluorescence. NIR ICG fluorography has been increasingly used in various clinical situations such as the assessment of lymphedema [11], real-time evaluation of blood flow in vascular [12] and intracranial surgeries [13], and sentinel lymph node (SLN) navigation in surgery for breast or gastrointestinal cancers [14, 15]. In hepatobiliary surgery, its usefulness has been reported in the intraoperative assessment of local anatomy [16], cholangiography for laparoscopic cholecystectomy [17], and bile leak test after hepatectomy [18].

To date, several authors have reported the usefulness of NIR ICG fluorography for the identification of HCC [19–21]. Along with other authors, we have incidentally found that HCC tumors showed very strong fluorescence in patients who received an intravenous injection of ICG for a routine liver function test before hepatectomy.

ICG Administration

ICG intravenously administered before surgery at a dose of 0.5 mg/kg of body weight was used to determine the operative indications and procedures. The median intervals between ICG administration and surgery in each of three studies were 4.8 days, 3 days, and 14.7 days, respectively. Fluorescence from preoperatively administered ICG can be detected inside or around the HCC for relatively long periods. In our unpublished data, we identified strong emission even beyond 30 days after administration, and the longest interval between ICG administration and surgery was 59 days. However, longer intervals may increase the false-negative rate, because ICG can be excreted into the diversionary bile duct and its fluorescence diminishes with time. Conversely, Ishizawa et al. reported that in cases of ICG injection within 24 h before surgery, the signal intensity of noncancerous liver parenchyma was high in patients with advanced cirrhosis. They recommended that the interval between ICG injection and surgery should be at least 2 days to obtain good lesion-to-liver contrast [20]. Although it is currently difficult to precisely define the best interval between ICG administration and surgery, an interval of 2–7 days seems to be appropriate.

NIR Fluorescence Imaging Systems

We and the other two study groups used the same commercially available NIR fluorescence imaging system (PDE™; Hamamatsu Photonics K.K., Hamamatsu, Japan), which activates ICG with emitted light (wavelength: 760 nm) and filters out light with a wavelength below 820 nm [19–21]. The light sources for the emission of ICG were 760-nm LEDs, and the detector was a charge-coupled device (CCD) camera. Several other commercially available NIR fluorescence imaging systems have been introduced for clinical use, such as the SPY imaging system (Novadaq Technologies, Concord, Canada), HyperEye Medical System (Mizuho Medical Co., Ltd., Tokyo, Japan), FLARE Imaging System Kit (The FLARE Foundation, Inc., Wayland, USA), Fluobeam (Fluoptics, Grenoble, France), and ICG system (KARL STORZ GmbH & Co. KG, Tuttlingen, Germany). Although differing in their technical applications such as the use of dual image or overlay of visible light and NIR, all of these systems provide an image of NIR fluorescence.

Real-Time Identification of HCC Using ICG Fluorography

For the real-time identification of either preoperatively identified or unidentified HCCs, it is important to observe the liver surface after mobilization of the liver. Although NIR is capable of relatively deep penetration into biological tissues, the depth at which NIR light can be observed is about 5–10 mm [14]. Only very limited regions of the liver can be observed using ICG fluorography just after laparotomy. In particular, attachment sites to diaphragma or retroperitoneum frequently fall into dead areas. Therefore, it is necessary to observe the liver surface from a number of different directions. Before liver transection, we routinely position the camera head about 20 cm above the liver surface with the surgical lights turned off. Surgeons can view the fluorescent images displayed on a television monitor or laptop computer connected to the NIR camera. Actual images of intraoperative and postoperative ICG fluorography can be seen in Video 15.1.

Imaging Appearance of HCC Using ICG Fluorography

The imaging appearance of HCCs is classified into three patterns: uniform (total) intra-lesion emission pattern (Fig. 15.1a), uneven (partial) intra-lesion emission pattern (Fig. 15.1b), and rim-like emission pattern (Fig. 15.1c). There is significant correlation between HCC differentiation and the imaging appearance. Well-differentiated carcinomas are commonly classified as uniform (total) intra-lesion emission pattern. Meanwhile, poorly differentiated carcinomas are commonly classified as rim-like emission pattern.

For the classification of emission patterns, it is vital to keep constant brightness in the room and a constant distance between the camera unit and the tumor. The strength of excitation light also influences the emission pattern. Emission patterns seem to be unreliable when the recording conditions are variable.

Fluorescent Microscopy

Emitted light can be seen even in formalin-fixed paraffin-embedded specimens. Fluorescent microscopy can be used to visualize fluorescence in the cytoplasm (Fig. 15.2a, b) and pseudogland (Fig. 15.2c, d) of HCC, or in the surrounding noncancerous liver tissues compressed by the tumor (Fig. 15.2e, f). ICG accumulations in the cytoplasm or pseudogland of HCCs cause macroscopic imaging appearances such as uniform (total) intra-lesion emission patterns or uneven (partial) intra-lesion emission patterns. Meanwhile, ICG accumulations in the surrounding noncancerous liver tissue cause a rim-like emission pattern. Hepatocytes express polarized transport systems at both the sinusoidal side and bile canalicular side. ICG uptake is mainly mediated by organ anion transporting polypeptide (OATP) 1B3 and Na^+-taurocholate co-transporting polypeptide (NTCP) [22]. OATP1B3 and NTCP expression are reported to be low in HCC [23, 24]. Meanwhile, ICG excretion is thought to be mainly mediated by multidrug resistance P-glycoprotein 3 (MDR3) [25, 26]. MDR3 translocates phosphatidylcholine from the inner to the outer hemileaflet of the canalicular membrane, which is followed by its release from the outer hemileaflet into bile. MDR3 expression levels in HCC are reported to vary between individuals [23]. The emission pattern and emission-positive or emission-negative observations might be influenced by the expression levels of these transporters. More recently, Ishizawa et al. reported detailed observations about the mechanistic background of ICG accumulation [27]. They reported that the expression levels of NTCP and OATP1B3 tended to be lower in patients with rim-like emission patterns. Since the expression of OATP1B3, a main uptake transporter of ICG [22], is reported to decrease according to the de-differentiation process (from low-grade dysplastic nodule to poorly differentiated HCC) [24], uptake of ICG by poorly differentiated HCCs would be limited. These results suggest that in well-differentiated HCC tissues, sinusoidal uptake is preserved, whereas biliary excretion of ICG is probably impaired because of morphological changes associated with cancer progression. Meanwhile, the precise molecular mechanisms leading to rim-like emission patterns remain unclear. As mentioned in other reports [20] [28], the rim-like emission pattern may be due to locally impaired biliary excretion. Further studies are needed to verify this possibility.

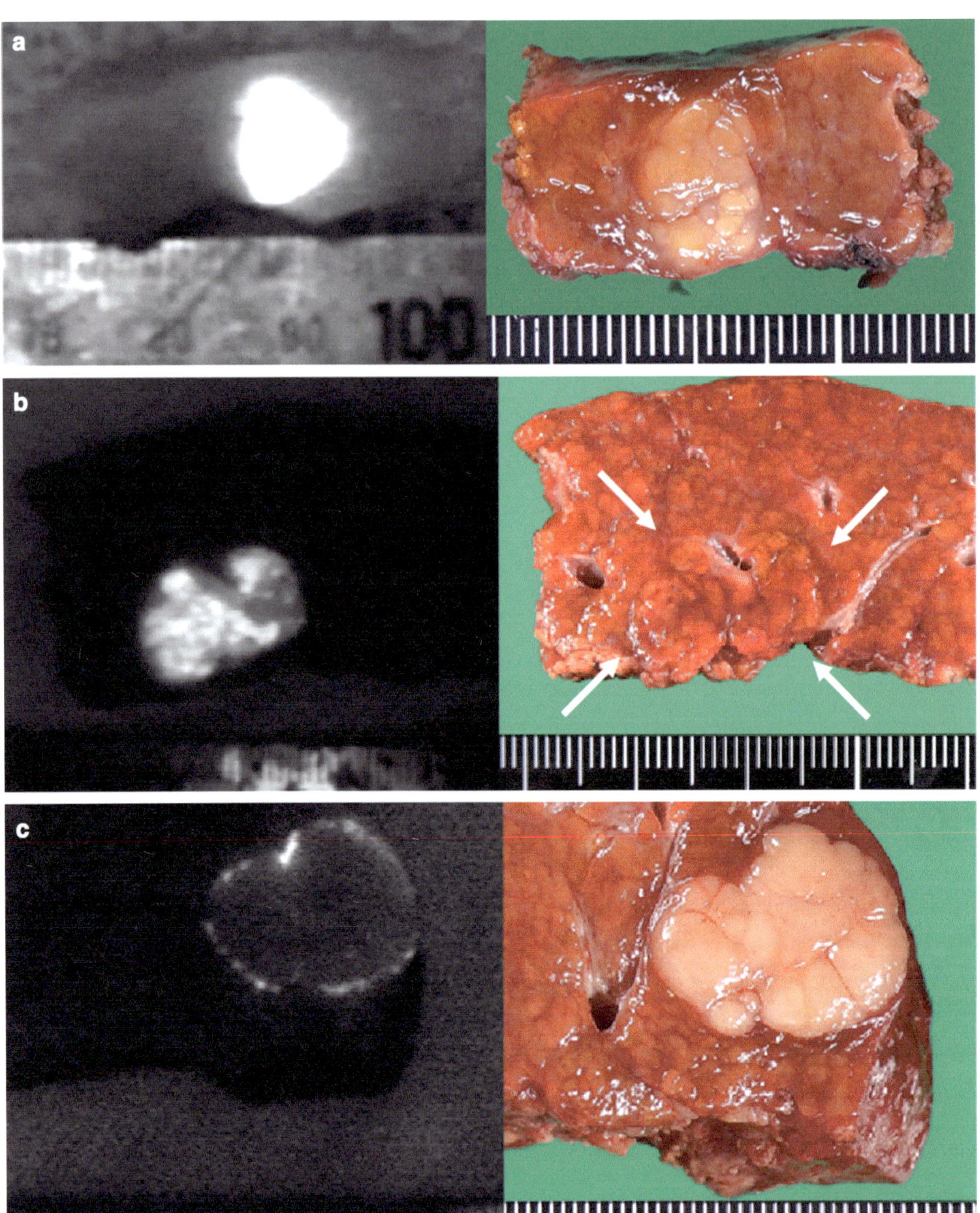

Fig. 15.1 Typical fluorescence imaging appearance of hepatocellular carcinoma. (**a**) Near-infrared (NIR) fluorescence imaging of a resected specimen showing uniform (total) intra-lesion emission pattern (*left panel*). Resected specimen contained grayish-white tumor 15 mm in diameter, with capsule and septum (*right panel*). (**b**) NIR fluorescence imaging of resected specimen showing uneven (partial) intra-lesion emission pattern, and its emission spots were heterogeneously distributed (*left panel*). Tumor was 15 mm in maximum diameter and its margin was indistinct (*arrow*, *right panel*). (**c**) NIR fluorescence imaging of resected specimen showing rim-like emission pattern. The emitted light arose from the rim of the tumor (*left panel*). Resected specimen contained grayish-white tumor 20 mm in diameter, with capsule and septum (*right panel*). With permission from Morita Y, Sakaguchi T, Unno N, Shibasaki Y, Suzuki A, Fukumoto K, et al. Detection of hepatocellular carcinomas with near-infrared fluorescence imaging using indocyanine green: its usefulness and limitation. Int J Clin Oncol 2013; 18(2):232–41© Springer [21]

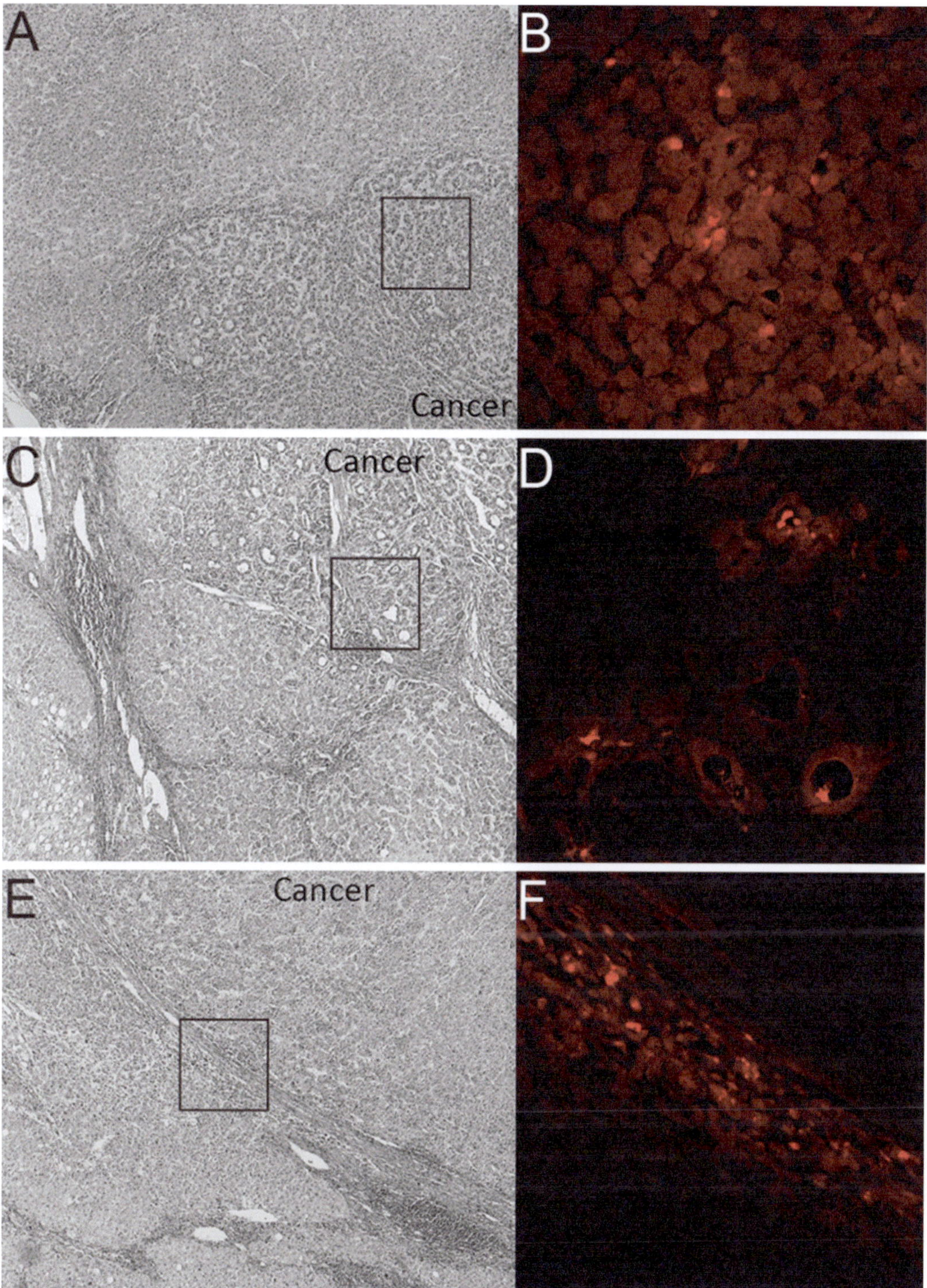

Fig. 15.2 Fluorescent microscopy findings of the resected specimen. (**a**) Photomicrograph of the resected specimen, which consisted of well-differentiated to moderately differentiated hepatocellular carcinoma (HCC) with trabecular structure. Hematoxylin staining, 40× at original magnification. (**b**) Fluorescent photomicrograph under near-infrared (NIR) showed clear emitted light (*red* in color) at the enclosed area in panel (**a**). Indocyanine green (ICG) localized in the cytoplasm of tumor cells, 200× at original magnification. (**c**) Photomicrograph of the resected specimen, which consisted of well-differentiated HCC with pseudoglandular structure. Hematoxylin staining, 40× at original magnification. (**d**) Fluorescent photomicrograph under NIR showed clear emitted light (*red* in color) at the enclosed area in panel (**c**). ICG localized in the pseudoglands of the tumor, 200× at original magnification. (**e**) Photomicrograph of the resected specimen, which consisted of moderately differentiated HCC with thick capsule. Hematoxylin staining, 40× at original magnification. (**f**) Fluorescent photomicrograph under NIR showed clear emitted light (*red* in color) at the enclosed area in panel (**e**). ICG localized in the cytoplasm of compressed adjacent parenchyma, 200× at original magnification

Intraoperative Detection of Tumors Using ICG Fluorography

In 3 reports [19–21], the intraoperative detection of preoperatively identified HCCs by surface observation with a NIR fluorescence imaging system was not always satisfactory. The detection rates in these studies using surface observation were 100 % (10/10), 51.2 % (21/41), and 61.8 % (47/76). Although NIR is capable of relatively deep penetration into biological tissues, the depth at which NIR light can be observed is about 5–10 mm [14]. According to Ishizawa's report [20], no tumor at a depth of more than 8 mm was identifiable. Therefore, the fluorescent detectability depends on the hepatic parenchymal thickness between the serosal surface and the tumors. This issue needs to be resolved to increase the intraoperative fluorescent detection rate of HCC. On the other hand, ICG fluorography enables the identification of new foci that may have been preoperatively and intraoperatively unidentifiable by conventional imaging modalities, visual inspection, and manual palpation. In particular, ICG fluorography is effective in shallow areas where ultrasonography is not practical. In cirrhotic livers, small HCCs, even those just beneath the surface, are not always palpable. Combining the modalities of intraoperative ICG fluorography and ultrasonography may improve detection rates.

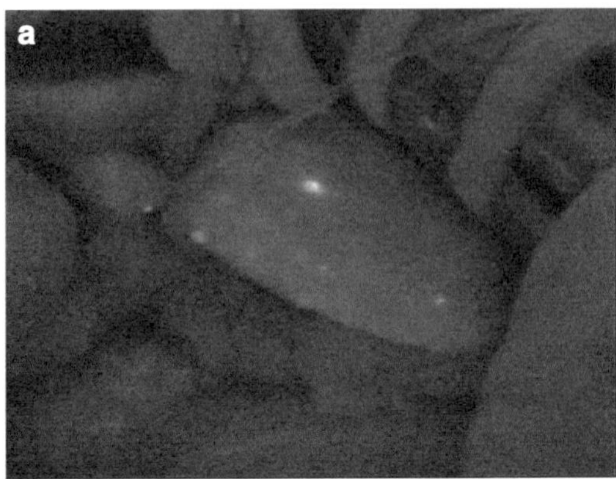

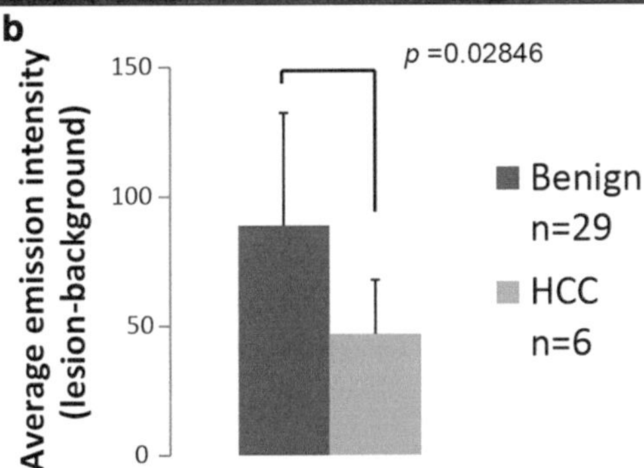

Fig. 15.3 Newly detected lesion under intraoperative NIR observation. (**a**) Newly detected emission spot, 3 mm in diameter, was intraoperatively detected on the surface of the liver under near-infrared (NIR) observation. (**b**) The calculated emission intensities in benign lesions and hepatocellular carcinomas (HCCs) using Scion image software are shown. With permission from Morita Y, Sakaguchi T, Unno N, Shibasaki Y, Suzuki A, Fukumoto K, et al. Detection of hepatocellular carcinomas with near-infrared fluorescence imaging using indocyanine green: its usefulness and limitation. Int J Clin Oncol 2013;18(2):232–41© Springer [21]

Newly Identified Foci During Hepatic Surface NIR Observation: Should We Resect?

We and two other study groups have reported that ICG fluorography could detect new lesions that could not be recognized using any other preoperative and intraoperative examinations (Fig. 15.3a) [19–21]. According to these reports, 26 (66.7 %) of 39 lesions larger than 3 mm were of the nodular lesions and 11 (47.8 %) of 23 larger than 5 mm were HCCs (Table 15.1). On the contrary, only 2 of 15 lesions smaller than 2 mm were histologically diagnosed as HCC. From our investigation, 21 of 29 benign lesions showed a homogenous intra-lesion emission pattern in the resected specimens. On the other hand, five of six newly identified HCCs showed an uneven intra-lesion emission pattern, and one had a homogeneous intra-lesion emission pattern. The uneven emission pattern rate was significantly higher in HCC ($p=0.01$) (Table 15.2). The calculated average emission intensities of HCCs were significantly lower than those of benign lesions (Fig. 15.3b). Interestingly, our results differed from those of Ishizawa [27]. They advocated that HCCs unidentifiable by gross inspection have higher emission intensities than non-cancerous lesions. Since these data were

Table 15.1 Newly detected lesions under ICG fluorography

Author (reference)	Histological diagnosis	Newly detected lesions (*N*)	Emission size (mm)
Gotoh [19]			
	HCC	4	3–6
	Benign lesion (DN, RN, normal liver)	4	3–5
Ishizawa [20]			
	HCC	8	2–19
	Benign lesion (RN, bile duct proliferation)	5	5–7
Morita [21]			
	HCC	6	2–5
	Benign lesion (DN, liver parenchyma, bile plug, cyst, inflammatory pseudotumor)	29	3–10

Histological diagnoses of newly detected small emission spots are listed

HCC hepatocellular carcinoma, *DN* dysplastic nodule, *RN* regenerative nodule

With permission from Gotoh K, Yamada T, Ishikawa O, Takahashi H, Eguchi H, Yano M, et al. A novel image-guided surgery of hepatocellular carcinoma by indocyanine green fluorescence imaging navigation. J Surg Oncol 2009;100(1):75–9. doi: 10.1002/jso.21272 © John Wiley & Sons, Inc [19], Ishizawa T, Fukushima N, Shibahara J, Masuda K, Tamura S, Aoki T, et al. Real-time identification of liver cancers by using indocyanine green fluorescent imaging. Cancer 2009;115(11):2491–504. doi: 10.1002/cncr.24291© John Wiley & Sons, Inc [20], and Morita Y, Sakaguchi T, Unno N, Shibasaki Y, Suzuki A, Fukumoto K, et al. Detection of hepatocellular carcinomas with near-infrared fluorescence imaging using indocyanine green: its usefulness and limitation. Int J Clin Oncol 2013;18(2):232–41© Springer [21]

Table 15.2 The emission pattern in newly detected HCCs and benign lesions

Emission pattern	*N*	Benign lesion	HCC
Uniform	22	21	1
Uneven	13	8	5
Total	35	29	6

HCC hepatocellular carcinoma, *N* number

With permission from Morita Y, Sakaguchi T, Unno N, Shibasaki Y, Suzuki A, Fukumoto K, et al. Detection of hepatocellular carcinomas with near-infrared fluorescence imaging using indocyanine green: its usefulness and limitation. Int J Clin Oncol 2013;18(2):232–41© Springer [21]

obtained from resected specimens, it is not clear whether these results are directly applicable to surface observation. At present, we cannot directly distinguish malignancies from small benign lesions by surface NIR observation. Further prospective studies are needed to determine whether incidentally found foci during hepatic surface NIR observation require resection.

Usefulness of ICG Fluorography for Detection of Extrahepatic Metastasis and Residual Tumors

Satou et al. presented the usefulness of ICG fluorography for detecting extrahepatic metastases of HCCs [29]. According to their report, intraoperative ICG fluorography identified 19 of 23 metastasis-suspected lesions, and all were pathologically HCCs. ICG fluorography can identify extrahepatic metastases because metastatic HCC cells can uptake ICG from the blood stream but cannot excrete it due to the lack of an adjacent biliary system. In addition, five preoperatively unidentified metastatic lesions were newly detected by intraoperative ICG fluorography in their report. Therefore, ICG fluorography may help avoid overlooking metastatic lesions in surgical fields and may benefit patients with HCC metastases. Presumably, through the application of ICG fluorography, the 1-year survival rate (85 %) in their study was relatively satisfactory.

References

1. Jemal A, Bray F, Center MM, Ferlay J, Ward E, Forman D. Global cancer statistics. CA Cancer J Clin. 2011;61(2):69–90.
2. Heiken JP, Weyman PJ, Lee JK, Balfe DM, Picus D, Brunt EM, et al. Detection of focal hepatic masses: prospective evaluation with CT, delayed CT, CT during arterial portography, and MR imaging. Radiology. 1989;171(1):47–51.
3. Quaia E, Calliada F, Bertolotto M, Rossi S, Garioni L, Rosa L, et al. Characterization of focal liver lesions with contrast-specific US modes and a sulfur hexafluoride-filled microbubble contrast agent: diagnostic performance and confidence. Radiology. 2004; 232(2):420–30.

4. Hammerstingl R, Huppertz A, Breuer J, Balzer T, Blakeborough A, Carter R, et al. Diagnostic efficacy of gadoxetic acid (Primovist)-enhanced MRI and spiral CT for a therapeutic strategy: comparison with intraoperative and histopathologic findings in focal liver lesions. Eur Radiol. 2008;18(3):457–67.
5. Mitsunori Y, Tanaka S, Nakamura N, Ban D, Irie T, Noguchi N, et al. Contrast-enhanced intraoperative ultrasound for hepatocellular carcinoma: high sensitivity of diagnosis and therapeutic impact. J Hepatobiliary Pancreat Sci. 2013;20(2):234–42.
6. Polom K, Murawa D, Rho YS, Nowaczyk P, Hunerbein M, Murawa P. Current trends and emerging future of indocyanine green usage in surgery and oncology: a literature review. Cancer. 2011;117(21):4812–22.
7. Hochheimer BF. Angiography of the retina with indocyanine green. Arch Ophthalmol. 1971;86(5):564–5.
8. Ganz W, Donoso R, Marcus HS, Forrester JS, Swan HJ. A new technique for measurement of cardiac output by thermodilution in man. Am J Cardiol. 1971;27(4):392–6.
9. Caesar J, Shaldon S, Chiandussi L, Guevara L, Sherlock S. The use of indocyanine green in the measurement of hepatic blood flow and as a test of hepatic function. Clin Sci. 1961;21:43–57.
10. Benson RC, Kues HA. Fluorescence properties of indocyanine green as related to angiography. Phys Med Biol. 1978;23(1):159–63.
11. Unno N, Inuzuka K, Suzuki M, Yamamoto N, Sagara D, Nishiyama M, et al. Preliminary experience with a novel fluorescence lymphography using indocyanine green in patients with secondary lymphedema. J Vasc Surg. 2007;45(5):1016–21.
12. Taggart DP, Choudhary B, Anastasiadis K, Abu-Omar Y, Balacumaraswami L, Pigott DW. Preliminary experience with a novel intraoperative fluorescence imaging technique to evaluate the patency of bypass grafts in total arterial revascularization. Ann Thorac Surg. 2003;75(3):870–3.
13. Raabe A, Beck J, Gerlach R, Zimmermann M, Seifert V. Near-infrared indocyanine green video angiography: a new method for intraoperative assessment of vascular flow. Neurosurgery. 2003;52(1):132–9. discussion 9.
14. Kitai T, Inomoto T, Miwa M, Shikayama T. Fluorescence navigation with indocyanine green for detecting sentinel lymph nodes in breast cancer. Breast Cancer. 2005;12(3):211–5.
15. Kusano M, Tajima Y, Yamazaki K, Kato M, Watanabe M, Miwa M. Sentinel node mapping guided by indocyanine green fluorescence imaging: a new method for sentinel node navigation surgery in gastrointestinal cancer. Dig Surg. 2008;25(2):103–8.
16. Mitsuhashi N, Kimura F, Shimizu H, Imamaki M, Yoshidome H, Ohtsuka M, et al. Usefulness of intraoperative fluorescence imaging to evaluate local anatomy in hepatobiliary surgery. J Hepatobiliary Pancreat Surg. 2008;15(5):508–14.
17. Ishizawa T, Bandai Y, Kokudo N. Fluorescent cholangiography using indocyanine green for laparoscopic cholecystectomy: an initial experience. Arch Surg. 2009;144(4):381–2.
18. Sakaguchi T, Suzuki A, Unno N, Morita Y, Oishi K, Fukumoto K, et al. Bile leak test by indocyanine green fluorescence images after hepatectomy. Am J Surg. 2010;200(1):e19–23.
19. Gotoh K, Yamada T, Ishikawa O, Takahashi H, Eguchi H, Yano M, et al. A novel image-guided surgery of hepatocellular carcinoma by indocyanine green fluorescence imaging navigation. J Surg Oncol. 2009;100(1):75–9. doi:10.1002/jso.21272.
20. Ishizawa T, Fukushima N, Shibahara J, Masuda K, Tamura S, Aoki T, et al. Real-time identification of liver cancers by using indocyanine green fluorescent imaging. Cancer. 2009;115(11):2491–504. doi:10.1002/cncr.24291.
21. Morita Y, Sakaguchi T, Unno N, Shibasaki Y, Suzuki A, Fukumoto K, et al. Detection of hepatocellular carcinomas with near-infrared fluorescence imaging using indocyanine green: its usefulness and limitation. Int J Clin Oncol. 2013;18(2):232–41.
22. de Graaf W, Hausler S, Heger M, van Ginhoven TM, van Cappellen G, Bennink RJ, et al. Transporters involved in the hepatic uptake of (99 m)Tc-mebrofenin and indocyanine green. J Hepatol. 2011;54(4):738–45.
23. Zollner G, Wagner M, Fickert P, Silbert D, Fuchsbichler A, Zatloukal K, et al. Hepatobiliary transporter expression in human hepatocellular carcinoma. Liver Int. 2005;25(2):367–79.
24. Kitao A, Matsui O, Yoneda N, Kozaka K, Shinmura R, Koda W, et al. The uptake transporter OATP8 expression decreases during multistep hepatocarcinogenesis: correlation with gadoxetic acid enhanced MR imaging. Eur Radiol. 2011;21(10):2056–66.
25. Huang L, Vore M. Multidrug resistance p-glycoprotein 2 is essential for the biliary excretion of indocyanine green. Drug Metab Dispos. 2001;29(5):634–7.
26. Kusuhara H, Sugiyama Y. Pharmacokinetic modeling of the hepatobiliary transport mediated by cooperation of uptake and efflux transporters. Drug Metab Rev. 2010;42(3):539–50.
27. Ishizawa T, Masuda K, Urano Y, Kawaguchi Y, Satou S, Kaneko J, et al. Mechanistic background and clinical applications of indocyanine green fluorescence imaging of hepatocellular carcinoma. Ann Surg Oncol. 2014;21(2):440–8. doi:10.1245/s10434-013-3360-4.
28. Harada N, Ishizawa T, Muraoka A, Ijichi M, Kusaka K, Shibasaki M, et al. Fluorescence navigation hepatectomy by visualization of localized cholestasis from bile duct tumor infiltration. J Am Coll Surg. 2010;210(6):e2–6.
29. Satou S, Ishizawa T, Masuda K, Kaneko J, Aoki T, Sakamoto Y, et al. Indocyanine green fluorescent imaging for detecting extrahepatic metastasis of hepatocellular carcinoma. J Gastroenterol. 2013;48(10):1136–43.

Identification of Malignant Tumors in the Liver

16

Henricus J.M. Handgraaf, Floris P.R. Verbeek, Cornelis J.H. van de Velde, Merlijn Hutteman, and Alexander L. Vahrmeijer

Introduction

Prognosis and survival of patients with cancer deteriorates dramatically when metastases in the liver are present. Most metastases in the liver originate from colorectal cancer. 14.5 % of colorectal cancer patients have liver metastases at the moment of diagnosis and 15.2 % will develop liver metastases within 5 years after curative resection of the primary tumor [1]. Metastases are the leading cause of cancer-related death, with no survivors at 5 years if left untreated [2]. Besides advancements in radiotherapy (neoadjuvant), chemotherapy and preoperative imaging, improvements in surgical techniques have led to improved prognosis and survival rate. In the past decades, hepatic metastasectomy has evolved from a high-risk procedure to a more commonly performed curative treatment. Surgical resection is currently the only potential and is regarded as standard-of-care, but is only possible in patients with sufficient functional liver reserve, without unresectable extrahepatic disease and when liver metastases can be resected with a tumor-free margin. Unfortunately, only 10–20 % of patients are potential candidates for curative resection. Even with strict patient selection, intrahepatic recurrence rates after hepatic metastasectomy remain high, ranging from 11 to 28 %, of which 78 % occurs within 1 year [3, 4]. Factors influencing the recurrence rate of colorectal cancer liver metastases are positive resection margins, extrahepatic disease, node-positive primary tumor, disease-free interval from primary tumor to metastases less than 12 months, more than one hepatic tumor, cases where the largest hepatic tumor is more than 5 cm, and a serum carcinoembryonic antigen level of more than 200 ng/ml [5]. The high recurrence rate could partly be explained by inability of available technologies to detect the presence of micrometastases during liver metastasectomy. In addition, positive resection margins (R1 and R2) are still a major issue, with reported rates ranging from 11 to 23 % [5, 6]. In the preoperative setting, imaging technologies such as ultrasonography, computed tomography (CT), magnetic resonance imaging (MRI), and positron emission tomography (PET) are used for surgical planning. However, these technologies are less suitable for intraoperative use and are unable to provide real-time surgical guidance. In addition, technologies available during surgery,

Electronic supplementary material: The online version of this chapter (doi:10.1007/978-3-319-15678-1_16) contains supplementary material, which is available to authorized users. Videos can also be accessed at http://link.springer.com/chapter/10.1007/978-3-319-15678-1_16.

H.J.M. Handgraaf, M.D. • F.P.R. Verbeek, M.Sc.
C.J.H. van de Velde, M.D., Ph.D. • M. Hutteman, M.D., Ph.D. • A.L. Vahrmeijer, M.D., Ph.D. (✉)
Department of Surgery, Leiden University Medical Centre, Albinusdreef 2, Leiden 2333ZA, The Netherlands
e-mail: h.j.m.handgraaf@lumc.nl; f.p.r.verbeek@lumc.nl; c.j.h.van_de_velde@lumc.nl; m.hutteman@lumc.n; a.l.vahrmeijer@lumc.nl

F.D. Dip et al. (eds.), *Fluorescence Imaging for Surgeons: Concepts and Applications*, DOI 10.1007/978-3-319-15678-1_16,

such as intraoperative ultrasonography (IOUS), visual inspection and palpation by the surgeon, are less suitable for the detection of small lesions [7]. Recently, the use of near-infrared (NIR) fluorescence imaging has emerged as an additional technique for real-time detection of small and superficial hepatic metastases (Video 16.1). This chapter focuses on the development and current applications of NIR fluorescence imaging in identifying metastases in the liver.

Conventional Technologies for Identification of Metastases in the Liver

Several imaging technologies can be used to detect hepatic abnormalities, though sensitivity and specificity can vary greatly. Among these technologies are ultrasound, CT, MRI, fluorodeoxyglucose positron emission tomography (FDG-PET), and FDG-PET/CT. CT is considered to be the preferred technology for staging of liver disease in the majority of hospitals, because it provides good coverage of the liver and, if necessary, the abdomen and thorax in the same scan [8]. However, CT is relatively expensive and requires ionizing radiation. In addition, lesions only several millimeters small are frequently missed [7]. MRI has a higher detection rate of lesions smaller than 10 mm compared to CT, with a sensitivity of 60.2 % and 47.3 % respectively [9]. However, for both modalities, this is still unacceptably low. Diagnostic accuracy of FDG-PET is strongly affected by neoadjuvant chemotherapy; treatment resulted in a drop in sensitivity from 81.3 to 54.5 % [10].

All these preoperative imaging modalities have great value for planning surgical procedures, but the hands and eyes of the surgeon combined with IOUS are still the most important tools during surgery. Hata et al. analyzed retrospectively a prospectively collected and recorded database of intraoperative detected colorectal liver metastases [11]. A total of 270 new metastases were detected in 183 consecutive patients. Intraoperative palpation and/or visual inspection detected 77 % (207/270) of the new lesions. IOUS also showed to be of great value. 12 % (33/270) of the new detected metastases was detected solely by IOUS. The remaining 11 % (30/270) were found in removed tissue. Most of the newly found metastases were located on or less than 1 cm underneath the liver surface. Although this obviously shows the benefit of IOUS, palpation and visual inspection, small tumors are still missed by all detection methods. Nomura et al. histologically examined resected liver specimen and concluded that IOUS misses 25 % of liver tumors smaller than 5 mm [7]. The same rate applies to palpation and visual inspection, although, sensitivity may vary greatly, e.g. due to the experience of surgeons and size and depth of tumors. In conclusion, there is a strong need for technologies which can intraoperatively detect small and superficial metastases in the liver. During minimally invasive liver resections, surgeons are deprived of tactile information and their visibility is hampered, increasing the need for additional imaging modalities.

Intraoperative NIR Fluorescence Imaging

Near-infrared (NIR) fluorescence imaging is a promising technique that can be used to identify tumors and vital structures during surgery [12]. It uses light with wavelengths between 700 and 900 nm, which is invisible to the naked eye. Advantages of NIR light include high tissue penetration and low autofluorescence. NIR imaging does not alter the surgical field, as it is invisible. It is safe to use as no ionizing radiation is used, and no tissue contact is needed. As NIR fluorescence images can be acquired in real-time (milliseconds), it allows surgeons to operate under direct image-guidance.

Several NIR fluorescence imaging systems are already commercially available for both open and laparoscopic surgery [13, 14]. The demand for minimally invasive liver resection increases due to exponential growth of patients eligible for hepatic metastasectomies. Laparoscopic liver resection shows a more favorable outcome than open liver resection with regard to complications, hospital stay and blood loss [15]. However, minimally invasive surgery also limits visualization and palpability of the liver surface. NIR fluorescence imaging could therefore contribute to this field of surgery by providing additional information. Several NIR fluorescence imaging systems can simultaneously acquire the NIR fluorescence signal and color video signal, enabling a real-time overlay of NIR fluorescence signal and enhance anatomical orientation. Ishizawa et al. were the first to demonstrate the safety and convenience of a prototype fluorescent imaging system during laparoscopic hepatectomy in a patient with hepatocellular carcinoma (HCC) and underlying chronic hepatitis C [16]. Laparoscopic NIR fluorescence imaging facilitated visual inspection by clearly delineating the tumor on the liver surface. Tummers et al. showed that laparoscopic NIR fluorescence imaging identified additional uveal melanoma metastases in the liver in 2 out of 3 patients [17].

Near-Infrared Fluorescent Contrast Agents

NIR fluorescence imaging of metastases in the liver can be challenging due to hepatic uptake and clearance of many fluorescent agents, resulting in fluorescent liver tissue and, hence, invisible metastases due to high background fluorescence. Furthermore, liver tissue absorbs a higher proportion of NIR fluorescent light compared to other human tissues, such as colon and breast, resulting in a lower signal [18]. Fluorophores emitting fluorescence with longer wavelenghts achieve deeper tissue penetration. However, wavelengths longer than 900 nm suffer from higher absorption by for example blood. An ideal fluorescent agent for hepatic metastases should be tumor-specific, cause low background fluorescence by being cleared renally, and emit light within the NIR fluorescence window (700–900 nm). To date, no such fluorescent agent is available for clinical use. Methylene blue (MB) and indocyanine green (ICG) are currently the only suitable NIR fluorescence agent approved by the Food and Drug Administration (FDA) and the European Medicines Agency (EMA). MB is cleared simultaneously by liver and kidneys, while ICG is cleared solely by liver. Hepatic metastases are visualized by non-tumor-targeted dyes when this dye is cleared by healthy hepatocytes, while it retains in cancerous tissue. The fluorescent properties of MB make it a less favorable fluorophore; its peak emitted fluorescence wavelength is 700 nm and therefore subject to higher background autofluorescence, more light absorption and less tissue penetration capacity. No study showed the capability of MB to identify metastases in the liver. The peak emitted fluorescence wavelength of ICG is 810 nm, making it a better candidate for NIR fluorescence imaging. In addition, since ICG is cleared exclusively by the liver, it results in a stronger signal in the liver compared to MB. Extensive medical experience with ICG already exists, due to its use in a broad spectrum of other clinical applications, e.g. assessing coronary artery bypass graft patency, retinal angiography and liver function. Since ICG contains iodine, iodine allergy and thyrotoxicosis are contraindications for its use. ICG is not metabolized, is cleared exclusively by the liver and does not undergo enterohepatic recirculation, making it an ideal candidate for detecting liver dysfunction [19].

Tumor-Targeting Dyes

Both ICG and MB are non-targeted dyes and the chemical structures do not readily allow conjugation to tissue-specific ligands. The lack of other clinically available NIR fluorescent agents is a considerable limitation, since ICG is only capable of imaging metastases inside the liver; extrahepatic metastases will not show any fluorescent signal. In contrast, tumor-targeted dyes do have this property and offer therefore great advantages. NIR fluorescence imaging of metastases in the liver is challenging due to hepatic uptake and clearance of many fluorescent dyes. This may result in unfavorable tumor-to-liver ratio (TLR). For identification, tumorous tissue has to be more fluorescent than its background. Several academic and commercial parties are currently developing tumor-targeting dyes. Integrin $\alpha_v\beta_3$ is a potential target, as it shows overexpression in various cancer types, such as colorectal, ovarian and breast cancer, but low expression in hepatocytes [20]. Several integrin $\alpha_v\beta_3$ targeting agents have been studied, such as IntegriSense680 (Perkin Elmer, Waltham, Massachusetts), targeting integrin $\alpha_v\beta_3$. Hutteman et al. demonstrated the feasibility for detecting colorectal metastases using IntegriSense680 in a syngeneic rat model [21]. Another potential target is matrix metalloproteinase 2 (MMP-2), which is associated with metastatic capacity of colorectal cancer. By using the NIR fluorescence probe Cy5.5-C6 in mice with induced colorectal cancer, tumors with increased expression of MMP-2 were successfully imaged [20]. Another excellent target for colorectal cancer is epidermal growth factor receptor (EGFR). Cetuximab, a human antibody binding specifically to EGFR, coupled with Cy5.5 (cetuximab-Cy5.5), resulted in the accurate detection of tumors [23]. Much more tumor-targeted dyes are currently under development and reviewed by Luo et al. [24]. However, as fluorescent dyes are excreted by liver and kidneys, this may result in high background signal. Therefore, optimal TLR should be studied for each new dye. Tumor-targeted dyes can potentially contribute during liver metastasectomy by visualizing not only intrahepatic, but, in contrast to non-targeted dyes, also extrahepatic tumors.

Dose and Timing of ICG Administration

After intravenous injection, ICG is absorbed by hepatocytes and eventually excreted into the bile. Therefore, the first period after intravenous administration results in highly fluorescent liver tissue, i.e. unfavorable TLR. Timing of administration before surgery is therefore crucial to reach an optimal TLR. Fluorescent intensity of the liver strongly decreases after 24 h, although liver dysfunction may influence this rate [25]. In patients with an unfavorable ICG retention rate, e.g. in cirrhosis, steatosis and after chemotherapy, the fluorescence signal of noncancerous liver parenchyma is higher, which makes it more challenging to obtain an adequate TLR [26]. To determine optimal dose and timing of ICG administration, van der Vorst et al. performed a preclinical study with syngeneic rats with colorectal liver metastases [25]. The highest TLR was achieved 72 h after intravenous injection. No significant effects were observed regarding doses, although a trend favoring 0.25 mg per kilogram body weight (extrapolated to humans) was shown. Ishizwa et al. suggested that the interval between ICG administration and surgery should be longer than at least 2 days to obtain an optimal TLR, especially in patients with advanced cirrhosis [24]. However, as in most other studies, the used dose was 0.5 mg per kilogram body weight. In a clinical trial, administration 24 and 48 h before surgery as well as doses of 10 and 20 mg ICG all showed sufficient TLR and no significant difference in TLR [27]. A dose of 10 mg ICG administered 24 h prior to surgery is therefore advised.

NIR Fluorescence Imaging of Metastatic Liver Cancer

NIR fluorescence imaging of hepatic tumors started after the incidental finding that hepatocellular carcinoma (HCC) shows a very strong fluorescent signal in patients who had been given ICG several days prior to surgery as a routine preoperative liver function test. In a subsequent study by Gotoh et al., all primary HCCs in 10 patients were identified as bright NIR fluorescent lesions and could be removed completely [28]. In addition, 4 new HCC nodules that were not detected by any preoperative examinations or IOUS were detected due to the use of NIR fluorescence imaging. Harada et al. were the first to demonstrate the feasibility during hepatectomy in a patient with colorectal hepatic metastasis [29]. Several other clinical studies followed, describing the detection of both HCC and metastatic liver cancer (Table 16.1) [26, 27, 29–34]. A total of 167 patients with colorectal or pancreatic liver metastases have been included in 8 studies. In these patients, ICG has been shown to accumulate as a rim around the tumor (Figs. 16.1 and 16.2). This pattern differs from primary hepatic cancers, e.g. HCC, where ICG shows

Table 16.1 List of studies using near-infrared fluorescence imaging in surgery for hepatic metastases

Study	Number of patients	Preoperative diagnosis	Imaging system	Dose of ICG	Injection site	Time between injection and imaging	Intraoperative IR (tumors)	Additional metastases identified	False-positive lesions	Smallest tumor size
Harada, 2010 [29]	3	ICC ($n=2$); CLM ($n=1$)	PDE	0.5 mg/kg	i.v.	4 days (1, 2) and 2 days (3)	3/3	–	0	20 mm
Ishizawa, 2009 [26]	49[a]	HCC ($n=37$); CLM ($n=12$)	PDE	0.5 mg/kg	i.v.	1–7 days for HCC and 1–14 days for CLM	21/41 HCCs and 16/16 CLM[b]	+	5	2 mm
Kasuya, 2011 [31]	1	CLM	PDE	500 μl mixed with ethanol	Locally	NA	NA	–	0	3 mm
Uchiyama, 2010 [33]	32	CLM	PDE	0.5 mg/kg	i.v.	<2 weeks	NA	+	2	4 mm
Yokoyama, 2012 [34]	49	PCM	PDE	25 mg	i.v.	1 day	NA	+	5	1.5 mm
Ishizuka, 2012 [30]	7	CLM	PDE	0.1 ml/kg	NA	NA	26/26	+	1	NA
Peloso, 2012 [32]	25	CLM	PDE	0.5 mg/kg	i.v.	24 h	NA/77	+	1	3 mm
van der Vorst, 2013 [27]	40	CLM	Mini-FLARE	10 and 20 mg	i.v.	24 and 48 h	71/97	+	0	1 mm\
Tummers, 2015 [17]	3	UMM	Karl Storz Fluorescence laparoscope	10 mg	i.v.	24h	NA	+	+	1 mm

HCC = Hepatocellular carcinoma; CLM = Colorectal Liver Metastases; ICC = Intrahepatic cholangiocarcinoma; PCM = pancreatic cancer metastasis; PDE = Photo Dynamic Eye; FLARE = Fluorescence-Assisted Resection and Exploration; NA = not available; IR = identification rate; i.v. = intravenous; UMM = Uveal Melanoma Metastases

[a]26 of 49 Patients (20 with HCC and 6 with CLM) underwent near-infrared fluorescence imaging during surgery

[b]Identification rate of 26 patients who were examined with near-infrared fluorescence imaging during surgery

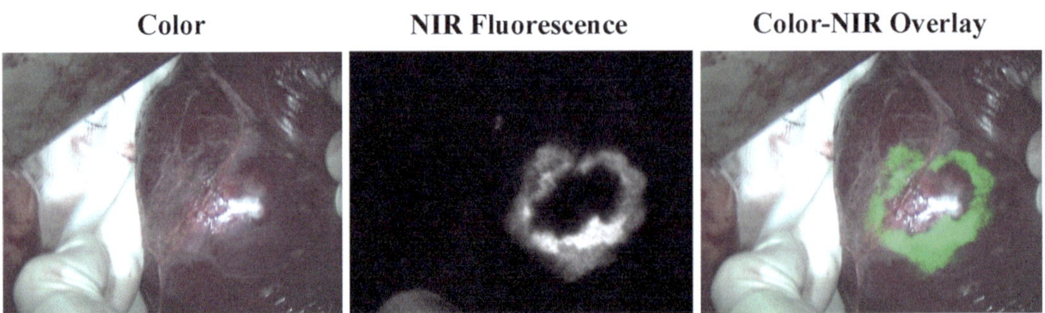

Fig. 16.1 In vivo near-infrared fluorescence imaging of a hepatic metastasis using the Mini-FLARE system (Beth Israel Deaconess Hospital, Boston, USA). A rim of fluorescence around the lesion is shown

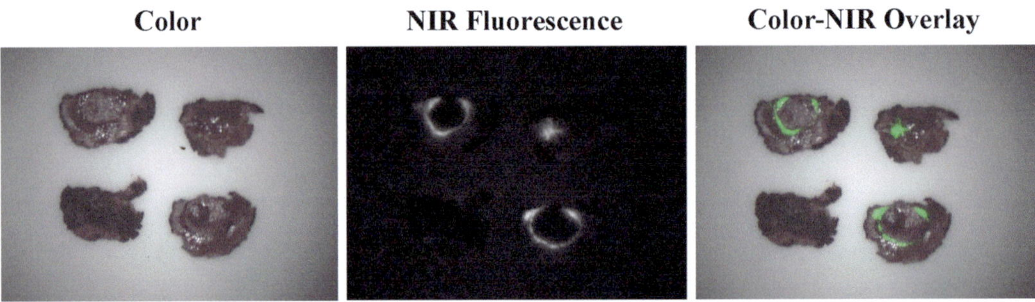

Fig. 16.2 Ex vivo near-infrared fluorescence imaging of resected and sliced lesion at pathology department using the FLARE imaging system (Beth Israel Deaconess Hospital, Boston, USA)

total or partial fluorescence of the tumor [26]. Since ICG is removed from circulation exclusively by the liver, the rim pattern is influenced by clearance by the liver and biliary drainage. Ishizawa et al. showed microscopically that fluorescence did not exist in metastatic tissue itself, but in surrounding noncancerous liver tissue compressed by the tumor [26]. Compression by malignant liver tumors not only leads to obstructed bile canaliculi, but also to changes in the liver parenchyma due to inflammation, ductular transformation and increased presence of immature hepatocytes [35]. Compared to well-differentiated hepatocytes, immature hepatocytes display less expression of organic anion transporters [36]. Multidrug resistance P-glycoprotein 2 (Mdr2), an organic anion transporter in the hepatocyte canalicular membrane, is essential for the transport of certain hydrophobic organic anions such as ICG and thus for its excretion. In its absence, ICG excretion is reduced by 90 % [37]. The rim pattern of fluorescence can therefore be explained by the fact that ICG can be transported from circulation into immature hepatocytes, but is retained intracellular and not cleared into the bile canaliculi. The rim pattern is specific for malignant lesions. Van der Vorst et al. could differentiate 25 benign lesions (8 hemangiomas, 13 cysts, and 4 bile duct hamartomas) from malignant lesions by a lack of a fluorescent signal rim (Fig. 16.3). False-negative results have not been reported so far, resulting in a sensitivity of 100 % on resected tissue ex vivo. However, several studies describe a combined total of 14 false-positive lesions, among which 4 large regenerative nodules and 1 bile duct proliferation [26, 30, 32–34]. The incidence and characteristics of false-positive lesions should be clarified in larger study populations.

One of the great challenges in NIR fluorescence imaging is still its limited capability to penetrate human tissue. None of the above-described

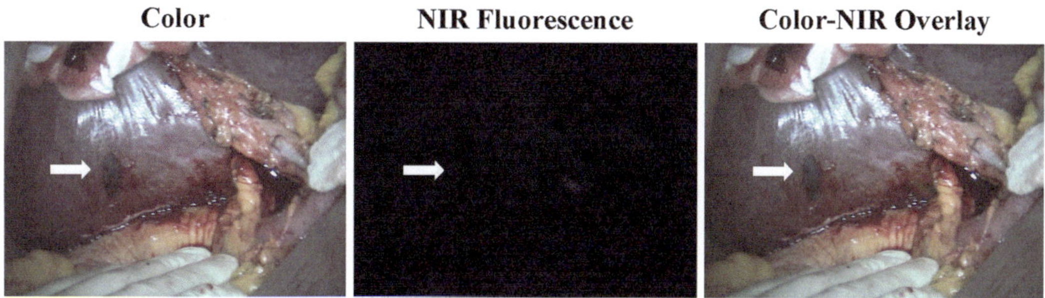

Fig. 16.3 Benign lesions (*arrow*) can be differentiated from malignant lesions by a lack of a near-infrared fluorescent rim around the lesion

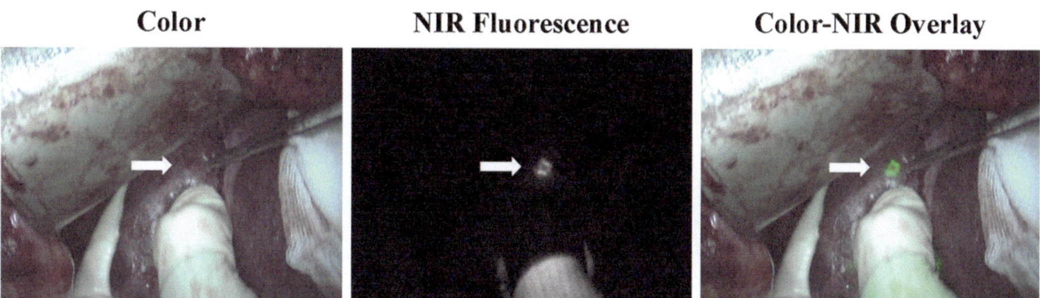

Fig. 16.4 Small, superficial, otherwise occult metastases are identified by near-infrared fluorescence imaging (*arrow*)

studies reported the ability to detect metastases more than 8 mm below the liver capsule. For colorectal liver metastases, however, this technique is very useful, as colorectal liver metastases are mostly located on the surface of liver parenchyma. Deeper localized metastases also show a fluorescent rim after resection and sectioning (Fig. 16.2) [38], but detection of these tumors still needs conventional technologies. Although this is a great limitation, NIR fluorescence imaging also has major advantages. Conventional imaging technologies easily miss superficial liver metastases smaller than 10 mm [7, 9]. In NIR fluorescence imaging, the bright signal enables surgeons to detect lesions as small as 1 mm in real-time (Fig. 16.4) [7, 27]. Indeed, additional, otherwise undetectable metastases were identified in 7 studies [26–28, 30, 32–34]. Van der Vorst et al. report identification of otherwise undetectable liver metastases in 5 of 40 patients (12.5 %, 95 % CI = 5.0–26.6) [27]. Combining contrast-enhanced IOUS and NIR fluorescence imaging together with CT and MRI, Uchhiyama et al. improved diagnostic sensitivity in 32 consecutive patients from 88.5 % to 98.1 % (P=0.050) [33]. Yokoyama et al. showed the potential of using NIR fluorescence imaging by screening the hepatic surface during pancreatic surgery with curative intent [34]. In 49 patients without preoperative detected hepatic metastases, 13 abnormal fluorescent lesions were detected without any apparent tumor, and 8 of them contained micrometastases. Within 6 months after surgery, 10 patients with abnormal fluorescence developed hepatic metastases, versus 1 of 36 patients in the fluorescence negative group, resulting in a positive predictive value of 77 % and a negative predictive value of 97 %. This outcome could possibly help surgeons select patients for either curative or palliative treatment.

Other Applications of NIR Fluorescence Imaging During Hepatectomy

Besides tumor demarcation, excretion of ICG into the bile can be used for real-time cholangiography of the biliary anatomy. Iatrogenic damage of bile ducts is a major issue in liver surgery. Bile leakage after hepatic resection is associated with high risk for liver failure and postoperative mortality [39]. During cholangiography, the feasibility is shown by Ishizwa et al. [40]. Potentially, bile leakage could also be detected during resection of hepatic metastasis, though no study reports the feasibility. Another novel and intraoperative technique is the use of ICG for image-guided liver segment identification for anatomical hepatic resection. Aoki et al. performed a study in 35 patients who underwent hepatectomy for hepatic malignancy [41]. After administration of 5 mg ICG into the portal vein, stained subsegments and segments of the liver were identified in 33 of 35 patients. Both cholangiography and identification of liver segments are very usable during hepatic metastectomy and are described in Chap. 27.

Conclusion

Surgical resection is currently the only potentially curative therapy and standard-of-care for metastases in the liver in selected patients. The high recurrence rate demands a technology which is able to detect lesions that are otherwise undetected by current visualization methods. Current available literature suggests an important complementary role for intraoperative NIR fluorescence imaging in the detection of metastases and primary tumors in the liver. It enables the identification of otherwise undetectable metastases as small as 1 mm and localized on or up to 8 mm below the liver surface. Although several studies show promising results, larger clinical trials are required to truly validate the benefit for patient requiring partial liver resection. In addition, optimization of NIR fluorescent-contrast agents and imaging systems is necessary before NIR fluorescence imaging becomes standard-of-care.

References

1. Manfredi S, Lepage C, Hatem C, Coatmeur O, Faivre J, Bouvier AM. Epidemiology and management of liver metastases from colorectal cancer. Ann Surg. 2006;244(2):254–9.
2. Goslin R, Steele Jr G, Zamcheck N, Mayer R, MacIntyre J. Factors influencing survival in patients with hepatic metastases from adenocarcinoma of the colon or rectum. Dis Colon Rectum. 1982;25(8): 749–54.
3. Abdalla EK, Vauthey JN, Ellis LM, Ellis V, Pollock R, Broglio KR, et al. Recurrence and outcomes following hepatic resection, radiofrequency ablation, and combined resection/ablation for colorectal liver metastases. Ann Surg. 2004;239(6):818–25.
4. Karanjia ND, Lordan JT, Fawcett WJ, Quiney N, Worthington TR. Survival and recurrence after neoadjuvant chemotherapy and liver resection for colorectal metastases: a ten year study. Eur J Surg Oncol. 2009;35(8):838–43.
5. Fong Y, Fortner J, Sun RL, Brennan MF, Blumgart LH. Clinical score for predicting recurrence after hepatic resection for metastatic colorectal cancer: analysis of 1001 consecutive cases. Ann Surg. 1999;230(3):309–18.
6. Tranchart H, Chirica M, Faron M, Balladur P, Lefevre LB, Svrcek M, et al. Prognostic impact of positive surgical margins after resection of colorectal cancer liver metastases: reappraisal in the era of modern chemotherapy. World J Surg. 2013;37(11): 2647–54.
7. Nomura K, Kadoya M, Ueda K, Fujinaga Y, Miwa S, Miyagawa S. Detection of hepatic metastases from colorectal carcinoma: comparison of histopathologic features of anatomically resected liver with results of preoperative imaging. J Clin Gastroenterol. 2007; 41(8):789–95.
8. Bipat S, Niekel MC, Comans EF, Nio CY, Bemelman WA, Verhoef C, et al. Imaging modalities for the staging of patients with colorectal cancer. Neth J Med. 2012;70(1):26–34.
9. Niekel MC, Bipat S, Stoker J. Diagnostic imaging of colorectal liver metastases with CT, MR imaging, FDG PET, and/or FDG PET/CT: a meta-analysis of prospective studies including patients who have not previously undergone treatment. Radiology. 2010; 257(3):674–84.
10. van Kessel CS, Buckens CF, van den Bosch MA, van Leeuwen MS, Hillegersberg H, Verkooijen HM. Preoperative imaging of colorectal liver metastases after neoadjuvant chemotherapy: a meta-analysis. Ann Surg Oncol. 2012;19(9):2805–13.
11. Hata S, Imamura H, Aoki T, Hashimoto T, Akahane M, Hasegawa K, et al. Value of visual inspection, bimanual palpation, and intraoperative ultrasonography during hepatic resection for liver metastases of colorectal carcinoma. World J Surg. 2011;35(12): 2779–87.

12. A. L. Vahrmeijer, M. Hutteman, J. R. van der Vorst, C. J. van de Velde, J. V. Frangioni, Nat. Rev. Clin. Oncol. 10, 507 (2013).
13. Schaafsma BE, Mieog JS, Hutteman M, van der Vorst JR, Kuppen PJ, Lowik CW, et al. The clinical use of indocyanine green as a near-infrared fluorescent contrast agent for image-guided oncologic surgery. J Surg Oncol. 2011;104(3):323–32.
14. Gioux S, Choi HS, Frangioni JV. Image-guided surgery using invisible near-infrared light: fundamentals of clinical translation. Mol Imaging. 2010;9(5):237–55.
15. Topal B, Fieuws S, Aerts R, Vandeweyer H, Penninckx F. Laparoscopic versus open liver resection of hepatic neoplasms: comparative analysis of short-term results. Surg Endosc. 2008;22(10):2208–13.
16. Ishizawa T, Bandai Y, Harada N, Muraoka A, Ijichi M, Kusaka K, et al. Indocyanine green-fluorescent imaging of hepatocellular carcinoma during laparoscopic hepatectomy: an initial experience. Asian J Endosc Surg. 2010;3:42.
17. Tummers QR, Verbeek FP, Prevoo HA, Braat AE, BaetenCI, Franioni JV et al. First experience on laparoscopic near-infrared fluorescence imaging of heaptic uveal melanoma metastases using indocyanine green. Surg Innov. 2015 Feb; 22(1):20–5.
18. Stolik S, Delgado JA, Perez A, Anasagasti L. Measurement of the penetration depths of red and near infrared light in human "ex vivo" tissues. J Photochem Photobiol B. 2000;57(2–3):90–3.
19. Faybik P, Hetz H. Plasma disappearance rate of indocyanine green in liver dysfunction. Transplant Proc. 2006;38(3):801–2.
20. A. J. Beer, S. Lorenzen, S. Metz, K. Herrmann, P. Watzlowik, H. J. Wester, C. Peschel, F. Lordick, M. Schwaiger, Comparison of integrin alphaVbeta3 expression and glucose metabolism in primary and metastatic lesions in cancer patients: a PET study using 18F-galacto-RGD and 18F-FDG. J.Nucl.Med. 49, 22–29 (2008).
21. Hutteman M, Mieog JS, van der Vorst JR, Dijkstra J, Kuppen PJ, van der Laan AM, et al. Intraoperative near-infrared fluorescence imaging of colorectal metastases targeting integrin alpha(v)beta(3) expression in a syngeneic rat model. Eur J Surg Oncol. 2011;37(3):252–7.
22. Lee CM, Jang D, Cheong SJ, Jeong MH, Kim EM, Kim DW, et al. Optical imaging of MMP expression and cancer progression in an inflammation-induced colon cancer model. Int J Cancer. 2012;131(8):1846–53.
23. Rosenthal EL, Kulbersh BD, King T, Chaudhuri TR, Zinn KR. Use of fluorescent labeled anti-epidermal growth factor receptor antibody to image head and neck squamous cell carcinoma xenografts. Mol Cancer Ther. 2007;6(4):1230–8.
24. Luo S, Zhang E, Su Y, Cheng T, Shi C. A review of NIR dyes in cancer targeting and imaging. Biomaterials. 2011;32(29):7127–38.
25. van der Vorst JR, Hutteman M, Mieog JS, de Rooij KE, Kaijzel EL, Lowik CW, et al. Near-infrared fluorescence imaging of liver metastases in rats using indocyanine green. J Surg Res. 2012;174(2): 266–71.
26. Ishizawa T, Fukushima N, Shibahara J, Masuda K, Tamura S, Aoki T, et al. Real-time identification of liver cancers by using indocyanine green fluorescent imaging. Cancer. 2009;115(11):2491–504.
27. van der Vorst JR, Schaafsma BE, Hutteman M, Verbeek FP, Liefers GJ, Hartgrink HH, et al. Near-infrared fluorescence-guided resection of colorectal liver metastases. Cancer. 2013;119(18):3411–8.
28. Gotoh K, Yamada T, Ishikawa O, Takahashi H, Eguchi H, Yano M, et al. A novel image-guided surgery of hepatocellular carcinoma by indocyanine green fluorescence imaging navigation. J Surg Oncol. 2009; 100(1):75–9.
29. Harada N, Ishizawa T, Muraoka A, Ijichi M, Kusaka K, Shibasaki M, et al. Fluorescence navigation hepatectomy by visualization of localized cholestasis from bile duct tumor infiltration. J Am Coll Surg. 2010;210(6):e2–6.
30. Ishizuka M, Kubota K, Kita J, Shimoda M, Kato M, Sawada T. Intraoperative observation using a fluorescence imaging instrument during hepatic resection for liver metastasis from colorectal cancer. Hepatogastroenterology. 2012;59(113):90–2.
31. Kasuya K, Sugimoto K, Kyo B, Nagakawa Y, Ikeda T, Mori Y, et al. Ultrasonography-guided hepatic tumor resection using a real-time virtual sonography with indocyanine green navigation (with videos). J Hepatobiliary Pancreat Sci. 2011;18(3):380–5.
32. Peloso A, Franchi E, Canepa MC, Barbieri L, Briani L, Ferrario J, et al. Combined use of intraoperative ultrasound and indocyanine green fluorescence imaging to detect liver metastases from colorectal cancer. HPB (Oxford). 2013;15:928.
33. Uchiyama K, Ueno M, Ozawa S, Kiriyama S, Shigekawa Y, Yamaue H. Combined use of contrast-enhanced intraoperative ultrasonography and a fluorescence navigation system for identifying hepatic metastases. World J Surg. 2010;34(12):2953–9.
34. Yokoyama N. Real-time detection of hepatic micrometastases from pancreatic cancer by intraoperative fluorescence imaging: preliminary results of a prospective study. Cancer. 2012;118(11):2813–9. doi:10.1002/cncr.26594.
35. Marchal GJ, Pylyser K, Tshibwabwa-Tumba EA, Verbeken EK, Oyen RH, Baert AL, et al. Anechoic halo in solid liver tumors: sonographic, microangiographic, and histologic correlation. Radiology. 1985;156(2):479–83.
36. Oshima H, Kon J, Ooe H, Hirata K, Mitaka T. Functional expression of organic anion transporters in hepatic organoids reconstructed by rat small hepatocytes. J Cell Biochem. 2008;104(1):68–81.
37. Huang L, Vore M. Multidrug resistance p-glycoprotein 2 is essential for the biliary excretion of indocyanine green. Drug Metab Dispos. 2001;29(5):634–7.
38. Verbeek FP, van der Vorst JR, Schaafsma BE, Hutteman M, Bonsing BA, van Leeuwen FW, et al. Image-guided hepatopancreatobiliary surgery using

near-infrared fluorescent light. J Hepatobiliary Pancreat Sci. 2012;19(6):626–37.
39. Yamashita Y, Hamatsu T, Rikimaru T, Tanaka S, Shirabe K, Shimada M, et al. Bile leakage after hepatic resection. Ann Surg. 2001;233(1):45–50.
40. Ishizawa T, Bandai Y, Ijichi M, Kaneko J, Hasegawa K, Kokudo N. Fluorescent cholangiography illuminating the biliary tree during laparoscopic cholecystectomy. Br J Surg. 2010;97(9):1369–77.
41. Aoki T, Yasuda D, Shimizu Y, Odaira M, Niiya T, Kusano T, et al. Image-guided liver mapping using fluorescence navigation system with indocyanine green for anatomical hepatic resection. World J Surg. 2008;32(8):1763–7.

Prediction for Postoperative Intrahepatic Recurrence of Hepatocellular Carcinoma

17

Kunihito Gotoh, Shigeru Marubashi, Terumasa Yamada, Hirofumi Akita, Hidenori Takahashi, Masahiko Yano, Osamu Ishikawa, and Masato Sakon

Abbreviations

AFP	α-Fetoprotein
ALT	Alanine aminotransferase
AS	Aspartate aminotransferase
HCC	Hepatocellular carcinoma
ICG	Indocyanine green
IM	Intrahepatic metastasis
MC	Multicentric carcinogenesis
NIR	Near-infrared
PIVKA-II	Protein induced by vitamin K absence/antagonist II
RFS	Recurrence-free survival
TACE	Transcatheter arterial chemoembolization

Background

Hepatocellular carcinoma (HCC) is one of the most common malignancies worldwide.

K. Gotoh, M.D., Ph.D. (✉) • S. Marubashi, M.D., Ph.D.
H. Akita, M.D., Ph.D. • H. Takahashi, M.D. M. Yano, M.D., Ph.D. • O. Ishikawa, M.D., Ph.D.
M. Sakon, M.D., Ph.D.
Department of Surgery, Osaka Medical Center for Cancer and Cardiovascular Diseases, 1-3-3, Nakamichi, Higashinari-ku, Osaka 537-8511, Japan
e-mail: gotou-ku@mc.pref.osaka.jp

T. Yamada, M.D., Ph.D.
Department of Surgery, Higashiosaka City General Hospital, Osaka, Japan

With the improvements in preoperative imaging modalities, surgical techniques, and perioperative management, long-term survival has been achieved in some patients who undergo hepatic resection [1]. However, even when a curative resection is performed, 70–80 % of patients experience tumor recurrence within 5 years postoperatively [2–4]. Many predictors of intrahepatic HCC recurrence after resection were reported previously [2, 3, 5]. These reports showed that the main predictors of recurrence were the presence of vascular invasion, multiple primary tumors, large tumors, less tumor differentiation, liver cirrhosis, hepatitis C virus (HCV) infection, and a high serum α-fetoprotein (AFP) level. Postoperative HCC recurrence is thought to occur in two ways: intrahepatic metastasis (IM) in the residual liver and metachronous, multicentric carcinogenesis (MC) based on chronic hepatitis [6]. These recurrent types have been determined based on histopathological analysis according to the Liver Cancer Study Group of Japan [6].

Differential diagnosis between IM and MC using radiological examination is difficult. Several studies, including genetic analyses, revealed that MC plays a considerable role in tumor recurrence, comprising approximately 50 % of intrahepatic recurrences [7–10]. Early tumor recurrences arise mainly from tiny IM or MC lesions that may have existed in the remnant liver at the time of surgery even when no macroscopically detectable tumors are found by preoperative or intraoperative examination [7, 11].

F.D. Dip et al. (eds.), *Fluorescence Imaging for Surgeons: Concepts and Applications*,
DOI 10.1007/978-3-319-15678-1_17, © Springer International Publishing Switzerland 2015

Therefore, more sensitive and accurate intrahepatic imaging evaluation of these lesions is essential for determining the extent of hepatic resection and selecting patients at high risk of early intrahepatic recurrence. Although sensitive optical sensors and probes for labeling cancer such as nanoprobes [12], fluorescent proteins [13], or near-infrared (NIR) fluorescent molecules [14, 15] have been developed, they are not yet ready for use in clinical applications.

ICG Fluorescence Liver Imaging

ICG is an NIR fluorescent dye that was approved by the Food and Drug Administration for cardiocirculatory and liver function diagnostic uses. The absorption and emission peaks of ICG dye (805 and 835 nm, respectively) lie within the "optical window" of the tissue, where the wavelength is barely absorbed by hemoglobin or water [16]. The fluorescence of ICG in the NIR wavelength can deeply penetrate living tissue and is advantageous for obtaining visual information [17]. ICG fluorescence imaging has been widely used in various surgical situations such as intraoperative angiography [18, 19] and sentinel lymph node navigation [20–23]. This method was recently applied to hepatobiliary surgery [24, 25]. ICG fluorescence liver imaging was described previously [26]. Briefly, ICG (Diagnogreen Inj.; Daiichi Pharmaceutical, Tokyo, Japan) was intravenously injected at a dose of 0.5 mg/kg body weight several days before hepatectomy. On the day of surgery, after laparotomy, the liver was examined by a commercially available NIR camera system (PDE; Hamamatsu Photonics K. K., Hamamatsu, Japan). The light source was a light-emitting diode, and the detector was a charge-coupled device camera. The camera unit of the device was directly handled and real-time fluorescence images were observed on the monitor in the operating room.

We previously reported that ten primary HCC tumors were detected as fluorescent signals by this method [26]. We could perform hepatic resection or enucleation under navigation with this device. Moreover, we could detect four new HCC nodules that were not detected by the use of any preoperative examinations including intraoperative ultrasonography. At the same time, Ishizawa et al. reported that 63 primary HCC tumors and six new HCC nodules were detected using this same technique [27]. All of the new HCC nodules detected in our study were very small (3–6 mm in diameter) and most were well-differentiated HCC. In some cases, tiny fluorescent signals (<2 mm in diameter) were observed in the residual liver parenchyma after hepatic resection. Since these lesions were so small, it is difficult to make a diagnosis pathologically. Therefore, we focused on these tiny fluorescence signals as cancer-potential lesions and followed them prospectively. The patients were grouped according to the pattern of intraoperative ICG fluorescence liver imaging. When tiny fluorescent signals (<2 mm in diameter) were identified throughout the residual planned liver after hepatic resection, the patient was considered to belong to the scattered pattern (SP) group (Fig. 17.1a). When no spotty fluorescent lesions were identified, the patient was considered to belong to the Non-SP group (Fig. 17.1b). In this chapter, we examine whether intraoperative ICG fluorescence liver imaging is associated with intrahepatic HCC recurrence after curative resection.

Correlations Between Intraoperative ICG Fluorescence Liver Imaging and Clinicopathological Factors

Between February 2007 and January 2008, 32 patients underwent curative hepatic resection for HCC at Osaka Medical Center for Cancer and Cardiovascular Diseases and were followed up for more than 2 years after surgery. Informed consent was obtained from every patient before surgery, and the study was approved by the Human Ethics Review Committee of Osaka Medical Center for Cancer and Cardiovascular Diseases. All data are expressed as mean ±SD. Comparisons between the two groups were performed using Student's *t*-test for continuous

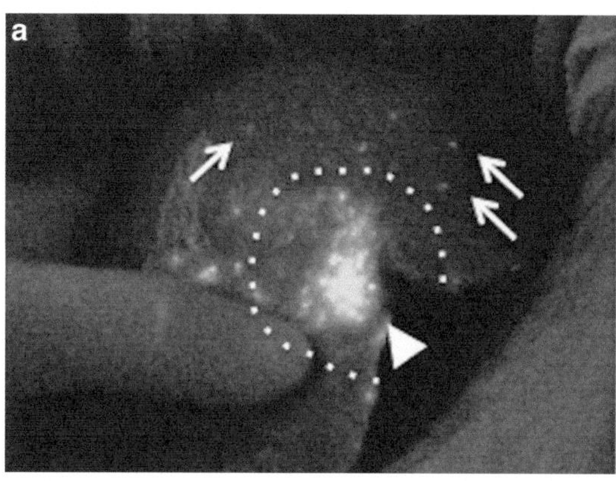

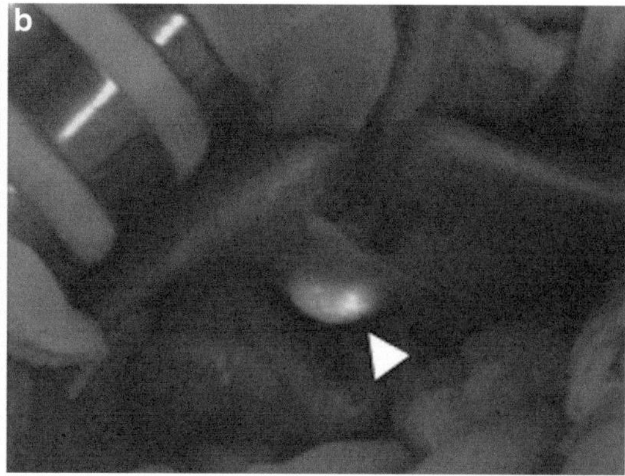

Fig. 17.1 Intraoperative ICG fluorescence liver imaging technique shows bright signals in the primary HCC nodule (*arrowheads*). *White broken lines* show the planned resection line of liver. (**a**) A photograph of scattered pattern (SP) by ICG fluorescence liver imaging. Tiny fluorescent signals (*thin arrows*) are identified throughout the residual planned liver. (**b**) A photograph of Non-SP. No spotty fluorescent lesions are identified

variables and the χ^2 test or Fisher's exact test for categorical variables. A P value <0.05 was considered to be statistically significant. All statistical analyses in this study were performed using SPSS software (SPSS Inc., Chicago, IL, USA).

The 27 men and 5 women had a mean age of 67 ± 10 years (range, 44–80 years). Twelve patients (37.5 %) were positive for hepatitis C virus (HCV) antibody, five patients (15.6 %) were positive for hepatitis B surface antigen, and 15 patients (46.9 %) were negative for both markers. The mean maximum tumor size was 28 ± 12 mm (range, 12–50 mm), and 27 patients (84.4 %) had a solitary tumor. Thirteen patients (40.6 %) underwent preoperative Transcatheter Arterial Chemoembolization (TACE). Liver cirrhosis was present in ten patients (31.3 %). Of the 32 patients, 17 (53.1 %) had a scattered pattern (SP) according to intraoperative ICG fluorescence liver imaging. Comparisons of patient clinicopathological characteristics according to presence or absence of SP are shown in Table 17.1.

Preoperative serum aspartate aminotransferase (AST) and alanine aminotransferase (ALT) activities were significantly higher in the SP than in the non-SP group, while other clinical factors such as gender, age, etiology (HBV and HCV), and background liver function (total bilirubin, albumin, ICG rate at 15 min, and prothrombin time) were similar in the two groups. In addition, there was no difference in tumor factors (number of nodules, tumor diameter, AFP, or protein induced by vitamin K absence/antagonist II [PIVKA-II]), surgical factors (preoperative TACE, hepatic resection, and resection volume), and pathological factors (IM, microscopic portal vein invasion, and liver cirrhosis) between the two groups.

Predictors of Intrahepatic Recurrence After Hepatic Resection

All patients were followed at 3–4-month intervals after surgery. The diagnosis of intrahepatic tumor recurrence was based on computed tomography scanning or ultrasonography and supplemented with the detection of serum AFP or PIVKA-II level elevation. Recurrence-free survival (RFS) rates and curves were analyzed by the Kaplan–Meier method, and the differences between the two groups were compared using the log rank test. Multivariate analysis of the prognostic factors was performed using the Cox proportional hazard stepwise model. RFS curves of patients with and without SP are shown in Fig. 17.2. The RFS rates for patients with and without SP were 41.2 and 86.7 % at 2 years, respectively. The RFS of patients with SP was

Table 17.1 Patient demographics ($n=32$)

	SP group ($n=17$)	Non-SP group ($n=15$)	p-value
Clinical factors			
Gender (M/F)	15/2	12/3	0.52
Age (years)	65±11	67±10	0.60
HBV Ag (positive/negative)	3/14	2/13	0.74
HCV Ab (positive/negative)	7/10	5/10	0.65
AST (IU/L)	35±14	24±9	0.02
ALT (IU/L)	36±15	24±16	0.04
Total bilirubin (mg/dL)	0.8±0.4	0.7±0.4	0.49
Albumin (g/dL)	3.8±0.4	3.8±0.3	0.89
1CG-R15 (%)	17.6±10.4	13.3±6.7	0.19
Prothrombin time (%)	94±13	94±10	0.91
Tumor factors			
Number of nodules (single/multiple)	14/3	13/2	0.74
Tumor diameter (mm)	50±24	41±22	0.32
AFP (ng/mL)	1,870±6,830	1,580±4,020	0.89
P1VKA-II (mAU/mL)	3,060±10,000	940±2,610	0.43
Surgical factors			
Preoperative TACE (yes/no)	8/9	5/10	0.49
Hepatic resection (anatomical/partial)	8/9	9/6	0.50
Volume of resection (g)	253±50	187±53	0.37
Pathological factors			
Intrahepatic metastasis (present/absent)	3/14	1/14	0.60
Portal vein invasion (present/absent)	3/14	1/14	0.60
Liver cirrhosis (present/absent)	8/9	2/13	0.06

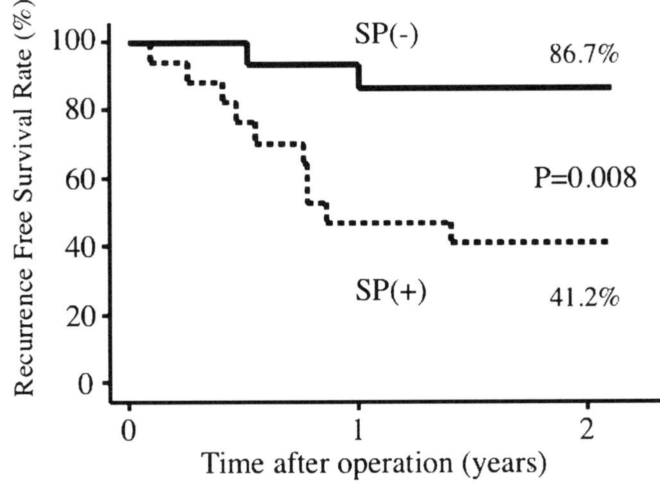

Fig. 17.2 Comparison of recurrence-free survival (RFS) rates in patients with and without scattered pattern (SP). The 2-year RFS rates are shown. The RFS of patients with SP was significantly shorter compared with survival of patients without SP ($P=0.008$)

significantly shorter than that of patients without SP ($P=0.008$). Preoperative AFP (≥8 ng/mL) and microscopic portal vein invasion (present) were also significantly correlated with decreased RFS rates, while other clinicopathological parameters were not. By multivariate analysis, a scattered ICG fluorescence liver imaging pattern, preoperative AFP (≥8 ng/mL), and microscopic portal vein invasion (present) were independent predictors of postoperative recurrence (Table 17.2).

Although it is hypothesized that the mechanisms of ICG accumulation in HCC nodules are associated with ICG transporter expression levels, those in HCC are reported to vary among individual tumors [28–31]. Since ICG is not a cancer-specific dye, false-positive results including dysplastic nodules were obtained in earlier reports [26, 27, 32]. It is difficult to distinguish HCC from benign lesions using ICG fluorescence imaging, especially in cases of tiny fluorescent lesions. Morita et al. reported that one of 14 fluorescent lesions <2 mm was neoplastic [32]. However, they resected all small surface-located emission spots (including those <2 mm in diameter) after employing ICG fluorescence liver imaging, and the short-term HCC recurrence rate within 12 months in their hospital tended to decrease. This result may support our hypothesis that tiny fluorescence signals by ICG fluorescence liver imaging are cancer-potential lesions. Therefore, further basic studies, such as molecular, genetic, and immunohistochemical research, are needed to understand the mechanism and pattern of ICG accumulation in HCC nodules during carcinogenesis.

Table 17.2 Univariate and multivariate analysis of recurrence free survival ($n=32$)

Variable		Patients No.	2-year DFS (%)	Univariate *P*	Multivariate *P*
Clinical factor					
Age	<65	15	73.3		
	≥65	17	52.9	0.17	
Gender	Male	27	66.7		
	Female	5	40.0	0.16	
HBV Ag	Negative	27	63.0		
	Positive	5	60.0	0.89	
HCV Ab	Negative	20	65.0		
	Positive	12	58.3	0.77	
AST (IU/L)	<28	15	66.7		
	≥28	17	58.8	0.55	
ALT (IU/L)	<28	16	62.5		
	≥28	16	62.5	0.97	
T. bilirubin (mg/dL)	<0.7	14	50.0		
	≥0.7	18	72.2	0.18	
Albumin (g/dL)	≥3.8	20	70.0		
	<3.8	12	50.0	0.20	
1CG-R15 (%)	<14.1	16	68.8		
	≥14.1	16	56.3	0.53	
Prothrombin time (%)	≥93	17	70.6		
	<93	15	53.3	0.40	
Tumor factor					
Number of nodules	Single	27	63.0		
	Multiple	5	60.0	0.81	

(continued)

Table 17.2 (continued)

Variable		Patients No.	2-year DFS (%)	Univariate *P*	Multivariate *P*
Tumor diameter (nun)	<40	16	68.8		
	≥40	16	56.3	0.35	
AFP (ng/mL)	<8	16	81.3		
	≥8	16	43.8	0.025	0.030
PIVKA-II	<55	16	62.5		
	≥55	16	62.5	0.81	
Surgical factor					
Preoperative TACE	Yes	13	68.4		
	No	19	53.8	0.30	
Hepatic resection	Anatomical	17	64.7		
	Partial	15	60.0	0.94	
ICGF pattern	Non-SP	15	86.7		
	SP	17	41.2	0.008	0.045
Pathological factor					
Intrahepatic metastasis	Absent	28	67.9		
	Present	4	25.0	0.059	
Portal vein invasion	Absent	28	67.9		
	Present	4	25.0	0.026	0.014
Liver cirrhosis	Absent	22	72.7		
	Present	10	40.0	0.12	

Closing Remarks

This preliminary study indicates that intraoperative ICG fluorescence liver imaging may predict early intrahepatic recurrence after curative hepatic resection for HCC. This technique is quite a simple and promising tool for routine intraoperative imaging during hepatic resection. It is hoped that further basic research and device innovation in this field will provide insight into the diagnosis and treatment of HCC in the near future.

References

1. Poon RT, Fan ST, Lo CM, et al. Improving survival results after resection of hepatocellular carcinoma: a prospective study of 377 patients over 10 years. Ann Surg. 2001;234:63–70.
2. Llovet JM, Fuster J, Bruix J. Intention-to-treat analysis of surgical treatment for early hepatocellular carcinoma: resection versus transplantation. Hepatology. 1999;30:1434–40.
3. Sasaki Y, Yamada T, Tanaka H, et al. Risk of recurrence in a long-term follow-up after surgery in 417 patients with hepatitis B- or hepatitis C-related hepatocellular carcinoma. Ann Surg. 2006;244:771–80.
4. Katz SC, Shia J, Liau KH, et al. Operative blood loss independently predicts recurrence and survival after resection of hepatocellular carcinoma. Ann Surg. 2009;249:617–23.
5. Imamura H, Matsuyama Y, Tanaka E, et al. Risk factors contributing to early and late phase intrahepatic recurrence of hepatocellular carcinoma after hepatectomy. J Hepatol. 2003;38:200–7.
6. Liver Cancer Study Group of Japan. General rules for the clinical and pathological study of primary liver cancer (3rd English edn). Tokyo: Kanehara Shuppan; 2010.
7. Sakon M, Umeshita K, Nagano H, et al. Clinical significance of hepatic resection in hepatocellular carcinoma: analysis by disease-free survival curves. Arch Surg. 2000;135:1456–9.
8. Nakashima O, Kojiro M. Recurrence of hepatocellular carcinoma: multicentric occurrence or intrahepatic metastasis? A viewpoint in terms of pathology. J Hepatobiliary Pancreat Surg. 2001;8:404–9.

9. Morimoto O, Nagano H, Sakon M, et al. Diagnosis of intrahepatic metastasis and multicentric carcinogenesis by microsatellite loss of heterozygosity in patients with multiple and recurrent hepatocellular carcinomas. J Hepatol. 2003;39:215–21.
10. Li Q, Wang J, Juzi JT, et al. Clonality analysis for multicentric origin and intrahepatic metastasis in recurrent and primary hepatocellular carcinoma. J Gastrointest Surg. 2008;12:1540–7.
11. Shimada M, Takenaka K, Taguchi K, et al. Prognostic factors after repeat hepatectomy for recurrent hepatocellular carcinoma. Ann Surg. 1998;227:80–5.
12. le Masne de Chermont Q, Chaneac C, Seguin J, et al. Nanoprobes with near-infrared persistent luminescence for in vivo imaging. Proc Natl Acad Sci U S A. 2007;104:9266–71.
13. Adusumilli PS, Stiles BM, Chan MK, et al. Real-time diagnostic imaging of tumors and metastases by use of a replication-competent herpes vector to facilitate minimally invasive oncological surgery. FASEB J. 2006;20:726–8.
14. Veiseh M, Gabikian P, Bahrami SB, et al. Tumor paint: a chlorotoxin:Cy5.5 bioconjugate for intraoperative visualization of cancer foci. Cancer Res. 2007;67:6882–8.
15. Ogawa M, Kosaka N, Choyke PL, Kobayashi H. In vivo molecular imaging of cancer with a quenching near-infrared fluorescent probe using conjugates of monoclonal antibodies and indocyanine green. Cancer Res. 2009;69:1268–72.
16. Frangioni JV. In vivo near-infrared fluorescence imaging. Curr Opin Chem Biol. 2003;7:626–34.
17. Kitai T, Miwa M, Liu H, Beauvoit B, Chance B, Yamaoka Y. Application of near-infrared time-resolved spectroscopy to rat liver–a preliminary report for surgical application. Phys Med Biol. 1999;44: 2049–61.
18. Rubens FD, Ruel M, Fremes SE. A new and simplified method for coronary and graft imaging during CABG. Heart Surg Forum. 2002;5:141–4.
19. Unno N, Suzuki M, Yamamoto N, et al. Indocyanine green fluorescence angiography for intraoperative assessment of blood flow: a feasibility study. Eur J Vasc Endovasc Surg. 2008;35:205–7.
20. Kitai T, Inomoto T, Miwa M, Shikayama T. Fluorescence navigation with indocyanine green for detecting sentinel lymph nodes in breast cancer. Breast Cancer. 2005;12:211–5.
21. Miyashiro I, Miyoshi N, Hiratsuka M, et al. Detection of sentinel node in gastric cancer surgery by indocyanine green fluorescence imaging: comparison with infrared imaging. Ann Surg Oncol. 2008;15:1640–3.
22. Noura S, Ohue M, Seki Y, et al. Feasibility of a lateral region sentinel node biopsy of lower rectal cancer guided by indocyanine green using a near-infrared camera system. Ann Surg Oncol. 2010;17:144–51.
23. Uchiyama K, Ueno M, Ozawa S, et al. Combined intraoperative use of contrast-enhanced ultrasonography imaging using a sonazoid and fluorescence navigation system with indocyanine green during anatomical hepatectomy. Langenbecks Arch Surg. 2011;396:1101–7.
24. Aoki T, Yasuda D, Shimizu Y, et al. Image-guided liver mapping using fluorescence navigation system with indocyanine green for anatomical hepatic resection. World J Surg. 2008;32:1763–7.
25. Kaibori M, Ishizaki M, Matsui K, Kwon AH. Intraoperative indocyanine green fluorescent imaging for prevention of bile leakage after hepatic resection. Surgery. 2011;150:91–8.
26. Gotoh K, Yamada T, Ishikawa O, et al. A novel image-guided surgery of hepatocellular carcinoma by indocyanine green fluorescence imaging navigation. J Surg Oncol. 2009;100:75–9.
27. Ishizawa T, Fukushima N, Shibahara J, et al. Real-time identification of liver cancers by using indocyanine green fluorescent imaging. Cancer. 2009;115: 2491–504.
28. de Graaf W, Hausler S, Heger M, et al. Transporters involved in the hepatic uptake of (99 m)Tc-mebrofenin and indocyanine green. J Hepatol. 2011;54:738–45.
29. Kusuhara H, Sugiyama Y. Pharmacokinetic modeling of the hepatobiliary transport mediated by cooperation of uptake and efflux transporters. Drug Metab Rev. 2010;42:539–50.
30. Kitao A, Matsui O, Yoneda N, et al. The uptake transporter OATP8 expression decreases during multistep hepatocarcinogenesis: correlation with gadoxetic acid enhanced MR imaging. Eur Radiol. 2011;21:2056–66.
31. Zollner G, Wagner M, Fickert P, et al. Hepatobiliary transporter expression in human hepatocellular carcinoma. Liver Int. 2005;25:367–79.
32. Morita Y, Sakaguchi T, Unno N, et al. Detection of hepatocellular carcinomas with near-infrared fluorescence imaging using indocyanine green: its usefulness and limitation. Int J Clin Oncol. 2013;18:232–41.

Detection of Bile Leakage After Hepatic Resection by Intraoperative Indocyanine Green Fluorescent Imaging

18

Masaki Kaibori, Kosuke Matsui, Morihiko Ishizaki, and Masanori Kon

Introduction

Improvements in surgical techniques and perioperative care have increased the safety of hepatic surgery over recent years, resulting in reduced operative mortality [1–4]. Despite the overall reduction in postoperative complications, the incidence of postoperative bile leakage has not changed and has a reported range of 3.6 to 33 % [5–13], making it one of the most common complications after hepatic surgery. Bile leakage is associated with increased risks of postoperative sepsis, liver failure, and mortality; and a longer hospital stay [6]. It is therefore important to minimize the occurrence of this complication.

Recently, intraoperative fluorescent angiography has been performed after intravenous injection of indocyanine green (ICG) to assess the patency of coronary artery bypass grafts [14–17]. Mitsuhashi et al. reported that intraoperative fluorescent imaging during hepatobiliary surgery leads to better understanding of the anatomy of the arteries, portal vein, and bile ducts [18]. ICG binds to plasma proteins, and protein-bound ICG emits near-infrared light [19, 20]. Human bile also contains proteins that bind with ICG [21], and Ishizawa et al. recently reported that fluorescent images of the biliary tract could be obtained after intrabiliary injection of ICG [22]. We hypothesized that fluorescent cholangiography using ICG would be able to detect minor bile leakage from the cut surface of the remnant liver after hepatic resection. The aim of this clinical trial was to evaluate the usefulness of fluorescent cholangiography for preventing postoperative bile leakage.

Materials and Methods

Patients

All 132 patients who were scheduled for liver resection at Hirakata Hospital of Kansai Medical University (Osaka, Japan) between August 2010 and December 2012 were included in this study. Of these, 72 had hepatocellular carcinoma, 9 had intrahepatic cholangiocellular carcinoma, 47 had colorectal liver metastasis, and 4 had another pathology. The patients were 108 males and 24 females with a mean age of 69.3 years. The operative procedure was extended hemihepatectomy in 14 patients, hemihepatectomy in 35, sectionectomy in 38, segmentectomy in 12, and limited resection in 33.

Electronic supplementary material: The online version of this chapter (doi:10.1007/978-3-319-15678-1_18) contains supplementary material, which is available to authorized users. Videos can also be accessed at http://link.springer.com/chapter/10.1007/978-3-319-15678-1_18.

M. Kaibori, M.D. (✉) • K. Matsui, M.D., Ph.D.
M. Ishizaki, M.D., Ph.D. • M. Kon, M.D., Ph.D.
Department of Surgery, Hirakata Hospital,
Kansai Medical University, 2-3-1 Shinmachi,
Hirakata, Osaka 573-1191, Japan
e-mail: kaibori@hirakata.kmu.ac.jp

F.D. Dip et al. (eds.), *Fluorescence Imaging for Surgeons: Concepts and Applications*,
DOI 10.1007/978-3-319-15678-1_18,

Surgical Techniques

Transection of the hepatic parenchyma was performed using the Cavitron Ultrasonic Surgical Aspirator (CUSA) system (Valleylab, Boulder, CO). After hepatic resection, a disposable cholangiography catheter was inserted into the cystic duct and ligated in place. The common bile duct was clamped below the cystic duct, and 10 ml of dilute ICG solution (Dianogreen 2.5 mg/ml; Daiichi Sankyo Co., Tokyo, Japan) was injected into the bile duct. Any sites of major dye leakage from the cut surface of the remnant liver were repaired by z-suturing with 6-0 non-absorbable sutures or by ligation with absorbable sutures, after which the absence of further leakage was confirmed by injecting 5–7 ml of saline. Only one injection of ICG solution was administered to each patient. Fluorescent imaging was performed with a Photodynamic Eye (PDE) imaging system (Hamamatsu Photonics K.K., Shizuoka, Japan) (Fig. 18.1) after the injection of ICG solution. The PDE system includes a control unit (322 × 283 × 55 mm, 2.8 kg) and a camera unit (80 × 181 × 80 mm, 0.5 kg). The camera unit contains a charge-coupled device camera that filters out light with a wavelength of less than 820 nm, and 36 light-emitting diodes with a wavelength of 760 nm. The camera head was positioned 20 cm above the remnant liver, and the operating lights were turned off (the room lights were left on). The operating field was therefore visible to the surgeons directly and on the television monitor. The fluorescent images of the cut surface of the liver were displayed on the monitor, and showed white spots representing potential bile leaks. The areas showing white spots were compressed with gauze, and leakage was tested by additional injection of 3–5 ml of saline (Fig. 18.2 and Video 18.1). If fluorescence was detected through the gauze, the white spots were considered to indicate minor leaks on the cut surface of the liver, and the presumably injured bile ducts were repaired by z-suturing or ligation.

Postoperative bile leakage (Fig. 18.3) was diagnosed by the following findings: detection of bile leakage from the wound or through the drainage tube (total bilirubin level in the drainage fluid >3 times the serum level), intra-abdominal accumulation of bile confirmed by drainage, or demonstration of bile leakage on postoperative cholangiography.

Results

Thirty-two of the 132 patients (24 %) had major leakage of ICG, defined as leakage of green dye without fluorescent imaging. Repair was performed by z-suturing with 6-0 non-absorbable sutures in 28 patients, and by ligation with 4-0 absorbable sutures in 4 patients. The leaks were all successfully repaired during surgery, and were further tested with fluorescent imaging.

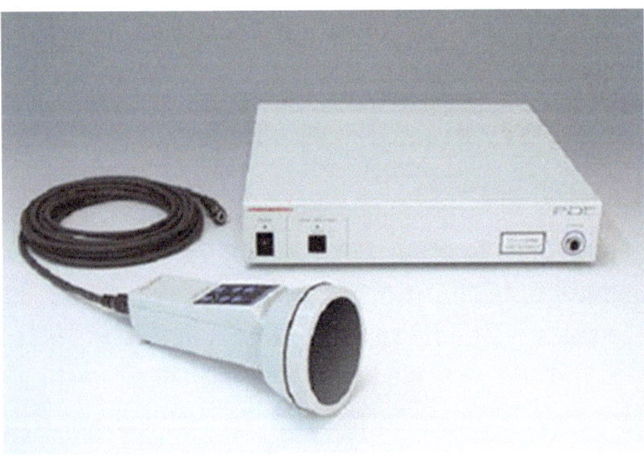

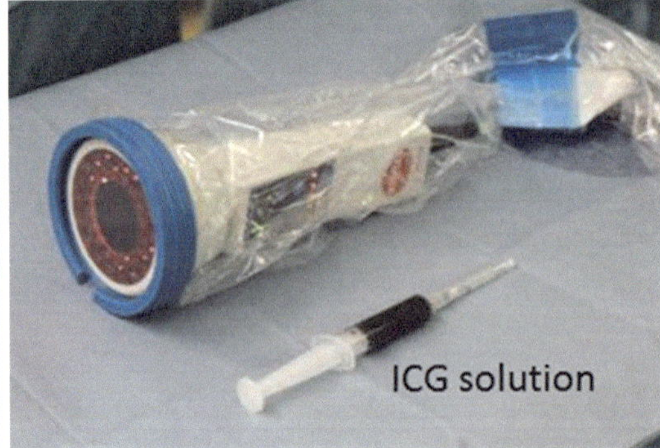

Fig. 18.1 PDE imaging system

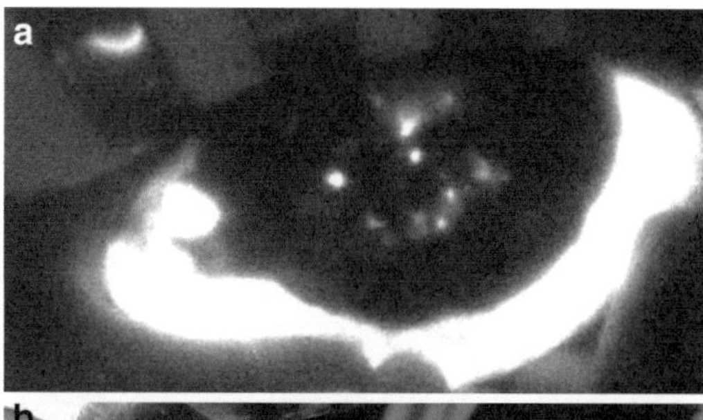

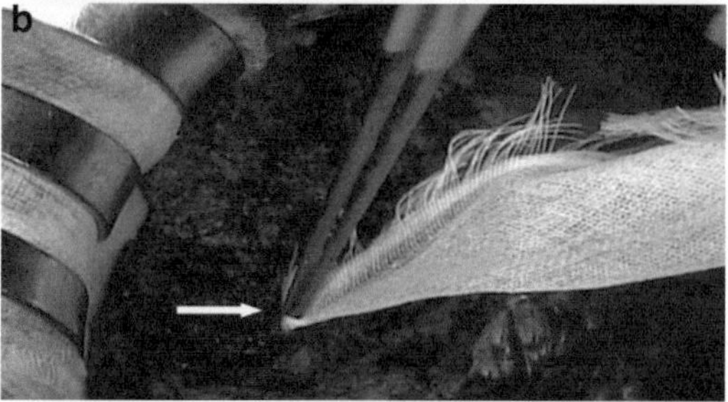

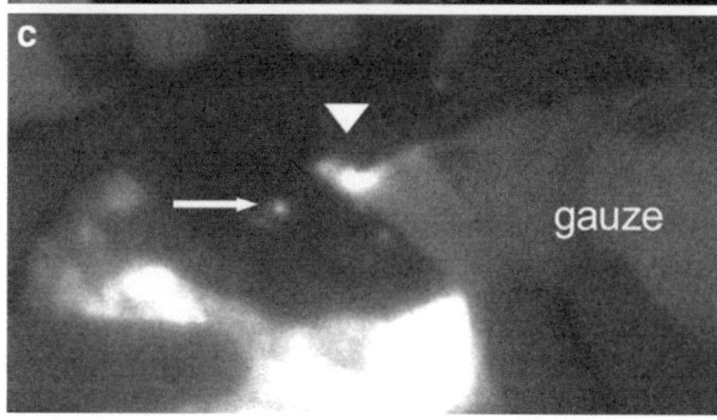

Fig. 18.2 Fluorescent images obtained using the PDE system. (**a**) *White spots* (possible bile leakage) on the cut surface of the liver. (**b**) Minor leakage (leaking duct) that is not visible to the surgeons (*arrow*) is compressed with gauze. (**c**) Fluorescent imaging after the application of gauze shows minor leakage on the cut surface of the liver (*arrow*), as fluorescence is detected through the gauze (*arrowhead*). With permission from Kaibori M, Ishizaki M, Matsui K, Kwon AH. Intraoperative indocyanine green fluorescent imaging for prevention of bile leakage after hepatic resection. Surgery 2011; 150: 91–8 (25)

Patterns of Fluorescence

In 37 of the 132 patients (19 %), no fluorescence was detected on the cut surface of the remnant liver on ICG fluorescent cholangiography, suggesting absence of bile ducts at the surgical margin (Type A pattern of fluorescence). In the remaining 95 patients, the pattern of fluorescence was classified into the following three types: intact bile duct type (Type B, fluorescence showed one or more intact bile ducts on the cut surface of the liver; $n=51$) (Fig. 18.4), injured bile duct type (Type C, leakage of dye from one or more bile duct stumps on the cut surface; $n=31$) (Fig. 18.5), and unconfirmed type (Type D, leakage of dye from the cut surface, but the source was unclear; $n=13$) (Fig. 18.6 and Video 18.2). In Type D pattern, the minor dye leaks could not be visualized by the surgeon, and could only be detected by viewing the fluorescent images on the monitor.

No repair was performed in patients with Type A pattern of fluorescence. In patients with Type B pattern, repair was performed by z-suturing with 6-0 non-absorbable sutures in 42 patients (1.5 ± 0.8 (mean ± SD) sutures per patient) and by ligation with 4-0 absorbable sutures in the remaining nine patients (1.0 ± 0.2 ligatures per patient). In the three patients with Type D pattern, fibrin sealant was applied to the area showing leakage (two pieces of 2.0 × 1.5 cm sealant per patient). The fluorescence imaging findings were used to guide treatment in the patients with Type D pattern.

Postoperative Bile Leakage

Postoperative bile leakage occurred in 7 of the 132 hepatectomy patients (5 %) and persisted for a median period of 6 weeks (range, 2–15 weeks). Table 18.1 shows the incidence of each pattern of fluorescence. The rate of Type D pattern differed from the rates of the other three patterns.

Discussion

Postoperative bile leakage remains a challenging problem in patients undergoing hepatic surgery, especially major hepatectomy, as it is associated with serious complications such as sepsis and liver failure [7]. The objective of the bile leakage test is to detect insufficiently closed bile duct stumps on the cut surface of the liver. However, such a test cannot completely prevent postoperative bile leakage, as leakage may occur from small ducts that are not in communication with the main part of the biliary tree [23]. There is no standard method of preventing postoperative

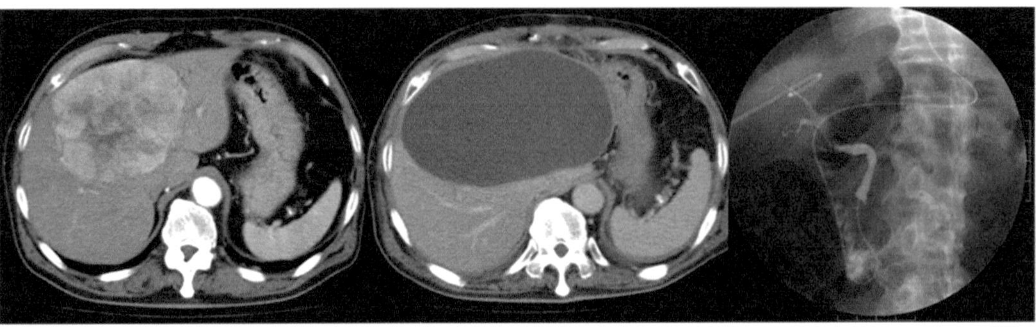

Fig. 18.3 Patient with postoperative bile leakage. A 67-year-old male underwent extended left hemihepatectomy for hepatocellular carcinoma. He was discharged uneventfully on postoperative day 12, and was readmitted on postoperative day 22 with severe abdominal pain and persistent fever. His bile leakage was treated by abdominal paracentesis and endoscopic biliary drainage, and was resolved on postoperative day 72

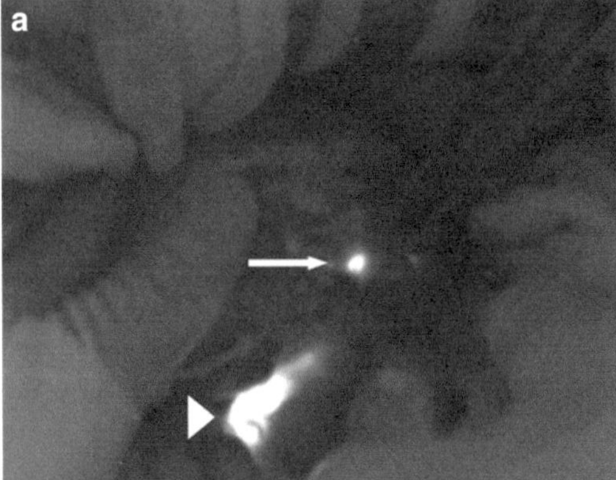

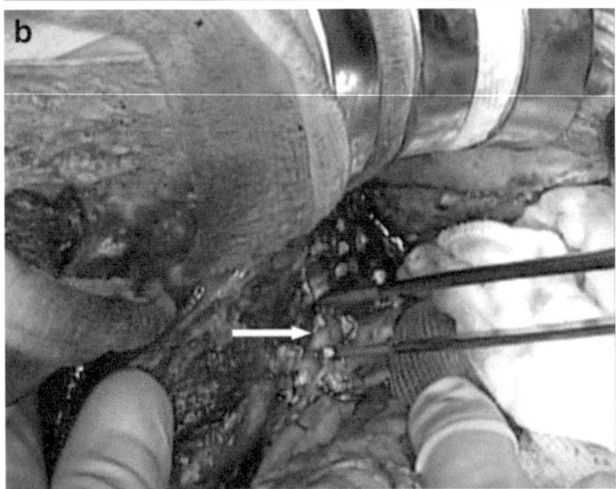

Fig. 18.4 Type B pattern of fluorescence (intact bile duct type). (**a**) Fluorescent imaging showed one fluorescent duct on the cut surface of the liver (*arrow*). *Arrowhead*: common bile duct. (**b**) There was an intact bile duct corresponding to the fluorescent area on the cut surface (*arrow*). With permission from Kaibori M, Ishizaki M, Matsui K, Kwon AH. Intraoperative indocyanine green fluorescent imaging for prevention of bile leakage after hepatic resection. Surgery 2011; 150: 91–8 (25)

bile leakage. We suspected that small bile duct stumps on the cut surface of the liver might be missed by the usual bile leakage test.

ICG fluorescent cholangiography is safe and feasible. ICG is already used worldwide to evaluate liver function before surgery, and the incidence of adverse reactions after intravenous injection of ICG is very low (approximately 0.003 %) [24]. Intraoperative ICG fluorescent cholangiography showed that 32 of our 132 patients had insufficient closure of bile duct stumps on the cut surface of the liver, and these stumps were treated by z-suturing or ligation. The major limitation of ICG fluorescent cholangiography is that it is not possible to visualize deep intrahepatic bile ducts or extrahepatic bile ducts covered by surrounding organs using this technique, because of the limited tissue penetration of the near-infrared light emitted by the current imaging system. However, the results of this study show that ICG fluorescent cholangiography enables evaluation of the cut surface of the liver. In our previous study, postoperative bile leakage occurred in five of the 50 patients in the control group versus none of the 52 patients in the PDE group [25]. In this study, the incidence of postoperative bile leakage was 0 % in patients with Type A pattern of fluorescence, 2 % in patients with Type B pattern, 6 % in patients with Type C pattern, and 31 % in patients with Type D pattern. In Type D pattern, visualization of white spots on the whole cut surface of the remnant

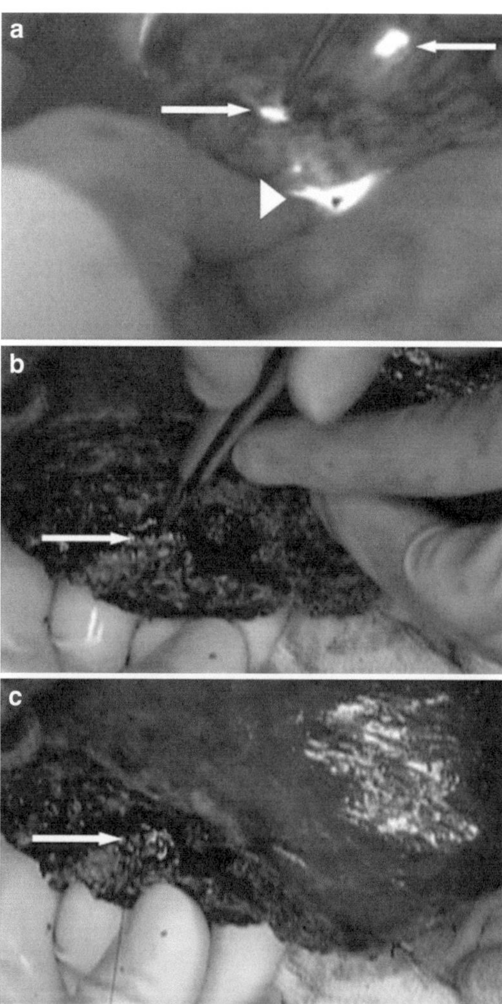

Fig. 18.5 Type C pattern of fluorescence (injured bile duct type). (**a**) Fluorescent imaging showed two areas of fluorescence corresponding to ducts on the cut surface of the liver (*arrows*). The upper bile duct was intact (*upper arrow*). *Arrowhead*: common bile duct. (**b**) The lower area of fluorescence corresponded to a partly closed bile duct stump (*arrow*). (**c**) The bile duct stump was repaired by z-suturing with 6-0 nonabsorbable sutures (*arrow*). With permission from Kaibori M, Ishizaki M, Matsui K, Kwon AH. Intraoperative indocyanine green fluorescent imaging for prevention of bile leakage after hepatic resection. Surgery 2011; 150: 91-8 (25)

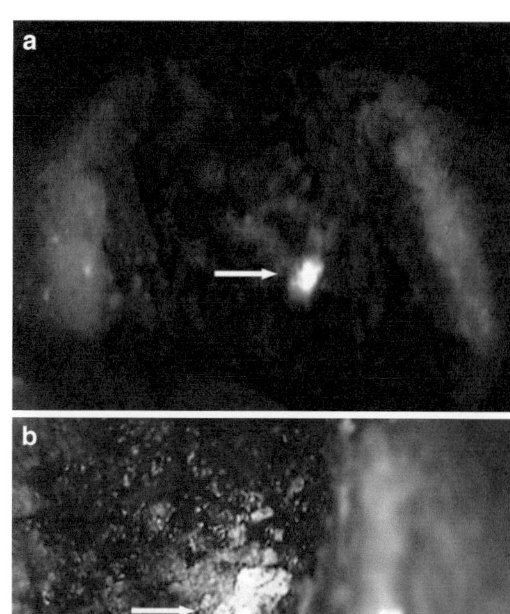

Fig. 18.6 Type D pattern of fluorescence (unconfirmed type). (**a**) Fluorescent imaging showed an area of fluorescence on the cut surface of the liver (*arrow*). (**b**) Fibrin sealant was applied to the fluorescent area (*arrow*). With permission from Kaibori M, Ishizaki M, Matsui K, Kwon AH. Intraoperative indocyanine green fluorescent imaging for prevention of bile leakage after hepatic resection. Surgery 2011; 150: 91–8 (25)

liver suggested leakage of a high volume of ICG dye from the stumps, or leakage of bile from ducts that were not in communication with the main part of the biliary tree. We consider that it is necessary to place intra-abdominal drainage catheters after hepatic resection to treat potential postoperative bike leakage. Patients with Type D pattern of fluorescence should be carefully monitored for bile leakage for several weeks after hepatic resection, because late-onset bile leakage can cause serious complications.

In conclusion, ICG fluorescent cholangiography enabled detection of leaking bile duct stumps that were missed by the conventional bile leakage test. ICG fluorescent cholangiography may be useful for the prevention of bile leakage after hepatic resection.

Table 18.1 Incidence of postoperative bile leakage for each pattern of fluorescence

	Indocyanine green fluorescent imaging	Number of patients	Number with postoperative bile leakage	%
Type A	No fluorescence was detected on the cut surface	37	0	0
Type B	Intact bile duct type	51	1	2
Type C	Injured bile ducts were detected and successfully repaired	31	2	6
Type D	Unconfirmed site of intraoperative bile leakage	13	4	31
Total		132	7	5

References

1. Imamura H, Seyama Y, Kokudo N, Maema A, Sugawara Y, Sano K, et al. One thousand fifty-six hepatectomies without mortality in 8 years. Arch Surg. 2003;138:1198–206.
2. Capussotti L, Polastri R. Operative risks of major hepatic resections. Hepatogastroenterology. 1998;45: 184–90.
3. Jarnagin WR, Gonen M, Fong Y, DeMatteo RP, Ben-Porat L, Little S, et al. Improvement in perioperative outcome after hepatic resection: analysis of 1,803 consecutive cases over the past decade. Ann Surg. 2002;236:397–407.
4. Poon RT, Fan ST, Lo CM, Liu CL, Lam CM, Yuen WK, et al. Improving perioperative outcome expands the role of hepatectomy in management of benign and malignant hepatobiliary diseases: analysis of 1222 consecutive patients from a prospective database. Ann Surg. 2004;240:698–710.
5. Lo CM, Fan ST, Liu CL, Lai EC, Wong J. Biliary complications after hepatic resection: risk factors, management, and outcome. Arch Surg. 1998;133:156–61.
6. Yamashita Y, Hamatsu T, Rikimaru T, Tanaka S, Shirabe K, Shimada M, et al. Bile leakage after hepatic resection. Ann Surg. 2001;233:45–50.
7. Nagano Y, Togo S, Tanaka K, Masui H, Endo I, Sekido H, et al. Risk factors and management of bile leakage after hepatic resection. World J Surg. 2003;27:695–8.
8. Terajima H, Ikai I, Hatano E, Uesugi T, Yamamoto Y, Shimahara Y, et al. Effectiveness of endoscopic naso-biliary drainage for postoperative bile leakage after hepatic resection. World J Surg. 2004;28:782–6.
9. Ijichi M, Takayama T, Toyoda H, Sano K, Kubota K, Makuuchi M. Randomized trial of the usefulness of a bile leakage test during hepatic resection. Arch Surg. 2000;135:1395–400.
10. Rudow DL, Brown Jr RS, Emond JC, Marratta D, Bellemare S, Kinkhabwala M. One-year morbidity after donor right hepatectomy. Liver Transpl. 2004;10:1428–31.
11. Tanaka S, Hirohashi K, Tanaka H, Shuto T, Lee SH, Kubo S, et al. Incidence and management of bile leakage after hepatic resection for malignant hepatic tumors. J Am Coll Surg. 2002;195:484–9.
12. Nakayama H, Masuda H, Shibata M, Amano S, Fukuzawa M. Incidence of bile leakage after three types of hepatic parenchymal transection. Hepatogastroenterology. 2003;50:1517–20.
13. Capussotti L, Ferrero A, Vigano L, Sgotto E, Muratore A, Polastri R. Bile leakage and liver resection: where is the risk? Arch Surg. 2006;141:690–5.
14. Rubens FD, Ruel M, Fremes SE. A new and simplified method for coronary and graft imaging during CABG. Heart Surg Forum. 2002;5:141–4.
15. Taggart DP, Choudhary B, Anastasiadis K, Abu-Omar Y, Balacumaraswani L, Pigott DW. Preliminary experience with a novel intraoperative fluorescence imaging technique to evaluate the patency of bypass grafts in total arterial revascularization. Ann Thorac Surg. 2003;75:870–3.
16. Reuthebuch O, Haussler A, Genoni M, Tavakoli R, Odavic D, Kadner A, et al. Intraoperative quality assessment in off-pump coronary artery bypass grafting. Chest. 2004;125:418–24.
17. Balacumaraswani L, Abu-Omar Y, Choudhary B, Pigott D, Taggart DP. A comparison of transit-time flowmetry and intraoperative fluorescence imaging for assessing coronary artery bypass graft patency. J Thorac Cardiovasc Surg. 2005;130: 315–20.
18. Mitsuhashi N, Kimura F, Shimizu H, Imamaki M, Yoshidome H, Ohtsuka M, et al. Usefulness of intra-operative fluorescence imaging to evaluate local anatomy in hepatobiliary surgery. J Hepatobiliary Pancreat Surg. 2008;15:508–14.
19. Landsman ML, Kwant G, Mook GA, Zijlstra WG. Light-absorbing properties, stability, and spectal stabilization of indocyanine green. J Appl Physiol. 1976;40:575–83.
20. Mordon S, Devoisselle JM, Soulie-Begu S, Desmettre T. Indocyanine green: physicochemical factors affecting its fluorescence in vivo. Microvasc Res. 1998;55:146–52.

21. Mulllock BM, Shaw LJ, Fitzharris B, Peppard J, Hamilton MJ, Simpson MT, et al. Sources of proteins in human bile. Gut. 1985;26:500–9.
22. Ishizawa T, Tamura S, Masuda K, Aoki T, Hasegawa K, Imamura H, et al. Intraoperative fluorescent cholangiography using indocyanine green: a biliary road map for safe surgery. J Am Coll Surg. 2009;208:e1–4.
23. Neuhaus P. Complications of liver surgery and their management. In: Lygidakis NJ, Tytgat GNJ, editors. Hepatobiliary and pancreatic malignancies: diagnosis, medical and surgical management. New York: Thieme-Stratton; 1989. p. 254–9.
24. Cherrick GR, Stein SW, Leevy CM, Davidson CS. Indocyanine green: observations on its physical properties, plasma decay, and hepatic extraction. J Clin Invest. 1960;39:592–600.
25. Kaibori M, Ishizaki M, Matsui K, Kwon AH. Intraoperative indocyanine green fluorescent imaging for prevention of bile leakage after hepatic resection. Surgery. 2011;150:91–8.

Near-Infrared Laser Photodynamic Therapy for Human Hepatocellular Carcinoma Cell Line Tumor with Indocyanine Green Fluorescence

19

Junichi Kaneko, Yoshinori Inagaki, Takeaki Ishizawa, and Norihiro Kokudo

Introduction

A goal of fluorescence imaging is treatment based on the fluorescent features of a photosensitizer and specific wavelength light source, so-called photodynamic therapy (PDT). Although it is effective for only a limited area, PDT is a well-established clinical treatment for various diseases, including cancer. PDT has been applied in a clinical setting for mainly skin cancer [1], Barrett's esophagus and/or esophageal carcinoma [2], superficial bladder cancer [3], and central—type early—stage lung cancer [4]. The major advantages of PDT compared to the other types of cancer treatment are its higher selectivity and lower toxicity due to the use of non-ionizing light. PDT is thus considered a safe and minimally invasive therapy. There are, however, two important obstacles. One is that the photosensitizer still has weak tumor selectivity. The other is the shallow light penetration. Consequently, PDT can be used to reliably treat only superficial lesions. PDT is therefore less optimal for the treatment of solid cancers.

Another light source, near-infrared (NIR) light, has important advantages. NIR light can penetrate deep into the tissue. The penetration depth of NIR light is reported to range from 7 to 10 mm [5, 6] up to several centimeters [7]. In NIR spectroscopy, NIR light passes through scalp skin and skull bone as incident and reflected light, respectively. In a study by Hock et al. [8], the NIR light emitter and detector were placed at a distance of 4 cm from human scalp skin.

Meanwhile, it has recently been found that human hepatocellular carcinoma (HCC) preferentially takes up indocyanine green (ICG) in the clinical setting [9]. Thereby, we previously developed a navigation method for identifying human HCC lesions during liver resection surgery using ICG fluorescence. ICG is administered preoperatively to evaluate liver function and is normally excreted

Electronic supplementary material: The online version of this chapter (doi:10.1007/978-3-319-15678-1_19) contains supplementary material, which is available to authorized users. Videos can also be accessed at http://link.springer.com/chapter/10.1007/978-3-319-15678-1_19.

J. Kaneko, M.D., Ph.D. • Y. Inagaki, Ph.D.
N. Kokudo, M.D., Ph.D. (✉)
Hepato-Biliary-Pancreatic Surgery Division, Department of Surgery, Graduate School of Medicine, University of Tokyo, 7-3-1 Hongo, Bunkyo-ku, Tokyo 113-8655, Japan

Artificial Organ and Transplantation Division, Department of Surgery, Graduate School of Medicine, University of Tokyo, 7-3-1 Hongo, Bunkyo-ku, Tokyo 113-8655, Japan
e-mail: kokudo-2su@h.u-tokyo.ac.jp

T. Ishizawa, M.D., Ph.D., F.A.C.S
Department of Gastroenterological Surgery, Cancer Institute Hospital, Japanese Foundation for Cancer Research, Koto-ku, Tokyo, Japan

Hepato-Biliary-Pancreatic Surgery Division, Department of Surgery, Graduate School of Medicine, University of Tokyo, 7-3-1 Hongo, Bunkyo-ku, Tokyo 113-8655, Japan

F.D. Dip et al. (eds.), *Fluorescence Imaging for Surgeons: Concepts and Applications*,
DOI 10.1007/978-3-319-15678-1_19,

into the bile. Because bile excretion is impaired in human HCCs, both ICG and bile stagnate in the tumor tissue. After ICG administration, we observed a contrast between the human HCC and the surrounding background liver parenchyma. Among these, the most important thing is that ICG absorbs NIR light [10].

Based on these previous findings, we examined whether the HuH-7 human HCC cell line preferentially takes up ICG and then assessed the effectiveness of PDT using ICG and NIR on a human HCC cell line transplanted into mice.

Methods

Animal Models Transplanted with Tumor Cell Lines

This pilot study protocol was approved by the University of Tokyo Animal Ethics Committee (P10-118, P12-42) in accordance with Japanese guidelines and regulations for scientific and ethical animal experimentation. Male BALB/c nude mice were purchased from Charles River Laboratories Japan, Inc. (Yokohama, Japan), and used at 5–6 weeks of age ($n=30$, mean weight 18 g, standard deviation 0.7 g). Animals were kept under specific pathogen-free conditions.

HuH-7 (a well-differentiated human HCC cell line, [11]) cells were obtained directly from Japanese Collection of Research Bioresources Cell Bank (JCRB, Osaka, Japan). These cells were maintained in Dulbecco's modified Eagle's medium (Invitrogen, Carlsbad, CA) with 10 % fetal bovine serum (Invitrogen, Carlsbad) at 37 °C in a humid atmosphere (5 % CO_2–95 % air) and harvested by brief incubation in Enzyme-free Cell Dissociation Solution (Millipore Co., Bedford, MA) [12].

Continuously cultured HuH-7 cells were harvested in tubes and re-suspended in serum-free medium after washing with phosphate-buffered saline. Each mouse was subcutaneously injected into the abdomen with 5×10^6 HuH-7 cells in 0.2 ml of serum-free medium containing 50 % Matrigel (Becton-Dickinson, Franklin Lakes, NJ).

Administration of ICG and Evaluation of Fluorescence Intensity

HuH-7

Mice with HuH-7 cell xenografts were randomized and divided into three groups: ICG administration only (ICG+ NIR− group, $n=8$), ICG administration and NIR laser exposure (ICG+ NIR+ group, $n=12$), and NIR laser exposure without ICG administration (ICG− NIR+ group, $n=5$).

ICG was administered intravenously via a tail vein when the maximum tumor size reached 50–1,000 mm^3 at approximately 10 days after transplantation. ICG (5 mg/kg Diagnogreen; Daiichi Sankyo Co. Ltd., Tokyo, Japan) was dissolved in sterile distilled water and injected into the lateral tail vein of the mice in the HuH-7, ICG+ NIR−, and ICG+ NIR+ groups. Total amount of injection solution was limited to less than 0.25 ml to avoid instability of hemodynamics [13]. Two randomly selected mice from the ICG+ NIR+ group were killed under general anesthesia for observation under fluorescence microscopy and histologic study.

Fluorescence Imaging System

We used a fluorescence imaging system (Photodynamic Eye, Hamamatsu Photonics, Hamamatsu, Japan) that was originally devised for identifying sentinel lymph nodes during breast surgery [14]. The fluorescence imaging system comprises a small control unit and a camera unit. The camera unit comprised a charge-coupled device camera that filtered out light with a wavelength below 820 nm, and 36 light-emitting diodes with a wavelength of 760 nm. The camera imaging head was positioned 20 cm above the mice. The light power density was 4 mW/cm^2.

Measurement of Signal Intensity on Fluorescent Images

The signal intensities of the tumor and background normal tissues were determined by calculating the mean brightness of an operator-defined region of interest on monochromatic fluorescent images of mice using image brightness analysis software (U11437, Hamamatsu Photonics). The brightness of each pixel was measured within a range of 0–255 arbitrary units (AU). To calculate the tumor-to-background ratio, the signal intensity of the background normal skin area was set to 1 AU in a darkroom. The tumor-to-background ratio was calculated by dividing the signal intensity of the background normal skin by that of the tumor.

Microscopic Fluorescence Evaluation and Pathologic Examination

We confirmed tumor fluorescence of two mice randomly selected from the ICG+ NIR+ group. Specimens with HuH-7 tumor were frozen in a dry ice and acetone slush, and sectioned at 20–40 μm. Fluorescence microscopic examinations were performed using an upright microscope system (upright microscope, Eclipse 80i; high pressure XBO lamp, Xenon 75 W, epi-fluorescence illuminator, D-FL; objective lens, CFI PlanApo 20×, Nikon Corporation, Tokyo, Japan) with an electron multiplying charge-coupled device camera (Rolera em-c2, QImaging, Surrey, Canada), filters (ICG single-band filter ICG-B-000, Semrock Inc., Rochester, NY), and imaging software (QCapture PRO 6, QImaging).

Irradiation with NIR Laser Light

The HuH-7 cells-transplanted mice were irradiated with 823 nm infrared laser light 24 h after (day 0) ICG administration (prototype NIR Diode Laser System, Hamamatsu Photonics). The power density was 160 mW/cm^2 for 3 min. Two other mice were randomly selected from the ICG+ NIR+ group and only parts of the tumors were irradiated with the NIR laser to allow for comparison of the histologic change between non-irradiated and irradiated tumor sites. At 36 h after partial NIR laser irradiation, the mice were killed under general anesthesia for histologic study with hematoxylin–eosin staining.

After laser treatment, the other mice were returned to the colony. HuH-7 tumor sizes were measured every 3 days using a caliper and tumor volume was monitored for a total of 9 days following treatment. Tumor volume (mm^3) was calculated by the formula $a \times b \times c$ (mm), where a, b, and c are the width, depth, and height of the tumor, respectively.

Statistical Analysis

The SPSS statistical software package (19.0; IBM, New York, NY) was used to perform the analysis of variance with repeated measures. Data are expressed as mean ± standard deviation. A P-value of 0.05 was considered statistically significant.

Results

Fluorescence Intensity

HuH-7

Developing HuH-7 tumors were observed in the center of abdomen of the transplanted mice under visible light (See Fig. 19.1a). Before ICG administration, the mouse body showed no fluorescence under the fluorescent imaging system in a darkroom. Immediately after ICG administration, the entire mouse body and HuH-7 tumor emitted strong fluorescence, 255 and 255 AU, respectively. Before NIR irradiation, all tumors in mice receiving ICG administration (ICG+ NIR− and ICG+ NIR+) exhibited uniform fluorescence under the fluorescence imaging system in 24 h after ICG administration (day 0) (See Fig. 19.1b and Video 19.1, (24 h after)). When the power density of the NIR light was increased from 0 to 4 mW/cm^2, HuH-7 tumor (24 h after ICG administration) showed increased fluorescence intensity. On the other hand, feces had fluorescence as previously reported [15]. When the NIR light was turned off, the fluorescence disappeared (See Video 19.1).

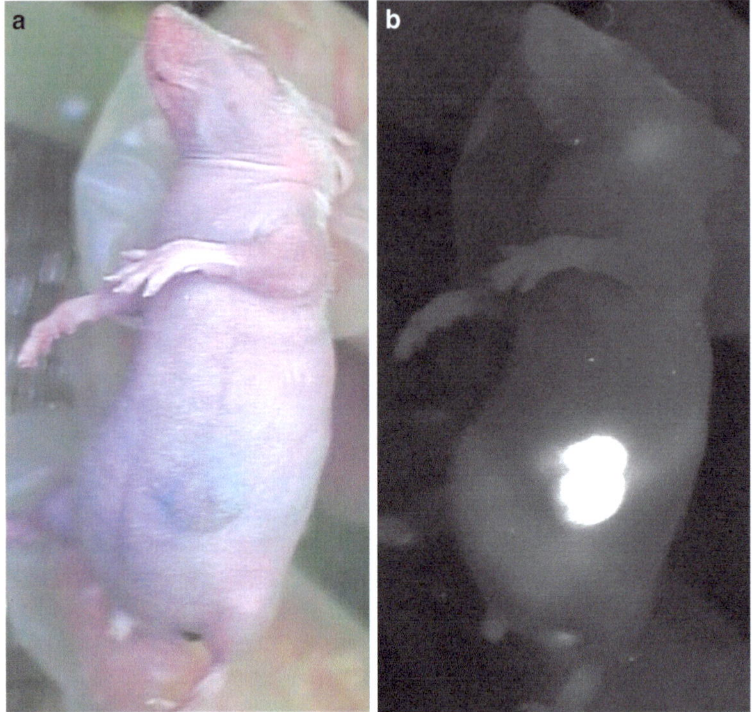

Fig. 19.1 (**a**) HuH-7 tumors were observed in the center of abdomen of the transplanted mice under visible light. (**b**) Twenty-four hour after ICG administration, HuH-7 tumors in mice exhibited uniform fluorescence under the fluorescence imaging system. When the fluorescence intensity of the background tissue was defined as 1 AU, the mean HuH-7 tumor-to-background ratio was 255 on 24 h after ICG administration

Fluorescence Histological Patterns and Attenuation Time of HuH-7 Cells

Under fluorescence microscopy, normal tissue surrounding HuH-7 tumors produced no fluorescence. All HuH-7 cells, however, were fluorescent. Fluorescence was observed in the cytoplasm of HuH-7 cells, but not in the nucleus (See Fig. 19.2a). Fluorescence of HuH-7 tumors gradually decreased over time, but retained an extremely high tumor-to-background ratio for up to 6 days. When the fluorescence intensity of the background tissue was defined as 1 AU, the mean HuH-7 tumor-to-background ratio (ICG+ NIR− and ICG+ NIR+ groups) was 255-, 100-, and 72-fold on days 0, 3, and 6, respectively (See Fig. 19.2b).

Pathologic Study of HuH-7 Cells Tumor 36 h After NIR Irradiation

Sections from the ICG+ NIR+ group (randomly selected and partially irradiated) were stained with hematoxylin and eosin 36 h after NIR laser exposure. At 100× magnification, a clear boundary was observed between areas with NIR laser exposure (right side) and those with no exposure (left side, See Fig. 19.3a). NIR laser exposure of HuH-7 tumors induced the following changes: pyknotic nuclei, increased intracellular spacing, and shrinkage of HuH-7 cells. These changes were more evident at 200× magnification (See Fig. 19.3b). These findings were observed not only superficially, but also in the deep parts of the tumor. HuH-7 cell behavior was not changed in the non-irradiated areas of the tumor.

HuH-7 Tumor Growth After NIR Laser Irradiation

Mean tumor volume in the ICG+ NIR− mice and mice receiving NIR only and no ICG (ICG− NIR+) increased steadily from 402 and 298 mm^3 on day 0 to 4,272 and 3,757 mm^3 by day 9, respectively (See Fig. 19.4a). In contrast, mean tumor volume in the ICG+ NIR+ group did not

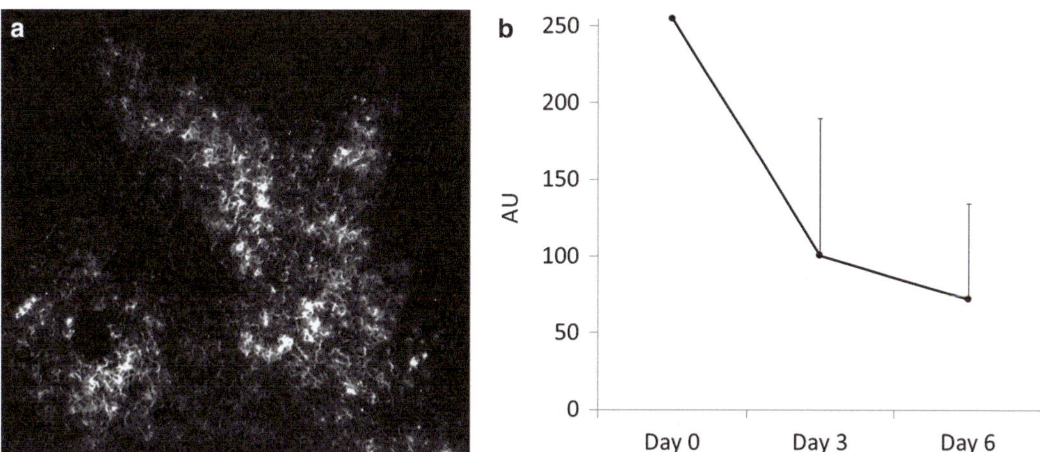

Fig. 19.2 (**a**) Under fluorescence microscopy, normal tissue surrounding HuH-7 tumors produced no fluorescence. All HuH-7 cells, however, were fluorescent. Fluorescence was observed in the cytoplasm of HuH-7 cells, but not in the nucleus. (**b**) Fluorescence in the ICG+ NIR− and ICG+ NIR+ groups gradually decreased over time, but retained an extremely high tumor-to-background ratio for up to 6 days. Under the fluorescence imaging, when the fluorescence intensity of the background tissue was defined as 1 AU, the mean HuH-7 tumor-to-background ratio was 255-, 100-, and 72-fold on days 0, 3, and 6, respectively

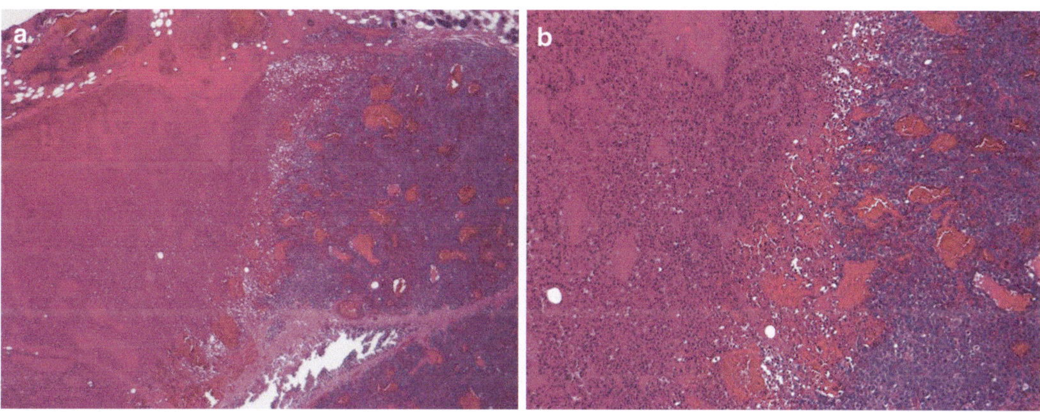

Fig. 19.3 (**a**) Sections from the ICG+ NIR+ group (randomly selected and partially irradiated) were stained with hematoxylin and eosin 36 h after NIR laser exposure. At 100× magnification, a clear boundary was observed between areas with NIR laser exposure (*right side*) and those with no exposure (*left side*). (**b**) NIR laser exposure (*right side*) of HuH-7 tumors induced the following changes: pyknotic nuclei, increased intracellular spacing, and shrinkage of HuH-7 cells. These changes were more evident at 200× magnification

change between days 0 and 3 (217 and 249 mm^3, respectively) and then gradually increased to 1,058 mm^3 by day 9 (See Fig. 19.4b). Mean tumor volume did not differ significantly between the ICG− NIR+ and ICG+ NIR− groups ($p=0.9$), but was significantly different between the ICG+ NIR+ group and both the ICG− NIR+ and ICG+ NIR− groups ($p<0.01$, both; Fig. 19.4c).

Discussion

The findings of the present study indicated that ICG is preferentially taken up by HuH-7 human HCC cell line tumors. The tumor-to-background ratio of HuH-7 tumors, in particular, was extremely high (255:1 at 24 h after ICG administration).

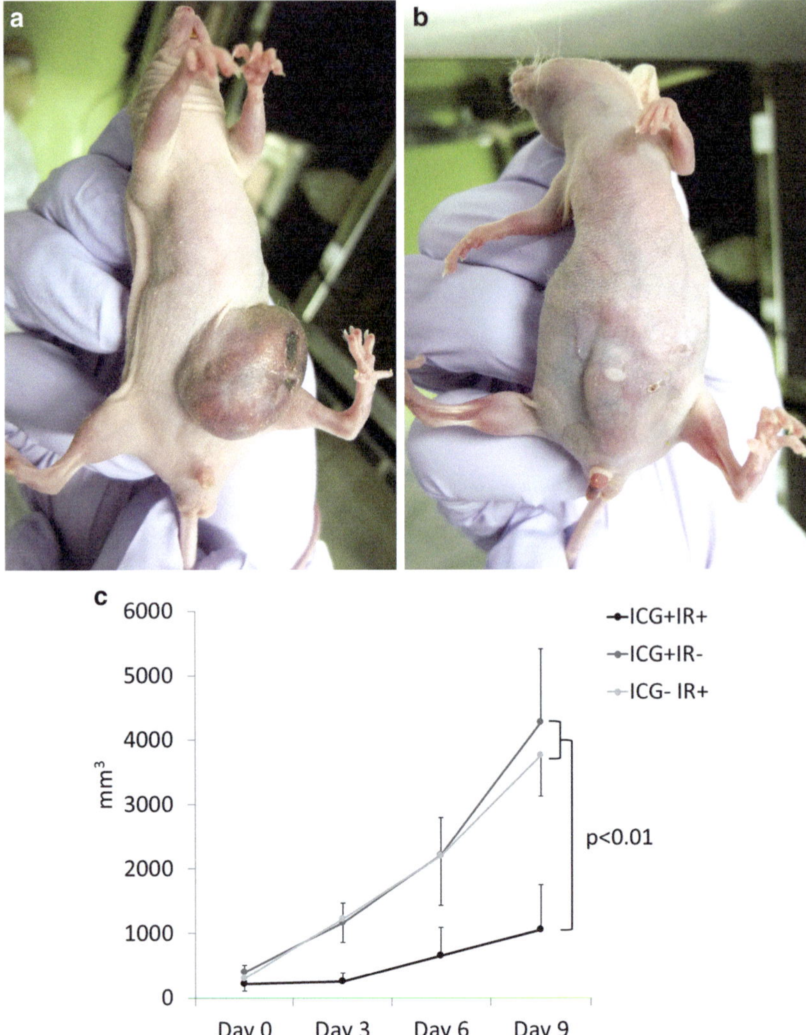

Fig. 19.4 (**a**) Mean tumor volume in the ICG+ NIR− and ICG− NIR+ groups increased steadily from 402 and 298 mm^3 on day 0 to 4,272 and 3,757 mm^3 by day 9, respectively. Large HuH-7 tumor was observed in the abdomen of the transplanted mice (ICG+ NIR− group) at day 9. (**b**) In contrast, mean tumor volume in the ICG+ NIR+ group did not change between days 0 and 3 (217 and 249 mm^3, respectively), and gradually increased to 1,058 mm^3 by day 9. Thin and obscure HuH-7 tumor was observed in the abdomen of the transplanted mice (ICG+ NIR+ group) at day 9. (**c**) Mean tumor volume did not differ significantly between the ICG− NIR+ and ICG+ NIR− groups ($p=0.9$), but was significantly different between the ICG+ NIR+ group and both the ICG− NIR+ and ICG+ NIR− groups ($p<0.01$, both)

In our previous clinical report, we hypothesized that the human HCC was fluorescent because it retained some of the preoperatively administered ICG due to a biliary excretion disorder in the cancerous tissue [9]. Here, we present an animal model that has similar characteristics and can therefore be used to clarify the pharmacokinetics of ICG in HCC cells. The other findings of the present study indicated that PDT with NIR light irradiation suppressed HuH-7 human HCC cell line tumor growth [16].

The detailed effects of combined ICG and NIR laser light PDT is not known. At least two possible mechanisms of action, however, have been reported. ICG absorbs around 800 nm NIR light and 88 % of the absorbed light is converted to heat [10]. Second, PDT generates singlet oxygen, which causes cancer cell death [17]. Although the mechanism was not determined in the present study, we confirmed in the preliminary study that HuH-7 cell behavior was altered and tumor growth was suppressed by the combined ICG and NIR PDT. Further study needs to clarify cell death mechanism [18], using the present study model. The analysis of impaired ICG excretion and correlation biliary transporters of HuH-7 cell line also needs to elucidate strong fluorescence of this cell line tumor [19].

Regarding photosensitizer ICG, in clinical setting, combination of ICG and NIR light is used in dermatology [20] for PDT treatment of port-wine stains. However, there were no reports of ICG and NIR light use for treatment of cancer in clinical situation. According to experimental report, ICG injected directly into a HeLa cell line tumor induced strong fluorescence, but the fluorescence was limited to only a portion of the tumor because the ICG did not infiltrate the entire tumor [21]. Other studies have reported limited success using this ICG and NIR combination for the PDT treatment of several non-HCC cancer cell lines, even though none of these tumor cell lines specifically take up ICG [5, 21, 22]. Those lacks of tumor selectivity of ICG, however, could result in the PDT inducing damage to the tissue surrounding the tumor.

The present PDT for human HCC model has several advantages. First, the fluorescence has a high tumor-to-background ratio. Previously evaluated combinations of other photosensitizers, light sources, and tumor cell lines showed weak tumor selectivity, with a tumor-to-background ratio ranging from 1.4:1 to 7:1 [7, 23–26], which is substantially lower than that obtained in the present HuH-7 tumor model. The selective high tumor-to-background ratio of the present model may enable us to perform safer PDT.

Second advantage of our model is that the ICG remained in the HuH-7 tumor for a long time, maintaining a high tumor-to-background ratio for at least 6 days after ICG administration. Shafirstein and colleagues reported ICG fluorescence in a murine mammary tumor mouse model. Their model showed a lack of tumor selectivity of ICG and the fluorescence continued for only around 15 min. Therefore, NIR light irradiation had to be performed immediately after intravenous ICG administration [5]. In the present study, we examined the effect of only a single session of NIR laser irradiation. As the HuH-7 human HCC cell line remained fluorescent for a long time, several cycles of NIR laser irradiation are possible and may be more effective. On the other hand, although PDT was not performed, Ogawa and colleagues reported an interesting animal model in which florescence continued for 1–4 days in ATAC4 tumor cells with ICG conjugates of a monoclonal antibody [26].

Third advantage of this model is clinical feasibility of ICG. ICG is a commonly used, safe, and inexpensive agent. ICG was first approved for clinical use by the United States Food and Drug Administration in 1956 [27] and is widely used in clinical settings to examine blood volume and cardiac output [28], liver function [29], and to visualize retinal and choroidal vasculature [30]. The reported incidence of adverse reactions after intravenous injection of ICG is quite small (approximately 0.003 %, [29]).

Unlike other light source, NIR has deep penetrating feature. By applying this property of NIR light, PDT may be effective for treating HCCs located in superficial as well as in deep areas of the liver using percutaneous or laparoscopic NIR laser exposure. However, the present study has some limitations. Further studies are needed to evaluate the tissue-penetration characteristics of NIR light and to determine the appropriate output power of NIR laser light and the mechanisms underlying the efficacy of PDT in this experimental model. Moreover, further studies are needed to determine whether other human HCC cell lines preferentially take up the photosensitizer ICG and the effectiveness of PDT.

Acknowledgments This work was supported by grants 21791271 (Kaneko) and 23249067 (Kokudo) from the Ministry of Education, Culture, Sports, Science, and Technology of Japan, and 2011 Tokyo Igakukai Medical Research Grants (Kaneko). We thank Dr. Yutaka Takazawa for helpful comments and suggestions regarding the pathology findings; Harukuni Tsuda for excellent technical assistance with the microscopic fluorescence evaluation; Yasuyuki Morishita for excellent technical assistance with histological preparation.

References

1. Christensen E, Mork C, Skogvoll E. High and sustained efficacy after two sessions of topical 5-aminolaevulinic acid photodynamic therapy for basal cell carcinoma: a prospective, clinical and histological 10-year follow-up study. Br J Dermatol. 2012;166(6):1342–8.
2. Dunki-Jacobs EM, Martin RC. Endoscopic therapy for Barrett's esophagus: a review of its emerging role in optimal diagnosis and endoluminal therapy. Ann Surg Oncol. 2012;19(5):1575–82.
3. Peng Q, Juzeniene A, Chen J, Svaasand LO, Warloe T, Giercksky K-E, et al. Lasers in medicine. Rep Prog Phys. 2008;71(5):056701.
4. Ikeda N, Usuda J, Kato H, Ishizumi T, Ichinose S, Otani K, et al. New aspects of photodynamic therapy for central type early stage lung cancer. Lasers Surg Med. 2011;43(7):749–54.
5. Shafirstein G, Bäumler W, Hennings LJ, Siegel ER, Friedman R, Moreno MA, et al. Indocyanine green enhanced near-infrared laser treatment of murine mammary carcinoma. Int J Cancer. 2012;130(5): 1208–15.
6. van Gemert MC, Welch AJ. Clinical use of laser-tissue interactions. IEEE Eng Med Biol Mag. 1989;8(4):10–3.
7. Mitsunaga M, Ogawa M, Kosaka N, Rosenblum LT, Choyke PL, Kobayashi H. Cancer cell–selective in vivo near infrared photoimmunotherapy targeting specific membrane molecules. Nat Med. 2011;17(12):1685–91.
8. Hock C, Villringer K, Muller-Spahn F, Wenzel R, Heekeren H, Schuh-Hofer S, et al. Decrease in parietal cerebral hemoglobin oxygenation during performance of a verbal fluency task in patients with Alzheimer's disease monitored by means of near-infrared spectroscopy (NIRS)–correlation with simultaneous rCBF-PET measurements. Brain Res. 1997;755(2):293–303.
9. Ishizawa T, Fukushima N, Shibahara J, Masuda K, Tamura S, Aoki T, et al. Real-time identification of liver cancers by using indocyanine green fluorescent imaging. Cancer. 2009;115(11):2491–504.
10. Reindl S, Penzkofer A, Gong SH, Landthaler M, Szeimies RM, Abels C, et al. Quantum yield of triplet formation for indocyanine green. J Photochem Photobiol A Chem. 1997;105(1):65–8.
11. Nakabayashi H, Taketa K, Miyano K, Yamane T, Sato J. Growth of human hepatoma cells lines with differentiated functions in chemically defined medium. Cancer Res. 1982;42(9):3858–63.
12. Inagaki Y. Effect of c-Met inhibitor SU11274 on hepatocellular carcinoma cell growth. Biosci Trends. 2011;5(2):52–6.
13. Diehl KH, Hull R, Morton D, Pfister R, Rabemampianina Y, Smith D, et al. A good practice guide to the administration of substances and removal of blood, including routes and volumes. J Appl Toxicol. 2001;21(1):15–23.
14. Kitai T, Inomoto T, Miwa M, Shikayama T. Fluorescence navigation with indocyanine green for detecting sentinel lymph nodes in breast cancer. Breast Cancer. 2005;12(3):211–5.
15. Kwon S, Sevick-Muraca EM. Non-invasive, dynamic imaging of murine intestinal motility. Neurogastroenterol Motil. 2011;23(9):881–e344.
16. Kaneko J, Inagaki Y, Ishizawa T, Gao J, Tang W, Aoki T, et al. Photodynamic therapy for human hepatoma-cell-line tumors utilizing biliary excretion properties of indocyanine green. J Gastroenterol. 2014;49(1): 110–6.
17. Brown SB, Brown EA, Walker I. The present and future role of photodynamic therapy in cancer treatment. Lancet Oncol. 2004;5(8):497–508.
18. Kessel D. Death pathways associated with photodynamic therapy. Med Laser Appl. 2006;21(4): 219–24.
19. Ishizawa T, Masuda K, Urano Y, Kawaguchi Y, Satou S, Kaneko J, et al. Mechanistic background and clinical applications of indocyanine green fluorescence imaging of hepatocellular carcinoma. Ann Surg Oncol. 2014;21(2):440–8.
20. Klein A, Szeimies RM, Baumler W, Zeman F, Schreml S, Hohenleutner U, et al. Indocyanine green-augmented diode laser treatment of port-wine stains: clinical and histological evidence for a new treatment option from a randomized controlled trial. Br J Dermatol. 2012;167(2):333–42.
21. Hirano T, Kohno E, Gohto Y, Obana A. Singlet oxygen generation by irradiation of indocyanine green and its effect to tissues. J Japan Soc Laser Surg Med. 2007;28:122–8.
22. Urbanska K, Romanowska-Dixon B, Matuszak Z, Oszajca J, Nowak-Sliwinska P, Stochel G. Indocyanine green as a prospective sensitizer for photodynamic therapy of melanomas. Acta Biochim Pol. 2002;49(2):387–91.
23. Thayer D, Unlu MB, Lin Y, Yan K, Nalcioglu O, Gulsen G. Dual-contrast dynamic MRI-DOT for small animal imaging. Technol Cancer Res Treat. 2010;9(1):61–70.
24. Kim TH, Mount CW, Dulken BW, Ramos J, Fu CJ, Khant HA, et al. Filamentous, mixed micelles of triblock copolymers enhance tumor localization of indocyanine green in a murine xenograft model. Mol Pharm. 2012;9(1):135–43.
25. Otake M, Nishiwaki M, Kobayashi Y, Baba S, Kohno E, Kawasaki T, et al. Selective accumulation of ALA-induced PpIX and photodynamic effect in chemically induced hepatocellular carcinoma. Br J Cancer. 2003;89(4):730–6.
26. Ogawa M, Kosaka N, Choyke PL, Kobayashi H. In vivo molecular imaging of cancer with a quenching near-infrared fluorescent probe using conjugates of monoclonal antibodies and indocyanine green. Cancer Res. 2009;69(4):1268–72.

27. Fox IJ, Wood EH. Applications of dilution curves recorded from the right side of the heart or venous circulation with the aid of a new indicator dye. Proc Staff Meet Mayo Clin. 1957;32(19):541–50.
28. Iijima T, Aoyagi T, Iwao Y, Masuda J, Fuse M, Kobayashi N, et al. Cardiac output and circulating blood volume analysis by pulse dye-densitometry. J Clin Monit. 1997;13(2):81–9.
29. Cherrick GR, Stein SW, Leevy CM, Davidson CS. Indocyanine green: observations on its physical properties, plasma decay, and hepatic extraction. J Clin Invest. 1960;39:592–600.
30. Hochheimer BF. Angiography of the retina with indocyanine green. Arch Ophthalmol. 1971;86(5):564–5.

Photodynamic Diagnosis of Gastric Cancer Using 5-Aminolevulinic Acid

20

Tsutomu Namikawa, Keiji Inoue, Taro Shuin, and Kazuhiro Hanazaki

Photodynamic Diagnosis

Protoporphyrin IX (PpIX) is an immediate precursor in the heme biosynthetic pathway that also has photoexcitable properties [1, 2]. Photodynamic diagnosis (PDD) is a relatively new modality that relies on the tumor specificity of photoactive PpIX accumulation, whereby the PpIX is photoexcited using blue light of 600–740 nm wavelength to emit a red fluorescence signal to detect cancer [2–4].

Our institution uses an endoscopic PDD system (Karl Storz, Tuttlingen, Germany) comprising a CCU Tricam SLII/3CCD CH Tricam-P PDD, D-Light C, and HOPKINSII Straight Forward Telescope 30° (Karl Storz) [3–6] (Fig. 20.1). The D-Light C light source (300 Watt xenon arc lamp; Karl Storz) is equipped with a band-pass filter designed to transmit blue light (excitation wavelength, 375–445 nm) and the CCU Tricam SLII/3CCD CHTricam-P PDD video camera system is equipped with a long-pass filter designed to exclude blue light for fluorescence imaging (fluorescence emission wavelength, 600–740 nm). This PDD system has the advantage that it can switch instantly between the blue light mode for fluorescence imaging and the white light mode for conventional observation.

Photosensitive Substance

PDD imaging systems were recently improved to enable detection of premalignant and early-stage malignant lesions in the lung and esophagus based on systemic administration of photosensitizing drugs, which are mostly porphyrin compounds such as Photofrin [7, 8]. However, the use of Photofrin has considerable disadvantages, such as strong phototoxic skin reactions that can persist for weeks.

5-Aminolevulinic acid (ALA) is a naturally occurring amino acid derivative that acts as an endogenous substrate and precursor to PpIX in the heme biosynthetic pathway. Although the underlying mechanism is not well known, exogenously administered ALA increases cellular levels of PpIX in a number of tissues, particularly at epithelial surfaces, and higher accumulation of PpIX occurs in tumor cells than in normal cells [5, 9, 10]. PDD using ALA (ALA-PDD) is clinically recognized as an effective procedure for

Electronic supplementary material: The online version of this chapter (doi:10.1007/978-3-319-15678-1_20) contains supplementary material, which is available to authorized users. Videos can also be accessed at http://link.springer.com/chapter/10.1007/978-3-319-15678-1_20.

T. Namikawa, M.D., Ph.D. (✉)
K. Hanazaki, M.D., Ph.D.
Department of Surgery, Kochi Medical School, Kohasu, Oko-cho, Nankoku, Kochi 783-8505, Japan
e-mail: tsutomun@kochi-u.ac.jp

K. Inoue, M.D., Ph.D. • T. Shuin, M.D., Ph.D.
Department of Urology, Kochi Medical School, Nankoku, Japan

F.D. Dip et al. (eds.), *Fluorescence Imaging for Surgeons: Concepts and Applications*,
DOI 10.1007/978-3-319-15678-1_20, © Springer International Publishing Switzerland 2015

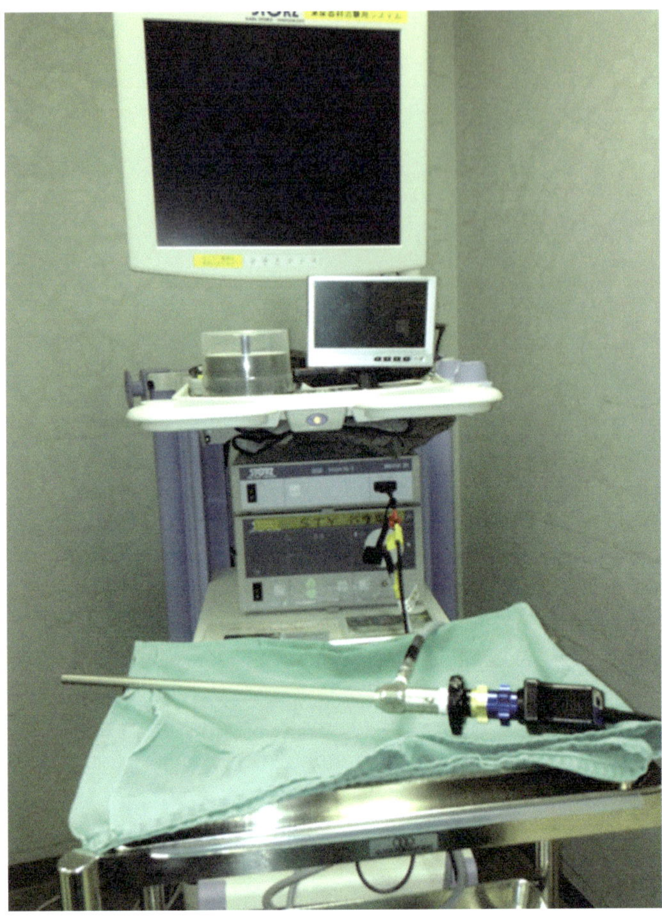

Fig. 20.1 Photodynamic diagnosis system. An endoscopic photodynamic diagnosis system comprising a straightforward telescope, light source, and a video camera system

detecting various cancers including brain tumor, bladder cancer, and prostatic cancer [3, 11, 12]. ALA offers several advantages as a photosensitizer for PDD, including the low risk of side effects; indeed, increases in serum aminotransaminase levels are only temporary and skin phototoxicity is limited to 1 or 2 days with ALA [2, 3, 13].

In our experience, no special precautions have been necessary during ALA-PDD, such as liver support or light shielding, and no adverse events were encountered. However, although ALA-PDD is an officially approved diagnostic procedure in Europe for detecting bladder cancer, many problems remain concerning the accuracy of diagnosis due to false-positive fluorescence and photobleaching during the procedure. Thus, we investigated using ALA as a photosensitizer for PDD in gastric cancer. ALA hydrochloride (Cosmo Bio Co., Ltd., Tokyo, Japan) was dissolved in 50 mL of 5 % glucose solution and 1.0 g of this solution was given orally through a stomach tube 3–4 h before the intraoperative PDD observation. Patients were shielded from direct sunlight for 24 h to avoid phototoxicity.

PDD for Gastric Cancer

Gastric cancer is the fourth most common cancer and second most common cause of cancer-related death worldwide [14]. It is the most prevalent form of cancer in Japan and East Asia, and the treatment of gastric cancer often requires gastrectomy with sufficient lymphadenectomy [15]. Surgical treatment achieves good rates of disease control, particularly in the early stages of gastric cancer [16], and the 5-year overall survival rate in Japanese gastrectomy patients with Stage I gastric cancer exceeds 90 % [17, 18]. Generally, diagnosis of gastric cancer relies on endoscopy and subsequent histological examination of tumor biopsies; however, several difficulties in diagnosis remain with regard to determining the extent of cancer infiltration in some carcinomas including the absence of typical morphological structures and indistinct margins [19]. Although esophagogastroduodenoscopy (EGD) is highly reliable in defining the area of cancer infiltration in early gastric cancer, it remains difficult to diagnose gastric cancer accurately in some cases such as gastric-type differentiated adenocarcinoma if biopsy specimens do not reveal the presence of adenocarcinoma [20, 21]. Only a few studies using ALA-PDD have been reported for cases of gastric cancer, and the clinical diagnostic accuracy and sensitivity of this approach as an alternative for diagnosing gastric tumors remain unclear [6, 22, 23].

Recently, we reported the feasibility and efficacy of intraoperative ALA-PDD for the identification of gastric cancer [4]. Intraoperative PDD using 1.0 g of ALA solution given orally through a stomach tube was carried out for 26 lesions of 21 patients with gastric cancer. Figure 20.2 shows a 71-year-old female who underwent proximal gastrectomy. Our ALA-PDD system revealed a slightly elevated lesion in this patient with a distinct margin demarcated by an intense fluorescent signal (Fig. 20.3 and Video 20.1). This lesion was pathologically diagnosed as well-differentiated adenocarcinoma confined to the mucosal layer.

Gastric carcinoma has been classified into two major histological types, intestinal and diffuse, according to the Lauren classification system [24]. In our studies, the incidence of intestinal-type tumor was significantly higher in PDD-positive than in PDD-negative cases (93.3 % vs. 27.3 %; $P<0.001$). Specifically, we found that significantly more ALA-PDD positive lesions were intestinal-type than diffuse-type gastric cancer. Interestingly, this finding could reflect differences in the porphyrin biosynthesis pathway at play in different histological types of gastric cancer [25–27].

Red fluorescence was detected in 15 lesions of 11 patients, and the sensitivity, specificity, positive predictive value, negative predictive value, and accuracy of ALA-PDD in detecting gastric cancer was measured at 57.7, 100, 100, 38.9, and 66.7 %, respectively (Table 20.1). The sensitivity and accuracy of ALA-PDD for detecting gastric

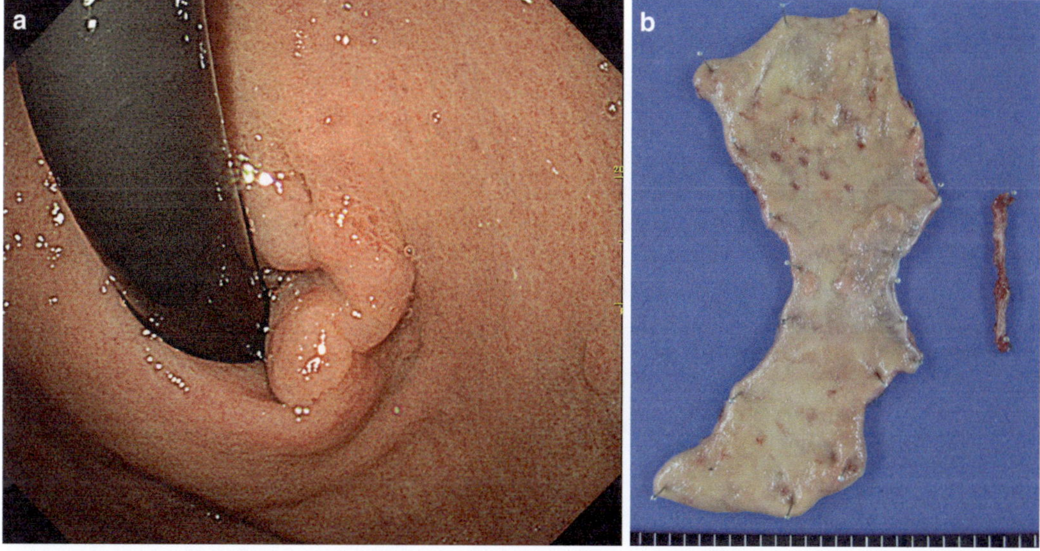

Fig. 20.2 Early gastric cancer based on photodynamic diagnosis. Esophagogastroduodenoscopy showed an early gastric cancer located in the cardia near the esophagogastric junction (**a**). The patient underwent proximal gastrectomy (**b**)

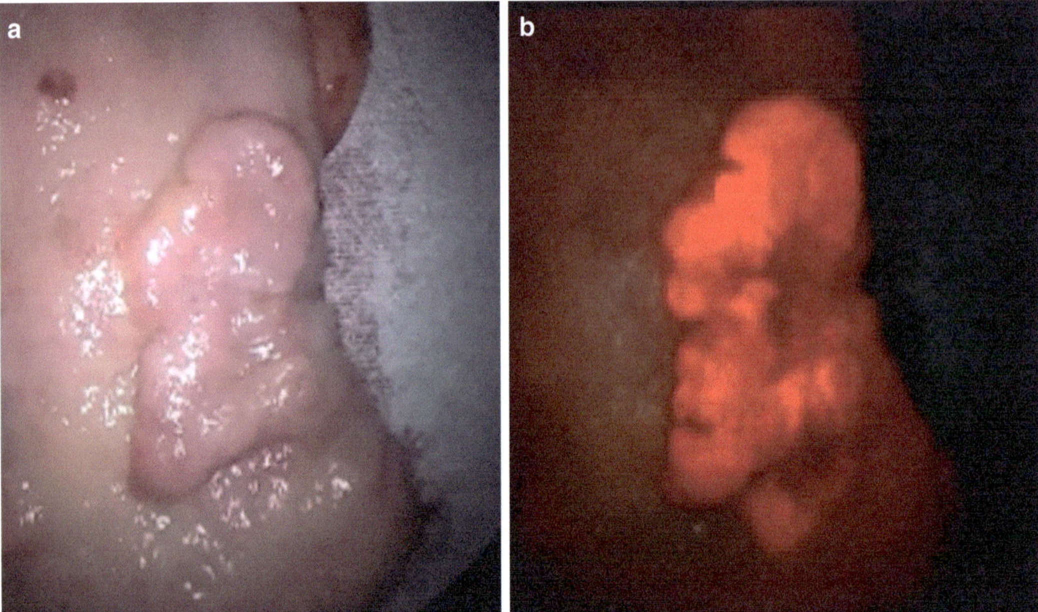

Fig. 20.3 Photodynamic diagnosis findings of gastric cancer compared to those using white light. Under white light, the tumor appeared as a slightly elevated lesion (**a**). Under ALA-induced fluorescence, the tumor emitted a red fluorescence signal that was consistent with the area of adenocarcinoma defined by pathological examination (**b**)

Table 20.1 Diagnostic ability for ALA-PDD in gastric cancer

	Sensitivity (%)	Specificity (%)	Accuracy (%)	PPV (%)	NPV (%)
Intestinal type	82.4	100	87.5	100	70
Total	57.7	100	66.7	100	38.9

PPV positive predictive value, *NPV* negative predictive value

tumors was therefore low in our clinical application of this modality to gastric cancer patients, and although the sensitivity of ALA-PDD for intestinal-type lesions was high, it was low for diffuse-type gastric cancer. Among only intestinal-type gastric tumors, the sensitivity, specificity, positive predictive value, negative predictive value, and accuracy of ALA-PDD in detecting gastric cancer was measured at 82.4, 100, 100, 87.5, and 70.0 %, respectively (Table 20.1).

Mechanism of PPIX Accumulation in Cancer Cells

The oligopeptide transporters (PEPT) such as peptide transporter 1 and 2 (PEPT 1 and 2) are reportedly involved in the cellular uptake of ALA [28–30]. Two molecules of ALA are converted to porphobilinogen, which is metabolized to porphyrinogen intermediates by porphobilinogen deaminase (PBG-D). The last step of metabolism for ALA is the incorporation of iron into PpIX, catalyzed by ferrochelatase (FC) [5]. Previous studies reported that some cancer cells have an increased PBG-D activity [31, 32] and a decreased FC activity [31, 33]. Therefore, ALA-induced PpIX accumulation is possibly due to the limited capacity of FC or the higher activity of PBG-D in tumors [34].

Recent studies also suggested that the expression of particular proteins could affect ALA-mediated PpIX accumulation in tumors, including PEPT1 and 2, FC, and ATP-binding cassette transporter G2 (ABCG2) [25, 26, 35]. Indeed, PEPT1 and ABCG2 showed pivotal roles in regulating intracellular PpIX levels and cellular photosensitivity in gastric cancer cells treated with ALA [25]. Experiments in lung adenocarcinoma, renal cancer, and colorectal cancer cell lines also indicated that mitochondrial ABCG2 might regulate ALA-mediated PpIX levels through exporting it from mitochondria to the cytosol [26]. In a gastric cancer cell line, high expression of PEPT1 and low expression of ABCG2 correlated with ALA-induced PpIX production and cellular photosensitivity [25]. Distribution of biomarkers including PEPT and ABCG depending on histological type might therefore underlie the observed differences in ALA-PDD positivity among gastric cancers.

Future Direction of PDD for Gastric Cancer

It is expected that PDD will be useful for the following future directions (Table 20.2). First, it could be used a tool for evaluating surgical resection margin, and thereby assisting the pathological diagnosis during surgery. Diagnostic ability without any adverse events is consistent with a recent study of ALA-PDD showing outstanding sensitivity for the detection of superficial bladder cancer, particularly for flat lesions such as dysplasia and carcinoma in situ without additional complication [3]. Superficial spreading-type cancer of the stomach also has some specific clinicopathological characteristics compared to the common type of gastric cancer [17]. For instance, in a case of duodenal invasion of early gastric cancer, the invasion was more extensive in superficial spreading lesions than in small cancer lesions [36, 37]. In such cases, ALA-PDD might be useful in determining the extent of gastric adenocarcinoma in cases with indistinct margins, and provide additional information when judging the sufficiency of proximal or distal margins during surgery.

Table 20.2 Feature direction of photodynamic diagnosis for gastric cancer

- Evaluation of surgical resection margin
  - Assist pathological diagnosis during surgery
- Evaluation of lesion extension
  - Supplemental modality to image-enhanced endoscopy and magnifying endoscopy
- Diagnosis of peritoneal metastasis during preoperative staging laparoscopy
  - Provide useful information for selecting a therapeutic modality

Moreover, it is expected that ALA-PDD could provide not only early detection of gastric cancer, but also a multimodal diagnostic tool for use during less invasive therapies such as endoscopic mucosal dissection (ESD). In such situations, ALA-PDD might prove valuable for evaluating the lesion extent as a supplemental modality to image-enhanced endoscopy and magnifying endoscopy. In Japan, ESD is commonly performed to treat early gastric cancer, allowing en bloc resection [38, 39]; however, it is important to detect any extension of the cancer [37, 40] and in this regard, ALA-PDD could provide additional information to that obtained with conventional endoscopic observation.

Another future direction for ALA-PDD is in the diagnosis of peritoneal metastasis during preoperative staging laparoscopy. Peritoneal metastasis caused by the seeding of exfoliated cancer cells from the primary gastric cancer is the most common type of spread in advanced gastric cancer with serosa-invading tumors [15, 41]. Although staging laparoscopy is used frequently in the management of patients with advanced gastric cancer to prevent unnecessary laparotomy [42, 43], it has limitations in visualizing dissemination of the cancer nest [44]. To this end, some investigators reported the effectiveness of ALA-PDD for early endoscopic detection of malignant esophageal lesions, and for the staging laparoscopy of advanced gastric cancer due to improvements in the detection sensitivity for peritoneal metastases [6, 13, 22]. Thus, ALS-PDD could also provide information that aids in selecting a therapeutic modality.

Conclusions

ALA-PDD is a promising and safe diagnostic modality for determining tumor extent and detecting metastatic lesions in gastric cancer. Fluorescence navigation by PDD provides good visualization and detection of gastric cancer, and could be particularly useful in cases with intestinal-type gastric lesions. Additionally, we have experienced no serious adverse events after administration of ALA, and ALA-PDD during gastric cancer surgery proved to be a safe procedure. However, further investigations such as multicenter studies with adequate statistical power and a larger number of patient subgroups, including a prospective randomized controlled trial, are needed to verify these conclusions.

References

1. Kelty CJ, Brown NJ, Reed MW, Ackroyd R. The use of 5-aminolaevulinic acid as a photosensitiser in photodynamic therapy and photodiagnosis. Photochem Photobiol Sci. 2002;1(3):158–68.
2. Inoue K, Karashima T, Kamada M, Shuin T, Kurabayashi A, Furihata M, et al. Regulation of 5-aminolevulinic acid-mediated protoporphyrin IX accumulation in human urothelial carcinomas. Pathobiology. 2009;76(6):303–14.
3. Inoue K, Fukuhara H, Shimamoto T, Kamada M, Iiyama T, Miyamura M, et al. Comparison between intravesical and oral administration of 5-aminolevulinic acid in the clinical benefit of photodynamic diagnosis for nonmuscle invasive bladder cancer. Cancer. 2012; 118(4):1062–74.
4. Namikawa T, Inoue K, Uemura S, Shiga M, Maeda H, Kitagawa H, et al. Photodynamic diagnosis using 5-aminolevulinic acid during gastrectomy for gastric cancer. J Surg Oncol. 2014;109(3):213–7. Epub 2013.
5. Hinnen P, de Rooij FW, van Velthuysen ML, Edixhoven A, van Hillegersberg R, Tilanus HW, et al. Biochemical basis of 5-aminolaevulinic acid-induced protoporphyrin IX accumulation: a study in patients with (pre)malignant lesions of the oesophagus. Br J Cancer. 1998;78(5):679–82.
6. Kishi K, Fujiwara Y, Yano M, Inoue M, Miyashiro I, Motoori M, et al. Staging laparoscopy using ALA-mediated photodynamic diagnosis improves the detection of peritoneal metastases in advanced gastric cancer. J Surg Oncol. 2012;106(3):294–8.
7. Lam S, Palcic B, McLean D, Hung J, Korbelik M, Profio AE. Detection of early lung cancer using low dose Photofrin II. Chest. 1990;97(2):333–7.
8. von Holstein CS, Nilsson AM, Andersson-Engels S, Willén R, Walther B, Svanberg K. Detection of adenocarcinoma in Barrett's oesophagus by means of laser induced fluorescence. Gut. 1996;39(5):711–6.
9. Loh CS, Vernon D, MacRobert AJ, Bedwell J, Bown SG, Brown SB. Endogenous porphyrin distribution induced by 5-aminolaevulinic acid in the tissue layers of the gastrointestinal tract. J Photochem Photobiol B. 1993;20(1):47–54.
10. Krieg RC, Messmann H, Rauch J, Seeger S, Knuechel R. Metabolic characterization of tumor cell-specific protoporphyrin IX accumulation after exposure to 5-aminolevulinic acid in human colonic cells. Photochem Photobiol. 2002;76(5):518–25.
11. Utsuki S, Miyoshi N, Oka H, Miyajima Y, Shimizu S, Suzuki S, et al. Fluorescence-guided resection of metastatic brain tumors using a 5-aminolevulinic acid-induced protoporphyrin IX: pathological study. Brain Tumor Pathol. 2007;24(2):53–5.
12. Fukuhara H, Inoue K, Satake H, Tamura K, Karashima T, Yamasaki I, et al. Photodynamic diagnosis of positive margin during radical prostatectomy: preliminary experience with 5-aminolevulinic acid. Int J Urol. 2011;18(8):585–91.
13. Mayinger B, Neidhardt S, Reh H, Martus P, Hahn EG. Fluorescence induced with 5-aminolevulinic acid for the endoscopic detection and follow-up of esophageal lesions. Gastrointest Endosc. 2001;54(5):572–8.
14. Jemal A, Bray F, Center MM, Ferlay J, Ward E, Forman D. Global cancer statistics. CA Cancer J Clin. 2011;61(2):69–90.
15. Nashimoto A, Akazawa K, Isobe Y, Miyashiro I, Katai H, Kodera Y, et al. Gastric cancer treated in 2002 in Japan: 2009 annual report of the JGCA nationwide registry. Gastric Cancer. 2013;16(1):1–27.
16. Sano T, Sasako M, Kinoshita T, Maruyama K. Recurrence of early gastric cancer. Follow-up of 1475 patients and review of the Japanese literature. Cancer. 1993;72(11):3174–8.
17. Namikawa T, Kobayashi M, Kitagawa H, Okabayashi T, Sugimoto T, Kuratani Y, et al. Differentiated adenocarcinoma with a gastric phenotype in the stomach: difficulties in clinical and pathological diagnoses. Clin J Gastroenterol. 2009;2(4):268–74.
18. Okabayashi T, Kobayashi M, Nishimori I, Sugimoto T, Namikawa T, Onishi S, et al. Clinicopathological features and medical management of early gastric cancer. Am J Surg. 2008;195(2):229–32.
19. Namikawa T, Hanazaki K. Clinicopathological features of early gastric cancer with duodenal invasion. World J Gastroenterol. 2009;15(19):2309–13.
20. Namikawa T, Hanazaki K. Mucin phenotype of gastric cancer and clinicopathology of gastric-type differentiated adenocarcinoma. World J Gastroenterol. 2010;16(37):4634–9.
21. Kabashima A, Yao T, Maehara Y, Tsuneyoshi M. Relationship between biological behavior and phenotypic expression in undifferentiated-type gastric carcinomas. Gastric Cancer. 2005;8(4):220–7.
22. Murayama Y, Ichikawa D, Koizumi N, Komatsu S, Shiozaki A, Kuriu Y, et al. Staging fluorescence laparoscopy for gastric cancer by using 5-aminolevulinic acid. Anticancer Res. 2012;32(12):5421–7.
23. Mayinger B, Reh H, Hochberger J, Hahn EG. Endoscopic photodynamic diagnosis: oral ami-

nolevulinic acid is a marker of GI cancer and dysplastic lesions. Gastrointest Endosc. 1999;50(2):242–6.
24. Lauren P. The two histological main types of gastric carcinoma: diffuse and so-called intestinal-type carcinoma: an attempt at a histoclinical classification. Acta Pathol Microbiol Scand. 1965;64:31–49.
25. Hagiya Y, Endo Y, Yonemura Y, Takahashi K, Ishizuka M, Abe F, et al. Pivotal roles of peptide transporter PEPT1 and ATP-binding cassette (ABC) transporter ABCG2 in 5-aminolevulinic acid (ALA)-based photocytotoxicity of gastric cancer cells in vitro. Photodiagnosis Photodyn Ther. 2012;9(3):204–14.
26. Kobuchi H, Moriya K, Ogino T, Fujita H, Inoue K, Shuin T, et al. Mitochondrial localization of ABC transporter ABCG2 and its function in 5-aminolevulinic acid-mediated protoporphyrin IX accumulation. PLoS One. 2012;7(11):e50082.
27. Inoue K, Fukuhara H, Kurabayashi A, Furihata M, Tsuda M, Nagakawa K, et al. Photodynamic therapy involves an antiangiogenic mechanism and is enhanced by ferrochelatase inhibitor in urothelial carcinoma. Cancer Sci. 2013;104(6):765–72.
28. Rodriguez L, Batlle A, Di Venosa G, MacRobert AJ, Battah S, Daniel H, et al. Study of the mechanisms of uptake of 5-aminolevulinic acid derivatives by PEPT1 and PEPT2 transporters as a tool to improve photodynamic therapy of tumours. Int J Biochem Cell Biol. 2006;38(9):1530–9.
29. Novotny A, Xiang J, Stummer W, Teuscher NS, Smith DE, Keep RF. Mechanisms of 5-aminolevulinic acid uptake at the choroid plexus. J Neurochem. 2000; 75(1):321–8.
30. Döring F, Walter J, Will J, Föcking M, Boll M, Amasheh S, et al. Delta-aminolevulinic acid transport by intestinal and renal peptide transporters and its physiological and clinical implications. J Clin Invest. 1998;101(12):2761–7.
31. el-Sharabasy MM, el-Waseef AM, Hafez MM, Salim SA. Porphyrin metabolism in some malignant diseases. Br J Cancer. 1992;65(3):409–12.
32. Navone NM, Polo CF, Frisardi AL, Andrade NE, Battle AM. Heme biosynthesis in human breast cancer–mimetic "in vitro" studies and some heme enzymic activity levels. Int J Biochem. 1990;22(12):1407–11.
33. Van Hillegersberg R, Van den Berg JW, Kort WJ, Terpstra OT, Wilson JH. Selective accumulation of endogenously produced porphyrins in a liver metastasis model in rats. Gastroenterology. 1992;103(2): 647–51.
34. Peng Q, Warloe T, Berg K, Moan J, Kongshaug M, Giercksky KE, et al. 5-Aminolevulinic acid-based photodynamic therapy. Clinical research and future challenges. Cancer. 1997;79(12):2282–308.
35. Robey RW, Steadman K, Polgar O, Bates SE. ABCG2-mediated transport of photosensitizers: potential impact on photodynamic therapy. Cancer Biol Ther. 2005;4(2):187–94.
36. Namikawa T, Kobayashi M, Kitagawa H, Okabayashi T, Dabanaka K, Okamoto K, et al. Early gastric cancer with widespread duodenal invasion within the mucosa. Dig Endosc. 2010;22(3):223–7.
37. Namikawa T, Kitagawa H, Iwabu J, Okabayashi T, Sugimoto T, Kobayashi M, et al. Clinicopathological properties of the superficial spreading type early gastric cancer. J Gastrointest Surg. 2010; 14(1):52–7.
38. Cao Y, Liao C, Tan A, Gao Y, Mo Z, Gao F. Meta-analysis of endoscopic submucosal dissection versus endoscopic mucosal resection for tumors of the gastrointestinal tract. Endoscopy. 2009;41(9):751–7.
39. Abe S, Oda I, Suzuki H, Nonaka S, Yoshinaga S, Odagaki T, et al. Short- and long-term outcomes of endoscopic submucosal dissection for undifferentiated early gastric cancer. Endoscopy. 2013;45(9): 703–7.
40. Abe N, Takeuchi H, Ooki A, Nagao G, Masaki T, Mori T, et al. Recent developments in gastric endoscopic submucosal dissection: towards the era of endoscopic resection of layers deeper than the submucosa. Dig Endosc. 2013;25 Suppl 1:64–70.
41. Koizumi W, Narahara H, Hara T, Takagane A, Akiya T, Takagi M, et al. S-1 plus cisplatin versus S-1 alone for first-line treatment of advanced gastric cancer (SPIRITS trial): a phase III trial. Lancet Oncol. 2008; 9(3):215–21.
42. Cardona K, Zhou Q, Gönen M, Shah MA, Strong VE, Brennan MF, et al. Role of repeat staging laparoscopy in locoregionally advanced gastric or gastroesophageal cancer after neoadjuvant therapy. Ann Surg Oncol. 2013;20(2):548–54.
43. Karanicolas PJ, Elkin EB, Jacks LM, Atoria CL, Strong VE, Brennan MF, et al. Staging laparoscopy in the management of gastric cancer: a population-based analysis. J Am Coll Surg. 2011;213(5):644–51.
44. Muntean V, Mihailov A, Iancu C, Toganel R, Fabian O, Domsa I, et al. Staging laparoscopy in gastric cancer. Accuracy and impact on therapy. J Gastrointestin Liver Dis. 2009;18(2):189–95.

Applications of ICG Fluorescence Imaging for Surgery in Colorectal Cancers

21

Shingo Shimada, Seiji Ohtsubo, and Mitsuo Kusano

Introduction

The usefulness of indocyanine green (ICG) fluorescence imaging (IFI) for fluorescence-guided surgery has been expanding [1–5]. In this study, we focus on the application of IFI in the surgery of colorectal cancers. Recently, gastrointestinal cancers including colorectal cancers have been identified as new targets for sentinel node navigation surgery (SNNS) [6–12]. In this section, we describe, first, the feasibility of sentinel lymph node (SLN) mapping guided by IFI in colorectal cancer and, second, the efficacy of the new tattooing method using IFI instead of India ink for tumor detection. Endoscopic tattooing with India ink is currently the agent of choice but has some complications [13–15]. Finally, we evaluated the IFI method of harvesting lymph nodes (LNs) from resected specimen from the viewpoint of accuracy of LN staging because the number of pathological LNs is very important for pathological staging.

S. Shimada, M.D. (✉) • M. Kusano, M.D., Ph.D.
Department of Surgery, Kushiro Rosai Hospital, Japan Labor Health and Welfare Organization, Hokkaido, Japan
e-mail: shingoshimada1979@true.ocn.ne.jp

S. Ohtsubo, M.D., Ph.D.
Department of Oral and Maxillofacial Surgery, Kushiro Rosai Hospital, Japan Labor Health and Welfare Organization, Hokkaido, Japan

Clinical Application

Sentinel Lymph Node Mapping by ICG Fluorescence Imaging

Background

SLN biopsy (SLNB) is often used and is a highly effective and accurate predictor of nodal metastasis in melanoma [16] and breast cancer [17]. Recently, several studies supporting the validity of the SLN concept for gastrointestinal cancer have been reported [6–12]. It has been suggested that SNNS could allow us to minimalize surgical invasion with individualized lymphadenectomy and to help improve postoperative outcomes and quality of life in several malignancies, including gastrointestinal cancer [6–12]. Here, we describe the feasibility of the SLN mapping guided by IFI in colorectal cancer.

Materials and Methods

Our series consisted of 26 patients with colorectal cancer who had undergone standard surgical resection. After laparotomy, 0.5 ml of an ICG solution prepared by dissolving 25 mg of powdered ICG in 10 ml of sterile physiologic saline was injected intraoperatively into the four points of the subserosa around the tumor. IFI was obtained by a charge-coupled device (CCD) camera with a light-emitting diode having a 760-nm wavelength as the light source and a cut filter to

F.D. Dip et al. (eds.), *Fluorescence Imaging for Surgeons: Concepts and Applications*,
DOI 10.1007/978-3-319-15678-1_21, © Springer International Publishing Switzerland 2015

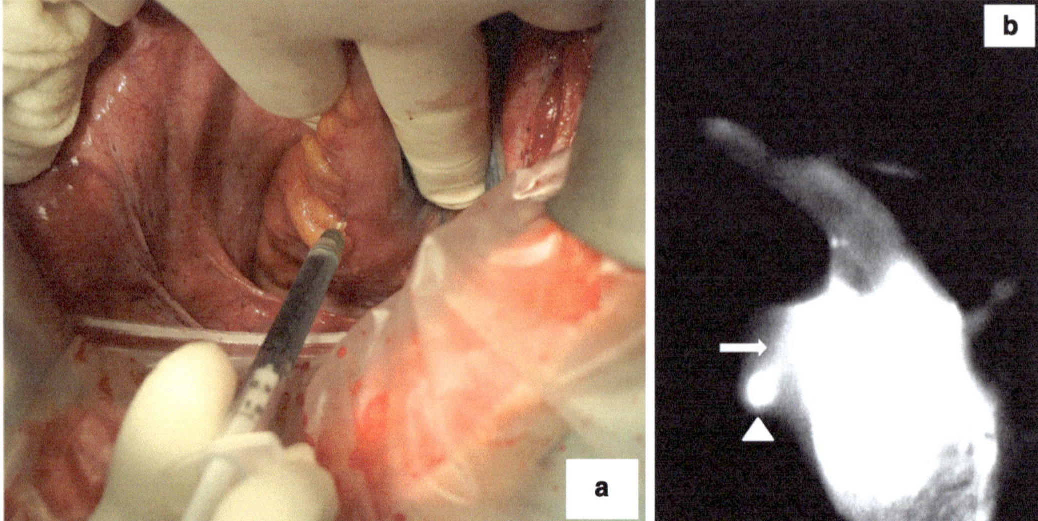

Fig. 21.1 Sentinel lymph node mapping. (**a**) ICG solution was injected intraoperatively into the four points of the subserosa around the tumor. (**b**) Lymph node (*arrowhead*) and lymph duct (*arrow*) were emitted

remove light with wavelengths below 820 nm (PDE system). We identified lymph nodes depicted by high brightness as SLNs and excised them. Then, all patients underwent standard operation with lymphadenectomy. The excised LNs (i.e., the SLNs) were evaluated by hematoxylin-eosin (HE) staining after operation.

Results

Immediately after the ICG injection (Fig. 21.1a), the lymphatic vessels draining the tumor and round-shaped SLNs were clearly visualized by their bright fluorescence (Fig. 21.1b). None of the SLNs were visible as green. However, even SLNs that were not green in color could be easily and clearly visualized by ICG fluorescence imaging. The SLN detection rate and mean number of SLNs were 88 % and 2.6.

Discussion

Here, we describe a way to detect SLNs of colorectal cancer using IFI. The methods for detecting SLNs can be categorized as dye methods, radioisotope (RI) methods, ICG fluorescence (IF) methods, and combination methods according to tracer. Dye methods offer only poor tissue contrast, which makes detection of SLN in deep, dark anatomical regions [18, 19] such as the lower rectum extremely difficult. RI methods are available without sighting of the lymphatic flow. However, they are expensive and there are limitations as to which institutions can use them [18, 19]. While IF methods require sighting of lymphatic flow, the visibility of the IF method is superior to the visibility of dye method, and the IF methods cost less than RI methods [2, 20, 21]. Hirche et al. reported that ICG guidance has been shown not to be influenced by the body mass index (BMI) or lymphatic invasion, whereas conventional methods are [21].

In this study, lymphatic vessels and SLN were clearly visualized by their bright fluorescence in real time.

As for colorectal cancer, the significance of SLNB itself is still controversial. But it may help reduce invasive operations and optimize lymph node dissection. SLNB using IFI is a promising tool that deserves further clinical exploration.

Tattooing of Tumor Location by ICG Fluorescence Imaging

Background

In early colorectal cancer, the precise detection of tumor location is often difficult due to unclear

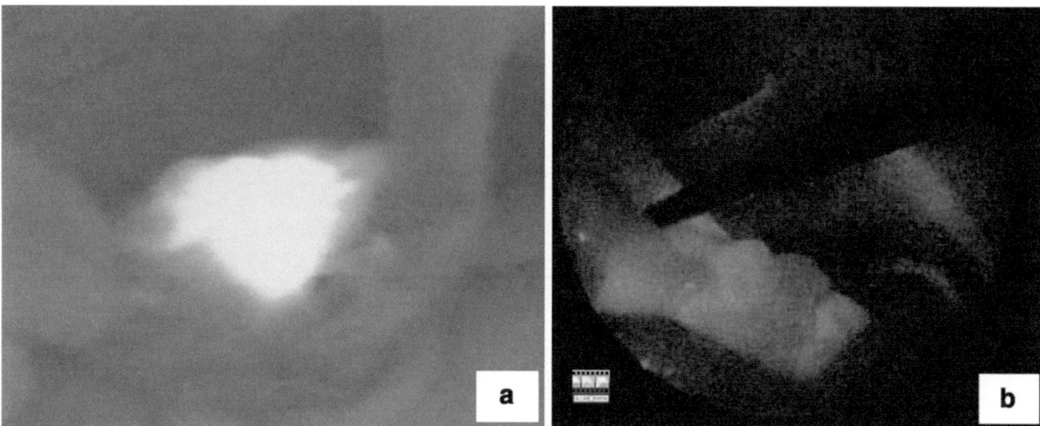

Fig. 21.2 Confirmation of tattooing of tumor location by ICG fluorescence imaging. (**a**) Open surgery. (**b**) Laparoscopic surgery

palpation with the fingers in colorectal surgery, particularly laparoscopic surgery. Therefore, endoscopic colorectal tattooing is frequently performed because it is a simple and economic technique to identify tumor localization [22]. Tattooing with India ink has been used since 1975, as first reported by Ponsky and King [23]. However, tattooing with India ink has been associated with complications, i.e., colonic abscess formation [13], peritonitis [13], inflammatory pseudotumor [14], and adhesion ileus [15]. Some institutions have reported the utility and safety of tattooing with IFI [24, 25]. Here, we describe our experience tattooing with IFI.

Materials and Methods

ICG solution was made by dissolving 25 mg of powdered ICG in 10 ml of sterile physiologic saline, and 0.2 ml volumes were endoscopically injected into 4–5 points of the submucosal areas around tumors in colon cancers. During subsequent laparotomy, the colon was first observed with the naked eye, and during subsequent laparotomy, IFI was observed by PDE (photodynamic emission) camera or by a prototype of fluorescence endoscopy for the detection of marking areas during laparoscopic operation.

Results

IFI showed the tumor localization clearly and accurately by bright fluorescence in all patients with open surgery (Fig. 21.2a) or laparoscopic surgery (Fig. 21.2b). The regions tattooed with ICG were hardly green at all in color.

There were no complications of LED-induced fluorescence, and no inflammatory signs were noted on the HE-stained slides for the identified injection sites in the resected specimens.

Discussion

Small cancer lesions in the colon can be difficult to palpate, and with lack of tactile sensation, it is essential to accurately localize them preoperatively. India ink tattooing is widely used for tumor localization [22]; however, this tattooing procedure is not yet standardized, and several complications associated with this technique such as colonic abscess formation [13], peritonitis [13], inflammatory pseudotumor [14], and adhesion ileus [15] have been reported.

As for the interval between injection and operation, it was reported that tattooing with ICG should be performed up to 5–8 days prior [24, 25]. However, tattooing with ICG seemed to disappear earlier than tattooing with India ink [25].

The regions tattooed with ICG were hardly green at all. ICG concentrations should be high enough to be recognized as green [24]. However, the ICG concentrations can be reduced for recognition of fluorescence by LED-induced fluorescence. Therefore, in spite of the lower ICG concentrations, tumor localization was clearly and accurately visualized. This may be related to the safety of tattooing with IFI.

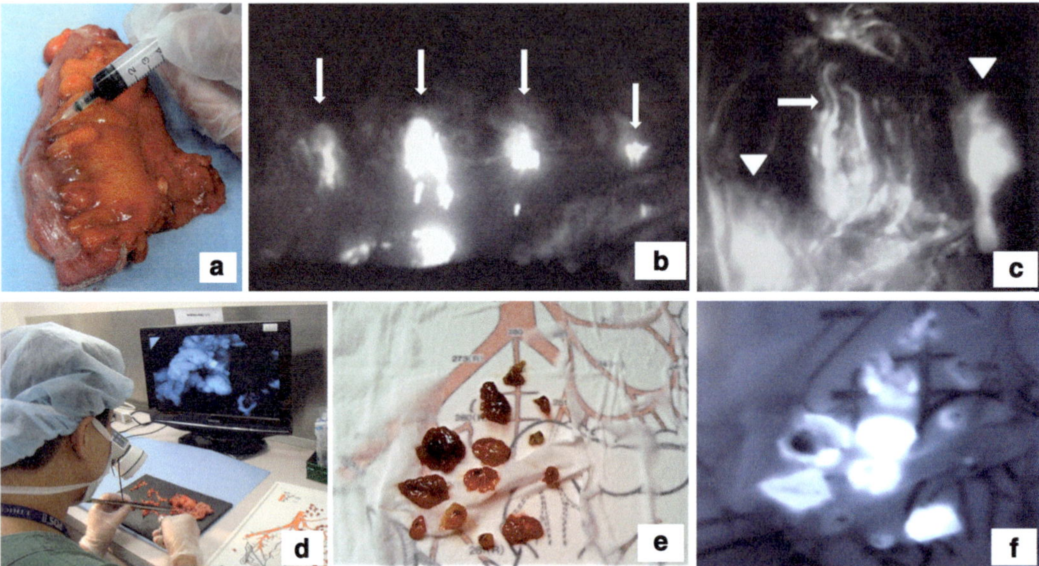

Fig. 21.3 Lymph node harvested from resected specimens using ICG fluorescence imaging. (**a**) Ex vivo ICG injection. (**b**) ICG injection point is emitted (*arrow*). (**c**) Lymph node (*arrowhead*) and lymph duct (*arrow*) were emitted. (**d**) Second harvest with ICG fluorescence imaging. (**e**) Harvested lymph nodes. (**f**) ICG fluorescence imaging of harvested lymph nodes

In this study, we could observe the tumor localization clearly and accurately. Moreover, there were no complications of LED-induced fluorescence and no inflammatory signs. The IFI tattooing method proved effective for tumor localization. Fine and clear tattooing points could be obtained, and the potential complications caused by the tattooing with India ink could be avoided by IFI technique.

LN Harvest from Resected Specimens by Using ICG Fluorescence Imaging

Background

The staging of lymph node factor is determined by the number of metastatic LNs rather than the stationary location of metastatic lymph nodes. Therefore, the number of harvested LNs is very significant for the exact staging of colorectal cancer. Here, we suggest one of the useful ways of harvesting LNs using IFI.

Materials and Methods

To harvest LNs, we injected 0.2 ml ICG solution made by dissolving 25 mg of powdered ICG in 10 ml of sterile physiologic saline at each of the four points around tumors intraoperatively or postoperatively (Fig. 21.3a–c). First, we harvested LNs by the conventional procedure, and then we reexamined whether there were LNs left behind by a PDE camera, which enabled us to detect the IFI absorbed by the LNs (second harvest) (Fig. 21.3d–f). Furthermore, we searched for IFI-labeled LNs in paraffin-embedded specimens (Fig. 21.4a–c). This study enrolled a total of 31 patients with colorectal cancer, including 24 with colon cancer and 7 with rectal cancer.

Results

We could observe lymph ducts (LDs) and nodes (LNs) by ICG injection into resected specimens (Fig. 21.3c). However, it was hard to observe IFI in cases with advanced cancer or ink tattooing. We observed IFI in harvested LNs in all 31 colorectal cancers (Fig. 21.3e). In 23 of the 31 cases (74 %), we could observe the LDs and LNs by IF ex vivo lymph node harvest procedures (Table 21.1). We tried to detect additional lymph nodes in 7 of 23 cases observed by IF ex vivo lymph node harvest procedures. In four of these seven cases (57 %) in which additional lymph nodes could be detected, we could find additional lymph nodes left in situ.

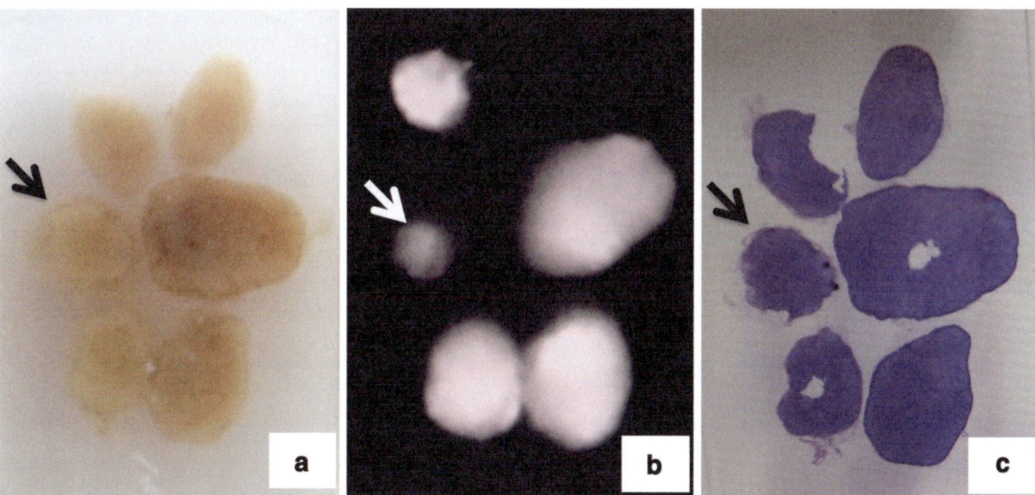

Fig. 21.4 Observation of paraffin block. *Arrow* shows metastatic lymph node. (**a**) Paraffin block of lymph nodes. (**b**) ICG fluorescence imaging of paraffin block of lymph nodes. (**c**) Hematoxylin-eosin staining of lymph nodes

Table 21.1 ICG fluorescence imaging observed by ICG ex vivo injection ($n=31$)

Stage	Possible	Impossible
I	6	2
II	1	3
IIIa	9	0
IIIb	6	3
IV	1	0
	23 (74 %)	8 (26 %)

ICG indocyanine green

Furthermore, in one of these four cases (25 %) at least one of the additionally harvested LNs was metastatic (Fig. 21.3e, f and 21.4a–c).

Discussion

In the UICC seventh edition, the cancer staging is determined by the number of metastatic LNs instead of the stationary location of the metastatic lymph nodes [26]. This suggests that the accurate detection and harvest of LNs are mandatory to determine the clinical staging of cancer patients. The ASCO guideline and ESMO guideline reported that having fewer than 12 dissected lymph node is a risk factor for recurrence [27, 28]. Moreover, the ratio of metastatic LNs, calculated as the number of metastatic LNs and divided by the total number of LNs harvested from the resected specimen, is considered another important parameter for the accuracy of clinical staging. A second harvest of LNs by IFI is recommended for the accurate determination of clinical staging in gastrointestinal cancers such as colorectal cancer.

Conclusions

Our preliminary results showed that IFI allows easy, highly sensitive, and real-time imaging-guided SLN mapping in patients with colorectal cancer. SLN mapping guided by IFI is a promising tool that deserves further clinical exploration.

Colonic tattooing using IFI is a new concept of colonic marking based on the fact that ICG is a near-infrared fluorescent dye and findings are useful, without any adverse effects, for perioperative tumor localization.

Harvesting LNs from resected specimens using IFI is a useful method with highly improved LN detection. We found IFI-labeled LNs even when ICG solution was injected ex vivo and observed them in paraffin-embedded specimens that could provide us a precise evaluation of the pathological status of LNs including sentinel LNs after surgery.

In our study, we confirmed the efficacy of IFI not only in the detection of SLNs in colon cancers but also for colonic tumor tattooing by open or laparoscopic surgery and for the accurate harvesting of LNs from resected specimens.

References

1. Kitai T, Inamoto T, Miwa M, et al. Fluorescence navigation with indocyanine green for detecting sentinel lymph nodes in breast cancer patients. Breast Cancer. 2005;12:211–5.
2. Kusano M, Tajima Y, Yamazaki K, et al. Sentinel node mapping guided by indocyanine green fluorescence imaging: a new method for sentinel node navigation surgery in gastrointestinal cancer. Dig Surg. 2008;25:103–8.
3. Parungo CP, Ohnishi S, Kimu SW, et al. Intraoperative identification of esophageal sentinel lymph nodes with near-infrared fluorescence imaging. J Thorac Cardiovasc Surg. 2005;129:844–50.
4. Rubens FD, Ruel M, Fremes SE. A new and simplified method for coronary and graft imaging during CABG. Heart Surg Forum. 2002;5:141–4.
5. Aoki T, Yasuda D, Shimizu Y, et al. Image-guided liver mapping using fluorescence navigation system with indocyanine green for anatomical hepatic resection. World J Surg. 2008;32:1763–7.
6. Ichikura T, Morita D, Uchida T, et al. Sentinel node concept in gastric carcinoma. World J Surg. 2002; 26:318–22.
7. Arigami T, Natsugoe S, Uenosono Y, et al. Evaluation of sentinel node concept in gastric cancer based on lymph node micrometastasis determined by reverse transcription-polymerase chain reaction. Ann Surg. 2006;243:341–7.
8. Joosten JJ, Strobbe LJ, Wauters CA, et al. Intraoperative lymphatic mapping and the sentinel node concept in colorectal carcinoma. Br J Surg. 1999;86:482–6.
9. Wiese DA, Saha S, Badin J, et al. Pathologic evaluation of sentinel node in colorectal carcinoma. Arch Pathol Lab Med. 2000;124:1759–63.
10. Merrie AE, van Rij AM, Phillips LV, et al. Diagnostic use of the sentinel node in colon cancer. Dis Colon Rectum. 2001;44:410–7.
11. Bertagnolli M, Miedema B, Redston M, et al. Sentinel node staging of resectable colon cancer: results of a multicenter study. Ann Surg. 2004;240:624–8.
12. Saha S, Seghal R, Patel M, et al. A multicenter trial of sentinel lymph node mapping in colorectal cancer: prognostic implications for nodal staging and recurrence. Am J Surg. 2006;191:305–10.
13. Park SI, Genta RS, Romeo DP, et al. Colonic abscess and focal peritonitis secondary to India ink tattooing of the colon. Gastrointest Endosc. 1991;37:68–71.
14. Coman E, Brandt LJ, Brenner S, et al. Fat necrosis and inflammatory pseudotumor due to endoscopic tattooing of the colon with India ink. Gastrointest Endosc. 1991;37:65–8.
15. Yano H, Okada K, Monden T. Adhesion ileus caused by tattoo-marking: unusual complication after laparoscopic surgery for early colorectal cancer. Dis Colon Rectum. 2003;46:987.
16. Morton DL, Wen DR, Wong JH, et al. Technical details of intraoperative lymphatic mapping for early stage melanoma. Arch Surg. 1992;127:392–9.
17. Veronesi U, Paganelli G, Galimberti V, et al. Sentinel-node biopsy to avoid axillary dissection in breast cancer with clinically negative lymph nodes. Lancet. 1997;349:1864–7.
18. Kitagawa Y, Fujii H, Kumai K, et al. Recent advances in sentinel node navigation for gastric cancer: a paradigm shift of surgical management. J Surg Oncol. 2005;90:147–51.
19. Aikou T, Kitagawa Y, Kitajima M, et al. Sentinel lymph node mapping with GI cancer. Cancer Metastasis Rev. 2006;25:269–77.
20. Ankersmit M, van der Pas MHGM, van Dam DA, et al. Near infrared fluorescence lymphatic laparoscopy of the colon and mesocolon. Colorectal Dis. 2011;13:70–3.
21. Hirche C, Mohr Z, Kneif S, et al. Ultrastaging of colon cancer by sentinel node biopsy using fluorescence navigation with indocyanine green. Int J Colorectal Dis. 2012;27:319–24.
22. Choi YS, Lee SI, Lee TG, et al. Economic outcomes of laparoscopic versus open surgery for colorectal cancer in Korea. Surg Today. 2007;37:127–32.
23. Ponsky JL, King JF. Endoscopic marking of colonic lesions. Gastrointest Endosc. 1975;22:42–7.
24. Watanabe M, Tsunoda A, Narita K, et al. Colonic tattooing using fluorescence imaging with light-emitting diode-activated indocyanine green : a feasibility study. Surg Today. 2009;39:214–8.
25. Miyoshi N, Ohue M, Noura S, et al. Surgical usefulness of indocyanine green as an alternative to India ink for endoscopic marking. Surg Endosc. 2009;23:347–51.
26. Sobin LH, Gospodarowicz MK, Wittekind C. TNM classification of malignant tumours. 7th ed. Hoboken, NJ: Wiley-Blackwell; 2010.
27. Benson 3rd AB, Schrag D, Somerfield MR, et al. American Society of Clinical Oncology recommendations on adjuvant chemotherapy for stage II colon cancer. J Clin Oncol. 2004;22:3408–19.
28. Van Cutsem EJ, Oliveira J, ESMO Guidelines Working Group. Colon cancer: ESMO clinical recommendations for diagnosis, adjuvant treatment and follow-up. Ann Oncol. 2008;19 Suppl 2:29–30.

Fluorophore-Conjugated Chimeric Anti-CEA Antibodies for Fluorescence-Guided Surgery of Gastrointestinal (GI) Tumors

22

Michael Bouvet and Robert M. Hoffman

Introduction

The ability of the surgeon to accurately visualize tumor margins and identify metastases is necessary for the success of cancer surgery. Fluorescence optical imaging, because of its high sensitivity, low cost, portability, and real-time capabilities, has great potential to improve surgical outcomes.

Fluorescence technology have already been used clinically. Frangioni et al. carried out the first-in-human clinical trial of fluorescent imaging in breast cancer sentinel lymph node mapping using indocyanine green (ICG) as a near-infrared (NIR) fluorescent lymphatic tracer [1]. Van Dam et al. used folate conjugated to fluorescein isothiocyanate for targeting folate receptor alpha (FR-α) in patients with ovarian tumors [2]. Fluorescence was detectable intraoperatively in all patients with a malignant tumor.

M. Bouvet, M.D. (✉)
Department of Surgery, Moores UCSD Cancer Center, 3855 Health Science Drive #0987, La Jolla, CA 92093-0987, USA
e-mail: mbouvet@ucsd.edu

R.M. Hoffman, Ph.D.
AntiCancer, Inc., 7917 Ostrow Street, San Diego, CA 92111, USA

Department of Surgery, University of California San Diego, 3855 Health Science Drive #0987, La Jolla, CA 92093-0987, USA
e-mail: all@anticancer.com

In preclinical studies, Tsien et al. used activatable cell-penetrating peptides (ACPPs), in which the fluorescently-labeled, polycationic cell-penetrating peptide (CPP) is coupled via a cleavable linker to a neutralizing peptide. Upon exposure to tumor proteases surgery, the linker is cleaved, dissociating the inhibitory peptide and allowing the CPP to bind to and enter tumor cells [3]. Kishimoto et al. selectively labeled tumors with a genetic reporter, GFP, using a telomerase-dependent adenovirus genetic OBP-401 [4]. They demonstrated that tumors recurred after fluorescence-guided surgery and maintained GFP expression [5]. Therefore, the detection of recurrence and future metastasis is possible with OBP-401 GFP labeling, since recurrent cancer cells stably express GFP, which is not possible with nongenetic labeling of tumors.

Negative tumor margins are especially necessary for curative pancreatic cancer surgery, For colorectal cancers the high mortality of this disease in the United States correlates with a high cancer incidence [6]. Colon cancer patients more often present with resectable disease [7] and have more surgical options than patients with pancreatic cancer [8, 9]. There is a clear advantage to compete resection of all primary and metastatic cancer at the time of surgery of colon cancer when clinically appropriate [10, 11].

Monoclonal antibodies specific for either CA19-9 or CEA were conjugated to a green fluorophores delivered to pancreatic or colon cancer-bearing mice as a single intravenous dose [12–18]. Using fluorescence imaging, the primary pancreatic

F.D. Dip et al. (eds.), *Fluorescence Imaging for Surgeons: Concepts and Applications*, DOI 10.1007/978-3-319-15678-1_22, © Springer International Publishing Switzerland 2015

or colon tumor was clearly visible at laparotomy as were small metastases in other organs.

The present chapter reviews the potential of using chimeric mouse-human antibodies conjugated with appropriate fluorophores for fluorescence-guided surgery (FGS) of metastatic pancreatic and colon cancer in orthotopic nude mouse models. The possibility of this technology to change the paradigm for surgical oncology is also discussed.

Carcinoembryonic Antigen

Carcinoembryonic antigen (CEA) was first described following immunization of xenogenic animals with human tumor tissue [19]. Human tissue specimens were positive for CEA expression from multiple cancers arising from endodermally-derived epithelium of the digestive tract [20] as well as in the human embryonic gut, pancreas, and liver tissue [21]. Although initially identified in adenocarcinoma of the colon [20], CEA is often also expressed in pancreatic ductal adenocarcinoma [22, 23].

CEA is considered is an oncofetal antigen, with expression in normal fetal tissues but only has a trace presence in normal adult human tissue [24, 25]. The CEA family is comprises a large class of complex glycoproteins of the immunoglobulin gene superfamily [26]. In addition to acting as a marker for human gastrointestinal cancers, CEA also functions as an intercellular adhesion molecule and may have some immunoregulatory function as well [27]. In clinical medicine, CEA is most commonly utilized as a serum marker in colorectal and pancreatic cancer for preoperative staging and follow-up of patient response outcome after surgery and chemotherapy [28, 29].

In Vitro Expression of CEA

In vitro, 70 % of the pancreatic cancer cell lines tested were positive for CEA immunostaining in culture. CEA-positive pancreatic cancer cell lines included MiaPaca-2, FG, ASPC-1, BxPC-3, CFPAC, Panc-1, and Capan-1. The cell lines tested that did not express CEA included XPA-1, XPA-3, and XPA-4. Of the human colon cancer cell lines that were tested in vitro, 67 % expressed CEA as identified by immunostaining. The CEA-positive colon cancer cell lines included LOVO, HCT-116, SW948, and LS174T. The colon cancer cell lines that did not express CEA were HT-29 and SW480. For each cell line tested, the cells were incubated with Alexa Fluor 488-labeled anti-CEA or IgG. Positive staining was indicated by fluorescence intensity above background staining with conjugated non-specific IgG (Table 22.1) [12].

Immunofluorescence Staining of Tissue for Binding with Anti-CEA Antibody

Screening of normal human tissue samples for binding to conjugated anti-CEA antibody used immunofluorescence staining of a human tissue array. This array contained two samples each of 19 different noncancerous adult human tissues including: salivary gland, liver, small intestine, stomach, kidney, skeletal muscle, skin, heart, placenta, breast, cervix, uterus, spleen, lung, brain, thyroid, pancreas, ovary, and adrenal gland. Human tumor grown in nude mice from the pancreatic cancer cell line ASPC-1 and the primary human colon cancer tumor Colo4104 were used as positive controls, and mouse axillary lymph node tissue was included as a negative control. Both the ASPC-1 pancreatic tumor and the Colo4104 colon tumor yielded positive staining for CEA. In noncancerous tissues the majority of samples did not demonstrate binding of conjugated anti-CEA above the isotype-control IgG background. A low level of staining above background was present within the small intestine and cervix. Notably, the normal colon and pancreas did not bind conjugated anti-CEA. Table 22.2 denotes the staining for all noncancerous human lists samples tested [12].

Table 22.1 CEA expression in vitro and in vivo

Human pancreatic cancer cell lines		
In vitro	+	–
Mia Paca-2	x	
FG	x	
BxPC-3	x	
CFPAC	x	
Panc-1	x	
Capan-1	x	
XPA-1		x
XPA-3		x
XPA-4		x
In vivo	+	–
ASPC-1	x	
BxPC-3	x	
CFPAC	x	
Panc-1	x	
Capan-1	x	
Human colon cancer cell lines		
In vitro	+	–
LOVO	x	
HCT-116	x	
SW948	x	
LS174T	x	
HT-29		x
SW480		x
In vivo	+	–
LS174T	x	
Colo4104	x	

Human pancreatic and colon cancer cell lines were tested for in vitro and in vivo expression of CEA and seven of ten (70 %) were positive. Four of 6 (67 %) colon cancer cell were positive. All seven pancreatic cancer cell lines and one colon cancer cell line tested in vivo positive. In addition, the primary patient human colon cancer tissue Col4104, grown in mice, was also positive [11].

Table 22.2 CEA expression in adult human tissues

Tissue	Staining
Salivary gland	–
Liver	–
Small intestine	+/–
Stomach	–
Kidney	–
Skeletal muscle	–
Skin	–
Heart	–
Placenta	–
Breast	–
Cervix	+
Uterus	–
Spleen	–
Lung	–
Brain	–
Thyroid	–
Pancreas	–
Ovary	–
Adrenal gland	–
ASPC-1 tumor[a]	+++
Colo4104 tumor[a]	++
Mouse axillary LN[b]	–

Staining of a tissue array of adult noncancerous human tissues demonstrated only a small amount of positive staining over background in cervix and small intestine tissues. In the small intestine, staining was primarily limited to cells on the mucosal surface. In the cervix, the staining was primarily on the luminal surface of glandular structures. The positive controls ASPC-1 and Colo4104 stained with conjugated anti-CEA both within the cytoplasm and on the cell membrane [11].

[a]Positive control

[b]Negative control

Fluorophore-Conjugated Anti-A Antibody for the Intraoperative Imaging and FGS of Pancreatic and Colorectal Cancer

In 2008, our team was the first to evaluate the use of a fluorophore-labeled anti-CEA monoclonal antibody to aid in primary and metastatic cancer visualization and FGS in nude mouse models of human colorectal and pancreatic cancer [12]. Anti-CEA conjugated with a green fluorophore, resulted in improved resection. Subcutaneous, orthotopic primary and metastatic human pancreatic and colorectal tumors were readily visualized with fluorescence imaging after administration of conjugated anti-CEA. The fluorescence signal was detectable 30 min after systemic antibody delivery and was stable for 2 weeks, with minimal in vivo photobleaching after exposure to standard operating room lighting. We demonstrated the principle that fluorophore-conjugated antibodies enabled successful FGS in orthotopic mouse models of pancreatic cancer [17].

Imaging of Subcutaneous Tumors with Fluorescent Anti-CEA Antibody

The human pancreatic cancer cell lines ASPC-1, BxPC-3, CFPAC, Panc-1, and Capan-1 were implanted subcutaneously in nude mice and evaluated for CEA expression. One colon cancer cell line (LS174T) and a primary human colon cancer patient specimen (Colo4104) , grown in nude mice, were also implanted subcutaneously. When the tumors had reached approximately 1–2 mm diameter, the animals were each given a single dose of Alexa Fluor 488-conjugated anti-CEA or IgG i.v. All 5 pancreatic cancer cell lines implanted bound the CEA as did the colon cancer cell line as well as the patient colon cancer model demostrated by fluorescence intensity above background IgG (Table 22.1).

Imaging Orthotopic Tumors with Fluorescent Anti-CEA Antibody

Tumors implanted orthotopically into the nude mouse pancreas and colon were evaluated for imaging using conjugated anti-CEA. The human pancreatic cancer cell lines ASPC-1 and BxPC-3 were used. For the colon cancer, the patient tumor Colo4014 was used in a model termed patient-derived orthotopic xenograft (PDOX®) [30, 31]. Orthotopic pancreatic or colon tumor-bearing mice were given a single dose of Alexa Fluor 488-conjugated anti-CEA or IgG via tail vein injection 7–10 days after tumor implantation. The animals were imaged under both bright field and fluorescence illumination using the Olympus variable-magnification OV100 Small Animal Imaging System [32]. Intravital fluorescence imaging revealed very small ASPC-1 and BxPC-3 tumors which were not visible under bright-field illumination, even at higher magnification (Fig. 22.1a–c). With fluorescence imaging it was clear that the extent of tumor invasion was much greater than that seen under standard bright light (Fig. 22.1d, e). The tumors in the colon cancer-bearing animals were visible under both bright field and fluorescence imaging at the magnification used (Fig. 22.2a–e). The animals which received control conjugated IgG had no green fluorescence in either their pancreatic (Fig. 22.1a, b) or colon (Fig. 22.2a, b) tumors [12].

Imaging Intra-abdominal Disseminated Tumor with Fluorescent Anti-CEA Antibody

Mouse models of intra-abdominal carcinomatosis of pancreatic and colorectal cancer were used to evaluate fluorophore-conjugated anti-CEA binding to these tumors in vivo. Animals were injected i.p. with human pancreatic (BxPC-3) or colorectal (Colo4104 or LS174T) cancer cells. After 1 week the mice were given a single 75 μg injection of Alexa Fluor 488-conjugated anti-CEA or IgG i.v. The mice were imaged 24 h later with the Olympus OV100 using both bright field and fluorescence illumination. At the time of imaging, these animals had very small peritoneal implants on the bowel and mesentery which were difficult to visualize using bright field imaging (Figs. 22.3a, b and 22.4a, b) but were very clearly visible under fluorescence illumination in mice given conjugated anti-CEA (Figs. 22.3c, d and 22.4c, d). The mice which received IgG had no visible fluorescence signal in their tumors (data not shown) [12].

Time-Course Imaging of Pancreatic Tumors in Nude Mice After Injection of Fluorescent Anti-CEA Antibody

Time-course imaging of human pancreatic tumors in nude mice labeled with conjugated anti-CEA demonstrated rapid binding of the antibody-fluorophore conjugate in vivo with very long signal duration. Mice with 1–2 mm diameter subcutaneous ASPC-1 tumors were given a single dose of Alexa Fluor 488-conjugated anti-CEA i.v. The mice were then imaged over 15 days after delivery of a single dose of

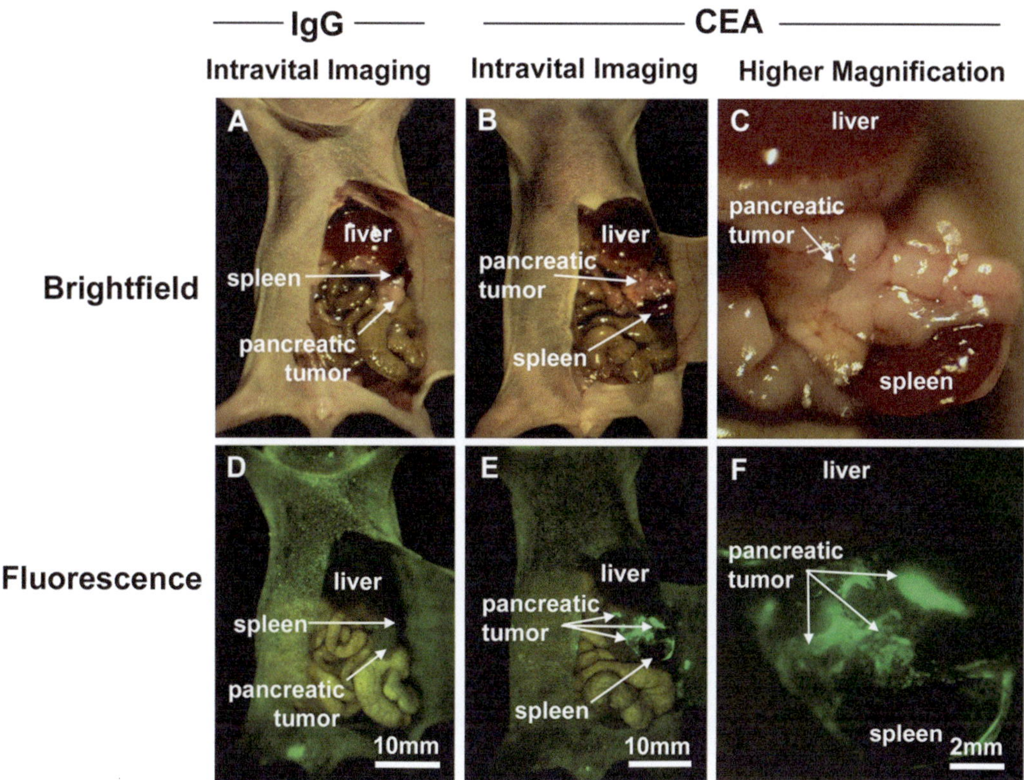

Fig. 22.1 Imaging of orthotopic human pancreas tumors in vivo had greatly improved primary tumor visualization at laparotomy. After treatment with fluorophore conjugated anti-CEA, mice with orthotopically-implanted BxPC-3 pancreatic tumors were imaged using both bright field (**a–c**) and fluorescence (**d–f**) illumination. Primary tumors were difficult to clearly distinguish under bright field imaging under both low (**a**, **b**) and high (**c**) magnification. In contrast, fluorescence illumination of anti-CEA-labeled tumors enabled identification of primary tumor (**e**, **f**), which was much more extensive than seen under bright field. Animals given conjugated non-specific IgG demonstrated no fluorescence signal in the orthotopic tumor (**d**). All tumors was confirmed by histology [12]

antibody. Two animals were imaged for each time point. A small amount of fluorescence signal could be seen at 30 min post-antibody injection, and the signal peaked at 24 h. This signal was stable over the next 24 h and then decreased over the following 6 days decaying to a low-level signal at 8 days post-injection. By 15 days there was undetectable signal remaining within the tumor tissue (Fig. 22.5) [12].

Use of Fluorescent Anti-CEA Antibody to Image Post-resection Residual Tumor

In animals bearing larger (3–10 mm diameter) subcutaneous tumors, we used fluorophore-conjugated anti-CEA to attempt a complete resection. Mice were given a single dose of Alexa Fluor 488-conjugated anti-CEA 24 h prior to

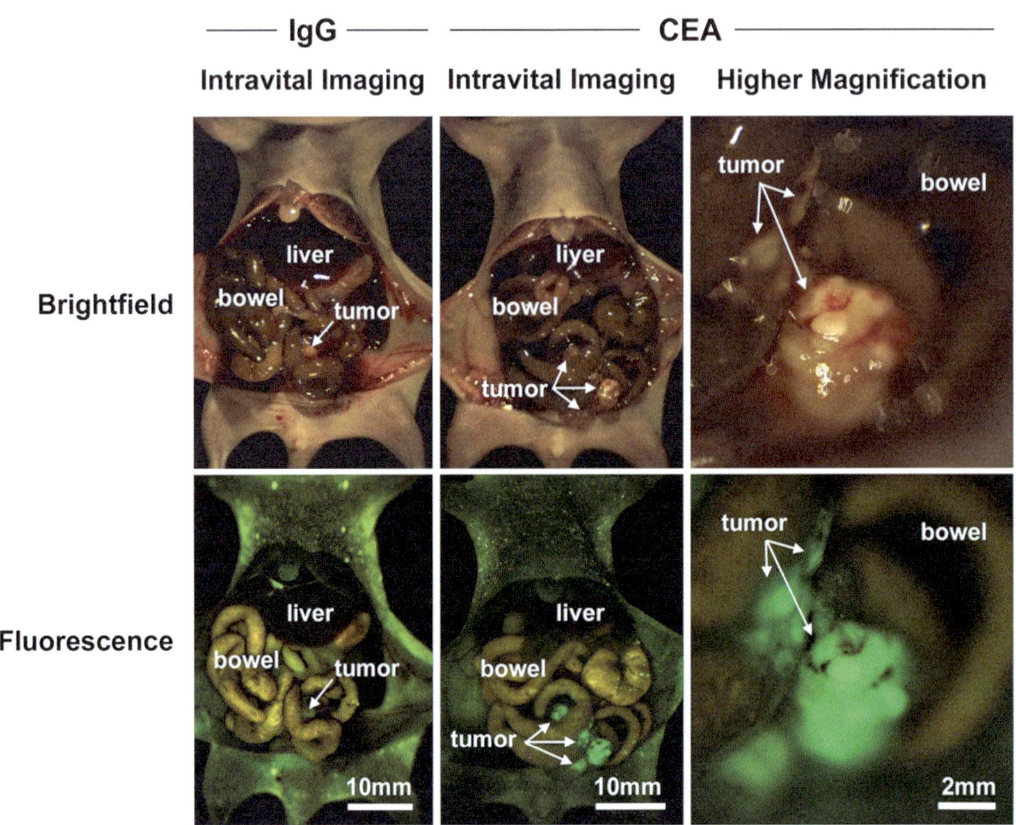

Fig. 22.2 Fluorescence imaging of patient-derived orthotopic xenograft (PDOX®) human colon tumors under fluorescence illumination improved primary tumor visualization at laparotomy. Mice with orthotopically-implanted low-passage AC4104 colon tumors originally derived from a patient were imaged under both bright field (**a–c**) and fluorescence (**d–f**) illumination. Primary tumors labeled with conjugated anti-CEA had bright green fluorescence (**e**, **f**). Mice given conjugated control IgG had no fluorescence signal in the orthotopic tumor (**d**). All tumors were confirmed by histology [12]

FGS. Before surgery, animals were anesthetized, and their tumors were imaged using bright field and fluorescence illumination (Fig. 22.6a, b). The tumors were then carefully resected with a dissecting microscope under bright field illumination with careful attention paid to removing all visible tumor tissue without adjacent normal skin or muscle (Fig. 22.6b–d). Following resection the operative bed was then imaged using fluorescence microscopy, with all remaining areas of fluorescence (Fig. 22.6e, f) were documented and biopsied. Of the three animals that underwent bright light surgery (BLS), all three had residual tumor present within the tumor bed which was not visible under bright field illumination. The presence of all tumor tissue was confirmed by histology (data not shown) [12].

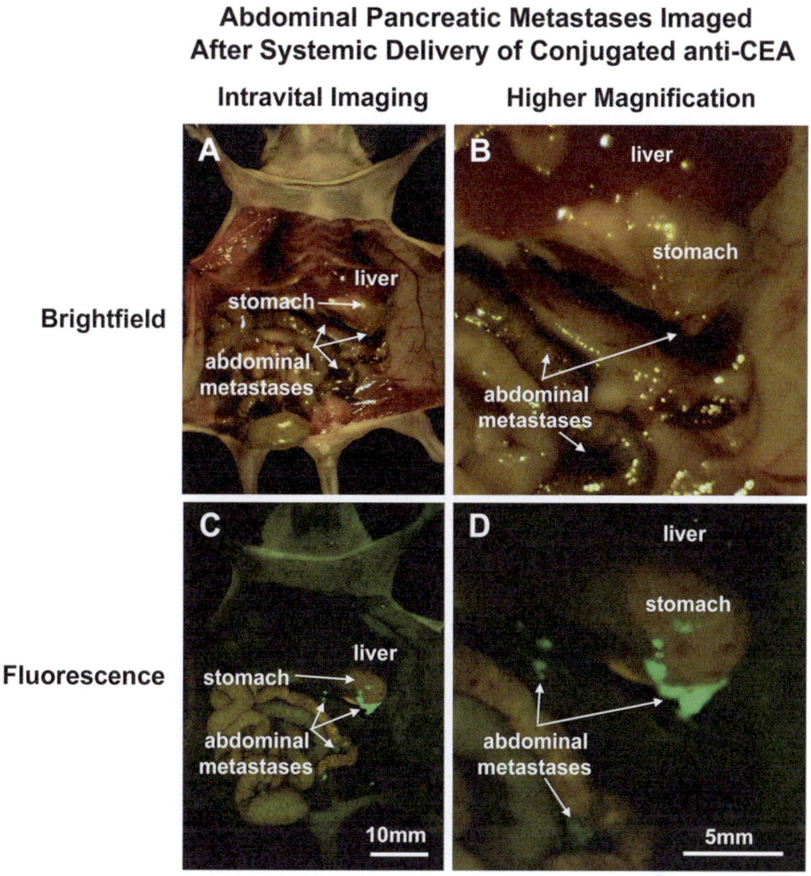

Abdominal Pancreatic Metastases Imaged After Systemic Delivery of Conjugated anti-CEA

Intravital Imaging | Higher Magnification

Brightfield | Fluorescence

Fig. 22.3 Fluorescence imaging of intra-abdominal metastases from human pancreatic cancer cell line BxPC-3 orthotopic models reveals greatly improved metastatic tumor visualization at laparotomy. Mice injected i.p. with BxPC-3 pancreatic cancer cells, were imaged using in bright field (**a**, **b**) and under fluorescence (**c**, **d**) illumination. Small metastatic implants on the bowel and mesentery were not seen with bright field imaging even under low (**a**) or high (**b**) magnification. In contrast, fluorescence illumination of anti-CEA-labeled tumors enabled easy identification of metastases (**c**, **d**) [12]

FGS with a Fluorophore-Conjugated Anti-CEA Antibody Improves Surgical Resection and Increases Survival in Orthotopic Mouse Models of Human Pancreatic Cancer

We showed that FGS with anti-CEA antibody conjugated to a green fluorophore improved outcomes in mouse models of pancreatic cancer [17]. Mouse models of human pancreatic cancer were established with surgical orthotopic implantation (SOI) of the human BxPC-3 pancreatic cancer cell line. Orthotopic tumors were allowed to develop for 2 weeks and the mice then underwent BLS or FGS 24 h after intravenous injection of anti-CEA-Alexa 488. Completeness of resection was assessed from postoperative fluorescence imaging. Mice were followed postoperatively to determine disease-free survival (DFS) and overall survival (OS). Complete resection was achieved in 92 % of FGS-treated mice as compared to 45.5 % in the BLS group ($p=0.001$). FGS resulted in a smaller postoperative tumors ($p=0.01$). Cure rates with FGS compared to BLS improved from 4.5 to 40 % ($p=0.01$). One year postoperative survival rates increased from 0 % with BLS to 28 % with FGS ($p=0.01$). Median DFS improved from 5 weeks with BLS to 11 weeks

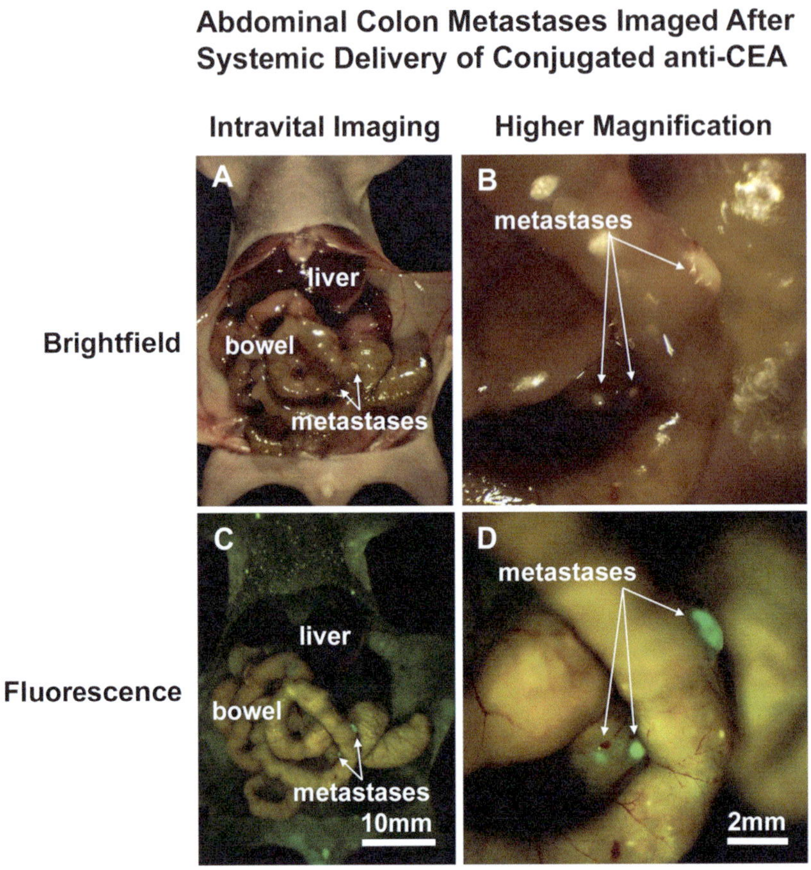

Fig. 22.4 Mice with metastatic human colon tumors had improved tumor visualization at laparotomy after injection of fluorophore-conjugated chimeric anti-CEA antibodies. Mice with intraperitoneally implanted Colo4104 colon cancer were imaged using both bright field (**a**, **b**) and fluorescence (**c**, **d**) illumination. The metastases were difficult to distinguish under bright field imaging under both low (**a**) and high (**b**) magnification. In contrast, fluorescence illumination of anti-CEA-labeled tumors enabled facile identification of metastases (**c**, **d**) [12]

with FGS ($p=0.0003$). Median OS improved from 13.5 weeks with BLS to 22 weeks with FGS ($p=0.001$) (Fig. 22.7) [17].

Fluorescently Labeled Chimeric Anti-CEA Antibody Improves Detection and Resection of Human Colon Cancer in a Patient-Derived Orthotopic Xenograft (PDOX®) Nude Mouse Model

A fluorescently labeled chimeric anti-CEA antibody improved detection and FGS of colon cancer [16]. Mouse monoclonal antibodies tend to evoke an immune reaction when administered to humans. Creating a chimeric "fusion" protein allows the introduction of segments of human constant domains while maintaining important properties from the "parent" mouse protein, thereby eliminating most of the potentially immunogenic portions of the antibody without compromising its specificity for the intended target (Fig. 22.8) [33]. Frozen tumor and normal human tissue samples were stained with chimeric and mouse antibody-fluorophore conjugates for comparison of tumor staining. PDOX® mice with human colon cancer underwent FGS or BLS 24 h after i.v. injection of fluorophore-conjugated chimeric anti-CEA

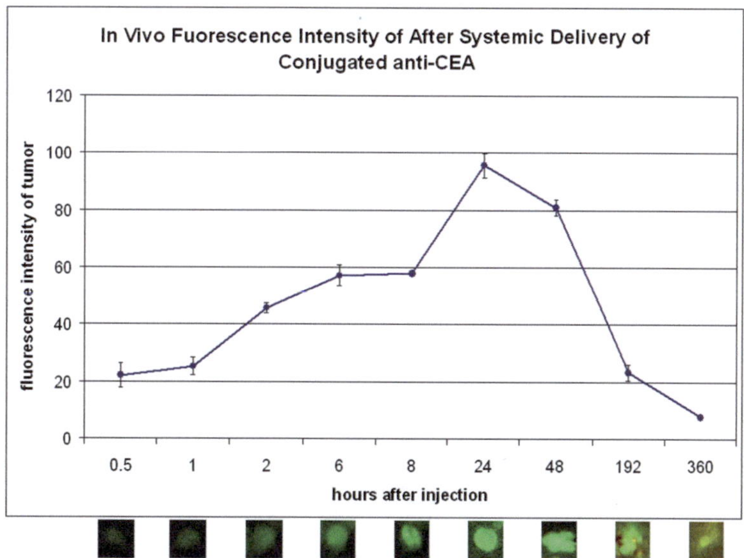

Fig. 22.5 Mice with small subcutaneous tumors were given a single dose of conjugated antibody and imaged at 30 min, 1 h, 2 h, 6 h, 8 h, 24 h, and 48 h and 8 and 15 days after. The fluorescence signal was detected at 30 min and peaked at 24 h after injection. The fluorescence signal remains high at 48 h but by 8 days (192 h) decayed to levels comparable to that seen at 30 min and by 15 days (360 h) was at background [12]

antibody. Completeness of resection was assessed using postoperative imaging. Mice were followed for 6 months to determine recurrence. The fluorophore conjugation efficiency (dye/mole ratio) improved from 3–4 to >5.5 with the chimeric anti-CEA antibody compared to mouse anti-CEA antibody. CEA-expressing tumors labeled with chimeric CEA antibody had a brighter fluorescence signal on frozen human tumor tissues ($p=0.046$). Normal human tissues had lower fluorescence with chimeric anti-CEA compared to mouse antibody. Chimeric CEA antibody accurately labeled PDOX® colon cancer in nude mice, enabling more effective FGS. The R0 resection rate improved from 86 to 96 % with FGS compared to BLS (Fig. 22.9).

Comparison of a Chimeric Anti-CEA Antibody Conjugated with Visible or Near-Infrared Fluorescent Dyes for Imaging Pancreatic Cancer in Orthotopic Nude Mouse Models

We evaluated a set of visible and near-infrared dyes conjugated to a tumor-specific chimeric antibody for their ability to improve high-resolution tumor imaging in orthotopic models of pancreatic cancer [14]. BxPC-3 human pancreatic cancer was orthotopically implanted into the pancreata of nude mice. The orthotopic models received a single i.v. injection of a chimeric anti-CEA antibody conjugated to one of the following fluorophores: the 488 nm group (Alexa Fluor 488 or DyLight 488),

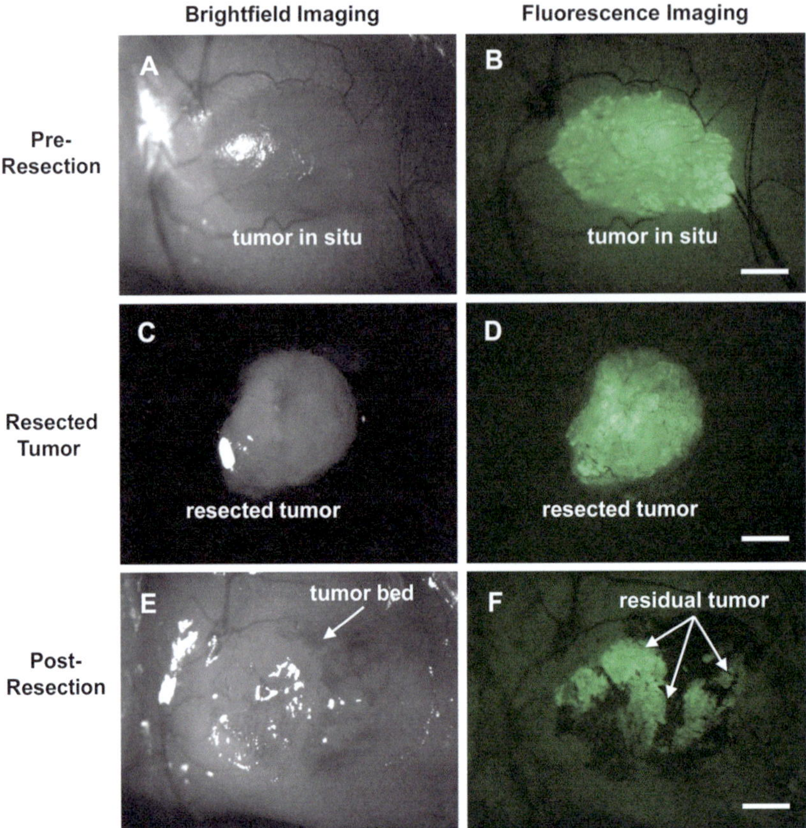

Fig. 22.6 Tumor resection under bright-light microscopy. Larger subcutaneous BxPC-3 tumors were imaged under a dissecting microscope via both bright field (**a**) and fluorescence (**b**) illumination. Under bright field microscopy, all visible tumor was resected, and the ex vivo tumor was imaged under bright field (**c**) and fluorescence (**d**) microscopy. The tumor resection bed (**e**) was then imaged under fluorescence microscopy for any evidence of residual fluorescence (**f**). In all animals resected, there was residual cancer cells visualized under fluorescence within the tumor bed. Resected and residual tumor was confirmed tumor by histology. All images taken at 20×, scale bars = 1 mm [12]

550 nm group (Alexa Fluor 555 or DyLight 550), 650 nm group (Alexa Fluor 660 or DyLight 650), and 750 nm group (Alexa Fluor 750 or DyLight 755). Twenty-four hours later, the Olympus OV100 small animal imaging system was used to compare the various dyes conjugated to anti-CEA for depth of imaging, resolution, tumor to background ratio, photobleaching, and hemoglobin quenching. The longer-wavelength dyes effected increased depth of penetration and ability to detect the smallest tumor deposits and provided the highest tumor to background ratios, resistance to hemoglobin quenching, and increased tumor specificity (Fig. 22.10). The shorter-wavelength dyes were more photostable [14].

Future Directions of Fluorescence-Guided Surgery and Fluorescence Laparoscopy

Fluorophore-conjugated tumor specific antibodies enabled labeling, detection, and subsequently FGS to improve surgical outcomes in mouse

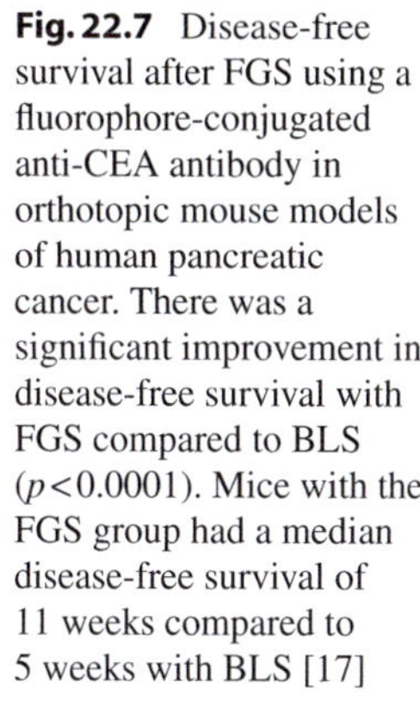
Fig. 22.7 Disease-free survival after FGS using a fluorophore-conjugated anti-CEA antibody in orthotopic mouse models of human pancreatic cancer. There was a significant improvement in disease-free survival with FGS compared to BLS ($p<0.0001$). Mice with the FGS group had a median disease-free survival of 11 weeks compared to 5 weeks with BLS [17]

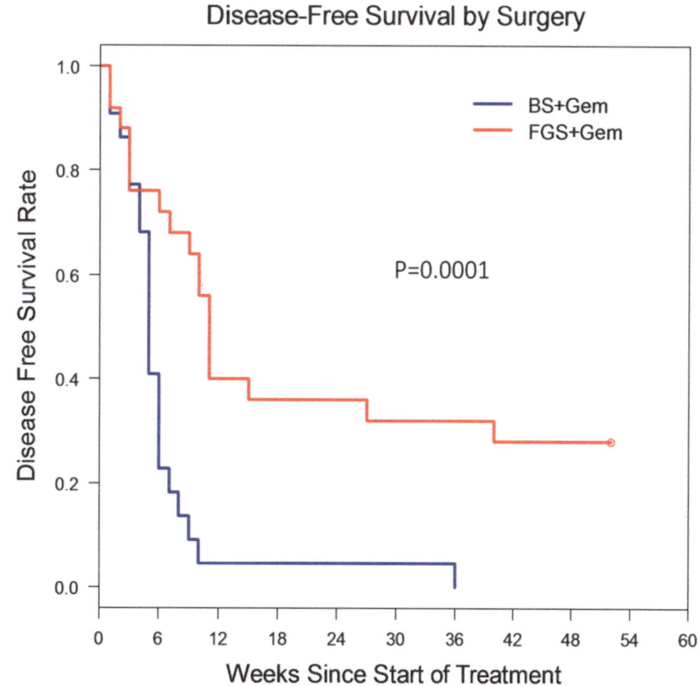

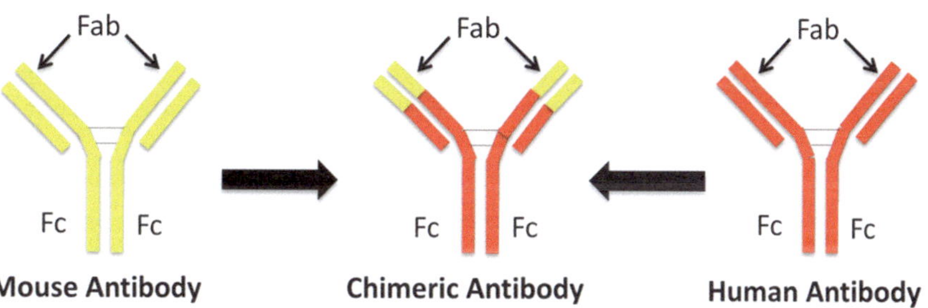

Fig. 22.8 Process of chimerization of anti-CEA antibody. This diagram represents the process of making a mouse-human chimeric anti-CEA antibody (Aragen Bioscience, Inc.) [16]

models of pancreatic and colon cancer. FGS significantly reduced the postoperative tumor burden in the recurrence. The improved R0 resection rate resulted in longer disease-free survival and overall survival. Fluorophore-conjugated antibodies were also be used to detect GI cancers using minimally invasive fluorescence laparoscopy [15, 18, 34].

The goal is to improve all methods of fluorescently labeling tumor and metastases for curative FGS. It is necessary to sterilize the tumor bed after FGS of all residual cancer cells using UVC irradiation, intraoperative chemotherapy, or photoimmunotherapy.

Conclusions

In this chapter, we have reviewed the use of a fluorophore-antibody conjugates specific for the oncofetal antigen CEA to effectively image both primary and metastatic colon and pancreatic tumors in clinically relevant orthotopic mouse models. This approach offers the advantage of a

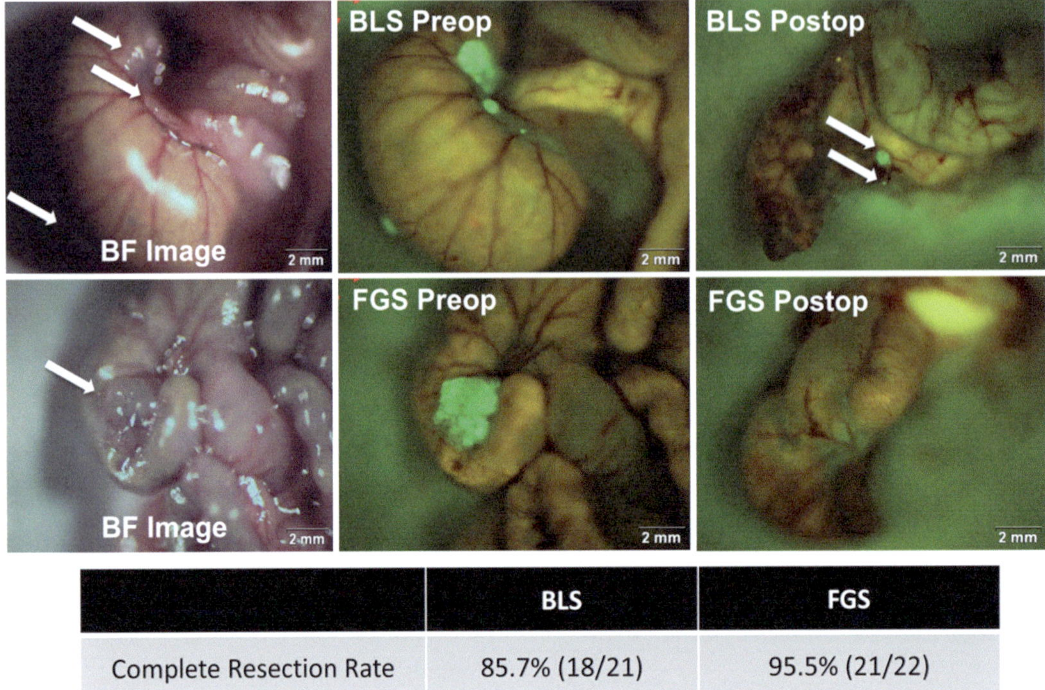

	BLS	FGS
Complete Resection Rate	85.7% (18/21)	95.5% (21/22)

Fig. 22.9 FGS with fluorophore-conjugated chimeric antibody improved real-time detection of tumor margins, thereby increasing complete resections with FGS- compared to BLS-treated mice from 85.7 % to 95.5 %, respectively. Small tumor deposits (indicated by *white arrows*) remained after BLS. The absence of any fluorescence signal in the mouse after FGS indicates complete resection. The preoperative bright field images are on the left to better illustrate the difficulty of identifying tumor margins without fluorescence labeling. *White arrows* indicate tumor lesions in the wall of the cecum [16]

single antibody delivery which improved the identification of residual tumor tissue at the time of resection thereby enabling FGS which significantly improved tumor resection, decreased recurrence, and lengthened overall survival.

Acknowledgment Work supported in part by grants from the National Cancer Institute CA142669 and CA132971 (to M.B. and AntiCancer, Inc.).

References

1. Troyan SL, Kianzad V, Gibbs-Strauss SL, Gioux S, Matsui A, Oketokoun R, et al. The FLARE intraoperative near-infrared fluorescence imaging system: a first-in-human clinical trial in breast cancer sentinel lymph node mapping. Ann Surg Oncol. 2009;16(10): 2943–52.
2. van Dam GM, Themelis G, Crane LM, Harlaar NJ, Pleijhuis RG, Kelder W, et al. Intraoperative tumor-specific fluorescence imaging in ovarian cancer by folate receptor-alpha targeting: first in-human results. Nat Med. 2011;17(10):1315–9.
3. Nguyen QT, Olson ES, Aguilera TA, Jiang T, Scadeng M, Ellies LG, et al. Surgery with molecular fluorescence imaging using activatable cell-penetrating peptides decreases residual cancer and improves survival. Proc Natl Acad Sci U S A. 2010;107(9): 4317–22.
4. Kishimoto H, Zhao M, Hayashi K, Urata Y, Tanaka N, Fujiwara T, et al. In vivo internal tumor illumination by telomerase-dependent adenoviral GFP for precise surgical navigation. Proc Natl Acad Sci U S A. 2009; 106(34):14514–7.
5. Kishimoto H, Aki R, Urata Y, Bouvet M, Momiyama M, Tanaka N, et al. Tumor-selective, adenoviral-mediated GFP genetic labeling of human cancer in the live mouse reports future recurrence after resection. Cell Cycle. 2011;10(16):2737–41.
6. Siegel R, Naishadham D, Jemal A. Cancer statistics, 2013. CA Cancer J Clin. 2013;63(1):11–30.

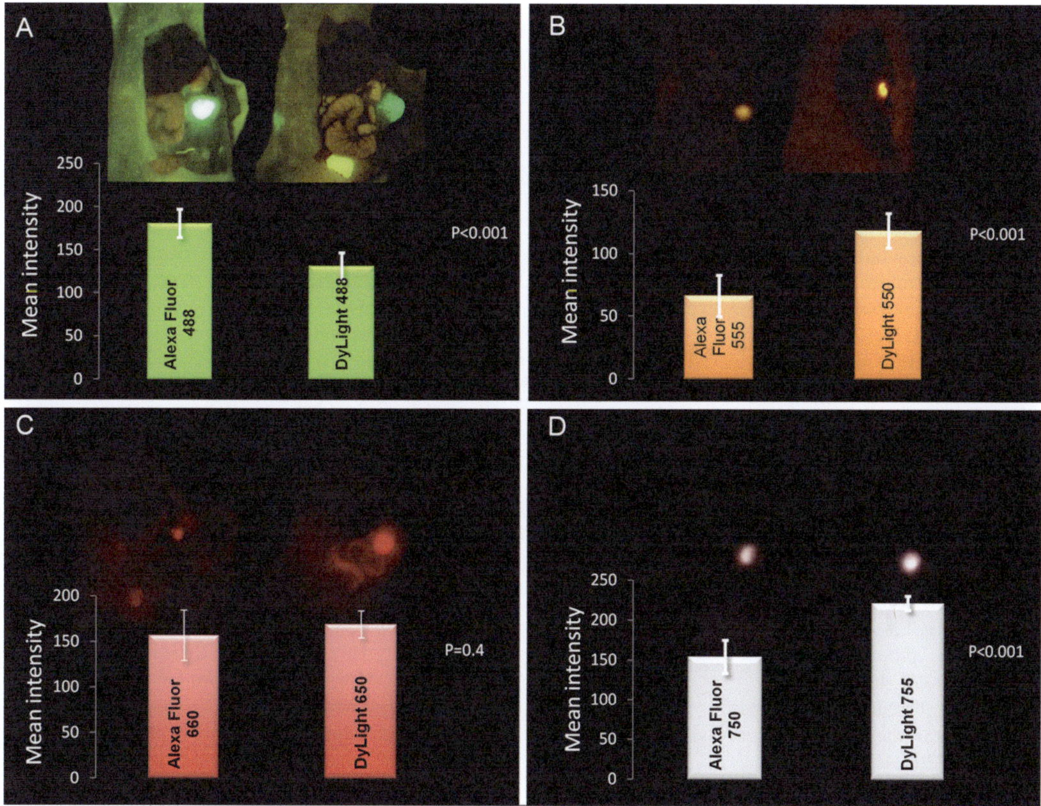

Fig. 22.10 Chimeric anti-CEA antibody conjugated with visible or near-infrared fluorescent dyes were used for imaging pancreatic cancer in orthotopic nude mouse models. (**a**) Alexa Fluor 488 was significantly brighter than DyLight 488 ($p<0.001$). (**b**) DyLight 550 was significantly brighter than Alexa Fluor 555 ($p<0.001$). (**c**) DyLight 650 was brighter than Alexa Fluor 660. (**d**) DyLight 755 was significantly brighter than Alexa Fluor 750 ($p<0.001$). There was decreasing ability to discern the background with increasing wavelength of the fluorophores (**a**–**d**) [14]

7. Robinson MH, Thomas WM, Hardcastle JD, Chamberlain J, Mangham CM. Change towards earlier stage at presentation of colorectal cancer. Br J Surg. 1993;80(12):1610–2.
8. Pawlik TM, Schulick RD, Choti MA. Expanding criteria for resectability of colorectal liver metastases. Oncologist. 2008;13(1):51–64.
9. Bonjer HJ, Hop WC, Nelson H, Sargent DJ, Lacy AM, Castells A, et al. Laparoscopically assisted vs open colectomy for colon cancer: a meta-analysis. Arch Surg. 2007;142(3):298–303.
10. Turrini O, Viret F, Guiramand J, Lelong B, Bege T, Delpero JR. Strategies for the treatment of synchronous liver metastasis. Eur J Surg Oncol. 2007;33(6):735–40.
11. Andreoni B, Chiappa A, Bertani E, Bellomi M, Orecchia R, Zampino M, et al. Surgical outcomes for colon and rectal cancer over a decade: results from a consecutive monocentric experience in 902 unselected patients. World J Surg Oncol. 2007;5:73.
12. Kaushal S, McElroy MK, Luiken GA, Talamini MA, Moossa AR, Hoffman RM, et al. Fluorophore-conjugated anti-CEA antibody for the intraoperative imaging of pancreatic and colorectal cancer. J Gastrointest Surg. 2008;12(11):1938–50.
13. McElroy M, Kaushal S, Luiken GA, Talamini MA, Moossa AR, Hoffman RM, et al. Imaging of primary and metastatic pancreatic cancer using a fluorophore-conjugated anti-CA19-9 antibody for surgical navigation. World J Surg. 2008;32(6):1057–66.
14. Maawy AA, Hiroshima Y, Kaushal S, Luiken GA, Hoffman RM, Bouvet M. Comparison of a chimeric anti-carcinoembryonic antigen antibody conjugated with visible or near-infrared fluorescent dyes for imaging pancreatic cancer in orthotopic nude mouse models. J Biomed Opt. 2013;18(12):126016.
15. Metildi CA, Kaushal S, Lee C, Hardamon CR, Snyder CS, Luiken GA, et al. An LED light source and novel fluorophore combinations improve fluorescence laparoscopic detection of metastatic pancreatic cancer in orthotopic mouse models. J Am Coll Surg. 2012; 214(6):997–1007.e2.

16. Metildi CA, Kaushal S, Luiken GA, Talamini MA, Hoffman RM, Bouvet M. Fluorescently labeled chimeric anti-CEA antibody improves detection and resection of human colon cancer in a patient-derived orthotopic xenograft (PDOX) nude mouse model. J Surg Oncol. 2014;109:451.
17. Metildi CA, Kaushal S, Pu M, Messer KS, Luiken GA, Moossa AR, et al. Fluorescence-guided surgery with a fluorophore-conjugated antibody to carcinoembryonic antigen (CEA) that highlights the tumor improves surgical resection and increases survival in orthotopic mouse models of human pancreatic cancer. Ann Surg Oncol. 2014;21:1405.
18. Tran Cao HS, Kaushal S, Metildi CA, Menen RS, Lee C, Snyder CS, et al. Tumor-specific fluorescence antibody imaging enables accurate staging laparoscopy in an orthotopic model of pancreatic cancer. Hepatogastroenterology. 2012;59(118):1994–9.
19. Gold P, Freedman SO. Demonstration of tumor-specific antigens in human colonic carcinomata by immunological tolerance and absorption techniques. J Exp Med. 1965;121:439–62.
20. Gold P, Shuster J, Freedman SO. Carcinoembryonic antigen (CEA) in clinical medicine: historical perspectives, pitfalls and projections. Cancer. 1978;42(3 Suppl):1399–405.
21. Gold P, Freedman SO. Specific carcinoembryonic antigens of the human digestive system. J Exp Med. 1965;122(3):467–81.
22. Albers GH, Fleuren G, Escribano MJ, Nap M. Immunohistochemistry of CEA in the human pancreas during development, in the adult, chronic pancreatitis, and pancreatic adenocarcinoma. Am J Clin Pathol. 1988;90(1):17–22.
23. Yamaguchi K, Enjoji M, Tsuneyoshi M. Pancreatoduodenal carcinoma: a clinicopathologic study of 304 patients and immunohistochemical observation for CEA and CA19-9. J Surg Oncol. 1991;47(3):148–54.
24. Prall F, Nollau P, Neumaier M, Haubeck HD, Drzeniek Z, Helmchen U, et al. CD66a (BGP), an adhesion molecule of the carcinoembryonic antigen family, is expressed in epithelium, endothelium, and myeloid cells in a wide range of normal human tissues. J Histochem Cytochem. 1996;44(1):35–41.
25. Nap M, Mollgard K, Burtin P, Fleuren GJ. Immunohistochemistry of carcino-embryonic antigen in the embryo, fetus and adult. Tumour Biol. 1988;9(2–3):145–53.
26. Fletcher RH. Carcinoembryonic antigen. Ann Intern Med. 1986;104(1):66–73.
27. Kammerer R, von Kleist S. CEA expression of colorectal adenocarcinomas is correlated with their resistance against LAK-cell lysis. Int J Cancer. 1994;57(3):341–7.
28. Locker GY, Hamilton S, Harris J, Jessup JM, Kemeny N, Macdonald JS, et al. ASCO 2006 update of recommendations for the use of tumor markers in gastrointestinal cancer. J Clin Oncol. 2006;24(33):5313–27.
29. Goldstein MJ, Mitchell EP. Carcinoembryonic antigen in the staging and follow-up of patients with colorectal cancer. Cancer Invest. 2005;23(4):338–51.
30. Fu XY, Besterman JM, Monosov A, Hoffman RM. Models of human metastatic colon cancer in nude mice orthotopically constructed by using histologically intact patient specimens. Proc Natl Acad Sci U S A. 1991;88(20):9345–9.
31. Kishimoto H, Maawy AA, Sato S, Murakami T, Uehara F, Miwa S, et al. Hand-held high-resolution fluorescence imaging system for fluorescence-guided surgery of patient and cell-line pancreatic tumors growing orthotopically in nude mice. J Surg Res. 2014;187:510.
32. Yamauchi K, Yang M, Jiang P, Xu M, Yamamoto N, Tsuchiya H, et al. Development of real-time subcellular dynamic multicolor imaging of cancer-cell trafficking in live mice with a variable-magnification whole-mouse imaging system. Cancer Res. 2006;66(8):4208–14.
33. Nelson AL, Dhimolea E, Reichert JM. Development trends for human monoclonal antibody therapeutics. Nat Rev Drug Discov. 2010;9(10):767–74.
34. Metildi CA, Hoffman RM, Bouvet M. Fluorescence-guided surgery and fluorescence laparoscopy for gastrointestinal cancers in clinically-relevant mouse models. Gastroenterol Res Pract. 2013;2013:290634.

Development of a Non-blurring, Dual-Imaging (X-Ray/ Fluorescence) Tissue Marker for Localization of Gastrointestinal Tumors

23

Hideki Hayashi, Taro Toyota, Shoichi Goto, Aki Oishi, Tao Gao, Lau Bik Ee, and Hisahiro Matsubara

Introduction

It is essential to secure minimum surgical margins from tumors in the treatments of gastrointestinal malignancies. However, achievements of better curability of the disease and improvements of postoperative quality of patients' life are of conflicting subjects in terms of extents of the resection margins from the tumor. For example, a gross margin of 2 cm is recommended for T1 gastric tumors, and a proximal margin of at least 3 cm is recommended for T2 or deeper tumors of type 1 or 2, according to the Japanese gastric cancer treatment guideline [1]. Generally, final decisions of the resection lines are made by the surgeons based on the information confirmed by their intraoperative palpation of the lesions or by that of marking clips which are preoperatively placed near the lesions [2, 3].

Recent advances of anastomotic techniques for alimentary tract reconstruction allow totally intracorporeal procedures for treatments of gastrointestinal malignancies, and these operations have been shown to have various clinical advantages over laparoscope-assisted operations [4–9]. However, such new procedures complicated the aforementioned issues about the resection lines, since the major targets of such laparoscopic surgeries have been early stage cancers that were not expected to cause serosal surface alterations and even palpations of target organs were no more available during surgery. Therefore, some countermeasures would be required to precisely localize the tumor. Although various techniques including preoperative or intraoperative gastrointestinal endoscopy to assist intraoperative localization of tumors had been proposed [10–20], these have yet to be satisfying remedies for this purpose.

In the previous report, we have shown that indocyanine green (ICG) mixed with giant vesicles of lecithin became optically stabilized and aggregate of ICG-containing giant vesicles with polyglycerol polyricinoleate (PGPR) was rather stable within biological tissues [21]. We have also shown that hydrophobic fluorophore, an ICG derivative tagged with alkyl chains (ICG-C18), hardly leaks from lecithin vesicles or liposomes in vivo, in contrast with conventional ICG molecule

H. Hayashi, M.D., Ph.D. (✉)
Center for Frontier Medical Engineering, Chiba University, 1-33 Yayoi-cho, Inage-ku, Chiba 263-8522, Japan
e-mail: hhayashi@faculty.chiba-u.jp

T. Toyota, Ph.D.
Department of Basic Science, University of Tokyo, Komaba, Tokyo, Japan

S. Goto, B.A. • A. Oishi
Faculty of Engineering, Chiba University, Chiba, Japan

T. Gao, M.S. • H. Matsubara, M.D., Ph.D.
Department of Frontier Surgery, Graduate School of Medicine, Chiba University, Chiba, Japan

L.B. Ee, B.A.
Graduate School of Engineering, Chiba University, Chiba, Japan

F.D. Dip et al. (eds.), *Fluorescence Imaging for Surgeons: Concepts and Applications*,
DOI 10.1007/978-3-319-15678-1_23,

showing some leaks into systemic circulation from liposomes when subcutaneously administered in mice [22]. Therefore, we have developed a tissue marker that consists of a giant lecithin vesicle with hydrophobic NIR fluorophore, ICG-C18, instead of ICG, expecting better stability in biological tissues.

Furthermore, we also tried to incorporate X-ray contrast medium in this giant vesicle in this study. Ordinary preoperative assessment of gastric cancer includes esophagogastroendoscopy and X-ray CT examination. Esophagogastroendoscopy is essential for pathologic diagnosis, and X-ray CT has an important role to detect nodal involvement and/or distant metastasis. However, assessment of tumor location by means of gastroendoscope is sometimes inaccurate [23], and early gastric cancer lesions are hardly visualized with X-ray CT [24]. Dual-imaging tissue markers (DI marker) that are preoperatively administered around the early gastric cancer lesions could help both preoperative simulation for surgery and intraoperative navigation with the use of NIR fluorescent laparoscope. The usefulness of this DI marker was tested in a porcine model.

Materials and Methods

Materials

Indocyanine green (Diagnogreen®) was purchased from Daiichi Sankyo Co., Ltd. (Tokyo, Japan). ICG-C18 was purchased from Yamada Chemical Co. Ltd. (Kyoto, Japan). Egg yolk phosphatidylcholine (EYPC), squalene, D-glucose, and sucrose were supplied by Kanto Chemical Co. Inc. (Tokyo, Japan). Ethiodized oil, Lipiodol®, as a radiographic contrast medium is purchased from Guerbet Japan (Tokyo, Japan). Polyglycerol polyricinoleate (Poem® PR-100) was provided by Riken Vitamin Co., Ltd. (Tokyo, Japan).

Preparation of the Tissue Marker

Detailed procedure for the preparation of DI marker was described elsewhere [21, 25]. Briefly, ICG-C18-containing giant vesicle (ICG-C18-GV) was formed by dispersion of pulverized ICG-C18 and EYPC in distilled water. ICG-C18-GV dispersion, 50 mM Tris–HCl (pH 7.8) with 1 M sucrose and Lipiodol®, was mixed with tapping. This mixture was suspended into the oil phase of squalene dissolving PGPR (15 w/w %) and Lipiodol® by agitation, and the resulting W/O emulsion was layered on a buffered solution of 50 mM Tris–HCl (pH 7.8) and 1 M D-glucose in a 15-mL plastic tube. The tube was centrifuged, and the precipitated PGPR polymersomes, containing ICG-C18-GVs and Lipiodol® emulsion, were collected using a micropipette.

In Vitro Analyses

NIR fluorescence camera system, HyperEye Medical System (Mizuho Co., Ltd., Tokyo, Japan), was used for macroscopic observation of NIR fluorescence. X-ray detectability of the marker was checked with multi-detector-row X-ray CT, Aquilion 16 (Toshiba Medical Systems Co., Tokyo, Japan). Microscopic observation of the marker was performed with an Olympus BX51 microscope (Tokyo, Japan) equipped with a xenon lamp and a filter unit (λex, 738–795 nm; λem, 814–862 nm). The micrographs were digitally recorded using an EM-CCD camera (iXon DU897E, Andor Technology plc., Belfast, UK) with bio-imaging analysis software, Lumina Vision (Mitani Co., Tokyo, Japan).

Animal Experiment

The animal experiment protocol was reviewed and approved by the Institutional Animal Care and Use Committee of Animal Experimentation, Chiba University. Four-month-old, male, castrated, specific pathogen-free domestic pigs were purchased from Takasugi Experimental Animals Inc. (Saitama, Japan). Each animal was kept in a cage with an appropriately warm and humid condition until the end of the experiments. They were fasted for 24 h before administration of the marker.

Under the general anesthesia, a metal clip for hemostasis [26] was placed on the mucosa of the middle stomach of the animal under the guidance of gastroendoscopy. Three hundred microliter of each marker was injected into the submucosal layer at four points on a circumference of radius 1 cm around the clip with the use of endoscopic injection needle. Consequently, we inserted the fluorescent laparoscope into the abdominal cavity of the animal and observed the serosal surface of the middle stomach. After that, a whole-body X-ray CT of the animal was performed, and then, let them awake from general anesthesia. The animal was anesthetized again on the next day, and fluorescent laparoscopy was performed. Animals were euthanized with intravenous injections of 20 mEq KCL solution after all analyses were completed.

In Vivo Analyses

NIR fluorescence laparoscope system used in this study was assembled in our laboratory. EM-CCD camera system (ADT-40S, Flovel, Co., Ltd, Tokyo, Japan), high-pass filter (λ>810 nm, LIX810, Asahi Spectra Co., Ltd., Tokyo, Japan), and telescopic rod lens with an optical angle of 30° (WA5005A, Olympus Medical Systems Corp., Tokyo, Japan) were used for the processing of NIR fluorescence images, and CLV-S20 (Olympus, Tokyo, Japan) was used as a light source. Aquilion 16 was also used for in vivo X-ray CT analysis. OsiriX DICOM viewer version 5.7 (Bernex, Switzerland) was used for volume rendering of X-ray CT images.

Results

In Vitro Analysis of the Marker

The DI marker showed gross appearance of viscous liquid with greenish white color (Fig. 23.1a). Bright fluorescence was observed with a default setting of the NIR fluorescence camera (Fig. 23.1b). X-ray CT with the setting for abdominal viscera clearly visualized the marker (Fig. 23.1c). Mean CT number of the marker was 627. Phase contrast microscopy of the marker revealed multiple compartments encapsulating various sizes of giant vesicles (Fig. 23.2a). NIR fluorescence from some of the encapsulated giant vesicles was detected by fluorescence microscopy with the NIR filter set (Fig. 23.2b).

In Vivo Analysis of the Marker

Next, we tested the usefulness of DI marker using a porcine model. Under the general anesthesia, gastroendoscopy of the pig was performed, and a metal clip was placed in each of the anterior and posterior sides of the middle stomach. Then, makers were injected into the submucosal layer at four points on a circumference of radius 1 cm around the clips. Approximately 300 μl of the marker was injected at each injection point. ICG aqueous solution of 64 μM was administered on the anterior side of the stomach and so did DI marker with the ICG-C18 concentration of 64 μM on the posterior side. Three hundred microliter of Lipiodol

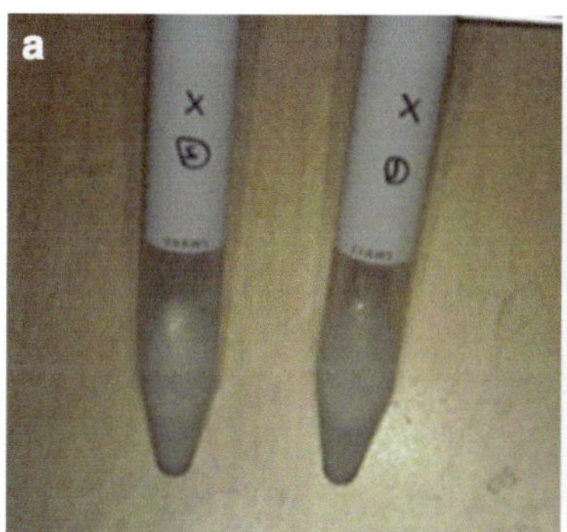

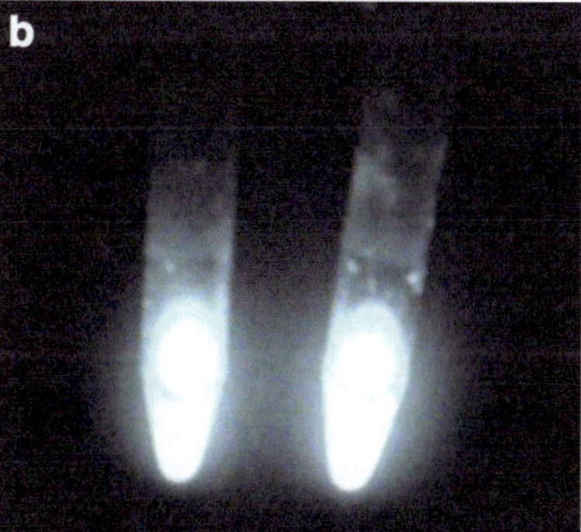

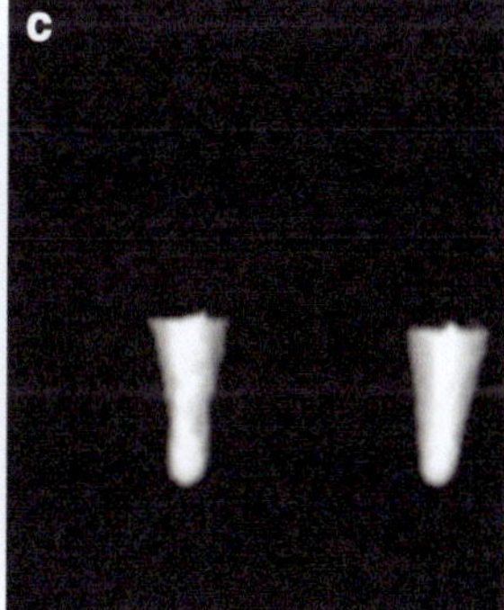

Fig. 23.1 Gross appearance (**a**), macroscopic NIR fluorescence image (**b**), and X-ray CT image of the marker

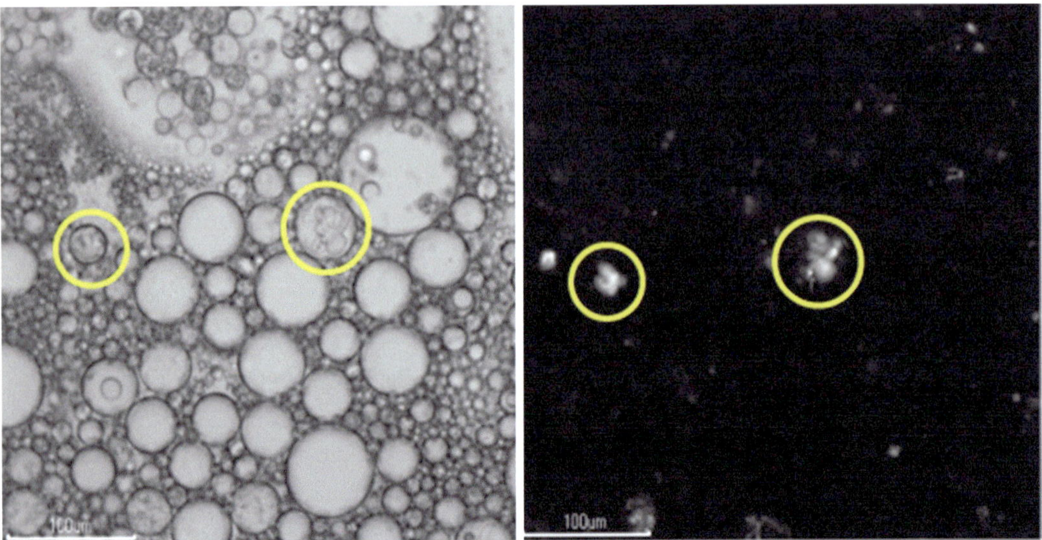

Fig. 23.2 Microscopic observation of the marker. (**a**) Phase contrast microscopy. (**b**) NIR fluorescence microscopy. *Circles* indicate vesicles which emit NIR fluorescence

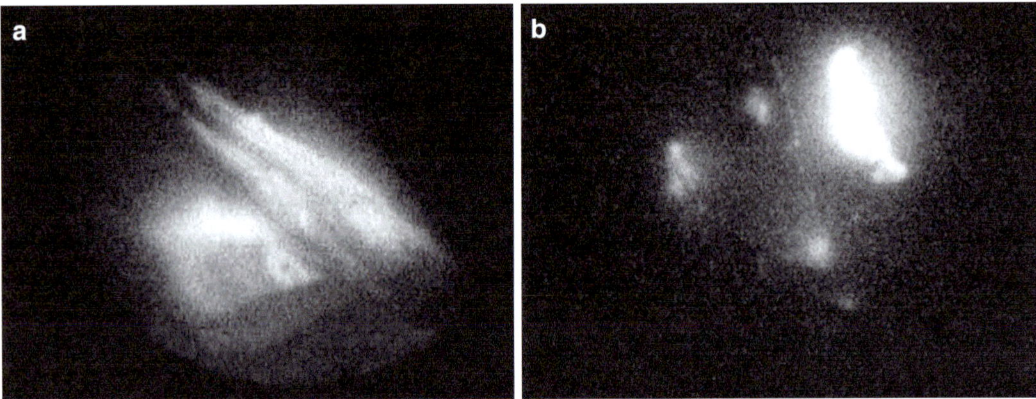

Fig. 23.3 NIR laparoscopic appearance of the stomach immediately after marker administration. (**a**) Anterior side where ICG aqueous solution was administered. (**b**) Posterior side where DI marker was administered

was injected at two points 1 cm apart from each other, in the lower stomach, as a positive control for X-ray CT examination.

Immediately after the marker injection, laparoscopic analysis of the animal was performed. Conventional laparoscopy could not reveal any alteration of the gastric serosa (data not shown). NIR laparoscopy showed bright fluorescence on each side of the stomach. The anterior side of the stomach showed some blurring already, and each injection point was barely distinguishable (Fig. 23.3a). In contrast, four injection points were independently observed on the posterior side of the stomach (Fig. 23.3b).

Then, an X-ray CT examination of the animal was followed. Nothing but a metal clip was observed on the anterior side of the stomach (data not shown); however, four injection points as well as a metal clip were individually detected in the posterior wall (Fig. 23.4a). Volume-rendered image of the X-ray CT examination showed each injection point and a metal clip separately (Fig. 23.4b).

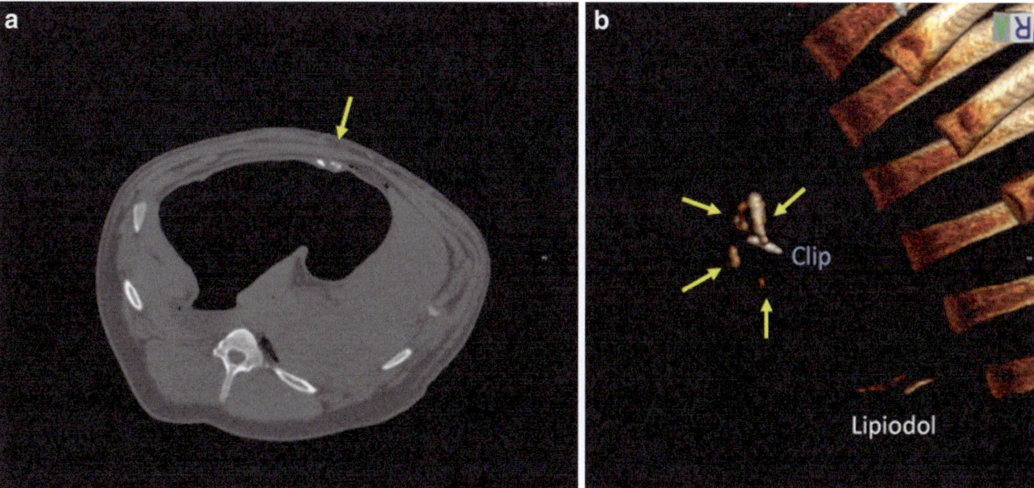

Fig. 23.4 X-ray CT images of the abdomen immediately after the marker administration. (**a**) Axial image for the level of the site of dual-imaging marker administration. (**b**) Volume-rendered images of the X-ray CT examination. *Arrows* indicate DI maker administered

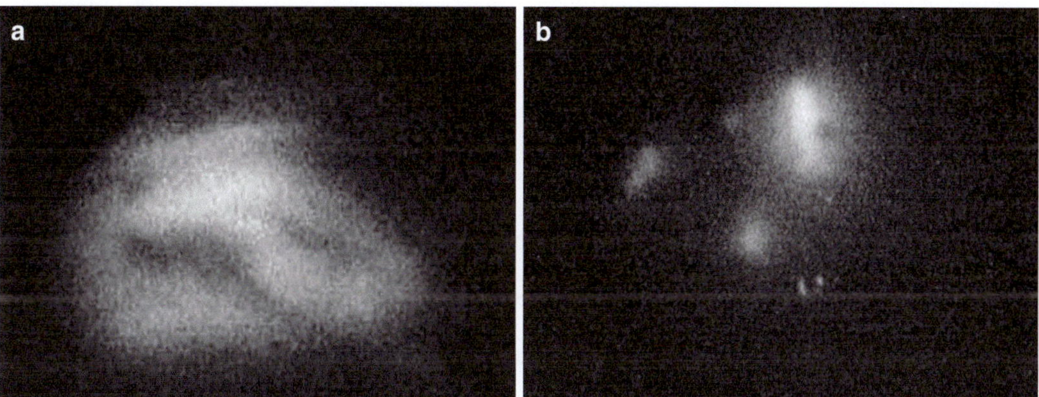

Fig. 23.5 NIR laparoscopic appearance of the stomach 18 h after marker administration. (**a**) Anterior side where ICG aqueous solution was administered. (**b**) Posterior side where DI marker was administered

Eighteen hours after marker administration, the animal was anesthetized and laparoscopy was performed again. NIR laparoscopy showed fluorescence of broad blurring over 7 cm on the anterior side of the stomach, and each injection point was not distinguishable (Fig. 23.5a). On the other hand, four injection points were still individually distinguishable on the other side of the stomach (Fig. 23.5b).

Discussions

Preoperative marking of target lesion is essential for gastrointestinal surgeries when they are not palpable. Tattooing with India ink has been long used for the surgeries of colon and rectum since the first report by Ponsky and King in 1975 [27]. This technique is useful for rough estimate of locations of interest within such a long thin organ.

However, it is not sufficient for the gastric surgeries that require precise setting of the resection line. Other dyes such as indocyanine green and indigo carmine and autologous blood have been tried to be administered preoperatively around early gastric cancer lesions to be localized intraoperatively [15, 17, 20]; however, a total volume of 2–5 ml of such markers was required to let them reach to the serosa and become visible. Therefore, these markers diffuse significantly through tissues, and resection margins varied among cases [20], suggesting these techniques remain not ideal to localize injection points in a pinpoint manner. Our technique required administration of only 0.3 ml of the marker in the submucosal layer to be visualized with NIR fluorescence laparoscope, and blurring was less than 1 cm even 18 h after administration.

So far, various other techniques to determine such precise resection line of the stomach during laparoscopic surgery have been reported. Park et al. first described the usefulness of intraoperative gastroscopy for laparoscopic gastric surgery. They detected target lesions by transillumination of scope or by pushing down the gastric wall with forceps and identifying the matching protrusion on the gastroscopy [11]. However, intraoperative gastroendoscopy is sometime cumbersome because of intratracheal intubation for general anesthesia and supine position which are not suitable for the procedure [19]. Furthermore, insufflation of the stomach and small bowel afterward with gastroendoscope might disturb subsequent laparoscopic procedures, and mobilization of the stomach sometimes causes acute alterations of the gastric mucosa such as redness and edema, hampering close observation of lesions.

In order to avoid such difficulties, combination of conventional imaging modalities such as intracorporeal ultrasonography [10] or portable radiograph [18] during laparoscopic surgery has been tried to detect preoperatively placed endoscopic marking clip. Although these techniques showed better results, they require complex procedure such as consideration of two image modalities (i.e., laparoscopic image and ultrasonographic or radiographic image) at the same time and the other space-occupying equipment and operating stuff in the OR.

The other new technologies have been also tested, as well. Ohdaira et al. introduced magnetic detection system to detect preoperatively placed marking clip [12]. This technique consisted of a small magnetic tag attached to marking clip with a nylon string and a magnetometer. It does not visualize the location of the clip; therefore, it requires a survey with the tip of a probe over the stomach. Choi et al. developed a novel fluorescent clip with NIR fluorochromes to be placed around target lesions [16]. Although it remains to be tested in an ex vivo animal model, it should be a fascinating method to detect preoperatively placed endoscopic clips in a pinpoint manner. However, these two techniques with marking clips should be considered with the risk to cause misfiring of linear staplers when they get stuck between the jaws of linear staplers during laparoscopic gastric resection.

Our novel tissue marker is provided as a form of viscous liquid which consists of biologically safe components, staying within biological tissues with minimum diffusion for a long time. It can be visualized with NIR fluorescence laparoscopic video systems without a risk of misfiring of linear staplers at gastric resections. Although it should be tested further in a clinical setting, it can be a useful tool to secure precise resection margin for the laparoscopic treatment of early stage tumors in the gastrointestinal tract.

References

1. Japanese Gastric Cancer Association. Japanese gastric cancer treatment guidelines 2010 (ver. 3). Gastric Cancer. 2011;14:113–23.
2. Hachisu T, Miyazaki S, Hamaguchi K. Endoscopic clip-marking of lesions using the newly developed HX-3L clip. Surg Endosc. 1989;3:142–7.
3. Ryu KW, Lee JH, Choi IJ, Bae JM. Preoperative endoscopic clipping: localizing technique of early gastric cancer. J Surg Oncol. 2003;82:75–7.
4. Ikeda O, Sakaguchi Y, Aoki Y, Harimoto N, Taomoto J, Masuda T, Ohga T, Adachi E, Toh Y, Okamura T, Baba H. Advantages of totally laparoscopic distal gastrectomy over laparoscopically assisted distal gastrectomy for gastric cancer. Surg Endosc. 2009;23:2374–9.
5. Kanaya S, Kawamura Y, Kawada H, Iwasaki H, Gomi T, Satoh S, Uyama I. The delta-shaped anastomosis in laparoscopic distal gastrectomy: analysis of the initial 100 consecutive procedures of intracorporeal gastroduodenostomy. Gastric Cancer. 2011;14:365–71.

6. Motoyama K, Kojima K, Hayashi M, Kato K, Inokuchi M, Sugihara K. Beta-shaped intracorporeal Roux-en-Y reconstruction after totally laparoscopic distal gastrectomy. Gastric Cancer. 2014;17:588. doi:10.1007/s10120-013-0311-5.
7. Kim HS, Kim MG, Kim BS, Lee IS, Lee S, Yook JH, Kim BS. Comparison of totally laparoscopic total gastrectomy and laparoscopic-assisted total gastrectomy methods for the surgical treatment of early gastric cancer near the gastroesophageal junction. J Laparoendosc Adv Surg Tech A. 2013;23: 204–10.
8. Jun G, Ping L, Jie C, Qi L, Tang D, Wang D. Totally laparoscopic vs. laparoscopically assisted distal gastrectomy for gastric cancer: a meta-analysis. Hepatogastroenterology. 2013;60:1530–4.
9. Gao J, Li P, Li QG, Chen J, Wang DR, Tang D. Comparison between totally laparoscopic and laparoscopically assisted distal gastrectomy for gastric cancer with a short follow-up: a meta-analysis. J Laparoendosc Adv Surg Tech A. 2013;23:693–7.
10. Hyung WJ, Lim JS, Cheong JH, Kim J, Choi SH, Song SY, Noh SH. Intraoperative tumor localization using laparoscopic ultrasonography in laparoscopic-assisted gastrectomy. Surg Endosc. 2005;19:1353–7.
11. Park DJ, Lee HJ, Kim SG, Jung HC, Song IS, Lee KU, Choe KJ, Yang HK. Intraoperative gastroscopy for gastric surgery. Surg Endosc. 2005;19:1358–61.
12. Ohdaira T, Nagai H. Intraoperative localization of early-stage upper gastrointestinal tumors using a magnetic marking clip-detecting system. Surg Endosc. 2007;21:810–5.
13. Nagata K, Endo S, Tatsukawa K, Kudo SE. Intraoperative fluoroscopy vs. intraoperative laparoscopic ultrasonography for early colorectal cancer localization in laparoscopic surgery. Surg Endosc. 2008;22:379–85.
14. Matsui H, Okamoto Y, Nabeshima K, Kondoh Y, Ogoshi K, Makuuchi H. Endoscopy-assisted gastric resection: a safe and reliable procedure for tumor clearance during laparoscopic high distal or proximal gastrectomy. Surg Endosc. 2009;23:1146–9.
15. Miyoshi N, Ohue M, Noura S, Yano M, Sasaki Y, Kishi K, Yamada T, Miyashiro I, Ohigashi H, Iishi H, Ishikawa O, Imaoka S. Surgical usefulness of indocyanine green as an alternative to India ink for endoscopic marking. Surg Endosc. 2009;23:347–51.
16. Choi Y, Kim KG, Kim JK, Nam KW, Kim HH, Sohn DK. A novel endoscopic fluorescent clip for the localization of gastrointestinal tumors. Surg Endosc. 2011;25:2372–7.
17. Jeong O, Cho SB, Joo YE, Ryu SY, Park YK. Novel technique for intraoperative tumor localization during totally laparoscopic distal gastrectomy: endoscopic autologous blood tattooing. Surg Endosc. 2012;26: 1778–83.
18. Kim HI, Hyung WJ, Lee CR, Lim JS, An JY, Cheong JH, Choi SH, Noh SH. Intraoperative portable abdominal radiograph for tumor localization: a simple and accurate method for laparoscopic gastrectomy. Surg Endosc. 2011;25:958–63.
19. Nakagawa M, Ehara K, Ueno M, Tanaka T, Kaida S, Udagawa H. Accurate, safe, and rapid method of intraoperative tumor identification for totally laparoscopic distal gastrectomy: injection of mixed fluid of sodium hyaluronate and patent blue. Surg Endosc. 2014;28:1371. doi:10.1007/s00464-013-3319-3.
20. Xuan Y, Hur H, Byun CS, Han SU, Cho YK. Efficacy of intraoperative gastroscopy for tumor localization in totally laparoscopic distal gastrectomy for cancer in the middle third of the stomach. Surg Endosc. 2013; 27:4364–70.
21. Hatayama H, Toyota T, Hayashi H, Nomoto T, Fujinami M. Application of a novel near infrared-fluorescence giant vesicle- and polymerasome-based tissue marker for endoscopic and laparoscopic navigation. Anal Sci. 2014;30:225.
22. Toyota T, Fujito H, Suganami A, Ouchi T, Ohishi A, Aoki A, Onoue K, Muraki Y, Madono T, Fujinami M, Tamura Y, Hayashi H. Near-infrared-fluorescence imaging of lymph nodes by using liposomally formulated indocyanine green derivatives. Bioorg Med Chem. 2014;22:721. doi:10.1016/j.bmc.2013.12.026.
23. Jeong SH, Bae K, Ha CY, Lee YJ, Lee OJ, Jung WT, Choi SK, Hong SC, Jung EJ, Ju YT, Jeong CY, Ha WS. Effectiveness of endoscopic clipping and computed tomography gastroscopy for the preoperative localization of gastric cancer. J Kr Surg Soc. 2013;84:80–7.
24. Yu JS, Choi SH, Choi WH, Chung JJ, Kim JH, Kim KW. Value of nonvisualized primary lesions of gastric cancer on preoperative MDCT. AJR Am J Roentgenol. 2007;189:W315–9.
25. Hayashi H, Toyota T, Goto S, Ooishi A, Gao T, Lau BE, Hatayama H, Nomoto T, Fujinami M, Matsubara H. Development of a non-blurring, dual-imaging tissue marker for gastrointestinal tumor localization. Surg Endosc. 2015;29:1445–51.
26. Hachisu T. Evaluation of endoscopic hemostasis using an improved clipping apparatus. Surg Endosc. 1988;2:13–7.
27. Ponsky JL, King JF. Endoscopic marking of colonic lesions. Gastrointest Endosc. 1975;22:42–3.

24 Sentinel Node Navigation Surgery by Infrared Imaging in Gastric Cancer

Yoshikazu Uenosono, Takaaki Arigami, Shigehiro Yanagita, Daisuke Matsushita, Sumiya Ishigami, and Shoji Natsugoe

Owing to the development of endoscopic screening, the incidence of early gastric cancer has been increased in nearly 60 % in Japan. Lymph node metastasis is one of the most important prognostic factors in patients with gastric cancer. Unfortunately, preoperative diagnosis by imaging means is not always reliable for detecting lymph node metastasis. Therefore, standard lymph node dissection for gastric cancer has been well accepted, with a resulting decrease in recurrence and improvement in survival rate. The 5-year survival rates of patients with mucosal and submucosal gastric cancer are 95–100 % and 85–95 %, respectively. However, the incidence of lymph node metastasis determined by histology in mucosal and submucosal gastric cancer is 2–4 % and 13–20 %, respectively.

Recently, minimally invasive surgery such as endoscopic submucosal dissection (ESD) and laparoscopic-assisted gastrectomy has been performed. According to the criteria of the Japanese Gastric Cancer Treatment Guidelines 2010 (ver. 3), the indication of ESD is tumors with differentiated type, mucosal cancer <2 cm in diameter, and no ulcer scar formation. Laparoscopic gastrectomy for early gastric cancer has been widely accepted, and many retrospective studies have demonstrated its advantages over open gastrectomy in terms of less pain, better cosmetic results, better postoperative respiratory function, and faster recovery. However, the long-term quality of life in the laparoscopic surgery is the same as an open surgery. This finding indicates that volume of residual stomach is the same in both operations. Standard gastrectomy such as distal gastrectomy and total gastrectomy causes postgastrectomy syndrome which remains to be a serious drawback for the cancer survivors [1]. In general, extensive gastrectomy is accompanied with a prophylactic lymphadenectomy. If lymph node metastasis can be confirmed by preoperative or intraoperative examination, the reduction of lymphadenectomy may be applicable for such patients.

The sentinel node (SN) concept has been clinically applied to melanoma and breast cancer. According to this concept, SN is the first lymph node receiving lymphatic flow from the primary tumor, and lymph node metastasis occurs first in SNs [2]. It is not necessarily the nearest lymph node from the primary tumor. To date, several studies have demonstrated the utility of the SN concept in patients with early stage gastric [3]. Radioisotope (RI) and/or blue dye was used as a tracer for detecting SNs in these studies. Nowadays, the SN detection by the infrared fluorescence imaging method using indocyanine

Y. Uenosono, M.D., Ph.D. (✉)
T. Arigami, M.D., Ph.D. • S. Yanagita, M.D., Ph.D.
D. Matsushita, M.D. • S. Ishigami, M.D., Ph.D.
S. Natsugoe, M.D., Ph.D.
Department of Digestive Surgery, Breast and Thyroid Surgery, Graduate School of Medicine, Kagoshima University, 8-35-1 Sakuragaoka, Kagoshima 890-8520, Japan
e-mail: uenosono@m3.kufm.kagoshima-u.ac.jp

F.D. Dip et al. (eds.), *Fluorescence Imaging for Surgeons: Concepts and Applications*, DOI 10.1007/978-3-319-15678-1_24, © Springer International Publishing Switzerland 2015

green (ICG) as a tracer is considered. In this chapter, we describe the SN detection for early gastric cancer, including infrared imaging (IRI) using ICG.

Sentinel Node Concept in Early Gastric Cancer

Recently, SN navigation surgery is one of the current trends in early gastric cancer. If the SN concept is applicable to patients with early gastric cancer, it would prove useful to identify the rational extent of lymph node dissection during surgery. The reduction of lymph node dissection enables extensive preservation of the stomach such as partial and segmental resection. However, clinical application of SN navigation surgery for gastric cancer remains controversial. In clinical application, an accurate diagnosis of SN is important preoperatively and during surgery. Usually, RI and/or dye methods are two major procedures for detection of SNs (Fig. 24.1).

We examined the value of SN navigation surgery, including estimation of nodal micrometastasis, in early gastric cancer patients [4]. The day before surgery, approximately 3 mCi ^{99m}Tc-radiolabeled tin colloid (Nihon Medi-Physics, Nishinomiya, Japan) was injected endoscopically into the gastric submucosa at four sites (0.5 ml each) near the tumor using a 23-gauge needle. During surgery, radioisotope uptake in each lymph node was measured using a Navigator GPS™ instrument (Dilon Diagnostics, USA). After laparotomy, SN excision was performed, followed by gastrectomy with lymphadenectomy. After resection, the absence of residual nodes with radioactivity in the abdomen was confirmed intraoperatively using the Navigator GPS. RI uptake in all dissected lymph nodes was again measured after surgery. All dissected lymph nodes including SNs were examined by hematoxylin and eosin (HE) staining, immunohistochemical (IHC) staining using AE1/AE2, and reverse transcription polymerase chain reaction (RT-PCR) using markers for CEA (Fig. 24.2) [5].

Feasibility study was performed in 201 patients with clinical T1N0 gastric cancer. Detection rate of SNs is 98 % (197/201), and the average number of SNs is 4.0. Metastasis was found in 9.1 %, and micrometastasis was found in 9.1 % by IHC and 10.1 % by RT-PCR, respectively. A total of 56 patients had lymph node metastasis including micrometastasis. The sensitivity and accuracy of SN mapping in HE staining were 94.4 % and 99.5 %, respectively. The sensitivity

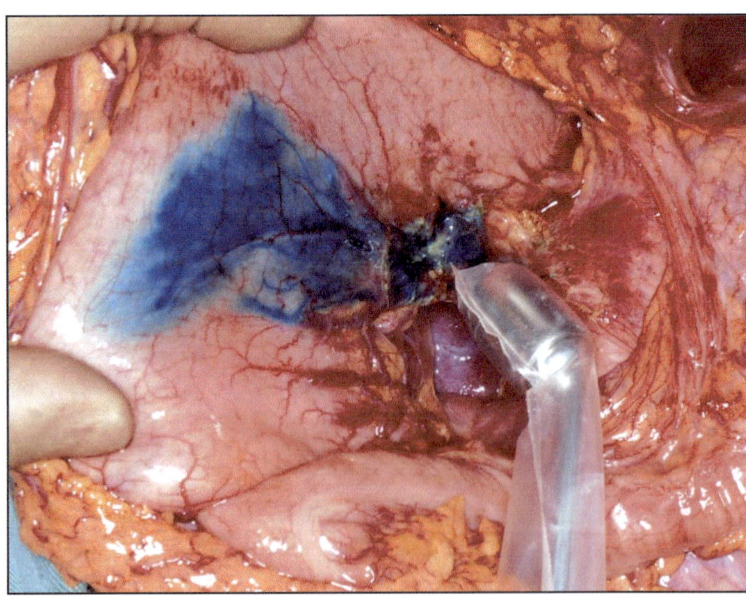

Fig. 24.1 Double tracers are recommended for SN detection of early gastric cancer. Navigator GPS is very useful for detection of SNs with RI

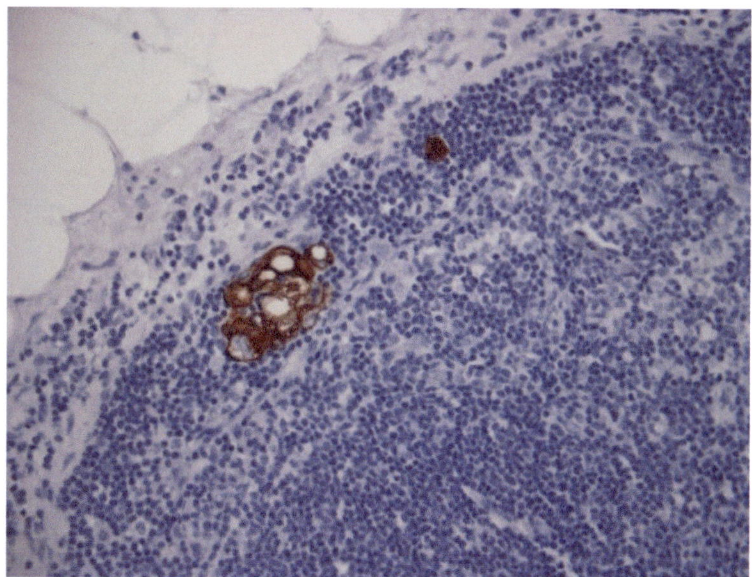

Fig. 24.2 Lymph node micrometastasis is detected by immunohistochemistry using AE1/AE3

and accuracy of SN mapping in IHC staining were 97.2 % and 99.5 %, respectively. Both sensitivity and accuracy of SN mapping by RT-PCR were 100 %. These results of feasibility study were thought to be promising for clinical application.

In 2013, the results of a prospective multicenter trial were reported from Japanese group for SN navigation surgery [6]. Three hundred ninety-seven patients were enrolled in this study. Detection rate, sensitivity, and accuracy were 97.5 %, 93 %, and 99 %, respectively, and the number of false negative cases was only four. In these cases, two had advanced tumors and lymph node metastases were included within the same lymphatic basin in three. Accordingly, we guess that SNNS by lymphatic basin dissection is feasible.

Infrared Imaging with ICG for SN Detection in Early Gastric Cancer

SN detection using infrared imaging with ICG has been clinically applied to melanoma and breast cancer. Infrared imaging may greatly contribute to the development of further SN detection in patients with gastric cancer. Lymphazurin and patent blue have been used for dye method in many facilities, but these are unauthorized for a drug of the lymph node identification in Japan. On the other hand, ICG is commonly and safely used for liver function test. The dye method is simple and easy, but there are some observation limitations due to the small size of the dye particles. RI method was used in our feasibility study, because of the difficulty in identifying blue nodes in a stomach with much fat and connective tissue.

Regarding infrared imaging for SN detection in gastric cancer, Nimura et al. reported the utility of infrared ray electronic endoscopy (IREE) combined with ICG in 2004 [7]. In this report, lymphatic vessel and SNs which took in ICG are found in absorbance imaging by IREE, but these are not found in white light. All of the 84 patients were detected SNs by IREE, and both sensitivity and accuracy are 100 %.

Clinical Application Based on SN Concept Using Infrared Red Imaging

Recently, IRI system was made by Olympus. Fluorescence imaging is clearer than absorbance imaging. This system is easily switchable in absorbance and fluorescence view, and we find

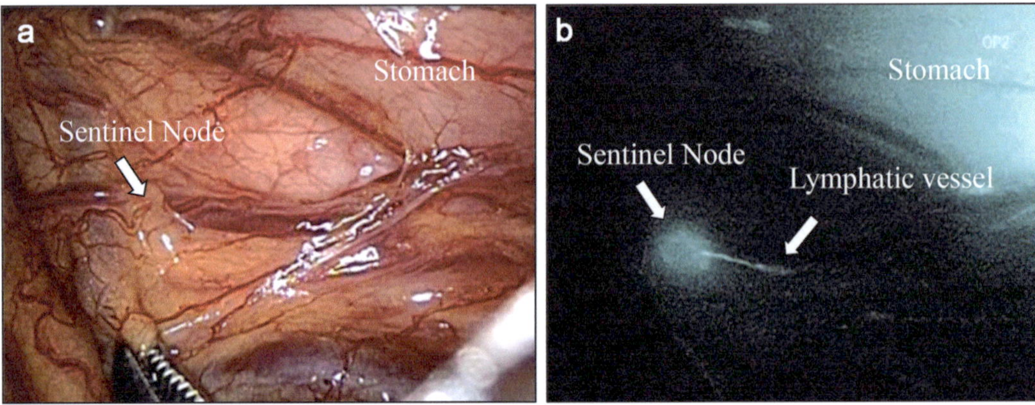

Fig. 24.3 ICG were injected 1 day before surgery. (**a**): Sentinel node with ICG is not found in white light. (**b**) Sentinel node is identified in fluorescence view using IRI system

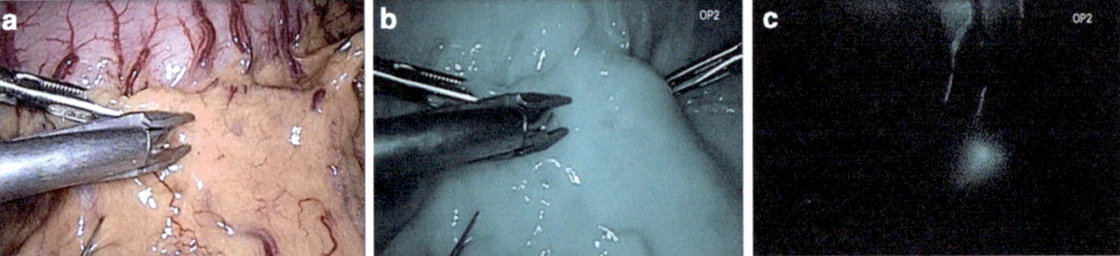

Fig. 24.4 Infrared imaging system is easily switchable in white light, absorbance, and fluorescence view, and we can find sentinel node easily and clearly. (**a**) White light view. (**b**) Absorbance view. (**c**) Fluorescence view

lymphatics clearly, and SNs are confirmed easily (Fig. 24.3). Recently, we used this system for SN detection in early gastric cancer. Technetium tin colloid and ICG are injected 1 day before surgery. ICG is used for 0.25 mg/ml in diluted form. There are some reports regarding intraoperative ICG injection. Since both lymphatic vessels and blood vessels are visualized in this procedure, it is difficult to discriminate between lymphatic and blood vessels. Therefore, we inject it 1 day before surgery. Fluorescence nodes (FNs) are identified using IRI system and marked by surgical clip during surgery (Fig. 24.4). A lot of FNs are identified in back table, due to the manipulation of stomach and omentum. The number of SNs has a limit for the pathological diagnosis during operation. Thus, it is important to confirm FNs carefully and mark it by surgical clip before the manipulation of gastric wall and omentum. FNs and hot nodes with RI were confirmed in back table, and SNs were examined by HE staining and RT-PCR during surgery [8]. The presence or absence of SN micrometastasis is an important issue, although the clinical significance of lymph node micrometastasis is controversial in gastric cancer [9]. Laparoscopic partial resection or segmental resection is applied to patients without metastasis in SNs determined by pathological examination and RT-PCR assay (Fig. 24.5).

Conclusion

SN detection by IRI in patients with early gastric cancer should allow performance of less invasive surgery such as laparoscopic partial gastrectomy with SN dissection and a reduction in the extent of lymphadenectomy. Infrared imaging is useful for SN navigation surgery in early gastric cancer. The reduction of lymphadenectomy based on SN

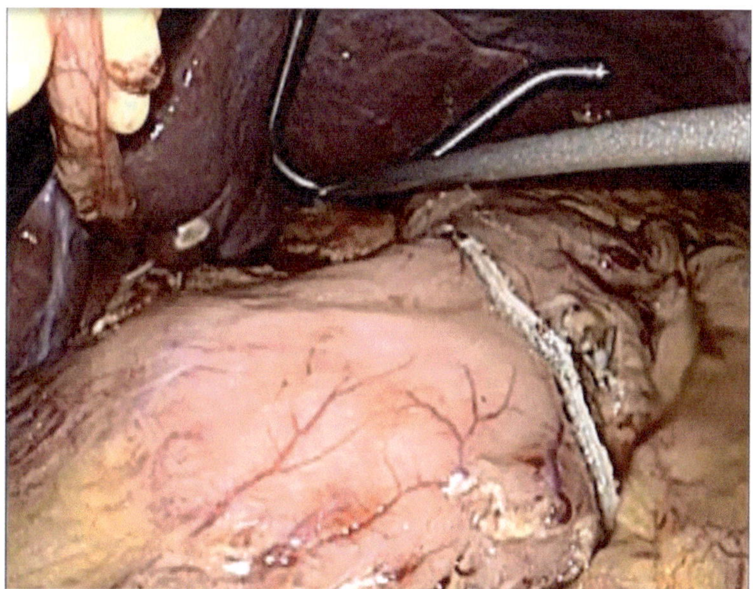

Fig. 24.5 There is little transformation of the stomach after laparoscopic partial resection

dissection using infrared imaging can be performed safely. IRI will be a promising method for the clinical application of SN concept.

References

1. Nakada K, et al. Characteristics and clinical relevance of postgastrectomy syndrome assessment scale (PGSAS)-45: newly developed integrated questionnaires for assessment of living status and quality of life in postgastrectomy patients. Gastric Cancer. 2015;18:147.
2. Morton DL, et al. Technical details of intraoperative lymphatic mapping for early stage melanoma. Arch Surg. 1992;127(4):392–9.
3. Kitagawa Y, et al. The validity of the sentinel node concept in gastrointestinal cancers. Nihon Geka Gakkai Zasshi. 2000;101(3):315–9.
4. Uenosono Y, et al. Detection of sentinel nodes and micrometastases using radioisotope navigation and immunohistochemistry in patients with gastric cancer. Br J Surg. 2005;92(7):886–9.
5. Arigami T, et al. Evaluation of sentinel node concept in gastric cancer based on lymph node micrometastasis determined by reverse transcription-polymerase chain reaction. Ann Surg. 2006;243(3):341–7.
6. Kitagawa Y, et al. Sentinel node mapping for gastric cancer: a prospective multicenter trial in Japan. J Clin Oncol. 2013;31(29):3704–10.
7. Nimura H, et al. Infrared ray electronic endoscopy combined with indocyanine green injection for detection of sentinel nodes of patients with gastric cancer. Br J Surg. 2004;91(5):575–9.
8. Yanagita S, et al. The utility of rapid diagnosis of lymph node metastasis in gastric cancer using a multiplex real-time reverse transcription polymerase chain reaction assay. Oncology. 2009;77(3–4):205–11.
9. Yanagita S, et al. Sentinel node micrometastases have high proliferative potential in gastric cancer. J Surg Res. 2008;145(2):238–43.

A Dual Infrared Ray Imaging System for Sentinel Node Mapping Against Early Gastric Cancer: Absorption and Florescence Methods by Infrared Ray Laparoscopy System Combined with Indocyanine Green

25

Naoto Takahashi, Hiroshi Nimura, Muneharu Fujisaki, Norio Mitsumori, and Katsuhiko Yanaga

Introduction

The sentinel node (SN) technique has been used in the management of variety of cancers to avoid unnecessary lymphadenectomy. In surgery for gastric cancer, acceptable detection rates of SNs and accuracy of intraoperative SN examination have been reported using the radioisotope plus dye method [1, 2]. We invented the use indocyanine green (ICG) and infrared ray laparoscopy system (IRLS) for the detection of SN for early gastric cancer with a high success rate and a high degree of accuracy [3]. Legal considerations and costs limit the use of radioactive substances in general hospitals; thus, the "ICG plus infrared ray method" is potentially safer, more convenient, and more cost-effective compared with the "radioisotope plus dye method."

There are two images in the ICG plus infrared ray method, an absorption imaging and a fluorescence imaging. Even though we reported that absorption imaging was useful for detection of SN [3], Kusano [4] and Miyashiro [5] recently proposed that SN mapping guided by ICG fluorescence imaging is a promising tool for SN navigation surgery for early gastric cancer. In our previous study, the sensitivity of fluorescence imaging by Photodynamic Eye® (Hamamatsu Photonics, Shizuoka, Japan) for detecting of ICG-stained SNs was very high, which however does not necessarily mean that the fluorescence method is superior to the absorption method. First, the latter procedure enables real-time observation of the lymphatic vessels and selective detection of SNs, while ICG injection 1 day before surgery in the fluorescence method eliminates such an advantage. Furthermore, dark environment necessary for the fluorescence method seems to interrupt surgical procedures, while the absorption method does not disturb such continuity. However, the absorption method has a limitation, such as loss of visibility in dense fat (Fig. 25.1).

We developed a new IRLS system (OLYMPUS, Tokyo, Japan) for detecting SN using both the

Electronic supplementary material: The online version of this chapter (doi:10.1007/978-3-319-15678-1_25) contains supplementary material, which is available to authorized users. Videos can also be accessed at http://link.springer.com/chapter/10.1007/978-3-319-15678-1_25.

N. Takahashi, M.D., Ph.D. (✉) • H. Nimura, M.D., Ph.D. • M. Fujisaki, M.D. • N. Mitsumori, M.D., Ph.D. • K. Yanaga, M.D., Ph.D.
Department of Surgery, The Jikei University School of Medicine, 3-25-8 Nishi-shinbashi, Minato-ku, Tokyo 105-8461, Japan
e-mail: ntakahashi@jikei.ac.jp

F.D. Dip et al. (eds.), *Fluorescence Imaging for Surgeons: Concepts and Applications*, DOI 10.1007/978-3-319-15678-1_25, © Springer International Publishing Switzerland 2015

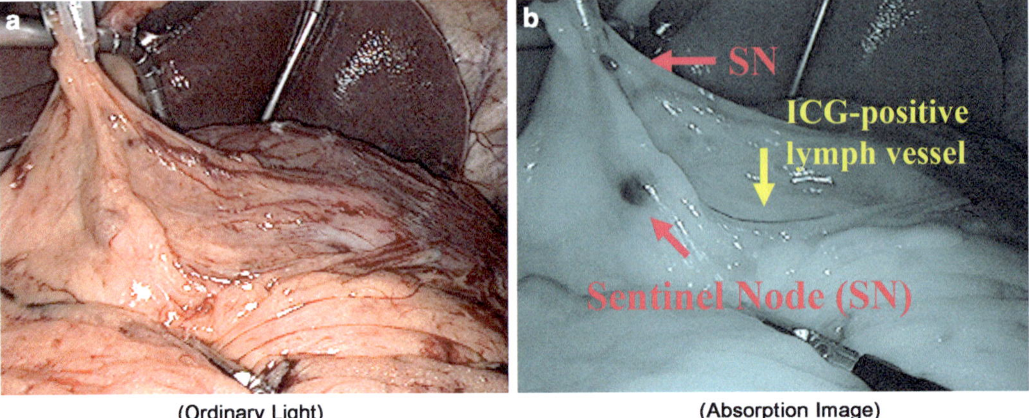

Fig. 25.1 Absorption imaging. Sentinel nodes and lymph vessel were clearly observed in the absorption imaging (**b**); however, they are not visualized in the ordinary light (**a**)

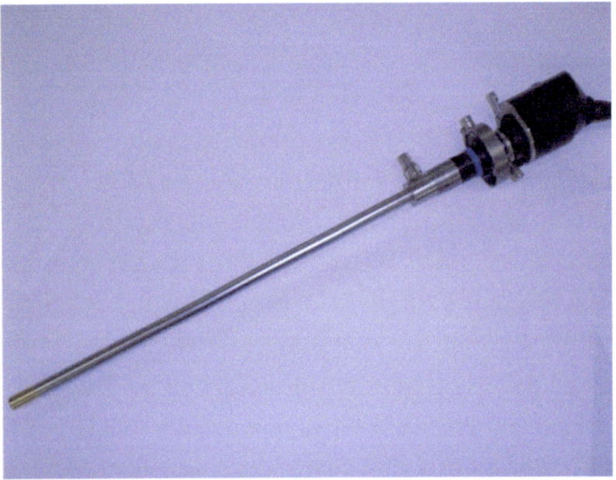

Fig. 25.2 New IRLS system. We developed a new IRLS system (OLYMPUS, Tokyo, Japan) for the detection of SN using both the absorption and the fluorescence imaging of ICG. One of the characteristics is instant switching of the observation mode between the absorption and the fluorescence imaging

absorption and the fluorescence imaging of ICG (Fig. 25.2). One of the characteristics is instant switching the observation mode between the light absorption and the fluorescence imaging. In this study, we determined the feasibility of ICG plus a new infrared ray laparoscopy system (n-IRLS) with dual imaging for SN of gastric cancer.

The following are the details of the principle of ICG and infrared ray: It is known that ICG has a maximum absorption of 805 nm wavelength when ICG is combined with plasma protein. In the irradiated light near the maximum absorption wavelength, the ICG area absorbs the light and becomes darker, and other areas without ICG as the background become brighter. This is the mechanism of the infrared light absorption imaging with ICG. On the other hand, ICG in itself emits a fluorescence which has a maximum of 830 nm wavelength. Compared to the reflected light, the intensity of the fluorescence is extremely weak. Thus, infrared fluorescence imaging with ICG is enabled by completely cutting the reflected light and receiving the light near the maximum fluorescence wavelength. With regards to the fluorescence observation, the difference from the above mentioned system is the changed spectral transmission wavelength of the infrared filter in the light source unit from the maximum light absorption wavelength (805 nm) to a shorter wavelength (690–790 nm). The light transmitted through the filter passes through the light guide and the laparoscope to irradiate the subject. Then ICG is excited and emits fluorescence. Inside the camera head, a special filter is used. The filter cuts the light (690–790 nm) reflected from the subject and allows only fluorescence (800 nm or longer) to pass, enabling fluorescence observation. During the absorption the light emitted from the light source unit has the same wavelength as that of the fluorescence. Also by collecting the reflected

Table 25.1 Patients characteristics and preoperative findings for sentinel navigation surgery

		Patients characteristics						Preoperative diagnosis		
Case	Age (Y)	Gender	Height (cm)	BW (kg)	BMI	Type	Size (mm)	Location	Depth	Pathology
1	60	M	170	73	25.26	0–IIc	10	M/less	M	Poor
2	77	F	145	79	37.57	0–IIc	10	M/less	M	Poor
3	57	M	177	66	21.07	0–IIc	10	U	SM	Well
4	74	M	166	51	18.5	0–IIc	30	M	SM	Well
5	78	M	168	62	21.97	0–IIc	10	MU/post	SM	Well
6	80	M	164	66	24.53	0–IIa	20	MU/post	SM	Well
7	63	M	169	60	21.01	0–IIc	20	M/less	M	Mod
8	66	F	148	48	21.91	0–IIc	40	MU/great	M	Sig
9	61	M	170	83	28.72	0–IIc	30	M/great	SM	Poor
10	61	M	170	83	28.71	0–IIc	30	M/great	SM	Poor
11	75	M	160	51	19.92	0–IIa	15	M/great	M	Moderate
12	73	M	163	82	30.86	0–IIc	18	U/post	M	Poor
13	69	M	169	59	20.66	0–IIa+IIc	15	U/ante	M	Well

BW=body weight; BMI=body mass index; M=mucosa; SM=submucosa

light at the camera head an absorption image is formed. The wavelength of the light for the absorption observation is changed from the maximum absorption wavelength (805 nm) to a shorter wavelength. However, the absorption images have practically no adverse effect for identifying the SN before clinical use (data not shown).

Materials and Methods

Patients

The study was approved by the Ethics Committee for Biomedical Research of the Jikei Institutional Review Board, and all patients provided informed consent. Patients admitted to the Jikei University Hospital with cT1 gastric cancer without obvious metastases were included prospectively in this study, which was diagnosed by preoperative abdominal CT scans and endoscopic ultrasonography. From February 2012 to December 2013, 13 patients were registered. Table 25.1 shows the patients characteristics and preoperative findings for SNNS for patients with gastric cancer.

Intraoperative SN Mapping Guided by a Newly Developed Dual Imaging Infrared Ray Laparoscopy System Combined with ICG

SN biopsy was conducted as described previously [3, 6]. In brief, 0.5 ml of indocyanine green (ICG) (5 mg/ml; Diagnogreen; Daiichi Pharmaceutical, Tokyo, Japan) was injected intraoperatively by endoscopy in four quadrants of the submucosa surrounding the tumor, using an endoscopic puncture needle. The concentration of ICG was determined in preclinical study in vivo (data not shown) (Fig. 25.3). At 20 min after the injection, SNs stained with ICG were observed with the naked eye as well as with a newly developed infrared ray laparoscopy system (OLYMPUS, Tokyo, Japan). For the first two patients, previous IRLS was used simultaneously comparing the absorption imaging by the new IRLS with that by the previous one. All lymph nodes that stained green node or black by the absorption imaging or bright by the fluorescence imaging were excised for a basin dissection before gastrectomy.

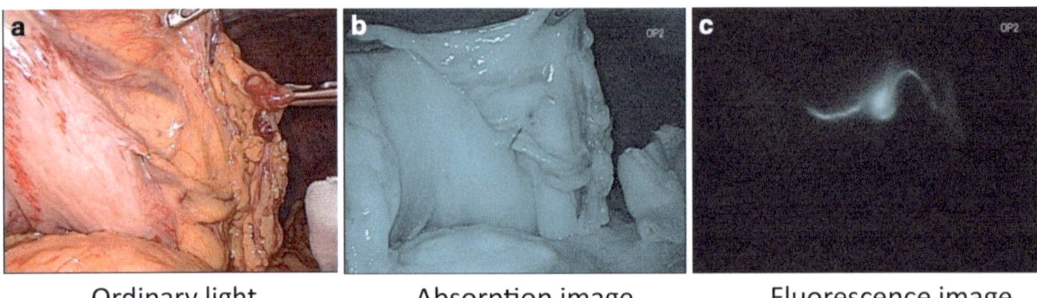

Fig. 25.3 Observation of No. 4d lymph node. Five milligram per milliliter of ICG was injected intraoperatively. (**a**) The green dye of ICG was difficult to be recognized in the ordinary light. (**b**) The ICG could be easily observed in the absorption imaging, which was along the right gastroepiploic artery (No. 4d). (**c**) The fluorescence of ICG was clearly observed in the dark view; however, the surgical procedure needs to be discontinued during the fluorescence imaging because of the dark operative view

Results

The average age of the patients was 63.4 years old. SNs were detected in all of 13 cases. Sensitivity was 100 % (8/8). SN in the absorption imaging was 6.6 per patient and that in the fluorescence imaging was 15.7 per patient. There was no difference in the SN mapping between two groups. For the first two patients, a previous IRLS was used to compare the absorption imaging by the new IRLS with that by the previous one. Because the wavelength used for the absorption image in the new IRLS is changed from the maximum absorption wavelength of 805 nm to a shorter wavelength. The intensity of lymph node and lymph vessels in the absorption imaging was weaker by the new IRLS than that by previous one. However, there was no difference in the SN mapping in both IRLS (data not shown).

Discussion

The present study shows that intraoperative submucosal injection of 0.5 ml × 4 of 5 mg/ml ICG is the adequate administration for detecting SNs. This 5 mg/ml of ICG concentration is as same as the former absorption method, which is too high to detect SNs by the fluorescence imaging in the literature. In the preclinical animal experiments, 0.05 mg/ml concentration of ICG was not useful (data not shown), because the low concentration of ICG did not spread adequately in 20 min and the ICG in the lymph system was not visualized in the absorption method. Although 5 mg/ml concentration of ICG may seem an excessive dose of ICG for the fluorescence imaging, there is no halation of the fluorescence imaging in the animal experiments. In this study, SNs were detected in all of the patients studied, and mean number of SNs in the fluorescence imaging was almost three times higher than that of the absorption imaging, which was not the adequate number of SNNS.

Though there have been several studies on the SN biopsy in gastric cancer, the technique and substance used for mapping differ a lot. Some groups use a dye combination with a radioactive probe [1, 2]. However, in some hospitals, the use of radioactive substances is limited for both logistic reasons and costs. A disadvantage of the use of radioactive substances is the shine-through effect of the injection site that can interfere with SN detection. In the past, we reported the results of our early SN series using absorption imaging by ICG and IRLS for gastric cancer [3, 6]. Recently there has been some discussion between our group and others about the use of the absorption imaging vs. the fluorescence imaging [7]. The absorption imaging enables real-time observation of lymphatic flow and detection of SN after ICG injection without disturbing the continuity of the operation. Especially, it is very easy to mark an SN and dissect SN basin under the absorption imaging, without the dark environment. However, the absorption imaging has a limitation, such as loss of visibility in dense fat, because infrared ray can penetrate fatty tissue up to a depth of 3 mm. No bright signal by the fluorescence imaging means that SN does not exist in this field.

It is very important to detect SNs correctly under the knowledge about the property between the absorption imaging and the fluorescence imaging, rather than to discuss the advantages and disadvantages of the two imaging.

A new IRLS system which has both the absorption and the fluorescence imaging of ICG can detect SN in real-time observation without disturbing the continuity of the operation and without difficulty in fatty patients.

References

1. Aikou T, Kitagawa Y, Kitajima M, Uenosono Y, Bilchik AJ, Martinez SR, Saha S. Sentinel lymph node mapping with GI cancer. Cancer Metastasis Rev. 2006;25:269–77.
2. Kitagawa Y, Fujii H, Kumai K, Kubota T, Otani Y, Saikawa Y, Yoshida M, Kubo A, Kitajima M. Recent advances in sentinel node navigation for gastric cancer: a paradigm shift of surgical management. J Surg Oncol. 2005;90:147–51.
3. Nimura H, Narimiya N, Mitsumori N, Yamazaki Y, Yanaga K, Urashima M. Infrared ray electronic endoscopy combined with indocyanine green injection for detection of sentinel nodes of patients with gastric cancer. Br J Surg. 2004;91:575–9.
4. Kusano M, Tajima Y, Yamazaki K, Kato M, Watanabe M, Miwa M. Sentinel node mapping guided by indocyanine green fluorescence imaging: a new method for sentinel node navigation surgery in gastrointestinal cancer. Dig Surg. 2008;25:103–8.
5. Miyashiro I, Miyoshi N, Hiratsuka M, Kishi K, Yamada T, Ohue M, Ohigashi H, Yano M, Ishikawa O, Imaoka S. Detection of sentinel node in gastric cancer surgery by indocyanine green fluorescence imaging: comparison with infrared imaging. Ann Surg Oncol. 2008;15:1640–3.
6. Kelder W, Nimura H, Takahashi N, Mitsumori N, van Dam GM, Yanaga K. Sentinel node mapping with indocyanine green (ICG) and infrared ray detection in early gastric cancer: an accurate method that enables a limited lymphadenectomy. Eur J Surg Oncol. 2010;36:552–8.
7. Takahashi N, Nimura H, Fujita T, Mitsumori N, Kashiwagi H, Yanaga K. Detection of sentinel node by fluorescence and infrared ray imaging system in gastric cancer. Ann Surg Oncol. 2009;16:1720.

26 Indocyanine Green Fluorescence-Navigated Sentinel Node Biopsy Showed Higher Sensitivity than the Conventional Radioisotope or Blue Dye Methods: It May Help to Reduce False-Negative Sentinel Lymph Node Biopsies in Skin Cancer

Yasuhiro Fujisawa

Introduction

Because subclinical lymph node metastases are found in nearly 20 % of patients with intermediate-thickness melanoma, wide excision of the primary tumor with prophylactic lymph node dissection (elective lymph node dissection (ELND)) has been recommended for the management of such patients [1]. On the other hand, others have recommended the excision of the primary site alone and therapeutic lymph node dissection (TLND) only when clinical nodal disease is present [2]. Several clinical trials were conducted, and although studies before 1990 showed no survival benefit from ELND [3, 4], the studies conducted after 1990 showed a tendency for ELND to improve the survival of patients with subclinical metastases by applying ELND [5, 6]. Thereafter, ELND was accepted as routine practice for managing intermediate-thickness melanomas, even though the majority of them (~80 %) do not have subclinical metastases and such patients are forced to suffer from the morbidity of unnecessary ELND.

To overcome this ELND-related morbidity, the technique of lymphatic mapping and sentinel lymph node biopsy (SLNB) was introduced as a minimally invasive method for the detection of microscopic regional lymph node metastases [7]. This SLNB is based on the concept that lymphatic drainage from the tumor to the regional lymph node basins runs a certain route and is captured at the first-tier nodes, so-called "sentinel" lymph nodes [8] as shown in Fig. 26.1. Therefore, we can dissect only the nodes harboring the direct lymphatic drainage from the tumor and determine pathologically who will benefit from ELND. Since this concept was introduced by Morton et al. [7], SLNB has become the most widely used procedure to determine the regional lymph node status of patients not only with melanoma but also with a variety of other skin cancers such as squamous cell carcinoma and extramammary Paget's disease [9–11].

SLNB was initially performed using blue dye (BD) [7]. However, this technique has several limitations, such as a loss of visibility when lymphatic channels are in deep fat and rapid washout of dye from the dyed node [12, 13]. A significant technical learning curve for SLN biopsy has been reported, and substantial experience is required to develop the technical skill necessary to achieve a high success rate [12, 13]. To overcome these

Y. Fujisawa, M.D., Ph.D. (✉)
Department of Dermatology, University of Tsukuba, 1-1-1 Tennodai, Tsukuba, Ibaraki 305-8575, Japan
e-mail: fujisan@md.tsukuba.ac.jp

F.D. Dip et al. (eds.), *Fluorescence Imaging for Surgeons: Concepts and Applications*, DOI 10.1007/978-3-319-15678-1_26, © Springer International Publishing Switzerland 2015

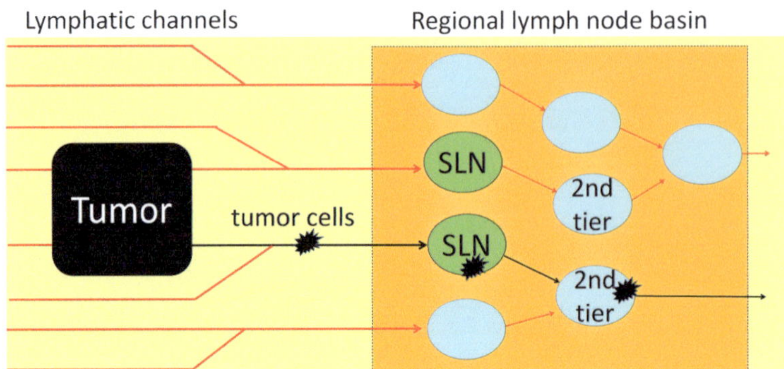

Fig. 26.1 Schematic image of sentinel lymph node (SLN) concept. SLN has direct flow from the tumor and the first metastasis via lymph channel should occur from here. If SLN harbors metastasis, second-tier node also has possibility to have metastasis. Inversely, if SLN shows no metastasis, no other lymph nodes in the regional basin will have metastasis

issues, BD combined with preoperative lymphoscintigraphy using radioisotope (RI) followed with an intraoperative handheld gamma-ray detector became the standard method to detect SLNs, providing a detection rate of nearly 98 % [14, 15]. Although the combined use of RI and BD affords a high SLN detection rate, 3–5 % of patients diagnosed as having negative SLN develop nodal metastasis (false-negative SLNB) [16, 17].

Kitai et al. [18] reported a new lymphatic mapping method using indocyanine green fluorescence imaging (ICG-FI) in breast cancer patients. Our attention was attracted by this ICG-FI because ICG fluorescence penetrates human tissue to a depth of 1–2 cm and we can transcutaneously detect the fluorescence in real time [19]. We started using this new ICG-FI combined with conventional RI and BD from 2010 [9]. After a series of cases in which ICG-FI was used, we became aware that some of the SLNs were detected only by ICG-FI. In this chapter, we would like to show our recent results obtained from SLNB using a combination of these three methods and to discuss the possibility of redetecting SLNs, which were overlooked by the conventional detection method.

Conventional Tracers for Sentinel Node Detection

Blue Dye Method

BD is safe, convenient, and cost-effective, but the detection rate was not satisfactory; BD by itself was reported to have a detection rate of 82 % [7]. As mentioned above, this technique has certain limitations, such as a loss of visibility in dense fat and rapid transit of the dye [12, 13]. However, BD is still useful in the operation because surgeons can confirm the blue lymphatic channel and lymph node by their own vision. Although a rare occurrence, we should remember possible allergic reaction to these dyes [20] and reduction in arterial oxygen saturation measured by pulse oximetry [21].

Radioisotope Method

To compensate for the limitations of BD, combining it with the use of radioisotope for tracers in detecting SLN is now a standard for SLNB. There are several different types of tracers depending on the conjugated particles. In general,

4–6 intradermal injections of 5–10 MBq of radiotracer are instilled closely around the tumor site. After injection, acquisition of dynamic images including the injection site to the regional lymph node basin continues until the RI drain into the SLN. This procedure is very important when using small-particle RI because they can easily pass through the SLN and accumulate at the second-tier nodes. At operation, the use of an intraoperative gamma-ray detector is required to search for "invisible" radioactive SLN. However, when the primary tumor is located near the regional lymphatic basin, the shine-through effect, which is a radiation from the injection site, interferes with the detection of SLN [22]. Thus, SLNB in such patients is sometimes difficult and leads to the failure to harvest SLN.

Indocyanine Green Fluorescence Imaging

Principle

ICG is widely used in a variety of examinations, including hepatic function, cardiac output, and retinal angiography [18], and is proven as a safe reagent. It binds with albumin and produces a peak wavelength of 840-nm near-infrared fluorescence when excited with 765-nm light (Fig. 26.2) [23]. This fluorescence is invisible to our eye, but a coupled-charged device (CCD) camera can visualize this fluorescence on a display. Unlike visible light, near-infrared light can penetrate human tissue and allows detection to a depth of 1–2 cm under the skin. Thus, we can observe real-time lymph flow under the skin.

Required Instruments

The basic ICG fluorescence detection system costs about $50,000 USD in Japan. However, the principles of ICG-FI being very simple, we made our own detection system using commercially available parts which cost only $1,600 as described in the previous report [9]. Briefly, we used a light-emitting diode that generates 760-nm wavelength light as the excitation light and as a detector a CCD camera with a long-wavelength pass filter to filter out light with a wavelength below 840 nm. The fluorescence image is sent to a TV monitor for real-time monitoring via a video recorder. When in use, the main body of the system is covered with a sterilized vinyl bag, and we can use it at the operation site.

The basic system including our handmade detector can only display fluorescence as "white lesion" in a dark black and white image (Fig. 26.3a). This black and white image sometimes made it difficult to localize the small fluorescence in the operative fields. Recently, a

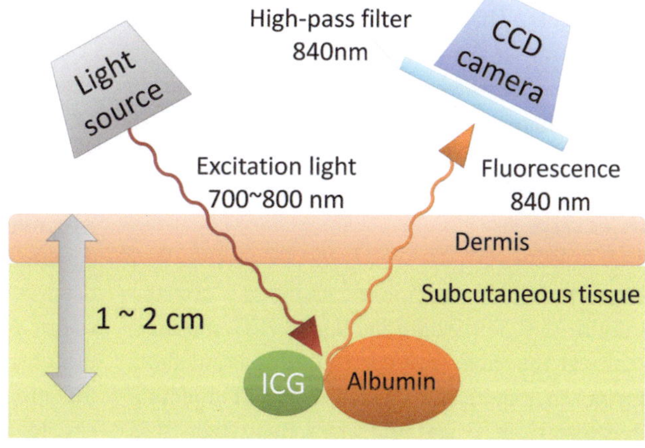

Fig. 26.2 Basic principle of indocyanine green fluorescence imaging (ICG-FI). The excitation wavelength of ICG that produces the maximum fluorescence is 765 nm, and the fluorescence occurs at 840 nm. This near-infrared fluorescence can be detected 1–2 cm beneath the skin

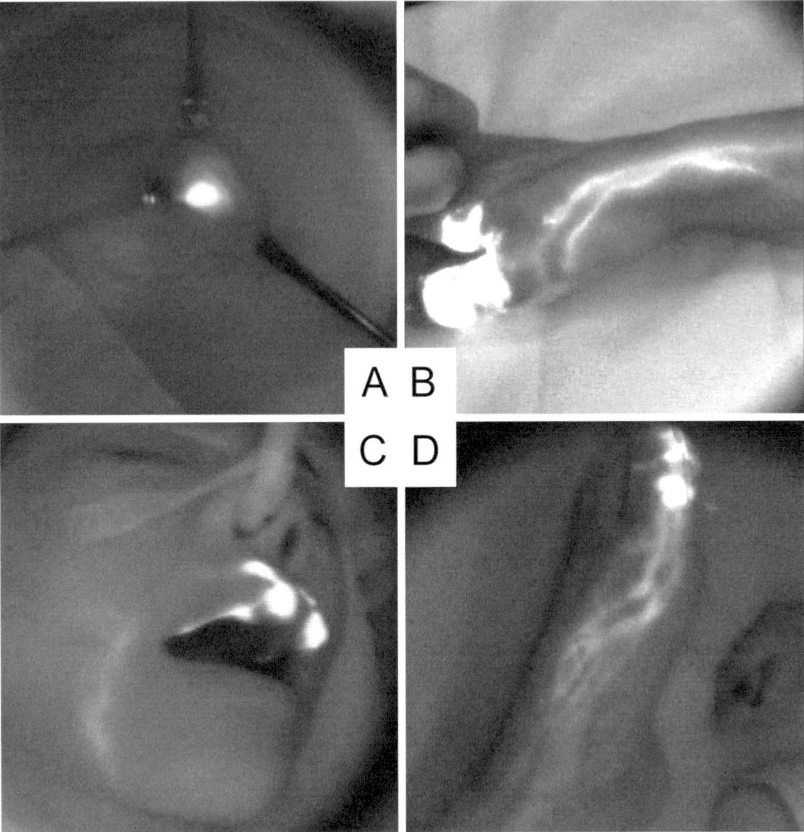

Fig. 26.3 Indocyanine green fluorescence images. (**a**) Lymph node stained with ICG displays as "white node". (**b**) Acral lentiginous melanoma of the right thumb. Several lymph channels to the axillae were evident. (**c**) Lentigo maligna melanoma of the upper lip. Single lymph channel to the submandibular region was evident. (**d**) Acral lentiginous melanoma of the left great toe. Several lymph channels form a netlike network and run toward the groin

new detector system, which can display fluorescence in a bright color image, has been introduced [24]. This new system may offer easier localization of fluorescence lymph channels and SLN, but it costs $120,000, about 70 times more than our handmade system [9, 24].

Injection

Dissolve ICG in physiologic saline at a concentration of 0.5 % solution. Inject ICG intradermally at the same spot where the RI is injected. From our experience, the sequence of injecting dye tracers, ICG and BD, does not affect SLN detection. Leaked ICG from the injection site should be wiped away using clean gauze, because ICG has very high fluorescence intensity and can easily contaminate the operation field.

Transcutaneous Localization of Sentinel Lymph Node(s)

Surgical light which includes wavelengths of 840 nm or longer should be turned off when detecting the fluorescence. However, there is no need to shut down all the lights in the operating room; the general ceiling lamp does not affect the observation of fluorescence. Soon after the injection of ICG, lymph flow from the injection site to the regional lymph node basin can be observed (Fig. 26.3b–d). The essential feature of this ICG-FI is real-time observation of lymph

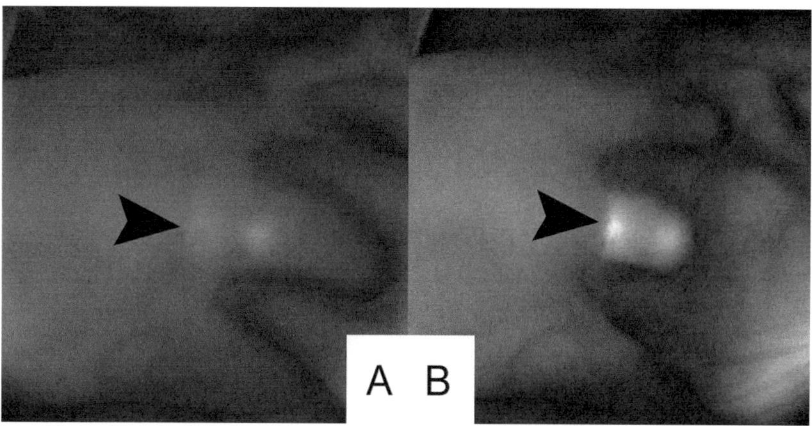

Fig. 26.4 Improvement of fluorescence detection by compression. *Arrowhead* indicates the location of the sentinel lymph node. (**a**) Fluorescence image before compression. (**b**) Improvement of fluorescence detection by compression

flow at operation; it enables us to see where the lymphatic channel is and localize the draining lymph node. In some cases, we can see the moving lymphatic fluid, which contains ICG moving toward the SLN. However, when a lymphatic channel is located under thick fat tissue or drains into solid tissue such as muscle or a salivary gland, transcutaneous observation of ICG fluorescence becomes difficult. In such cases, we can improve fluorescence observation by compressing the overlying skin and reducing the depth of fluorescence (Fig. 26.4a, b) [9, 25]. In our results, transcutaneous localization of SLNs by ICG-FI was possible for 139 of 190 SLNs (73 %).

Visualization of Lymphatic Flow to SLN Located in Unusual Sites

It was interesting for us to see how the lymphatic channels connected to SLNs are located in unusual sites, such as cubital and popliteal lymph node basins. We experienced two patients with SLNs in cubital nodes and three patients with them in popliteal lymph nodes. The patients with SLN in the cubital region both had two main lymph channels on the radial side of the forearm and the channel, which runs outside and made a sudden curve towards the lateral side of the upper arm and finally drained into the cubital LN. The patients with SLNs in the popliteal region had completely different lymph channels which drain into the groin region. The lymph flow toward the groin LN runs along with the greater saphenous vein, but the lymph flow toward the popliteal LN runs along the dorsal side of the lower leg where the lesser saphenous vein is.

Limitation of Indocyanine Green Fluorescence Imaging

ICG-FI has several limitations: (1) The maximum depth of detection is 2 cm from the surface of the skin, (2) ICG shows very high fluorescence intensity and only a small leak of ICG can contaminate the surgical field, (3) quantitative evaluation of fluorescence is difficult at present, and (4) a halogen surgical lamp contains light with wavelengths over 840 nm and must be turned off during fluorescence detection. Most of these issues can be solved by combined use of RI; however, assuming an RI-free situation, we suggest the following solutions.

For issues (1) and (2), we advise the use of preoperative ultrasonography and marking of potential lymph nodes. This marking helps to estimate the possible SLN location when the lymphatic channel cannot be traced as far as the draining SLN through the skin. Thus, the incision line can be drawn at the estimated SLN position

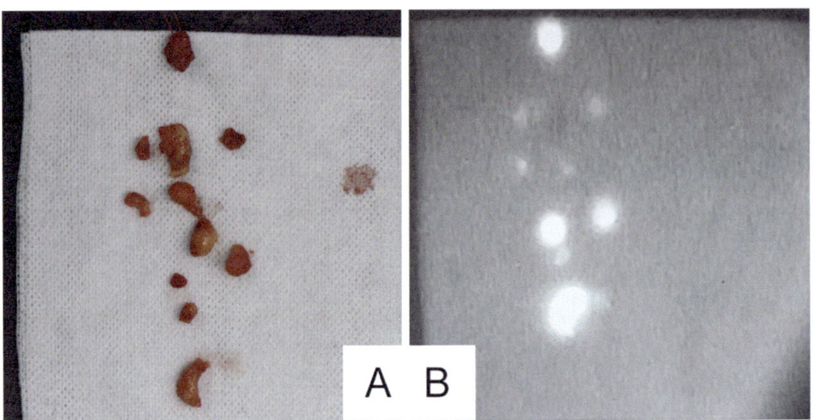

Fig. 26.5 Florescence detection in nonsentinel lymph nodes. (**a**) Photograph of dissected nonsentinel lymph nodes. (**b**) Fluorescence image of dissected nonsentinel lymph nodes. Most of the dissected nonsentinel lymph nodes showed fluorescence

and the risk of damaging unnecessary lymph channels reduced. However, the location of SLNs in the neck region varies, and fluorescence is difficult to detect through the skin when the SLN is located in a salivary gland or under the muscle. Therefore, SLN mapping in the neck region still requires the RI method.

At present, (3) quantitative evaluation of fluorescence is difficult, and this means that we cannot exclude the possibility of harvesting second-tier nodes because of their high fluorescence intensity (Fig. 26.5). To exclude second-tier nodes, observation of the dynamic image soon after the injection is essential; trace the fluorescent lymph channels toward the regional lymph node basin and locate the SLN.

With (4) the lighting issue, using a visible light source which does not contain 840-nm or longer wavelength is the best solution. Using a short-wavelength pass filter, which filters out long wavelengths from the conventional head light system, is one of the solutions [9]. Recently, surgical lighting systems are changing from using halogen lamps to light-emitting diodes (LED) because of their longer life, low energy consumption, and low generation of heat. LEDs produce light consisting of a very narrow range of wavelengths; therefore, it is theoretically possible to produce a surgical lighting system using LED which will not affect fluorescence imaging. Further development is required.

Comparison Between Indocyanine Green Fluorescence Imaging and the Conventional Detection Method in Skin Cancer: Experience with 76 Patients in Our Institute

Patients

We performed SLN biopsy using ICG-FI combined with the conventional methods from January 2010 through November 2013 on 75 patients with invasive skin cancer. Of these, 40 had malignant melanoma, 16 had squamous cell carcinoma, 15 had extramammary Paget's disease, 1 had sweat gland carcinoma, and 1 had Merkel cell carcinoma, 1 had eccrine porocarcinoma, and 1 had clear cell sarcoma. Tumors were located in the head or neck in 13, upper extremity in 13, trunk in 30, and lower extremity in 19.

Procedures

Before the operation, lymphoscintigraphy was performed. On the morning of the operation, 99mTc-tin colloid was intradermally injected around the primary tumor. We marked the injected sites for the subsequent injections of ICG and BD. After the induction of general anesthesia, 0.4–0.8 mL of 2 % patent blue solution and 0.5 % ICG were intradermally injected at the marked spot

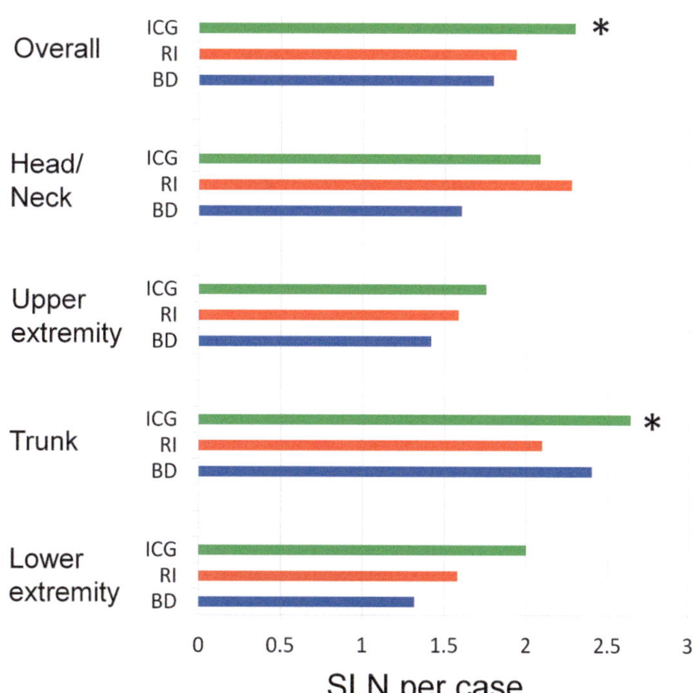

Fig. 26.6 Number of detected sentinel lymph nodes by methods and site of the primary tumor. Indocyanine green fluorescence imaging detected more SLNs than the conventional methods. (ICG: indocyanine green fluorescence imaging, RI: radioisotope method, BD: blue dye method, *: $p<0.05$ by Mann–Whitney *U*-test)

around the primary tumor, and ultrasonography was performed to locate potential lymph nodes. Basically, the skin incision line was determined according to the lymphoscintigraphy results. For lymphatic basins in which only the ICG flow was confirmed, the skin incision line was determined according to the results of the ICG-FI with ultrasonography support. An intraoperative gamma-ray detector was also used to detect SLNs. The lymph nodes were considered to be SLNs and removed when any of the following conditions were observed: (1) lymph flow from the primary site to the draining lymphatic basin and a fluorescent lymph node, (2) a radioactive lymph node (including the hottest node and nodes that showed higher than 10 % of the maximum value as measured by the gamma probe), and (3) accumulation of BD in the lymph nodes. SLNs identified only by ICG-FI were termed "occult" SLNs.

Identification of Sentinel Lymph Nodes

SLNs were successfully identified in 75 of the 76 patients (99 %) by at least one of the three methods. One patient with Merkel cell carcinoma on the upper arm failed to have SLNs detected. Total identified SLNs were 176 with 102 lymphatic basins (2.35 SLNs per case, 1.73 SLNs per basin). The ICG-FI identified more SLNs than conventional methods (Fig. 26.6). ICG failed to detect SLNs in 4:2 patients had SLNs in the external iliac node identified by preoperative lymphoscintigraphy, and we decided not to remove them. In another two patients, they were in the head and neck region, and we failed to remove some of the SLNs detected by preoperative lymphoscintigraphy. In both cases, we could not locate some SLNs even though we used an intraoperative gamma-ray detector.

Difference According to Tumor Site

As shown in Fig. 26.6, ICG detected more SLNs of patients with a primary tumor in the upper extremity, trunk, and lower extremity; but only tumors in the trunk showed a statistically significant increase in detection ($p<0.05$). Tumors in the trunk have multiple possible lymph node basins, e.g., bilateral groin or axillae plus groin. The increased detection was a

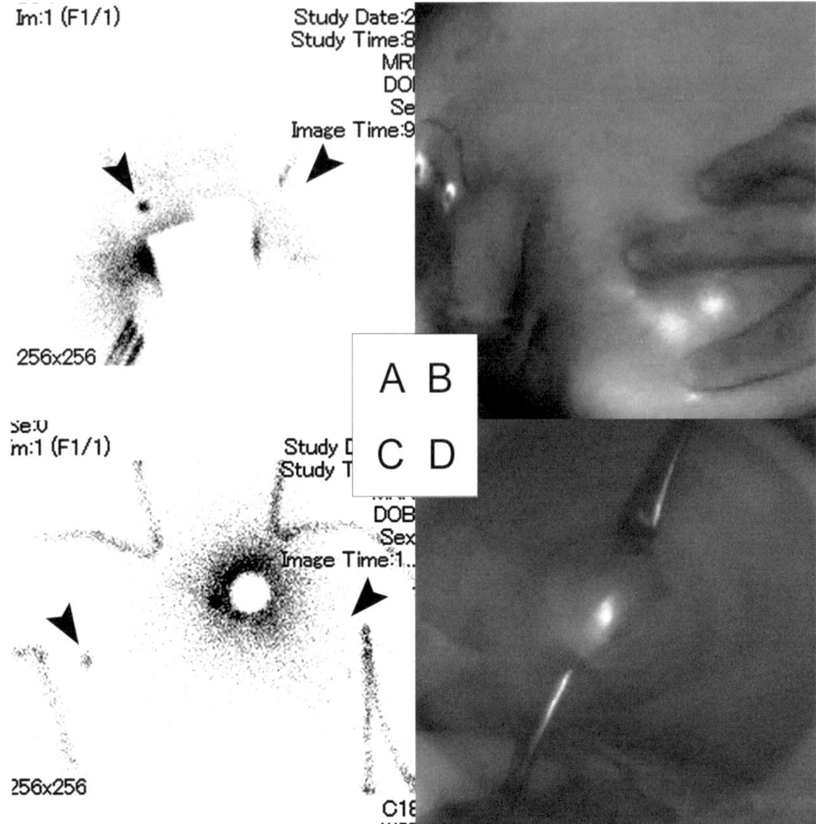

Fig. 26.7 Sentinel lymph node detected only by indocyanine green fluorescence imaging. (**a**) Image of preoperative lymphoscintigraphy. Extramammary Paget's disease located in the left penoscrotal region and only single sentinel lymph node detected in the right groin (*arrowheads*). (**b**) Image of indocyanine green fluorescence imaging. Two "occult" sentinel lymph nodes were detected in the left groin and confirmed by handheld gamma detector to have no radioactivity. (**c**) Image of preoperative lymphoscintigraphy. Superficial spreading melanoma of the left back and only single sentinel lymph node detected in the right axillae (*black arrowhead*). (**d**) Image of indocyanine green fluorescence imaging. One "occult" sentinel lymph node was detected in the left axillae and had metastasis

result not only of the greater number of SLNs detected per basin but also of the greater number of lymphatic basins detected per case. On the other hand, even though tumors in the head and neck region tend to have multiple SLNs, SLN location varies a lot in the neck region and fluorescence is difficult to detect through the skin when the SLN is located in a salivary gland or under the muscle. Therefore, we assumed that RI-free SLNB may be possible using ICG-FI with BD for patients with tumors except in the head and neck region.

Identification of "Occult" Sentinel Lymph Nodes

ICG-FI identified "occult" SLNs that conventional methods overlooked in a certain percent of the patients. In our cohort, 24 of the 75 patients (32 %) had SLNs that were detected only by ICG-FI. In most patients with an "occult" SLN, it is located within the same lymphatic basin where a conventional detection method identified an SLN. Of note, 7 of the 74 patients (9 %) had an "occult" SLN in another basin, which RI did not detect (Fig. 26.7).

Interestingly, we experienced a patient with malignant melanoma of the back who had metastasis only in an "occult" SLN (Fig. 26.7c, d) and he received TLND. In another patient, an "occult" SLN was found to have metastasis. Fortunately, however, this patient was found to be SLN positive by RI and received complete lymph node dissection. However, the improvement afforded by ICG-FI may be limited by the following situations: A large-sized metastasis in an SLN blocks the lymph flow, causing it to flow into a non-SLN that may not yet harbor tumor cells [26], and the tumor cells floating in the lymphatic channel at the time of the SLN biopsy end up later in the non-SLN and may become evident as nodal failure. Therefore, we assume that the ICG-FI will improve the SLN detection rate, but cannot completely eliminate false-negative cases.

Factors Influencing Sentinel Node Detection

There are several factors that are known to influence the time to detection, such as particle size of the tracers, site of injection, and the characteristics of the patients. The terminals of lymphatic capillaries are located at the level of the reticular dermis, and therefore, tracers injected in this layer might gain maximum incorporation into the lymphatic system [27, 28]. Another factor, particle size, determines the speed of tracer movement in the lymphatic system [29]. The last factor is the patient characteristics such as older age and high body mass index, which have been shown to be associated with SLN detection failure [30].

The reasons why ICG-FI has more sensitivity than the conventional methods might be explained by the following two factors: particle size of the tracer and fluorescence intensity/contrast. The distribution of administered tracers depends on the particle size, meaning that smaller-sized particles will be transported quickly [31]. ICG is a small substance (molecular weight, 774.96 Da), and after administration, it binds with albumin (2–3 nm) to show fluorescence [23, 31, 32]. On the other hand, most of the commonly used tracers such as tin colloid [33], phytate colloid [34], and sulfide colloid [35] are 100–200 nm in size. Smaller-sized tracers such as human serum albumin colloid (particle size, 5–80 nm) [35] and antimony trisulfide (particle size, 3–30 nm) [36] are sometimes used. However, the ICG–albumin complex is the smallest of all the available RI tracers.

Another explanation is its high fluorescence intensity and contrast [19]. Only a small amount of ICG is required to gain sufficient fluorescence intensity. This means that ICG can detect SLNs with faint lymph flow that may be undetected by RI. On the other hand, this feature may induce contamination of the operation field if the lymphatic tract is damaged. Thus, the operation should be performed very carefully.

Conclusion

SLNB using ICG-FI is now widely accepted for various cancers. The merit of this method is the detection of lymph flow through the skin in real time in the operation field. Seeing how the lymphatic channels run toward the regional lymphatic basin greatly helps surgeons to locate the SLNs. As mentioned above, 32 % of patients were found to have SLNs, which were only detected by ICG-FI, and such "occult" SLNs may explain the occurrence of false-negative SLNs. Therefore, we suggest the use of ICG-FI in SLNB to remedy the lack of detection of SLNs by conventional detection methods.

Acknowledgement We are grateful to Brian K. Purdue for his native English speaker manuscript editing.

Financial disclosure This work was partly supported by a National Cancer Center Research and Development Fund (23-A-22).

References

1. Schneebaum S, Briele HA, Walker MJ, et al. Cutaneous thick melanoma. Prognosis and treatment. Arch Surg. 1987;122:707–11.
2. Elder DE, Dt G, Van Horn M, et al. The role of lymph node dissection for clinical stage I malignant melanoma of intermediate thickness (1.51-3.99 mm). Cancer. 1985;56:413–8.
3. Sim FH, Taylor WF, Pritchard DJ, Soule EH. Lymphadenectomy in the management of stage I malignant melanoma: a prospective randomized study. Mayo Clin Proc. 1986;61:697–705.
4. Veronesi U, Adamus J, Bandiera DC, et al. Inefficacy of immediate node dissection in stage 1 melanoma of the limbs. N Engl J Med. 1977;297:627–30.
5. Cascinelli N, Morabito A, Santinami M, MacKie RM, Belli F. Immediate or delayed dissection of regional nodes in patients with melanoma of the trunk: a randomised trial. WHO Melanoma Programme. Lancet. 1998;351:793–6.
6. Balch CM, Soong S, Ross MI, et al. Long-term results of a multi-institutional randomized trial comparing prognostic factors and surgical results for intermediate thickness melanomas (1.0 to 4.0 mm). Intergroup Melanoma Surgical Trial. Ann Surg Oncol. 2000; 7:87–97.
7. Morton DL, Wen DR, Wong JH, et al. Technical details of intraoperative lymphatic mapping for early stage melanoma. Arch Surg. 1992;127:392–9.
8. Reintgen D, Balch CM, Kirkwood J, Ross M. Recent advances in the care of the patient with malignant melanoma. Ann Surg. 1997;225:1–14.
9. Fujisawa Y, Nakamura Y, Kawachi Y, Otsuka F. A custom-made, low-cost intraoperative fluorescence navigation system with indocyanine green for sentinel lymph node biopsy in skin cancer. Dermatology. 2011;222:261–8.
10. Fujisawa Y, Nakamura Y, Kawachi Y, Otsuka F. Indocyanine green fluorescence-navigated sentinel node biopsy showed higher sensitivity than the radioisotope or blue dye method, which may help to reduce false-negative cases in skin cancer. J Surg Oncol. 2012;106:41–5.
11. Tanaka R, Nakashima K, Fujimoto W. Sentinel lymph node detection in skin cancer using fluorescence navigation with indocyanine green. J Dermatol. 2009; 36:468–70.
12. Giuliano AE. Lymphatic mapping and sentinel node biopsy in breast cancer. JAMA. 1997;277:791–2.
13. Bostick P, Essner R, Glass E, et al. Comparison of blue dye and probe-assisted intraoperative lymphatic mapping in melanoma to identify sentinel nodes in 100 lymphatic basins. Arch Surg. 1999;134:43–9.
14. Sandrucci S, Casalegno PS, Percivale P, Mistrangelo M, Bombardieri E, Bertoglio S. Sentinel lymph node mapping and biopsy for breast cancer: a review of the literature relative to 4791 procedures. Tumori. 1999; 85:425–34.
15. Hung WK, Chan CM, Ying M, Chong SF, Mak KL, Yip AW. Randomized clinical trial comparing blue dye with combined dye and isotope for sentinel lymph node biopsy in breast cancer. Br J Surg. 2005; 92:1494–7.
16. Morton DL, Cochran AJ, Thompson JF, et al. Sentinel node biopsy for early-stage melanoma: accuracy and morbidity in MSLT-I, an international multicenter trial. Ann Surg. 2005;242:302–11. discussion 311–303.
17. Carlson GW, Page AJ, Cohen C, et al. Regional recurrence after negative sentinel lymph node biopsy for melanoma. Ann Surg. 2008;248:378–86.
18. Kitai T, Inomoto T, Miwa M, Shikayama T. Fluorescence navigation with indocyanine green for detecting sentinel lymph nodes in breast cancer. Breast Cancer. 2005;12:211–5.
19. Murawa D, Hirche C, Dresel S, Hunerbein M. Sentinel lymph node biopsy in breast cancer guided by indocyanine green fluorescence. Br J Surg. 2009;96:1289–94.
20. Keller B, Yawalkar N, Pichler C, Braathen LR, Hunger RE. Hypersensitivity reaction against patent blue during sentinel lymph node removal in three melanoma patients. Am J Surg. 2007;193:122–4.
21. Takahashi Y, Hara K, Sata T. A case of prolonged reduction in arterial oxygen saturation measured by pulse oximetry after administering patent blue in an elderly patient. Masui. 2013;62:442–4.
22. Wu CT, Morita ET, Treseler PA, et al. Failure to harvest sentinel lymph nodes identified by preoperative lymphoscintigraphy in breast cancer patients. Breast J. 2003;9:86–90.
23. Benson RC, Kues HA. Fluorescence properties of indocyanine green as related to angiography. Phys Med Biol. 1978;23:159–63.
24. Troyan SL, Kianzad V, Gibbs-Strauss SL, et al. The FLARE intraoperative near-infrared fluorescence imaging system: a first-in-human clinical trial in breast cancer sentinel lymph node mapping. Ann Surg Oncol. 2009;16:2943–52.
25. Kitai T, Kawashima M. Transcutaneous detection and direct approach to the sentinel node using axillary compression technique in ICG fluorescence-navigated sentinel node biopsy for breast cancer. Breast Cancer. 2012;19:343–8.
26. Leijte JA, van der Ploeg IM, Valdes Olmos RA, Nieweg OE, Horenblas S. Visualization of tumor blockage and rerouting of lymphatic drainage in penile cancer patients by use of SPECT/CT. J Nucl Med. 2009;50:364–7.
27. McMasters KM, Wong SL, Martin 2nd RC, et al. Dermal injection of radioactive colloid is superior to peritumoral injection for breast cancer sentinel lymph node biopsy: results of a multiinstitutional study. Ann Surg. 2001;233:676–87.
28. Borgstein PJ, Meijer S, Pijpers RJ, van Diest PJ. Functional lymphatic anatomy for sentinel node

biopsy in breast cancer: echoes from the past and the periareolar blue method. Ann Surg. 2000;232:81–9.
29. Jinno H, Ikeda T, Matsui A, et al. Sentinel lymph node biopsy in breast cancer using technetium-99m tin colloids of different sizes. Biomed Pharmacother. 2002;56 Suppl 1:213–6.
30. Pritsivelis C, Garcia Mendonca CA, Pinheiro Pessoa MC, Coelho-Oliveira A, Gutfilen B, Barbosa Da Fonseca LM. Failure predictors of the sentinel lymph node in patients with breast cancer using Tc-99m sulfur colloid and periareolar injection. Q J Nucl Med Mol Imaging. 2007;51: 189–93.
31. Wilhelm AJ, Mijnhout GS, Franssen EJ. Radiopharmaceuticals in sentinel lymph-node detection - an overview. Eur J Nucl Med. 1999;26:S36–42.
32. Miyashiro I, Miyoshi N, Hiratsuka M, et al. Detection of sentinel node in gastric cancer surgery by indocyanine green fluorescence imaging: comparison with infrared imaging. Ann Surg Oncol. 2008;15:1640–3.
33. Higashi H, Natsugoe S, Uenosono Y, et al. Particle size of tin and phytate colloid in sentinel node identification. J Surg Res. 2004;121:1–4.
34. Jankovic D, Maksin T, Djokic D, et al. Particle size analysis: 90Y and 99mTc-labelled colloids. J Microsc. 2008;232:601–4.
35. Lens M. Sentinel lymph node biopsy in melanoma patients. J Eur Acad Dermatol Venereol. 2010;24:1005–12.
36. Chen SL, Iddings DM, Scheri RP, Bilchik AJ. Lymphatic mapping and sentinel node analysis: current concepts and applications. CA Cancer J Clin. 2006; 56:292–309. quiz 316–297.

27 The Indocyanine Green Method Is Equivalent to the (99m) Tc-Labeled Radiotracer Method for Identifying the Sentinel Node in Breast Cancer: A Concordance and Validation Study

Bettina Ballardini, Germana Lissidini, and Paolo Veronesi

Sentinel Node Biopsy

The second major twentieth-century development in the surgical treatment of breast cancer—after the introduction of the new breast-conserving approach called QUART—was sentinel node biopsy (SNB). The sentinel node (SN) is the first node or nodes to receive lymph from the area around the tumor. The first randomized study on SNB was conducted at the European Institute of Oncology (EIO): it randomized 516 patients to compare SNB plus immediate axillary dissection, with SNB plus axillary dissection only if the SN was positive. After over 5 years of follow-up, no differences between the two arms were found, either in terms of axillary recurrences or distant metastases. Moreover, the trial reported decreased arm mobility, increased pain, and increased lymphedema rates in the axillary dissection arm [1].

This trial, and other subsequently published studies, validated the concept of SNB: if the SN is disease-free, then the whole axilla is also disease-free, and axillary dissection can be avoided. SNB has now supplanted axillary dissection as an accurate and less invasive axillary staging procedure in patients with clinically node-negative early-stage breast cancer in high-resource countries. In fact SNB alone (without axillary dissection) is associated with less than 1 % isolated axillary recurrence in patients with node-negative disease and provides excellent regional nodal control [2].

Although a recent review suggested that patients with an involved SN or clinically positive nodes should continue to be offered axillary dissection, the results of the recently published Z0011 randomized trial indicate that axillary dissection confers no advantage in patients with early breast cancer and no more than three involved SNs, provided they undergo standard whole-breast irradiation and adjuvant chemotherapy [3].

The findings of the International Breast Cancer Study Group 23.01 trial (IBCSG 23.01), a phase 3 randomized controlled trial, also indicate that axillary dissection confers no advantage in patients with only micrometastases in the SN (one or more foci up to 2 mm) and tumor of maximum 5 cm [4].

Thus, it appears that Z0011 and IBCSG 23.01 will change clinical practice, allowing many patients with early breast cancer and minimal

B. Ballardini, M.D. (✉) • G. Lissidini, M.D.
P. Veronesi, M.D.
Division of Senology, European Institute of Oncology, Via G. Ripamonti, 435, Milan 20141, Italy
e-mail: bettina.balliardi@ieo.it

F.D. Dip et al. (eds.), *Fluorescence Imaging for Surgeons: Concepts and Applications*,
DOI 10.1007/978-3-319-15678-1_27,

sentinel node involvement, to be spared axillary dissection, reducing the surgical complications related to axillary dissection with no adverse effect on survival. Patients with clinically evident axillary involvement should continue to receive axillary dissection.

Sentinel Node Identification Method

Several methods have been developed to identify SNs. The most common are peritumoral or periareolar injection of colloid labeled with the short-lived gamma emitter 99Tc and injection of blue dye. These two methods were evaluated in the study published by Cox et al. [5]: the SN identification rate with blue dye only was 80.3 % as compared to 88.6 % with radiolabeled colloid. Using blue dye and radiolabeled colloid together, the identification rate reached 95.7 %, and the false-negative rate was lower. However, both techniques remain widely employed. We use the radiotracer method exclusively at the European Institute of Oncology; however, there is a considerable interest in using a new SNB method that employs the fluorescent dye indocyanine green (ICG).

Indications for SNB

Large Tumors

SNB has generally been confined to patients with small tumors, and its use in large breast cancer remains controversial. However, published data show [6] that for large tumors, rates of SN identification and false negatives compare favorably with those in studies on patients with smaller cancer. Furthermore, in nearly a third of patients with large tumors, the SN is negative permitting avoidance of axillary dissection even in these cases.

Multifocal or Multicentric Invasive Breast Cancer

Multifocal and multicentric cancers were once considered relative contraindications for SNB, based on concerns that these cancers may involve multiple lymphatic tracts draining to multiple SNs [7].

However, there is now good evidence [8] that drainage from the breast occurs through relatively few common lymphatic channels to a common axillary SN, regardless of the precise location of the lesion. A study on SNB in 337 patients with unilateral multicentric breast cancer and a clinically negative axilla found an SN in all cases (identification rate of 100 %). A total of 138 patients with negative SN ($n=134$) or only isolated tumor cells (ITCs) in the SN ($n=4$) did not receive completion axillary lymph node dissection. During a median follow-up of 5 years, 3 (2.2 %) cases developed axillary disease, indicating that the use of SNB to stage the axilla in multicentric breast cancer provides comparable results to those obtained in monocentric cancer [9].

Ductal Intraepithelial Neoplasia

The role of SNB in the management of ductal intraepithelial neoplasia (DIN)—formerly known as ductal carcinoma in situ, DCIS—has not been definitively established. By definition pure DIN (no microinvasion) does not metastasize. However, the EIO investigated 854 patients with pure DIN who underwent SNB and found SN metastases in 12 (1.4 %) cases and ITCs in additional four patients [10].

Consistent with other reports, the SN was found to be the only metastatic node in all patients who received axillary dissection. These findings suggest that routine axillary dissection is unnecessary in pure DIN cases with a metastatic SN, particularly when only micrometastases are present. The EIO recommends that (a) SNB should not be a standard procedure for all patients with DIN: if the DIN is completely excised by radical surgery (free resection margins) or completely (macroscopically) removed by core needle biopsy or vacuum-assisted biopsy (limited clusters of microcalcifications or small solid lesions <2 cm), then diagnostic underestimation and the risk of an invasive component at final histology are unlikely and SNB is not indicated. (b) The principal reason for performing SNB in DIN is diagnostic uncertainty at final histological examination.

This happens when DIN is not completely excised by conservative surgery (positive resection margins or residual microcalcifications at postoperative mammogram) or in large tumors or diffuse, multicentric microcalcifications not macroscopically removed by core needle biopsy or vacuum-assisted biopsy. In these cases, the risk of finding invasion at final histology is much higher (10–20 %) depending on lesion size. (c) If mastectomy is indicated for DIN, then SNB is mandatory since upstaging to invasive cancer is reported in 28–48 % of patients after mastectomy for extensive DIN [11].

Previous Breast and Axillary Surgery

In order to accurately predict axillary status, SNB requires the presence of an intact lymphatic path from the site of the primary—or recurrence—to the axilla. Previous axillary surgery could partially or temporarily interrupt or modify this path, compromising the ability of SNB to identify the "real" SN (which reliably predicts axillary status). Data suggest, however, that a new lymphatic system develops shortly after surgery has removed tissues or disrupted lymphatic flow [12] so that subsequent SNB allows identification of a new SN for the new tumor. The EIO recommends that a second SNB be offered to selected women with ipsilateral breast cancer recurrence after previous conservative breast surgery and a negative SN followed by adjuvant radiotherapy. Previous mastectomy is generally considered an absolute technical contraindication to SNB. However, in selected mastectomized patients with recurrence, subdermal injection of radioisotope permits identification of an axillary SN [13], and there appear to be no anatomical or physiological reasons why this staging opportunity should be missed. Nevertheless, a large study with sufficient follow-up is necessary to determine predictive value of SNB in such patients.

Pregnancy

Studies indicate that SNB with radiotracer can be performed in pregnancy with negligible risk to the fetus [14, 15]. Since the injected radiotracer remains at the injection site, within the lymphatic ducts or the SN, fetal exposure should be zero. A pilot study carried out at the EIO indicated that SNB can be performed safely in pregnancy, at any phase of gestation, since doses to the uterus were well below threshold values for deterministic effects [16].

More recently, the EIO published experience on SNB in 12 pregnant patients with breast cancer [17]. Eleven healthy babies resulted from the 12 pregnancies. The 12th, delivered at 34 weeks by Cesarean section, was operated on at 3 months for a ventricular septal defect and at 43 months was in good health. The malformation was suspected at ultrasound at week 21 (well before SNB radiotracer was injected) and confirmed a posteriori by an independent observer. Precautions to further minimize fetal exposure include avoiding contact with other patients as potential sources of radioactivity (e.g., by scheduling pregnant patients as first procedure of the day and accommodating them in a single room), reducing the time between radiotracer injection and surgery, and injecting less activity.

Neoadjuvant Chemotherapy

Recent large studies indicate that SNB identification and false-negative rates in patients receiving primary chemotherapy are similar to those in patients who do not receive neoadjuvant treatment [18, 19].

Thus, SNB appears useful in women who have a clinically negative axilla before starting neoadjuvant chemotherapy and in whom the disease does not progress during treatment. SNB is not suggested for patients with a clinically suspicious axilla after primary chemotherapy. However, in women with a clinically involved axilla at presentation, which downstages to clinical negativity after systemic therapy, SNB might be an option after appropriate preoperative evaluation (e.g., PET or ultrasound) appears to confirm a negative axilla [20].

SNB in Men

There are no biological or anatomical reasons why lymphatic drainage of the breast should differ in men from that in women, so men with breast cancer and a clinically negative axilla should receive SNB: this hypothesis was tested in an EIO study. Between April 1999 and January 2005, 75 men with breast cancer were treated at the EIO, 32 of these (with a clinically negative axilla) SNB was performed, and the SN was identified after lymphoscintigraphy in all cases. Metastases were found in 4 patients, micrometastasis in 2 patients, and a negative SN in the remaining 26 patients. After a median follow-up of 30 months, there was no axillary recurrence and no other breast cancer-related events [21].

Biopsy of the Internal Mammary Chain

For patients with inner quadrant cancer, biopsy of the internal mammary chain lymph nodes can be performed during breast surgery with a minimum increase in operating time. Internal mammary nodes are accessed easily through the intercostal space. Fatty tissue containing the node to be examined is carefully freed from blood vessels, and care is taken to avoid damaging to the underlying pleura. If an internal mammary node is found to be involved (even if the axilla is disease-free), the disease is upstaged and adjuvant treatment (chemotherapy or RT to the internal mammary chain) is given [22].

When both an axillary SN and intramammary SN are observed on preoperative lymphoscintigraphy, the policy at the EIO is that both should be biopsied, but only a positive axillary SN justifies axillary dissection.

The SOUND Trial

The Z0011 and IBCSG 23.01 trials showed that for most patients with early breast cancer and an SN with limited metastatic involvement, axillary dissection confers no advantage. This raises the question as to whether even SNB is necessary in such patients, and this is the question being investigated by the SOUND trial (sentinel node biopsy vs. observation after axillary ultrasound).

SOUND is a multicentric randomized controlled trial, designed and promoted by the EIO, to compare SNB with no SNB (observation only) in patients with small breast cancer and negative axillar on preoperative assessment with US. Patients with breast cancer up to 2 cm, candidates for breast-conserving surgery, and with a clinically negative axilla undergo axillary US to compliment the clinical examination and reduce the probability of axillary involvement. Patients with single doubtful node on US undergo US-guided fine-needle aspiration biopsy. Patients with negative US, or with the single doubtful node negative on cytologic examination, are eligible for randomization to either SNB ± axillary dissection or no surgical staging of the axilla (i.e., no SNB and no axillary dissection). Those randomized to SNB will not receive axillary dissection if the SN is negative, is micrometastatic, or has ITCs; axillary dissection will be performed only if macrometastasis is present. The primary endpoint is distant disease-free survival. Secondary endpoints are cumulative incidence of distant metastasis, cumulative incidence of axillary disease, disease-free survival, and overall survival. Other endpoints are quality of life and assessment of influence of type of adjuvant treatment administered [23].

Sentinel Node Identification with Indocyanine Green (ICG)

Sentinel node biopsy (SNB) is the standard procedure for axillary staging in breast cancer. Randomized controlled trials have shown that 5-year overall survival in patients with a negative SN who do not undergo axillary dissection is indistinguishable for that in comparable patients who do undergo axillary dissection, making it possible to avoid axillary dissection in a considerable fraction of patients, for whom the adverse sequelae of axillary dissection are avoided and quality of life is improved [24–26].

Colloid labeled with technetium 99m(^{99m}Tc) is widely used to identify and localize the sentinel node (SN), either alone or in combination with blue dye [27]. A method employing ^{99m}Tc bound to human albumin was developed at the European Institute of Oncology, Milan, where it has been in use for over 15 years, with over 22,000 SNBs performed [28]. This method detects the sentinel node in a high (95–99 %) proportion of cases [29]. However radioisotopes are not available to all treatment centers, and their use requires licensing and a nuclear medicine department; furthermore, the time window between radiotracer injection and surgery is limited.

A growing body of evidence supports the feasibility and efficacy of using the fluorescent dye indocyanine green (ICG) to identify the SN [30–33]. The method involves injection of ICG subdermally close to the tumor or in the peri-areolar region and following its progress through the lymphatic ducts to the sentinel node (SN) using an excitation illumination system in combination with a high-sensitivity camera, which detects the emitted fluorescence. The method is additionally characterized by very low complication and adverse event rates [34]; however, it has not been formally and prospectively compared with the gold-standard radiotracer method in terms of SN detection rate.

At the European Institute of Oncology, therefore, we performed a study to assess the concordance between the ICG method and the ^{99m}Tc-labeled radiotracer method, to determine whether ICG can be effectively used alone to identify the SN.

Materials and Methods

Between June 2011 and January 2013, 134 patients with early breast cancer, confirmed by core or fine-needle biopsy, and a clinically negative axilla were enrolled in the present single-center study, approved by the Ethical Committee of the European Institute of Oncology, Milan, and registered as EudraCT No. 2010-021815-18. Patients gave written informed consent to treatment with ICG for SNB prior to enrolment. Those with cancer >3 cm, clinically positive lymph nodes, previous surgery for invasive breast cancer, thyroid dysfunction, hypersensitivity to iodine, and hepatic or renal insufficiency were excluded. Patients eligible for SNB according to ASCO guidelines [35] (excluding those with cancer >3 cm) were injected, on the afternoon before the day of surgery, with 12–15 MBq of ^{99m}Tc-labeled albumin particles (Nanocoll, GE Healthcare, Italy) in 0.2 ml saline subdermally close to the tumor or in the peri-areolar region. Planar anterior and anterior-oblique scintigraphic scans of the breast and axilla were taken 30 min after injection. If no nodes were visualized, a further scan was taken 3 h later [36, 37].

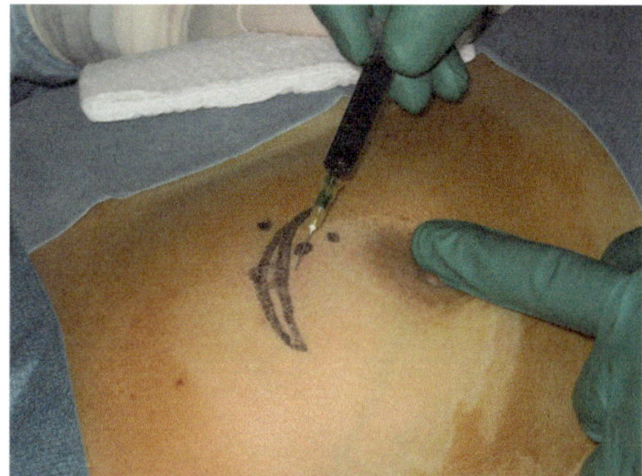

Fig. 27.1 ICG solution injection

Immediately before surgery, after disinfection of the operative field, 1 ml 0.5 % ICG solution was injected subdermally close to the tumor or into the subareolar region (Fig. 27.1). ICG movement in the lymph ducts was facilitated by massage. ICG fluorescence was elicited and detected by a photodynamic eye (PDE) camera (Hamamatsu Photonics, Hamamatsu, Japan) (Fig. 27.2), and the lymphatic drainage, made evident by the fluorescent dye, was visualized in real time on a monitor (Fig. 27.3). The fluorescence was followed from the site of injection toward the axilla, and where the fluorescence disappeared into the axilla, an incision was made to start the biopsy. Fluorescent lymph nodes (ICG positive) were then localized and excised and the axilla was inspected for any residual fluorescence (Fig. 27.4). Excised ICG-positive nodes were then tested for radioactivity using a gamma-detecting probe and classified as hot

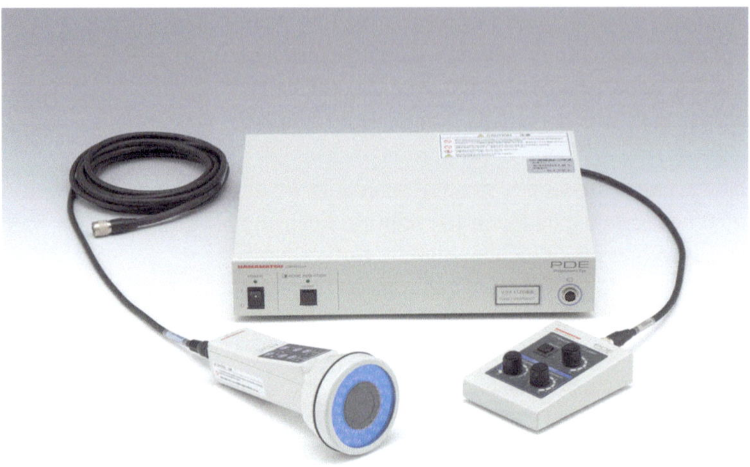

Fig. 27.2 Photodynamic eye (PDE) camera (Hamamatsu Photonics, Hamamatsu, Japan)

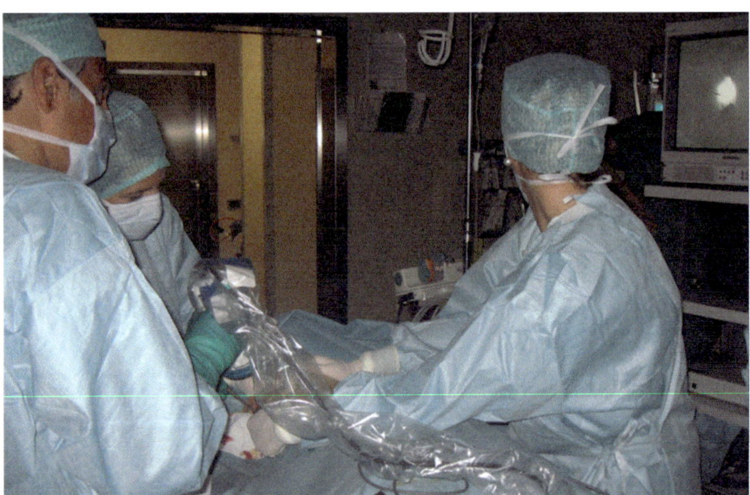

Fig. 27.3 Lymphatic drain real-time visualization

(^{99m}Tc positive) or cold (^{99m}Tc negative). Finally, the axillary region was checked with the gamma-detecting probe to determine whether any radioactivity was left in place. In the event of significant residual radioactivity, the hot spot (considered to be a ^{99m}Tc-positive SN) was removed and examined.

The number of sentinel nodes (ICG positive, ^{99m}Tc positive, or both) removed from each patient was noted. Patient characteristics were also recorded.

Statistical Methods

The study was designed to determine whether the ICG method was equivalent to the Tc method—considered the gold standard—in terms of its ability to detect SNs.

Let A = number of Tc-positive and ICG-positive SNs detected, B = number of Tc-positive and ICG-negative SNs detected, and C = number of Tc-negative and ICG-positive SNs detected. The total number (N) of SNs detected is therefore

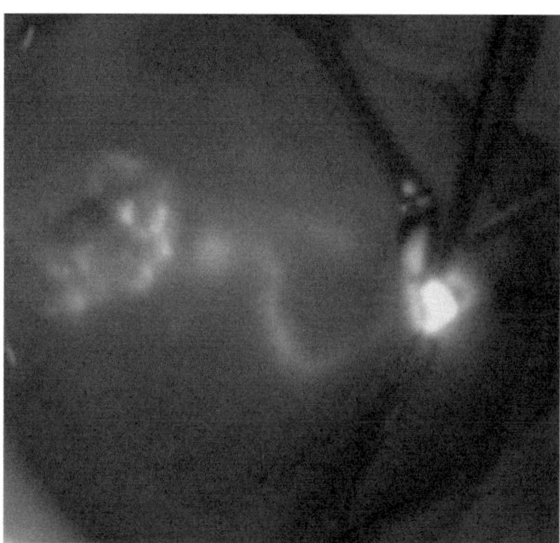

Fig. 27.4 Fluorescent lymph node visualization (permission by Kobe City Medical Center General Hospital, Japan—Hamamatsu)

Table 27.1 Characteristics of patients and their cancers

Characteristic	*N* (patients)	%
Tumor size (cm)		
≤0.5	5	3.7
>0.5, ≤1	33	24.6
>1, ≤1.5	44	32.8
>1.5, ≤2	39	29.1
>2, ≤3	11	8.2
>3	2	1.5
Histological type		
Invasive ductal	112	83.6
Invasive lobular	9	6.7
Other invasive	12	9.0
Ductal intraepithelial neoplasia grade 3	1	0.7
Grade		
G1	24	17.9
G2	58	43.3
G3	45	33.6
Not available	7	5.2
Median age (range)	56 (26–80)	years
Median BMI (range)	23 (18–40)	kg/m^2

$N=(A+B+C)$; the proportion of SNs (P_{Tc}) detected by the Tc method is $(A+C)/N$; and the proportion of SNs (P_{ICG}) detected by the ICG method is $(A+B)/N$.

The null hypothesis for the equivalence of the two methods is that the difference between the proportions of SNs detected by each method lies outside the interval $-\delta$ and $+\delta$, where δ was set as 5 %. Formally, the null hypothesis for equivalence is

$$P_{ICG}-P_{Tc}<-\delta \text{ and } P_{ICG}-P_{Tc}\geq+\delta$$

If the null hypothesis is rejected, we may conclude for the hypothesis of equivalence that

$$\delta<\left(P_{ICG}-P_{Tc}\right)<\delta$$

The equivalence hypothesis can be transformed into two one-sided hypotheses: (A) that the difference between the proportions is beyond the lower equivalence margin, null hypothesis A, $P_{ICG}-P_{Tc}<-\delta$, which, if rejected, permits the conclusion $P_{ICG}-P_{Tc}\geq-\delta$ and (B) that the difference between the proportions is greater than the upper equivalence margin, null hypothesis B, $P_{ICG}-P_{Tc}\geq+\delta$ which, if rejected, permits the conclusion $P_{ICG}-P_{Tc}\leq+\delta$.

Hypotheses A and B were tested by comparing the 95 % confidence interval (CI) of the percentage difference in the number of SNs detected between the two methods, with the 5 % equivalence margins ($-\delta$ and $+\delta$). A Wald-type sample-based test statistic, based on Lu and Bean algorithm, was used to determine the 95 % CI of the actual difference [38]. To conclude for equivalence, both null hypotheses had to be rejected.

To estimate the number of SNs that needed to be examined, we assumed a detection rate $(A+C)/N$ of 97 % for the gold-standard Tc method, a discordance rate $(B+C)/N$ of 6 %, and set δ at 5 %. We found that 255 SNs are needed to be examined to demonstrate equivalence between the two methods with 80 % power and type I error (α) of 5 % [39, 40]. Sample size was calculated by using PASS 2008 software [41]. Statistical analyses were performed using the SAS statistical software, version 9.2 [41].

Results

Patient and tumor characteristics are shown in Table 27.1. One hundred thirty-four patients provided a total of 246 SNs detected by one or both methods. One, 2, and 3 SNs, respectively, were detected, excised, and examined in 70 (52.2 %), 39 (29.1 %), and 17 (12.7 %) patients; 4–10 SNs were detected and examined in the remaining 8 patients. Tc detected 231 of the 246 (93.9 %) SNs found; ICG detected 245 (99.6 %) of the total SNs found (Table 27.2). The two methods were concordant for 230 of the 246 SNs (93.5 %) examined and discordant for 16 SNs (6.5 %). Fifteen (6.1 % of total) of the latter SNs were detected by ICG and not Tc; the remaining SN (0.4 %) were detected by Tc but not ICG. The difference between the two methods is therefore (15–1)/246 = +5.7 %. Application of the Lu and Bean [15] algorithm yielded 2.6–8.8 % as the 95 % confidence interval (CI) for this difference. Since this CI does not include −5 % (Fig. 27.5), we were able to reject one-sided null hypothesis A and conclude that ICG is not inferior to Tc. However, this 95 % CI does include +5 % (Fig. 27.5), so it was not possible to reject one-sided hypothesis B that ICG is superior to Tc. This does not mean, however, that the ICG method is superior.

The analysis based on patients was performed by classifying as "concordant" those for whom at least one SN was detected by both methods. As shown in Table 27.3, 121 patients were "concordant" because all SNs removed (range 1–10) were detected by both methods. Among the remaining 13 patients, all were concordant except one. For the single discordant patient, the single SN identified was detected by ICG but not Tc. In addition, a single ICG negative and Tc positive was found in a patient in whom two other SNs were detected by both methods (Table 27.3).

Table 27.2 Numbers of sentinel nodes identified with the indocyanine green (ICG) and technetium (TC) methods

	N	%
ICG+ Tc+	230	93.5
ICG+ Tc−	15	6.1
ICG− Tc +	1	0.4
Total SNs	246	100

For 93.5 % of the nodes removed, the two methods were concordant

Discussion

This was a prospective study to compare the promising ICG method with the gold-standard radiotracer method for finding and removing axillary SNs in early breast cancer. The comparison

Table 27.3 Numbers of sentinel nodes removed from patients according to whether all nodes were identified by both the ICG and TC methods ($n = 121$) and whether not all nodes were identified by both methods ($n = 13$)

N SNs removed	ICG+ Tc+	ICG+Tc−	ICG− Tc+	*N* patients
1	1	0	0	69
2	2	0	0	34
3	3	0	0	12
4	4	0	0	3
5	5	0	0	1
7	7	0	0	1
10	10	0	0	1
Patients in whom all sentinel nodes were identified by both ICG and TC methods (N = 121)				
1	0	1	0	1
2	1	1	0	5
3	2	1	0	2
3	2	0	1	1
3	1	2	0	2
5	4	1	0	1
8	6	2	0	1
Patients in whom not all removed nodes were identified by both methods (N = 13)				

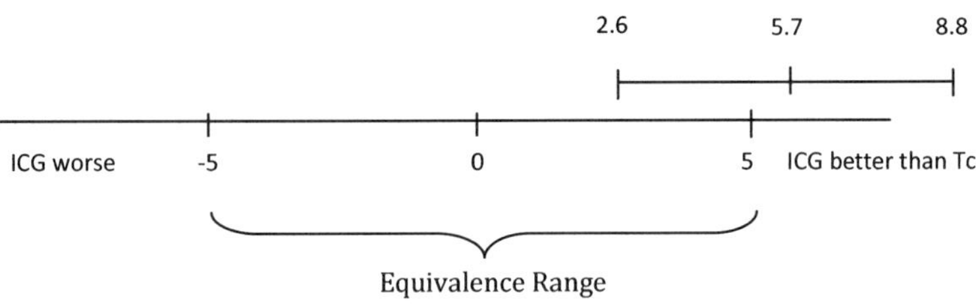

Fig. 27.5 Percentage difference between ICG and radiotracer methods with 95 % confidence interval, in relation to equivalence margins (−5 % and +5 %)

was designed to assess whether the ICG method can be used as a reliable alternative to the radiotracer method. We found that the ICG method detected 99.6 % of all SNs found, while the Tc method detected 93.9 % of all SNs. Furthermore, only 1 SN (0.4 % of total) was found by the Tc method and not the ICG method.

Considering patients, the concordance between the two methods was 99.3 %. For the single discordant patient, the single SN was detected by ICG but not Tc. Furthermore, SNs identified by Tc were also fluorescent except for one SN. The patient with this hot but nonfluorescent SN had two other SNs that were both hot and fluorescent. Thus, all patients with a hot SN also had an ICG-positive SN (even though these SNs did not perfectly coincide in one case). Furthermore, and in line with the findings of other studies, we found that the ICG method identified more SNs than the Tc method. Thus, of the 246 SNs identified, 15 (6.1 %) were Tc negative. The most likely reason for this is that, being a small molecule, ICG migrates more readily beyond the "first" SN than the much larger albumin particles. In support of this supposition, our experience is that fewer SNs are found if biopsy is started immediately after ICG reaches the axilla (disappears from view) than if started after some delay. Our analysis enabled us to reject null hypothesis A, to conclude that the ICG method is not inferior to the radiotracer method, but did not enable us to reject null hypothesis B and conclude that the ICG method is superior to the Tc method.

The present study is an adequately powered prospective study to assess the ability of the ICG method to detect SNs in comparison with the standard radiotracer method. Several previous studies have shown that the ICG method identifies the SN in a high proportion of cases (97–100 %) and is associated with low toxicity and few allergic reactions or other side effects [30–34, 42–47].

One of the earliest studies was performed by Murawa et al. [31]. They assessed the feasibility of SN detection using ICG and a PDE camera in 30 patients, 20 of whom also received the radiotracer method. All patients underwent axillary dissection. The fluorescent method detected SNs in 97 % of patients. Among the 20 patients in whom both techniques were used, ICG identified SNs in 20, and radiotracer identified SNs in 17. The study demonstrated that lymphatics were reliably imaged by the ICG, with an SN found in all but one case. Closely similar findings were reported by Hirsche et al. [32] on 43 patients. Hojo et al. [33] compared ICG with the blue dye and radiotracer methods, but only in 29 patients, the ICG and radiotracer methods were compared directly. The SN detection rate of the fluorescence method was 99.3 % compared to 92.9 % for the blue dye method and 100 % for the radiotracer method.

Apparently, the first study to investigate the feasibility of the ICG method in a large series of patients was that of Sugie et al. [43]. Their multi-institute study compared the ICG and blue dye methods in 411 breast cancer patients. At least one SN was identified and removed in 408 patients. The identification rate with ICG (99 %) was higher than with blue dye (83–93 %); as also found in a recent paper [48], the authors emphasized the need for a direct comparison between the radioisotope and ICG methods.

The 2012 feasibility study by Wishart et al. [49] compared ICG with blue dye, radiotracer, or both in 100 patients. They defined ICG sensitivity as the proportion of SNs detected by blue dye and/or radioisotope that were also fluorescent; they showed that fluorescence imaging using ICG was highly sensitive for detecting the SN (100 %) and that combined nodal sensitivity was higher for blue dye and ICG (95 %) than the combination of blue dye and radioisotope (73.1 %). Detection rates were 99 % for blue dye, 91.3 % for radioisotope, and 100 % for fluorescence.

Some studies have investigated ICG conjugated with human serum albumin to limit the mobility of the fluorescent molecule and thereby reduce the number of SNs found. However, no significant differences in the number of SNs identified by ICG and conjugated ICG have been found [50–52].

Conclusions

The ICG method has previously been shown to be surgically simple procedure that identifies axillary SNs in breast cancer patients, without the use of radiopharmaceutical. The present study has validated the ICG method by demonstrating that it is statistically non-inferior to the gold-standard method that used Tc-labeled albumin, allowing us to conclude that the ICG method can be used as a reliable and safe alternative to the radiotracer method. This finding is potentially of major importance in clinical practice since the ICG method has advantages over the current gold-standard radiotracer method including (a) direct implementation in the operating room and (b) no prior preparation by, or involvement of, nuclear medicine physicians. The lack of need for a nuclear medicine department will appeal to centers without such a department.

We did not investigate costs in the present study. However, it is likely that the ICG method will cost less than the radiotracer method since nuclear medicine personnel are not required and preoperative radiotracer injection and lymphoscintigraphic SN identification are eliminated. The disadvantages of the technique are that a separate incision is preferred to identify and isolate the SN and that the sentinel node is not always visualized transcutaneously before the incision, so sentinel node location is not always available before the incision. Furthermore, extra-axillary sentinel nodes (e.g., those in the intramammary chain) are not visualized as readily as with lymphoscintigraphy [53].

References

1. Veronesi U, Paganelli G, Viale G, et al. A randomized comparison of sentinel-node biopsy with routine axillary dissection in breast cancer. N Engl J Med. 2003;349(6):546–53.
2. Giuliano AE, Han SH. Local and regional control in breast cancer: role of sentinel node biopsy. Adv Surg. 2011;45:101–16. Review.
3. Giuliano AE, Hunt KK, Ballman KV, Beitsch PD, Whitworth PW, Blumencranz PW, Leitch AM, Saha S, McCall LM, Morrow M. Axillary dissection vs no axillary dissection in women with invasive breast cancer and sentinel node metastasis: a randomized clinical trial. JAMA. 2011;305(6):569–75.
4. Galimberti V, Cole BF, Zurrida S, Viale G, Luini A, Veronesi P, Baratella P, Chifu C, Sargenti M, Intra M, Gentilini O, Mastropasqua MG, Mazzarol G, Massarut S, Garbay JR, Zgajnar J, Galatius H, Recalcati A, Littlejohn D, Bamert M, Colleoni M, Price KN, Regan MM, Goldhirsch A, Coates AS, Gelber RD, Veronesi U, International Breast Cancer Study Group Trial 23–01 investigators. Axillary dissection versus no axillary dissection in patients with sentinel-node micrometastases (IBCSG 23–01): a phase 3 randomised controlled trial. Lancet Oncol. 2013;14(4):297–305.
5. Cox CE, Salud C, Whitehead GF, Reintgen DS. Sentinel lymph node biopsy for breast cancer: combined dye-isotope technique. Breast Cancer. 2000; 7(4):389–97.
6. Spillane AJ, Brennan ME. Accuracy of sentinel lymph node biopsy in large and multifocal/multicentric breast carcinoma—a systematic review. Eur J Surg Oncol. 2011;37(5):371–85. Epub 2011 Feb 3. Review.
7. Schwartz GF, Giuliano AE, Veronesi U, et al. Proceedings of the consensus conference on the role of sentinel lymph node biopsy in carcinoma of the breast, April 19–22, Philadelphia, Pennsylvania. Cancer. 2001;94:2542–7.
8. Ferrari A, Dionigi P, Rovera F, et al. Multifocality and multicentricity are not contraindications for sentinel lymph node biopsy in breast cancer surgery. World J Surg Oncol. 2006;4:79.
9. Gentilini O, Veronesi P, Botteri E, et al. Sentinel lymph node biopsy in multicentric breast cancer: five-year results in a large series from a single institution. Ann Surg Oncol. 2011;18:2879–84.
10. Intra M, Rotmensz N, Veronesi P, et al. Sentinel node biopsy is not a standard procedure in ductal carcinoma in situ of the breast: the experience of the European Institute of Oncology on 854 patients in 10 years. Ann Surg. 2008;247:315–9.
11. Patani N, Khaled Y, Al Reefy S, Mokbel K. Ductal carcinoma in situ: an update for clinical practice. Surg Oncol. 2011;20(1):e23–31. Epub 2010 Nov 24.
12. Maaskant-Braat AJ, de Bruijn SZ, Woensdregt K, Pijpers H, Voogd AC, Nieuwenhuijzen GA. Lymphatic mapping after previous breast surgery. Breast. 2012; 21(4):444–8.
13. Intra M, Veronesi P, Gentilini OD, et al. Sentinel lymph node biopsy is feasible even after total mastectomy. J Surg Oncol. 2007;95(2):175–9.
14. Nicklas A, Baker M. Imaging strategies in pregnant cancer patients. Semin Oncol. 2000;27:623–32.
15. Khera SY, Kiluk JV, Hasson DM, et al. Pregnancy-associated breast cancer patients can safely undergo lymphatic mapping. Breast J. 2008;14(3):250–4.
16. Gentilini O, Cremonesi M, Trifirò G, et al. Safety of sentinel node biopsy in pregnant patients with breast cancer. Ann Oncol. 2004;15(9):1348–51.
17. Gentilini O, Cremonesi M, Toesca A, et al. Sentinel lymph node biopsy in pregnant patients with breast

cancer. Eur J Nucl Med Mol Imaging. 2010;37: 78–83.

18. Piñiero A, Giménez J, Vidal-Sicart S, Intra M. Selective sentinel lymph node biopsy and primary systemic therapy in breast cancer. Tumori. 2010 Jan-Feb;96(1):17–23.
19. Tan VK, Goh BK, Fook-Chong S, Khin LW, Wong WK, Yong WS. The feasibility and accuracy of sentinel lymph node biopsy in clinically node-negative patients after neoadjuvant chemotherapy for breast cancer – A systematic review and meta-analysis. J Surg Oncol. 2011;104(1):97–103. doi:10.1002/jso.21911. Epub 2011 Mar 24. Review.
20. Beatty JD, Precht LM, Lowe K, Atwood M. Axillary-conserving surgery is facilitated by neoadjuvant chemotherapy of breast cancer. Am J Surg. 2009;197(5):637–41. discussion 641–2. Epub 2009 Mar 24.
21. Gentilini O, Chagas E, Zurrida S, et al. Sentinel lymph node biopsy in male patients with early breast cancer. Oncologist. 2007;12:512–5.
22. Veronesi U, Arnone P, Veronesi P, et al. The value of radiotherapy on metastatic internal mammary nodes in breast cancer. Results on a large series. Ann Oncol. 2008;19(9):1553–60.
23. Gentilini O, Veronesi U. Abandoning sentinel lymph node biopsy in early breast cancer? A new trial in progress at the European Institute of Oncology of Milan (SOUND: Sentinel node vs Observation after axillary UltrasouND). Breast. 2012;21:678–81.
24. Veronesi U, Viale G, Paganelli G, Zurrida S, Luini A, Galimberti V, Veronesi P, Intra M, Maissoneuve P, Zucca F, Gatti G, Mazzarol G, De Cicco C, Vezzoli D. Sentinel node biopsy in breast cancer: ten year results of a randomized controlled study. Ann Surg. 2010;251(4):595–600.
25. Veronesi U, Galimberti V, Paganelli G, Maisonneuve P, Viale G, Orecchia R, Luini A, Intra M, Veronesi P, Caldarella P, Renne G, Rotmensz N, Sangalli C, De Brito LL, Tulli M, Zurrida S. Axillary metastases in breast cancer patients with negative sentinel nodes: a follow up of 3548 cases. Eur J Cancer. 2009;45(8): 1381–8.
26. Krag DN, Anderson SJ, Julian TB, Brown AM, Harlow SP, Costantino JP, Ashikaga T, Weaver DL, Mamounas EP, Jalovec LM, Frazier TG, Noyes RD, Robidoux A, Scarth HM, Wolmark N. Sentinel-lymph-node resection compared with conventional axillary-lymph-node dissection in clinically node-negative patients with breast cancer: overall survival findings from the NSABP B-32 randomised phase 3 trial. Lancet Oncol. 2010;11(10):927–33.
27. Cody III HS, Fey J, Akhurst T, Fazzari M, Mazumdar M, Yeung H, Yeh SD, Borgen PI. Complementarity of blue dye and isotope in sentinel node localization for breast cancer: univariate and multivariate analysis of 966 procedures. Ann Surg Oncol. 2001;8(1):13–9.
28. Paganelli G. Lymphoscintigraphy and sentinel node biopsy in breast cancer: where are we after 10 years? Eur J Nucl Med Mol Imaging. 2007;34(12):2152–3.
29. Trifirò G, Viale G, Gentilini O, Travaini LL, Paganelli G. Sentinel node detection in pre-operative axillary staging. Eur J Med Mol Imaging. 2004;31 Suppl 1:46–55.
30. Kitai T, Inomoto T, Miwa M, Shikayama T. Fluorescence navigation with indocyanine green for detecting sentinel nodes in breast cancer. Breast Cancer. 2005;12(3):211–5.
31. Murawa D, Hirche C, Dresel S, Hunerbein M. Sentinel node biopsy in breast cancer guided by indocyanine green fluorescence. Br J Surg. 2009;96:1289–94.
32. Hirche C, Murawa D, Mohr Z, Kneif S, Hunerbein M. ICG fluorescence-guided sentinel node biopsy for axillary nodal staging in breast cancer. Breast Cancer Res Treat. 2010;121(2):373–8.
33. Hojo T, Nagao T, Kikuyama M, Akashi S, Kinoshita T. Evaluation of sentinel node biopsy by combined fluorescent and dye method and lymph flow for breast cancer. Breast. 2010;19(3):210–3.
34. Alford R, Simpson HM, Duberman J, Hill GC, Ogawa M, Regino C, Kobayashi H, Choyke PL. Toxicity of organic fluorophores used in molecular imaging: literature review. Mol Imaging. 2009;8(6):341–54.
35. Lyman GH, Giuliano AE, Somerfield MR, Benson 3rd AB, Bodurka DC, Burstein HJ, Cochran AJ, Cody III HS, Edge SB, Galper S, Hayman JA, Kim TY, Perkins CL, Podoloff DA, Sivasubramaniam VH, Turner RR, Wahl R, Weaver DL, Wolff AC, Winer EP. American Society of Clinical Oncology guideline recommendations for sentinel lymph node biopsy in early-stage breast cancer. J Clin Oncol. 2005;23(30): 7703–20.
36. Paganelli G, De Cicco C, Cremonesi M, Prisco G, Calza P, Luini A, Zucali P, Veronesi U. Optimized sentinel node scintigraphy in breast cancer. Q J Nucl Med. 1998;42(1):49–53.
37. De Cicco C, Cremonesi M, Luini A, Bartolomei M, Grana C, Prisco G, Galimberti V, Calza P, Viale G, Veronesi U, Paganelli G. Lymphoscintigraphy and radioguided biopsy of the sentinel axillary node in breast cancer. J Nucl Med. 1998;39:2080–4.
38. Lu Y, Bean JA. On the sample size for one-sided equivalence of sensitivities based upon McNemar's test. Stat Med. 1995;14(16):1831–9.
39. Lui KJ. Interval estimation of generalized odds ratio in data with repeated measurements. Stat Med. 2002;21(20):3107–17.
40. Nam JM. Comparison of validity of assessment methods using indices of adjusted agreement. Stat Med. 2007;26(3):620–32.
41. Hintze J. PASS: power analysis and sample size. Kaysville, UT: NCSS, LLC; 2008.
42. Tagaya N, Aoyagi H, Nakagawa A, Abe A, Iwasaki Y, Tachibana M, Kubota K. A novel approach for sentinel node identification using fluorescence imaging and image overlay navigation surgery in patients with breast cancer. World J Surg. 2010;35(1):154–8.
43. Sugie T, Kassim KA, Tekeuchi M, Hashimoto T, Yamagami K, Masai Y, Toi M. A novel method for sentinel node biopsy by indocyanine green fluorescence

technique in breast cancer. Cancers. 2010;2: 713–20.
44. Hirche C, Mohr Z, Kneif S, Murawa D, Hunerbein M. High rate of solitary sentinel node metastases identification by fluorescence-guided lymphatic imaging in breast cancer. J Surg Oncol. 2012;105(2): 162–6.
45. Aoyama K, Kamio T, Nishizawa M, Ohchi T, Kameoka S. Sentinel node biopsy for breast cancer patients using fluorescence navigation with indocyanine green. World J Surg Oncol. 2011;9(1):157.
46. Kitai T, Kawashima M. Transcutaneous detection and direct approach to the sentinel node using axillary compression technique in ICG fluorescence-navigated sentinel node biopsy for breast cancer. Breast Cancer. 2012;19(4):343–8.
47. Abe H, Mori T, Umeda T, Tanaka M, Kawai Y, Shimizu T, Cho H, Kubota Y, Kurumi Y, Tani T. Indocyanine green fluorescence imaging system for sentinel node biopsies in early breast cancer patients. Surg Today. 2011;41(2):197–202.
48. Sugie T, Sawada T, Tagaya N, Kinoshita T, Yamagami K, Suwa H, Ikeda T, Yoshimura K, Niimi M, Shimizu A, Toi M. Comparison of the indocyanine green fluorescence and blue dye methods in detection of sentinel lymph nodes in early-stage breast cancer. Ann Surg Oncol. 2013;20(7):2213–8.
49. Wishart GC, Loh SW, Jones L, Benson JR. A feasibility study (ICG-10) of indocyanine green (ICG) fluorescence mapping for sentinel lymph node detection in early breast cancer. Eur J Surg Oncol. 2012;38(8): 651–6.
50. Polom K, Murawa D, Nowaczyk P, et al. Breast cancer sentinel lymph node mapping using near infrared guided indocyanine green and indocyanine green-human serum albumin in comparison with gamma emitting radioactive colloid tracer. Eur J Surg Oncol. 2011;38:137–42.
51. Mieog JSD, Troyan SL, Hutteman M, et al. Toward optimization of imaging system and lymphatic tracer for near-infrared fluorescence sentinel lymph node mapping in breast cancer. Ann Surg Oncol. 2011;18: 2438–91.
52. Hutteman M, Mieog JS, van der Vorst JR, Liefers GJ, Putter H, Löwik CW, Frangioni JV, van de Velde CJ, Vahrmeijer AL. Randomized, double-blind comparison of indocyanine green with or without albumin premixing for near-infrared fluorescence imaging of sentinel lymph nodes in breast cancer patients. Breast Cancer Res Treat. 2011;127(1):163–70.
53. Ballardini B, Santoro L, Sangalli C, Gentilini O, Renne G, Lissidini G, Pagani GM, Toesca A, Blundo C, del Castillo A, Peradze N, Caldarella P, Veronesi P. The indocyanine green method is equivalent to the ^{99}mTc-labeled radiotracer method for identifying the sentinel node in breast cancer: a concordance and validation study. Eur J Surg Oncol. 2013;39(12): 1332–6.

Photodynamic Detection of Lymph Node Metastases in Gastrointestinal Cancer by Using 5-Aminolevulinic Acid

28

Takeo Minamikawa, Yoshinori Harada, and Tetsuro Takamatsu

Introduction

Minimally invasive approaches in cancer surgery are now becoming increasingly important to improve postoperative quality of life of recurrence-free long-term survivors. The accurate detection of metastasis is a key component of minimally invasive surgery because accurate staging and appropriate treatments for patients with cancer are based on the extent and/or presence of metastasis. Especially in gastrointestinal cancer, the detection of lymph node (LN) metastasis after lymphadenectomy is generally applied [1–4]. LN metastasis is a common feature of metastasis and serves as one of the most important prognostic indicators for gastrointestinal cancer patients. Accurate and rapid diagnosis of LN metastasis is thus essential for designing therapeutic strategies and assessing outcomes of patients.

Computed tomography (CT) and fluorodeoxyglucose-positron emission tomography (FDG-PET) are widely used for the detection of LN metastasis before surgery. The sensitivities of these methods in the detection of LN metastasis are, however, no more than about 67–77 % percent for CT and 29–85 % for FDG-PET [5–8]. Histopathological diagnosis using frozen sections of LNs is regarded as the gold standard for the detection of the LN metastasis during surgery. LN metastasis is examined with imaging of a few hematoxylin-eosin (HE)-stained sections per LN. Although histopathological diagnosis is the most reliable method for the detection of LN metastasis, the limitation of the number of evaluated sections (generally 2–3 sections with 4-μm thickness) due to the time-consuming nature of the preparation of the stained sections (20–30 min) results in diagnostic errors of small metastases that are sometimes not included in the evaluated sections. A reverse transcriptase-polymerase chain reaction (RT-PCR)-based detection system of LN metastasis was recently developed [9–11]. Marker genes of LN metastasis (e.g., CK19) are detected with the RT-PCR or its alternate methods. The sensitivity of the RT-PCR-based methods is more than 90 %; however, these LNs are lost due to their solubilization for the examination, and thus, postoperative histopathological diagnosis for precise evaluation of disease cannot be applied using solubilized LNs. The lengthy examination time (30–40 min) is also a problematic aspect of the RT-PCR-based detection of LN metastasis. Therefore, a method of detection of LN metastasis in removed whole LN is desirable for ultrarapid and accurate

T. Minamikawa, Ph.D. • Y. Harada, M.D., Ph.D.
T. Takamatsu, M.D., Ph.D. (✉)
Department of Pathology and Cell Regulation, Graduate School of Medical Science, Kyoto Prefectural University of Medicine, 456 Kajii-cho, Kawaramachi-Hirokoji, Kamigyo-ku, Kyoto 602-8566, Japan
e-mail: yoharada@koto.kpu-m.ac.jp; ttakam@koto.kpu-m.ac.jp

F.D. Dip et al. (eds.), *Fluorescence Imaging for Surgeons: Concepts and Applications*,
DOI 10.1007/978-3-319-15678-1_28, © Springer International Publishing Switzerland 2015

diagnosis of LN metastasis. Here, we introduce a novel and rapid method of the detection of LN metastasis by fluorescence imaging with exogenous administration of 5-aminolevulinic acid (5-ALA).

Mechanism of Detection of LN Metastasis by Using 5-ALA

5-ALA is an endogenous amino acid, which is synthesized with glycine and succinyl-CoA by aminolevulinic acid synthase in mitochondria and which is metabolized to heme via intermediates (Fig. 28.1) [12, 13]. In cancer cells, exogenous 5-ALA administration causes selective intracellular accumulation of a fluorescent intermediate protoporphyrin IX (PpIX), due to the altered activity of enzymes, the increase in porphobilinogen deaminase activity, and the reduction in ferrochelatase activity, while PpIX is rapidly metabolized into non-fluorescent heme in normal nucleated cells [14–18]. Figure 28.2 shows the fluorescence spectrum of PpIX. Strong red fluorescence is observed at around 635 nm on blue light excitation. Observation of the fluorescence of PpIX allows selective detection of cancer cells.

Figure 28.3 shows fluorescence images of various colorectal cancer cell lines (HT-29, CACO-2, DLD-1, and HCT-15) and isolated normal LN cells [19]. Primary-culture normal LN cells were negative for 5-ALA-derived PpIX fluorescence at 1 and 3 h; in contrast, all colorectal cancer cell lines tested were positive after the start of incubation. Speckled PpIX-fluorescence signal was detected in the cytoplasm of the colorectal cancer cells.

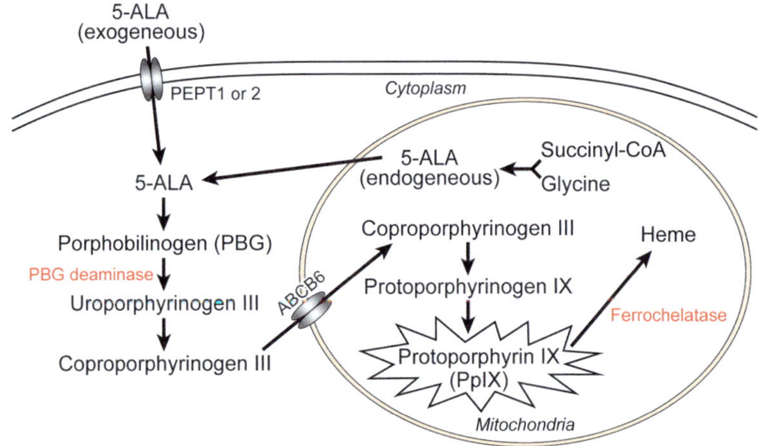

Fig. 28.1 Metabolic pathway of 5-ALA

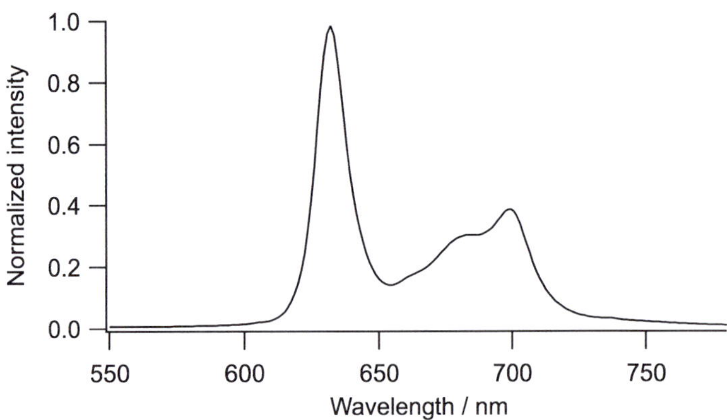

Fig. 28.2 Fluorescence spectrum of PpIX excited with 405 nm

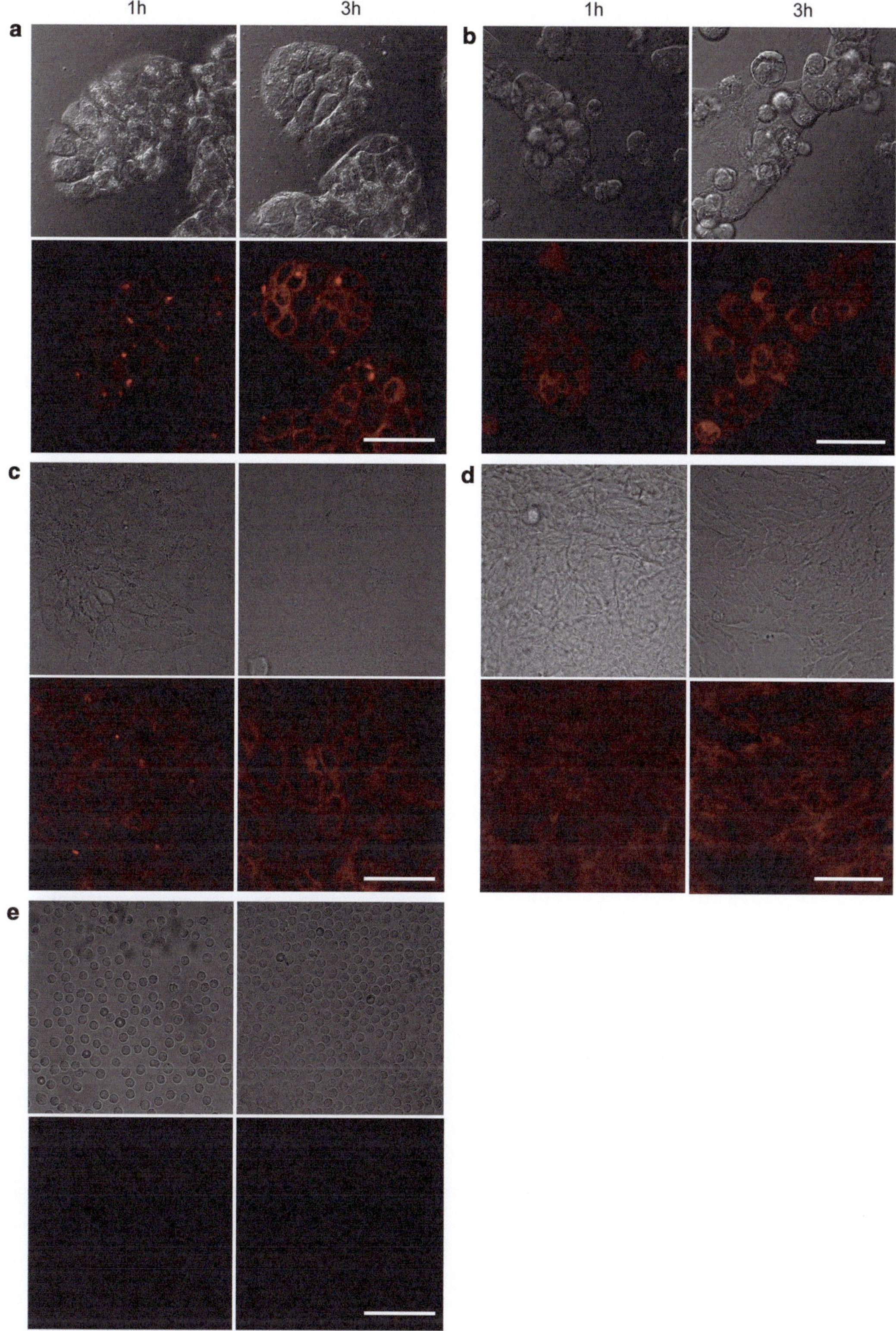

Fig. 28.3 Fluorescence detection of cancer cells by using 5-ALA administration. (**a**) HT-29, (**b**) CACO-2, (**c**) DLD-1, and (**d**) HCT-15 colorectal cancer cells and (**e**) isolated normal LN cells (**e**) were incubated with 15 mM 5-ALA for 30 min. With permission from Murayama Y, Harada Y, Imaizumi K, Dai P, Nakano K, Okamoto K, et al. Precise detection of lymph node metastases in mouse rectal cancer by using 5-aminolevulinic acid. Int J Cancer 2009;15;125(10):2256–63. © 2009 John Wiley and Sons [19]

Since the exogenous administration of the 5-ALA has few side effects in patients without metabolic disease of heme synthesis, such as porphyria, 5-ALA has been widely applied clinically for the detection of primary malignant tumors such as brain tumor, urinary bladder tumor, and gastrointestinal tumor [20–25]. Recently, several studies revealed that exogenous 5-ALA also accumulates in metastatic cancer cells at LNs to an extent sufficient to allow fluorescence detection of PpIX [19, 26–30]. Regarding its application to gastrointestinal cancer, the feasibility of PpIX-based fluorescence detection of LN metastasis was initially investigated by using a mouse model bearing rectal cancer [19], and the first clinical application to patients with gastrointestinal cancer was conducted in gastric cancer patients [27].

Fluorescence Detection of LN Metastasis Using 5-ALA in Xenograft Mouse Model of Gastrointestinal Cancer

The feasibility of fluorescence-based diagnosis of LN metastasis using 5-ALA has been preliminarily evaluated by using a xenograft mouse model of gastrointestinal cancer [19]. Figure 28.4 shows images from a feasibility study of the fluorescence diagnosis of LN metastasis carried out with an orthotopic mouse model of human rectal cancer [19]. This model develops LN metastases, around the abdominal aorta, from the aortic bifurcation to the renal arteries, 5 weeks after the inoculation of the rectal tumor [31].

After injection of 5-ALA via the tail vein, 5-ALA-derived PpIX fluorescence could be detected, as shown in Fig. 28.4b. The PpIX-fluorescence-positive region coincided with metastatic lesions. In contrast, PpIX-fluorescence-negative LNs contained no metastatic lesion. In this model, sufficient PpIX fluorescence was observed from 6 to 9 h after the 5-ALA injection. In general, the presence of metastasis in LNs is often identified based on the shape and hardness of the LNs, i.e., the metastatic LNs are generally enlarged and hardened due to infiltration and proliferation of metastatic cancer cells in the LNs. In Fig. 28.4b, c, although the metastatic lesions were present on the LNs but remained unrecognized on the white-light imaging, they could be clearly and accurately identified by the PpIX-fluorescence diagnostic method.

The PpIX-fluorescence diagnostic method provides high sensitivity and accuracy for the diagnosis of LN metastasis in this model. Figure 28.5 shows scatterplots of the fluorescence intensity ratios of metastatic and nonmetastatic LNs of the orthotopic mouse model of human rectal cancer at 6 and 9 h after 5-ALA injection. The mean intensity ratios were 3.67 ± 1.38 in the metastatic LNs and 0.99 ± 0.15 in the nonmetastatic LNs. The PpIX-fluorescence diagnostic method identified all metastatic and nonmetastatic LNs when the cutoff of the intensity ratio was set at 1.7.

In histological analysis of surgical specimens, simplified examination of HE-stained sections is generally performed, i.e., a few slides obtained from each LN with spacing of a few millimeter are analyzed for rapid diagnosis. Although precise HE staining analysis of whole LNs with narrow spacing of serial sections (<100 μm) provides definitive diagnosis of LN metastasis, the simplified HE staining analysis sometimes yields false-negative result due to the finite number of sections of metastatic LNs [32–34]. The PpIX-fluorescence detection method requires no thin sectioning of LNs and no time-consuming preparation for diagnosis and thus has the potential to allow detection of LN metastasis including small metastatic lesions that might otherwise lead to diagnostic error in the analysis with the simplified HE staining analysis.

Procedure of Fluorescence Detection of LN Metastasis Using 5-ALA in Human Gastrointestinal Cancer Patients

Appropriate Patients

Since 5-ALA is naturally generated in the human body, it is considered to be relatively safe even after exogenous administration. Indeed, 5-ALA is commercially distributed worldwide as a

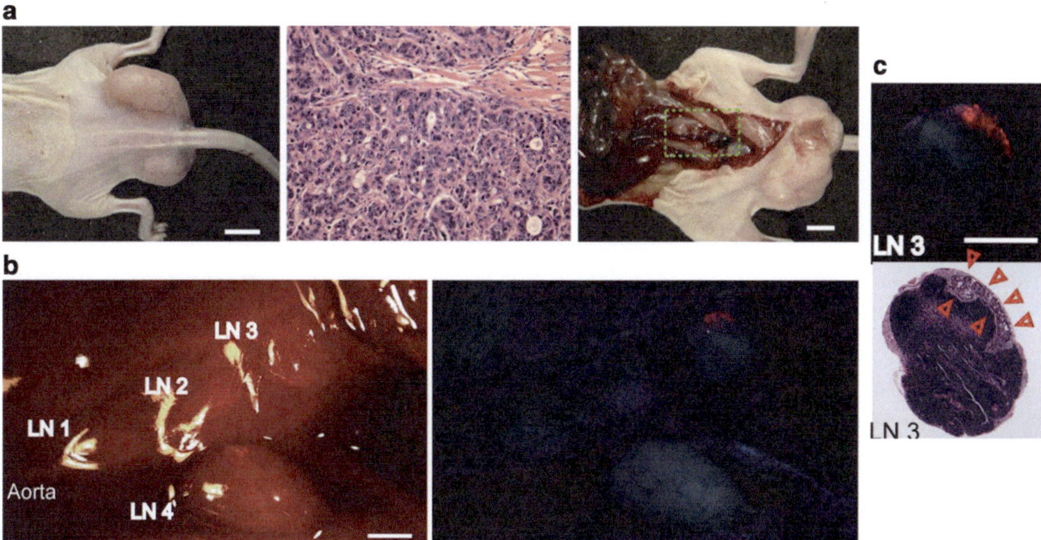

Fig. 28.4 Feasibility study of the fluorescence diagnosis of LN metastasis conducted with an orthotopic mouse model of human rectal cancer. (**a**) White-light image of a BALB/c nude mouse after the injection of human rectal cancer (HT-29) (*left*), image of HE-stained specimen of primary tumor (*center*), and white-light image of abdominal cavity of the nude mouse (*right*). (**b**) PpIX-fluorescence diagnosis of LN metastasis. White-light image of LNs (*left*) and fluorescence image of PpIX (*right*). (**c**) Fluorescence image and image of HE-stained specimen of excised LN3. With permission from Murayama Y, Harada Y, Imaizumi K, Dai P, Nakano K, Okamoto K, et al. Precise detection of lymph node metastases in mouse rectal cancer by using 5-aminolevulinic acid. Int J Cancer 2009;15;125(10):2256–63. © 2009 John Wiley and Sons [19]

medicinal drug for photodynamic diagnosis of bladder cancer and malignant glioma and for photodynamic therapy of actinic keratosis and condyloma acuminatum.

Since 5-ALA is metabolized to heme, its use in patients with metabolic diseases of heme synthesis such as acute or chronic porphyria is inappropriate. Hypersensitivity to 5-ALA or porphyrins is also a major contraindication. The liver and kidneys are two of the main accumulation sites for the excretion of 5-ALA; 5-ALA should be used with caution in patients with renal or hepatic insufficiency.

Administration of 5-ALA

For the detection of LN metastasis in gastrointestinal cancer patients, oral administration of 5-ALA is generally performed [26, 27, 35]. In general, 5-ALA hydrochloride dissolved in water with 50 % glucose is applied orally at a concentration ranging from 15 to 20 mg per kg of body weight. Although the concentration of 5-ALA for oral administration is based on comparison with the detection of primary tumors such as glioma, the optimal administration concentration of 5-ALA for the detection of LN metastasis of various cancers is now under investigation. After the administration of 5-ALA, patients should be protected from strong light such as direct sunlight for 24–48 h to avoid photosensitivity reaction.

Various groups reported that the half-life of 5-ALA in the human body is approximately 45 min, and the maximum concentration of PpIX in blood is reportedly approximately 6–9 h after oral administration of 5-ALA [36–38]. Therefore, in general surgery for gastrointestinal cancer, in which LNs are excised in about 3–5 h after general anesthesia, 5-ALA should be administered to the patients 2–4 h prior to induction of general anesthesia.

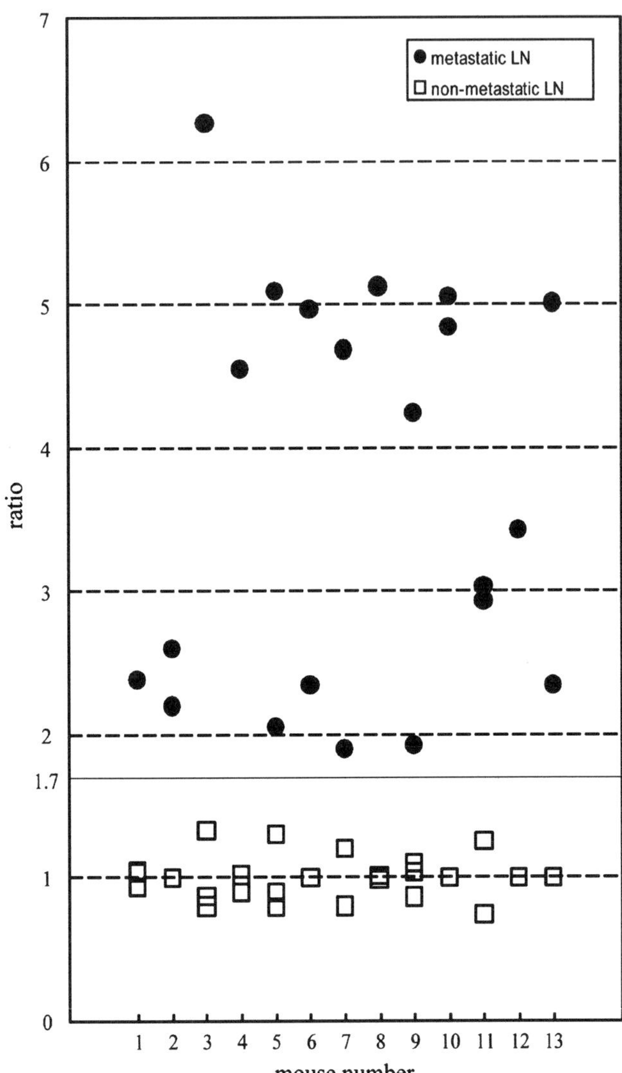

Fig. 28.5 Scatterplots of the fluorescence intensity ratio between excised metastatic and nonmetastatic LNs of mice with rectal cancer 6 and 9 h after administration of 5-ALA. The decision line (1.7) yields sensitivity of 100 % and specificity of 100 % for differentiating metastatic from nonmetastatic LNs. Filled circles and open rectangles denote metastatic LNs and nonmetastatic LNs, respectively. With permission from Murayama Y, Harada Y, Imaizumi K, Dai P, Nakano K, Okamoto K, et al. Precise detection of lymph node metastases in mouse rectal cancer by using 5-aminolevulinic acid. Int J Cancer 2009;15;125(10):2256–63. © 2009 John Wiley and Sons [19]

Side Effects of 5-ALA Administration

The side effects of the administration of 5-ALA have been investigated in patients in several studies [35, 38, 39]. Although substance-specific side effects, such as hypotension, nausea, photosensitivity reaction, and photodermatosis, are reported, these effects are uncommon reactions (less than 1 % incidence). Other side effects such as anemia, thrombocytopenia, leukocytosis, and impairment of liver function have also developed after surgery; however, these seem to be related to procedure involved in the surgery.

Sample Preparation

Tissues including LNs at several sites at which metastatic lymph nodes are suspected to be present are carefully removed during gastrointestinal surgery. The LNs are isolated from the resected tissues and are analyzed immediately, avoiding unnecessary light exposure. For precise analysis of PpIX fluorescence from the LNs, if necessary, the LNs are cut into a few parts with a fine blade due to the limited penetration depth of excitation light. Fluorescence imaging of the cut surface and diagnostic evaluation of LN metastasis are performed. After the PpIX-fluorescence diagnostic evaluation, the LNs are subjected to a postoperative histopathological diagnosis for precise evaluation of disease.

Fluorescence Detection of LN Metastasis Using 5-ALA for Human Gastrointestinal Cancer Patients

Fluorescence Detection of LN Metastasis for Gastric Cancer Patients

In this section, we introduce some applications of fluorescence diagnosis of LN metastasis using 5-ALA in human gastrointestinal cancer patients. Firstly, the detection of LN metastasis of gastric cancer is presented.

Figure 28.6 shows representative images of metastatic and nonmetastatic LNs from gastric cancer patients [27]. 5-ALA hydrochloride at 15 mg per kg of body weight was dissolved in 20 ml of 50 % glucose solution and administered orally to the patients 2 h prior to induction of general anesthesia. Isolated lymph nodes were cut in half longitudinally one by one with a fine blade. Fluorescence images of the cut surface were obtained by using a stereomicroscope that comprised a 405-nm band-pass filter for excitation and a color CCD camera equipped with a 430-nm long-pass filter for emission. In practice, the peak wavelength of PpIX fluorescence is about 635 nm, and the fluorescence from PpIX must be filtered with its peak wavelength to obtain the optimum contrast of the PpIX image. Since the PpIX fluorescence observed in the red channel of the color CCD camera is relatively strong compared to the background autofluorescence, the color CCD camera is widely used for the observation of PpIX fluorescence owing to its convenience for use. Furthermore, blue and green fluorescence predominantly represents the distribution of flavins, collagens, and fats, and the analysis of RGB channels of the color CCD camera provides us data for not only the distribution of PpIX but also the distribution of parenchyma and stroma of LNs.

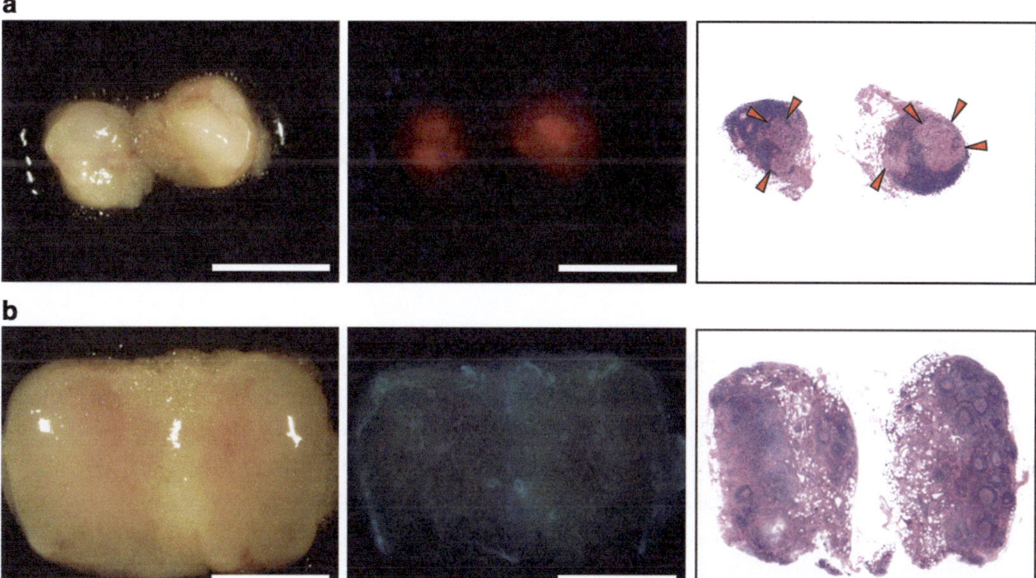

Fig. 28.6 Representative images of (**a**) metastatic and (**b**) nonmetastatic LNs from gastric cancer patients. (*left*) White-light and (*center*) fluorescence images obtained with a color CCD camera and (*right*) image of HE-stained specimens from each lymph node are shown. *Arrowheads* indicate metastatic lesions. Scale bar = 3 mm. With permission from Harada K, Harada Y, Beika M, Koizumi N, Inoue K, Murayama Y, et al. Detection of lymph node metastases in human colorectal cancer by using 5-aminolevulinic acid-induced protoporphyrin IX fluorescence with spectral unmixing. Int J Molecular Sci 2013;14(11):23140–52.© Springer [26]

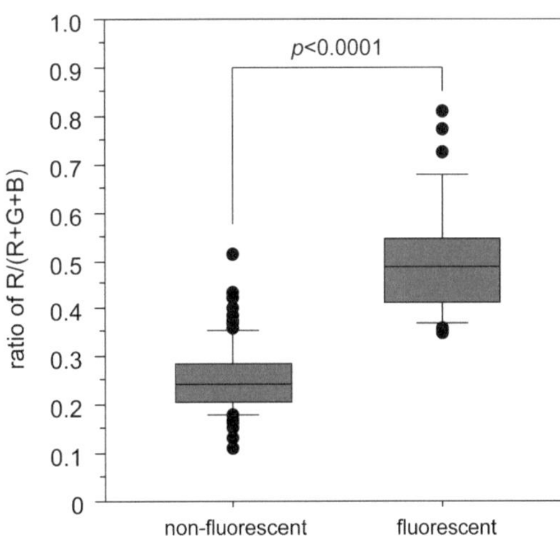

Fig. 28.7 Quantitative analysis of fluorescence intensity of metastatic and nonmetastatic LNs observed in the R channel of the color CCD camera. The fluorescence intensity of the R channel is normalized with total fluorescence of RGB channels to eliminate individual variation of autofluorescence. The normalized fluorescence intensity of the R channel between metastatic lymph nodes and nonmetastatic lymph nodes without follicular accumulation showed significant difference ($p<0.0001$). With permission from Harada K, Harada Y, Beika M, Koizumi N, Inoue K, Murayama Y, et al. Detection of lymph node metastases in human colorectal cancer by using 5-aminolevulinic acid-induced protoporphyrin IX fluorescence with spectral unmixing. Int J Molecular Sci 2013;14(11):23140–52.© Springer [26]

In the metastatic LNs, strong red fluorescence was observed, and the distribution coincided with malignant lesions that were confirmed by HE stain image. In contrast, red fluorescence was weak in nonmetastatic LNs. Results for quantitative analysis of fluorescence intensity of LNs observed on the R channel of the color CCD camera are shown in Fig. 28.7. The fluorescence intensities of the metastatic and nonmetastatic LNs without follicular accumulation are significantly different.

Diagnostic accuracy of the PpIX-fluorescence diagnosis was evaluated with 144 LNs. The PpIX-fluorescence diagnosis was compared with histopathological diagnosis as the gold standard for the detection of LN metastasis. Totals of 17 out of 24 metastatic LNs and 106 out of 120 nonmetastatic LNs were detected by the PpIX-fluorescence diagnostic method with agreement with the histopathological diagnosis. The overall diagnostic accuracy of gross fluorescence inspection was 85.4 % (123/144).

For the PpIX-fluorescence diagnosis, there were false-positive results for 14 nonmetastatic LNs that were deemed to be metastatic and false-negative results for 7 metastatic LNs that were deemed to be nonmetastatic. False-positive results were mainly due to the accumulation of PpIX in normal lymphoid follicles [40, 41]. The normal lymphoid follicles exhibit specific accumulation patterns of PpIX, and the diagnostic specificity and accuracy are, respectively, improved from 88.3 to 96.7 % and from 85.4 to 92.4 % after special consideration of the fluorescence patterns of the malignant lesions and the normal lymphoid follicles. In false-negative results, it is considered that extraneous fluorescence from nonmetastatic regions such as that from collagen, flavins, and fats obscures the PpIX fluorescence from metastatic LNs; in particular, some of the metastatic LNs of patients with scirrhous carcinoma could not be detected owing to strong autofluorescence emitted from increased amount of fibrous tissues. To eliminate the autofluorescence from nonmetastatic regions, detailed analysis of fluorescence spectra of PpIX and nonmetastatic regions is required. One technique of spectral analysis to eliminate autofluorescence of nonmetastatic regions is presented in the next subsection. Individual variation of absorption capacity of 5-ALA may also be in part responsible for false-negative results. Because 5-ALA is mainly absorbed at the upper gastrointestinal tract, patients with diseased stomach may take in 5-ALA of insufficient dose.

Fluorescence Detection of LN Metastasis for Diagnosis of Colorectal Cancer Patients

Autofluorescence from the nonmetastatic region sometimes leads to false-negative results, as discussed above. To improve the diagnostic accuracy of the PpIX-fluorescence-based diagnosis, spectral analysis of PpIX fluorescence is essential. Here, we introduce an application of fluorescence detection of LN metastasis for diagnosis of

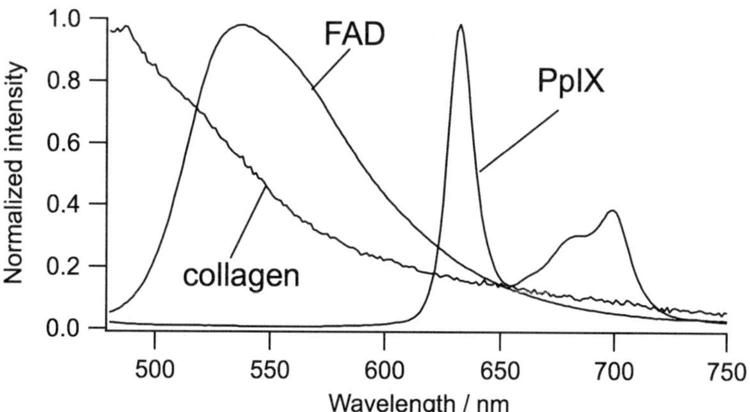

Fig. 28.8 Fluorescence spectra of PpIX, collagen type I, and flavins (FAD, flavin adenine dinucleotide) excited with 405 nm. The spectra were normalized at the peak intensity

colorectal cancer patients employing a spectral analysis of spectral unmixing method [26].

The spectral unmixing method is a useful technique in fluorescence imaging for separating multiple fluorescence signals derived from several materials [42]. The spectral unmixing method assumes that the fluorescence spectra at each pixel in the image can be expressed by linear superposition of fluorescence spectra of tissue species [43]. When two-dimensional spectral images are used for spectral unmixing, the following algorithm is generally applied: two-dimensional spectral images of LNs are obtained with a fluorescence microscope; three-dimensional data set (two dimensions of space and wavelength) is reshaped to a two-dimensional matrix to apply the spectral unmixing algorithm. It is assumed that the two-dimensional data set X is expressed as

$$X = CS,$$

where C and S represent concentration matrix and user-defined spectral data set, respectively. The user-defined spectral data set is composed of such elements as the fluorescence spectra of PpIX, collagen, and flavins (Fig. 28.8). The concentration matrix can be calculated

$$C = XS^t \left(SS^t\right)^{-1}.$$

The calculated concentration matrix composes concentration maps of the user-defined spectra such as PpIX, and therefore, more accurate PpIX-fluorescence diagnosis can be performed without the influence of autofluorescence.

Figure 28.9 shows representative unmixed images of metastatic and nonmetastatic LNs from a colorectal cancer patient. 5-ALA hydrochloride at 15 mg per kg body weight was dissolved in 20 ml of 50 % glucose solution and administered orally to the patients 2 h prior to induction of general anesthesia. Isolated lymph nodes were cut in half longitudinally one by one with a fine blade. Fluorescence images of the cut surface were obtained by using a macrozoom microscope with a 405-nm band-pass filter for excitation and a monochrome CCD camera via a liquid-crystal tunable filter for emission. In the spectral unmixing, fluorescence spectra of pure PpIX and pure collagen type I were used for the user-defined spectra. The distribution of the each chemical was estimated by plotting the score of each chemical in the calculated concentration matrix.

By using the spectral unmixing method, the distributions of PpIX and collagen were clearly visualized in the metastatic LN with the agreement of the histopathological observation, while the nonmetastatic LN showed no apparent spectral unmixed PpIX fluorescence. The diagnostic power of the spectral unmixing method in colorectal cancer patients was estimated by using receiver-operating characteristic (ROC) analysis (Fig. 28.10). The ROC curve of the spectral unmixing method was compared with the conventional method of PpIX-intensity-based diagnosis observed at 640 nm. The area under the curve (AUC), which is an index in evaluating the inherent capacity of the diagnostic methods, was 0.95 for the spectral unmixing

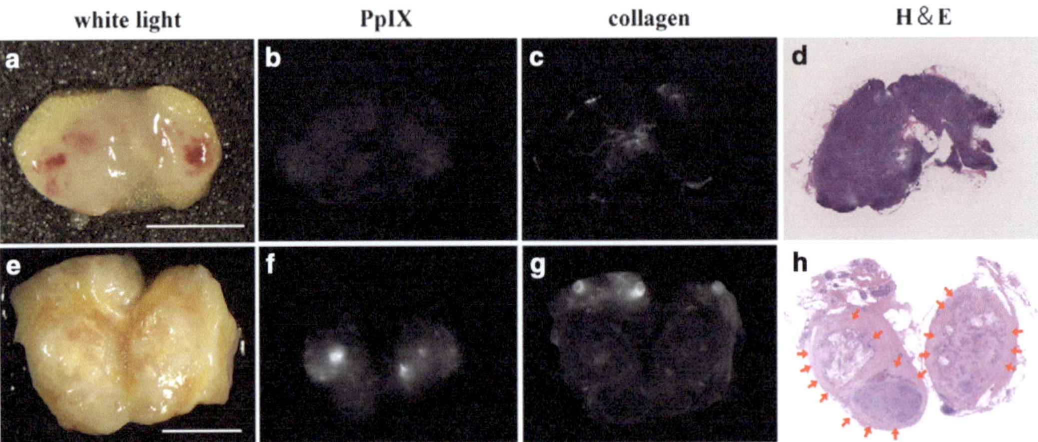

Fig. 28.9 Representative images of (**a**–**d**) nonmetastatic and (**e**–**h**) metastatic LNs of a colorectal cancer patient by using spectral unmixing method. (**a**, **e**) White-light images, (**b**, **f**) spectral unmixed fluorescence images of PpIX, (**c**, **g**) spectral unmixed fluorescence images of collagen, and (**d**, **h**) HE stain images. *Red arrows* indicate metastatic lesions. Scale bar = 5 mm. With permission from Harada K, Harada Y, Beika M, Koizumi N, Inoue K, Murayama Y, et al. Detection of lymph node metastases in human colorectal cancer by using 5-aminolevulinic acid-induced protoporphyrin IX fluorescence with spectral unmixing. Int J Molecular Sci 2013;14(11):23140–52.© Springer [26]

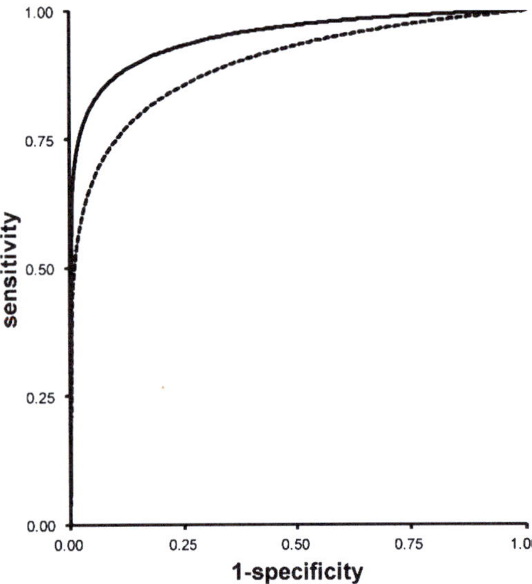

Fig. 28.10 Fitted ROC curves of PpIX-fluorescence diagnosis. (*solid curve*) ROC curve derived by analyzing with the spectral unmixing method. (*dashed line*) ROC curve derived by analyzing with the fluorescence intensity at 640 nm, which includes the peak fluorescence of PpIX (conventional method). With permission from Harada K, Harada Y, Beika M, Koizumi N, Inoue K, Murayama Y, et al. Detection of lymph node metastases in human colorectal cancer by using 5-aminolevulinic acid-induced protoporphyrin IX fluorescence with spectral unmixing. Int J Molecular Sci 2013;14(11):23140–52.© Springer [26]

method and 0.9 for the conventional PpIX-intensity-based diagnostic method. This result indicated that the unfavorable autofluorescence of the nonmetastatic region was eliminated, and the diagnostic accuracy was improved, by using the spectral unmixing.

Conclusion

The fluorescence detection of LN metastases of gastrointestinal cancer by using 5-ALA is now becoming a promising technique in gastrointestinal cancer surgery. The most important advantage of this method is that rapid intraoperative detection can be attained without inducing any serious side effects of 5-ALA and its metabolites. The intraoperative detection of LN metastasis will lead to improved postoperative quality of life of gastrointestinal cancer patients by affording design of minimally invasive treatment. Although it is necessary to examine this method based on larger population studies, this rapid and feasible method has the potential of becoming an indispensable technique in gastrointestinal surgery.

References

1. Kitagawa Y, Kitano S, Kubota T, Kumai K, Otani Y, Saikawa Y, et al. Minimally invasive surgery for gastric cancer–toward a confluence of two major streams: a review. Gastric Cancer. 2005;8(2):103–10. PubMed PMID: 15864717.
2. Koeda K, Nishizuka S, Wakabayashi G. Minimally invasive surgery for gastric cancer: the future standard of care. World J Surg. 2011;35(7):1469–77. PubMed PMID: 21476116.
3. Nicholl M, Bilchik A. Is routine use of sentinel node biopsy justified in colon cancer? Ann Surg Oncol. 2008;15(1):1–3. PMID: 17929100.
4. Cohen AM, Kelsen D, Saltz L, Minsky BD, Nelson H, Farouk R, et al. Adjuvant therapy for colorectal cancer. Curr Probl Surg. 1997;34(8):601–76. PubMed PMID: 9251585.
5. Yun M, Lim JS, Noh SH, Hyung WJ, Cheong JH, Bong JK, et al. Lymph node staging of gastric cancer using (18)F-FDG PET: a comparison study with CT. J Nucl Med. 2005;46(10):1582–8. PubMed PMID: 16204706.
6. Kim S-K, Kang K, Lee J, Kim H, Chang H, Choi J, et al. Assessment of lymph node metastases using 18F-FDG PET in patients with advanced gastric cancer. Eur J Nucl Med Mol Imaging. 2006;33(2):148–55.
7. Mukai K, Ishida Y, Okajima K, Isozaki H, Morimoto T, Nishiyama S. Usefulness of preoperative FDG-PET for detection of gastric cancer. Gastric Cancer. 2006;9(3):192–6.
8. Kim EY, Lee WJ, Choi D, Lee SJ, Choi JY, Kim B-T, et al. The value of PET/CT for preoperative staging of advanced gastric cancer: comparison with contrast-enhanced CT. Eur J Radiol. 2011;79(2):183–8.
9. Mejia A, Waldman SA. Previstage™ GCC test for staging patients with colorectal cancer. Expert Rev Mol Diagn. 2008;8(5):571–8.
10. Tsujimoto M, Nakabayashi K, Yoshidome K, Kaneko T, Iwase T, Akiyama F, et al. One-step nucleic acid amplification for intraoperative detection of lymph node metastasis in breast cancer patients. Clin Cancer Res. 2007;13(16):4807–16. PubMed PMID: 17699859.
11. Le Frere-Belda MA, Bats AS, Gillaizeau F, Poulet B, Clough KB, Nos C, et al. Diagnostic performance of one-step nucleic acid amplification for intraoperative sentinel node metastasis detection in breast cancer patients. Int J Cancer. 2012;130(10):2377–86. PubMed PMID: 21780107.
12. Ishizuka M, Abe F, Sano Y, Takahashi K, Inoue K, Nakajima M, et al. Novel development of 5-aminolevulinic acid (ALA) in cancer diagnoses and therapy. Int Immunopharmacol. 2011;11(3):358–65. PubMed PMID: 21144919.
13. Berlin NI, Neuberger A, Scott JJ. The metabolism of delta-aminolevulinic acid. 2. Normal pathways, studied with the aid of 14C. Biochem J. 1956;64(1):90–100. PubMed PMID: 13363810, Pubmed Central PMCID: 1199694.
14. Van Hillegersberg R, Van den Berg JW, Kort WJ, Terpstra OT, Wilson JH. Selective accumulation of endogenously produced porphyrins in a liver metastasis model in rats. Gastroenterology. 1992;103(2):647–51. PubMed PMID: 1386052.
15. Schoenfeld N, Epstein O, Lahav M, Mamet R, Shaklai M, Atsmon A. The heme biosynthetic pathway in lymphocytes of patients with malignant lymphoproliferative disorders. Cancer Lett. 1988;43(1–2):43–8. PubMed PMID: 3203329.
16. Dailey HA, Smith A. Differential interaction of porphyrins used in photoradiation therapy with ferrochelatase. Biochem J. 1984;223(2):441–5.
17. Kennedy JC, Pottier RH. Endogenous protoporphyrin IX, a clinically useful photosensitizer for photodynamic therapy. J Photochem Photobiol B. 1992; 14(4):275–92. PubMed PMID: 1403373.
18. Peng Q, Warloe T, Berg K, Moan J, Kongshaug M, Giercksky KE, et al. 5-Aminolevulinic acid-based photodynamic therapy. Clinical research and future challenges. Cancer. 1997;79(12):2282–308. PubMed PMID: 9191516.
19. Murayama Y, Harada Y, Imaizumi K, Dai P, Nakano K, Okamoto K, et al. Precise detection of lymph node metastases in mouse rectal cancer by using 5-aminolevulinic acid. Int J Cancer. 2009;125(10):2256–63. PubMed PMID: 19569177.
20. Kriegmair M, Baumgartner R, Knuchel R, Stepp H, Hofstadter F, Hofstetter A. Detection of early bladder cancer by 5-aminolevulinic acid induced porphyrin fluorescence. J Urol. 1996;155(1):105–9. discussion 9–10, PubMed PMID: 7490803.
21. Jichlinski P, Forrer M, Mizeret J, Glanzmann T, Braichotte D, Wagnieres G, et al. Clinical evaluation of a method for detecting superficial surgical transitional cell carcinoma of the bladder by light-induced fluorescence of protoporphyrin IX following the topical application of 5-aminolevulinic acid: preliminary results. Lasers Surg Med. 1997;20(4):402–8. PubMed PMID: 9142679.
22. Stummer W, Stocker S, Wagner S, Stepp H, Fritsch C, Goetz C, et al. Intraoperative detection of malignant gliomas by 5-aminolevulinic acid-induced porphyrin fluorescence. Neurosurgery. 1998;42(3):518–25. discussion 25–6, PubMed PMID: 9526986.
23. Friesen SA, Hjortland GO, Madsen SJ, Hirschberg H, Engebraten O, Nesland JM, et al. 5-Aminolevulinic acid-based photodynamic detection and therapy of brain tumors (review). Int J Oncol. 2002;21(3):577–82. PubMed PMID: 12168102.
24. Mayinger B, Reh H, Hochberger J, Hahn EG. Endoscopic photodynamic diagnosis: oral aminolevulinic acid is a marker of GI cancer and dysplastic lesions. Gastrointest Endosc. 1999;50(2):242–6. PubMed PMID: 10425420.
25. Murayama Y, Ichikawa D, Koizumi N, Komatsu S, Shiozaki A, Kuriu Y, et al. Staging fluorescence

laparoscopy for gastric cancer by using 5-aminolevulinic acid. Anticancer Res. 2012;32(12):5421–7. PubMed PMID: 23225446.

26. Harada K, Harada Y, Beika M, Koizumi N, Inoue K, Murayama Y, et al. Detection of lymph node metastases in human colorectal cancer by using 5-aminolevulinic acid-induced protoporphyrin IX fluorescence with spectral unmixing. Int J Mol Sci. 2013;14(11):23140–52. PubMed PMID: 24284403, Pubmed Central PMCID: 3856110.
27. Koizumi N, Harada Y, Murayama Y, Harada K, Beika M, Yamaoka Y, et al. Detection of metastatic lymph nodes using 5-aminolevulinic acid in patients with gastric cancer. Ann Surg Oncol. 2013;20(11):3541–8. PubMed PMID: 23846777.
28. Kato S, Kawamura J, Kawada K, Hasegawa S, Sakai Y. Fluorescence diagnosis of metastatic lymph nodes using 5-aminolevulinic acid (5-ALA) in a mouse model of colon cancer. J Surg Res. 2012;176(2):430–6. PubMed PMID: 22221602.
29. Takizawa H, Kondo K, Toba H, Kenzaki K, Sakiyama S, Tangoku A. Fluorescence diagnosis of lymph node metastasis of lung cancer in a mouse model. Oncol Rep. 2009;22(1):17–21. PubMed PMID: 19513499.
30. Asai J, Harada Y, Beika M, Takenaka H, Katoh N, Takamatsu T. Photodynamic diagnosis of metastatic lymph nodes using 5-aminolevulinic acid in mouse squamous cell carcinoma. J Dermatol Sci. 2014;74:171.
31. Tsutsumi S, Kuwano H, Morinaga N, Shimura T, Asao T. Animal model of para-aortic lymph node metastasis. Cancer Lett. 2001;169(1):77–85. PubMed PMID: 11410328.
32. Fisher ER, Swamidoss S, Lee CH, Rockette H, Redmond C, Fisher B. Detection and significance of occult axillary node metastases in patients with invasive breast cancer. Cancer. 1978;42(4):2025–31. PubMed PMID: 213191.
33. Cochran AJ, Wen DR, Morton DL. Occult tumor cells in the lymph nodes of patients with pathological stage I malignant melanoma. An immunohistological study. Am J Surg Pathol. 1988;12(8):612–8. PubMed PMID: 3041850.
34. Trojani M, de Mascarel I, Bonichon F, Coindre JM, Delsol G. Micrometastases to axillary lymph nodes from carcinoma of breast: detection by immunohistochemistry and prognostic significance. Br J Cancer. 1987;55(3):303–6. PubMed PMID: 3552017, Pubmed Central PMCID: 2001759.
35. European Assessment Report – Gliolan. Available from: URL: http://www.ema.europa.eu/ema/index.jsp?curl=pages/medicines/human/medicines/000744/human_med_000807.jsp&mid=WC0b01ac05800 1d124
36. Dalton JT, Yates CR, Yin D, Straughn A, Marcus SL, Golub AL, et al. Clinical pharmacokinetics of 5-aminolevulinic acid in healthy volunteers and patients at high risk for recurrent bladder cancer. J Pharmacol Exp Ther. 2002;301(2):507–12. PubMed PMID: 11961050.
37. Rick K, Sroka R, Stepp H, Kriegmair M, Huber RM, Jacob K, et al. Pharmacokinetics of 5-aminolevulinic acid-induced protoporphyrin IX in skin and blood. J Photochem Photobiol B. 1997;40(3):313–9. PubMed PMID: 9372622.
38. Webber J, Kessel D, Fromm D. Side effects and photosensitization of human tissues after aminolevulinic acid. J Surg Res. 1997;68(1):31–7. PubMed PMID: 9126192.
39. Hinnen P, de Rooij FW, Terlouw EM, Edixhoven A, van Dekken H, van Hillegersberg R, et al. Porphyrin biosynthesis in human Barrett's oesophagus and adenocarcinoma after ingestion of 5-aminolaevulinic acid. Br J Cancer. 2000;83(4):539–43. PubMed PMID: 10945504, Pubmed Central PMCID: 2374652.
40. Bjorkman DJ, Samowitz WS, Brigham EJ, Peterson BJ, Straight RC. Fluorescence localization of early colonic cancer in the rat by hematoporphyrin derivative. Lasers Surg Med. 1991;11(3):263–70. PubMed PMID: 1861565.
41. Jones BB, Jessop LD, Samowitz WS, Bjorkman DJ. Computer-assisted fluorescence identification of colon cancer in rats. Am J Gastroenterol. 1993;88(10):1724–8. PubMed PMID: 8213714.
42. Gao X, Cui Y, Levenson RM, Chung LW, Nie S. In vivo cancer targeting and imaging with semiconductor quantum dots. Nat Biotechnol. 2004;22(8):969–76. PubMed PMID: 15258594.
43. Garini Y, Young IT, McNamara G. Spectral imaging: principles and applications. Cytometry A. 2006;69(8):735–47. PubMed PMID: 16969819.

Application of Photodynamic Treatment

29

Tomás G. Núñez and Tamara Portas

Introduction

Lung cancer is the most common cause of cancer death among both men and women. It was estimated that in the U.S. in 2013, there would be 228,190 new cases of lung cancer and 159,480 deaths. Only 15 % of these tumors were diagnosed at a localized stage [1].

It is well known that lung cancer can be cured when diagnosed and resected at its earliest stage; however, many lung cancer patients cannot undergo curative resection because of either poor pulmonary reserve or advanced disease.

Obstruction of the large central airways is not an uncommon complication in these patients, and hence strategies to preserve functional lung tissue and to palliate hemoptysis or dyspnea are of importance. Treating these tumors with a minimally invasive therapeutic modality should be highly beneficial in patients who are often in poor general condition due to a previous radical surgery, aggressive therapies such as chemo or radiotherapy or to comorbidities (cardiovascular, respiratory or metabolic).

A wide range of bronchoscopic therapeutic modalities are available to treat central airway obstruction. These include Nd-YAG laser, cryotherapy, electrocautery, and endobronchial stenting. Because of its selectivity, photodynamic therapy is an important adjunctive modality in the management of lung cancer.

Photodynamic therapy (PDT), which is based on the combined effects of a photosensitizing drug and visible light, has the potential to treat such endobronchial tumors. This therapeutic modality has the distinct advantage that it can be repeated multiple times without inducing any resistance or cumulative toxicity.

PDT can also be utilized in association with other procedures, and it does not preclude the use of eventual more aggressive future treatments in case of failure. Introduction of photodynamic diagnosis (fluorescence and autofluorescence) has improved the available endoscopic imaging techniques that can diagnose and determine

Electronic supplementary material: The online version of this chapter (doi:10.1007/978-3-319-15678-1_29) contains supplementary material, which is available to authorized users. Videos can also be accessed at http://link.springer.com/chapter/10.1007/978-3-319-15678-1_29.

T.G. Núñez, M.D. (✉)
Thoracic Surgery Division, Hospital de Clínicas, Universidad de Buenos Aires, Córdoba 2351 8th Floor, Buenos Aires 1120, Argentina

Membrillar 112 14 "B", Buenos Aires C1406DPD, Argentina
e-mail: tgnunez@yahoo.com

T. Portas
Thoracic Surgery Division, Hospital de Clínicas, Universidad de Buenos Aires, Córdoba 2351 8th Floor, Buenos Aires 1120, Argentina

F.D. Dip et al. (eds.), *Fluorescence Imaging for Surgeons: Concepts and Applications*, DOI 10.1007/978-3-319-15678-1_29, © Springer International Publishing Switzerland 2015

extension of precancerous and early-stage lesions of central airway. In this setting, PDT is used with a high rate of clinical response.

Mechanism of Action

The photodynamic effect is essentially a photochemical oxidative reaction that uses a photosensitizing agent to kill tumor cells. Specific wavelength laser light activates the drug and, in the presence of oxygen, generates highly reactive singlet oxygen species that lead to a cytotoxic reaction [2]. Currently, the most commonly used agent is porfimer sodium (Photofrin II; Sanofi Pharmaceuticals, New York, NY, USA) which is a dihematoporphyrin ether consisting of a complex mixture of porphyrinoids including hematoporphyrin, hydroxyethylvinyldeuteroporphyrin, and a hydrophobic fraction responsible for tumor localization and photosensitization. The half-life is usually 20–30 h and it is cleared from most organ systems within 72 h. However, it is retained for a longer period of time in tumors, liver, spleen, and skin.

The mechanisms involved in the tissue selectivity of dihematoporphyrin ether are unknown, but possible explanations include variations in the route of delivery, binding to lipoproteins, pH changes within tumor stroma and cells, tumor angiogenesis factors, and poor tumor lymphatic drainage [3, 4]. The available photosensitizing drugs and their specific wavelengths of activation are listed in Table 29.1.

Endobronchial Lung Cancer

Photodynamic therapy (PDT) treatment of non-small lung cancer (NSCLC) is primarily employed as an endobronchial therapy to treat patients in basically two settings: (a) advanced disease with central airway obstruction, to palliate symptoms and (b) occult central airway tumors that were detected and evaluated using photodynamic diagnosis (PDD) and other endoscopic diagnostic modalities [5]. Current indications of endobronchial PDT are summarized in Table 29.2. Common contraindications and disadvantages are those of other endobronchial procedures. The specific ones are listed in Table 29.3.

Table 29.1 Photosensitizing agents and specific wavelengths

Agent	Wavelength (in nm)
Porfimer sodium (Photofrin)	630
2-[1-hexyloxyethyl]-2-devinylpyropheophorbide-a (HPPH)	665
Temoporfin (*m*-tetrahydroxyphenylchlorin, Foscan)	652
Talaporfin (mono-(L)-aspartylchlorin-e6, NPe6)	664

Table 29.2 Current PDT options in NSCLC

Localized/early stage (non-metastatic)	Advanced (metastatic/non-surgical)
Occult central airway carcinoma	Palliation of central airway obstruction
Recurrent (metachronous) central tumors after definitive therapy.	Neoadjuvant therapy to improve performance status to convert non-surgical candidates into operable
Reduce extent of surgical resection	Palliation of hemoptysis

PDT = Photodynamic treatment; NSCLC = non-small lung cancer

Table 29.3 Contraindications and disadvantages

Contraindications	Disadvantages
Tumors adjacent to major blood vessel	Skin photosensitivity
Trachco-bronchocsophageal fistula	Cannot be made as emergency procedure

PDT for Palliative Treatment

Endoscopic palliative treatment in advanced lung cancer patients has a very important role. Because of the poor prognosis and often poor performance status, management of these patients is quite complicated and improving their quality of life may require a multidisciplinary approach [6]. PDT has been used in combination with chemo or radiotherapy without increasing the toxicity of

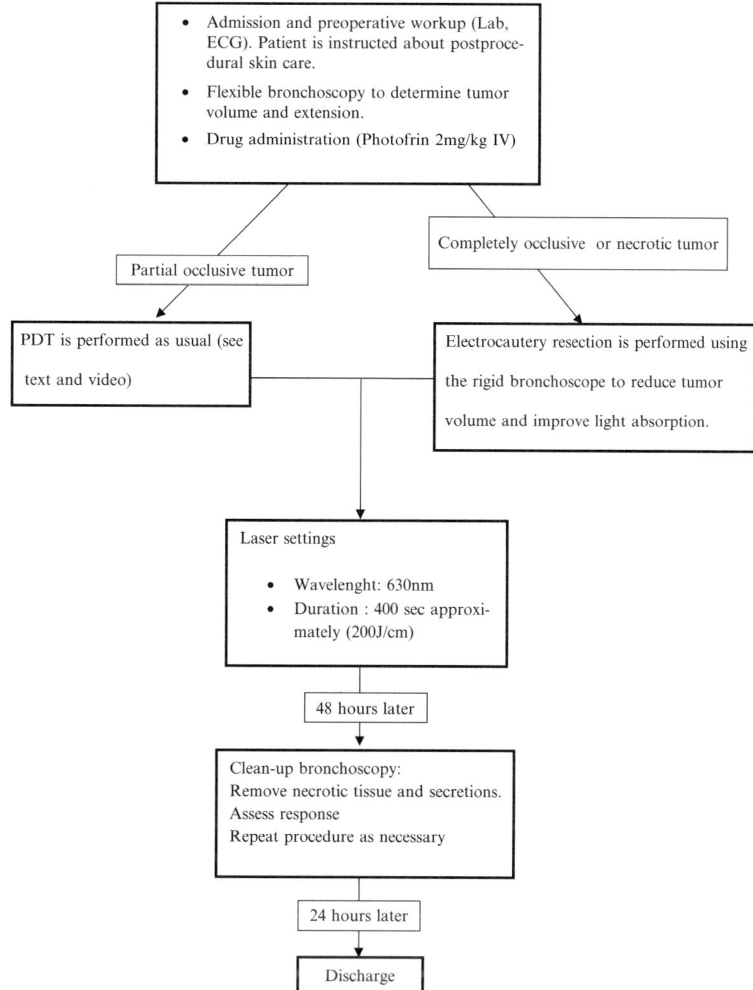

Fig. 29.1 Steps of typical PDT procedure for main stem bronchus obstruction

these therapies to treat airway obstruction and recurrent hemoptysis [7, 8].

The typical PDT procedure can be done on an outpatient basis but we prefer to perform it as an inpatient setting because of the usual need to "clean-up" procedures. As shown in Video 29.1, we follow the procedure guidelines as found elsewhere [9]. The steps are outlined in Fig. 29.1.

We mostly use rigid bronchoscopy under general anesthesia as part of the occluding tumor can be removed using electrocautery and core resection, particularly when necrotic tissue is visible. We did not see any complications due to the immediate PDT after electrofulguration of the gross part of the tumor; even better, this resection may enhance light penetration. The laser fiber is introduced via a steerable bronchoscopic laser guide adjacent or even inside the tumor. After the fiber was properly placed, laser activation of the drug is performed. Operating room dim lights are dimmed to avoid skin lesions. During the procedure, laser fiber movement should be kept to a minimum and be checked frequently by turning off the laser. Usual laser safety measures are used like proper glasses. After the laser dose is completed, debris and secretions are aspired. The patient is then sent to the surgery ward where vital signs are monitored. Supplemental humidified oxygen and mucolytic drugs are provided as usually. In case of difficult clearance of secretions, respiratory physical therapy is started.

Two days after initial treatment a new rigid bronchoscopy is performed and PDT is repeated as needed. Again, we prefer the easier management of necrotic tissue and mucous plugs provided by larger-bore rigid instruments. The day after this second bronchoscopy, the patient is discharged if another clean-up procedure is not needed. Follow-up procedures are scheduled in an individual basis and outcome is assessed by patient interview about symptoms, flow-volume spirometry, and flexible bronchoscopy under local anesthesia. Prompt referral to clinical oncology for base treatment is mandatory.

Special Issues

Life-threatening situations like critical airway obstruction and postobstructive pneumonia cannot be treated by PDT because of its delayed mechanism of action. In these cases we prefer to perform electrocautery resection, start antibiotics, and schedule PDT when the patient has overcome the acute episode. Many reports have been published comparing effectiveness of PDT and Nd-YAG laser therapy for endoluminal airway obstruction [10–12] and their results are contradictory, but I think that PDT and thermal ablation (Nd-YAG, electrocautery) are complemental, as shown in the video. Moreover, in our experience and other reports, electrocautery is more cost-effective and can be used during the same procedure [13], unlike Nd-YAG [9].

PDT for Early-Stage Central Lung Cancer

Since the decade of 1980, a growing interest in treating early-stage carcinomas of the central airways has been developed since the initial use of photodynamic diagnosis (PDD). Early clinical experience with PDD was published by Kato, Cortese and Lam working with hematoporphyrin-induced fluorescence [14, 15]. Since then, several autofluorescence devices (AFB) for observing bronchial mucosa allowed the detection of precancerous lesions as severe dysplasia and preinvasive lesions (carcinoma in situ [CIS]) in patients with sputum cytological atypia. Different techniques of illumination have been developed and all proved better than white light alone in detecting early-stage carcinomas [16–18]. The use of different device across published studies produced equivalent results [19].

Although treatment with curative intention of these lesions has been accomplished successfully in several centers [20, 21], it is not clear that such interventions improved survival among patients with localized epithelial atypia. Moreover, these subjects are seldom found in clinical practice because of the recommendation against screening with sputum cytology [22].

In our experience with AFB, we were not able to detect severe dysplasia and/or CIS in a cohort of high-risk patients selected on a risk-factor evaluation basis (unpublished data, presented at the 2006 World Congress for Bronchology). Long-term follow-up of these patients showed that none developed central lung cancer in a 4-year period.

Newer endoscopic imaging techniques like AFB can be used, if available, as an adjunct to white-light bronchoscopy to evaluate high-risk individual with preinvasive lesions, delineate surgical margins in invasive carcinomas and even identify synchronous lesions in patients being evaluated for surgical resection [23].

Treatment of early lesions needs information about depth of invasion since lasers cannot penetrate beyond the cartilage wall. The use of endobronchial ultrasound (EBUS) may provide information about wall invasions and lymph node involvement. In the case series published by Kato and Furuse [20, 21] curation of preinvasive cancers, most of them squamous cell carcinomas, has been achieved in 80–100 % of patients but some experienced local recurrence.

Conclusions

PDT has been used to treat central airway carcinomas since the decade of 1980 and still seems to be underutilized. It is well tolerated, with low toxicity and side effects particularly when new agents are used.

In advanced-stage central non-small cell lung cancer, PDT may be used to palliate symptoms as obstruction and hemoptysis, alone or in a multimodality treatment, without added toxicity and can be repeated safely.

In surgical candidates, PDT can be used as a neoadjuvant treatment to diminish extent of surgical resection, increase operability and does not compromise future treatment options [24].

In early stage carcinomas has a high rate of clinical response but close follow up is needed. Autofluorescence bronchoscopy, when available, is an important tool to assess extension and response of central lesions. EBUS can determine the depth of invasion and rule out cartilage compromise. Newer endoscopic imaging techniques and photosensitizers may extend the use of this technique.

References

1. American Cancer Society. Cancer facts and figures 2013. Accessed on December 15, 2013. http://www.cancer.org/acs/groups/content/@epidemiologysurveilance/documents/document/acspc-036845.pdf
2. Agostinis P, Berg K, Cengel KA, Foster TH, Girotti AW, Gollnick SO, et al. Photodynamic therapy of cancer: an update. CA Cancer J Clin. 2011;61:250–81.
3. Okunaka T, Kato H, Konaka C, Kawate N, Bonaminio A, Yamamoto H, et al. Photodynamic therapy for multiple primary bronchogenic carcinoma. Cancer. 1991;68:253–8.
4. Pass HI. Photodynamic therapy for lung cancer. Chest Surg Clin N Am. 1991;1:135–51.
5. Kato H. Our experience with photodynamic diagnosis and photodynamic therapy for lung cancer. J Natl Compr Canc Netw. 2012;10 Suppl 2:S3–8.
6. Hopwood P, Stephens RJ. Symptoms at presentation for treatment in patients with lung cancer: implications for the evaluation of palliative treatment. The Medical Research Council (MRC) Lung Cancer Working Party. Br J Cancer. 1995;71:633–6.
7. Hugh-Jones P, Gardner WN. Laser photodynamic therapy for inoperable bronchogenic squamous carcinoma. Q J Med. 1987;64:565–81.
8. McCaughan Jr JS, Hawley PC, LaRosa JC, Thomas JH, Hicks WJ. Photodynamic therapy to control life-threatening hemorrhage from hereditary hemorrhagic telangiectasia. Lasers Surg Med. 1996;19:492–4.
9. Mehrishi S, Ost D. How I do it, Photodynamic therapy. J Bronchol. 2002;9:218–21.
10. Furukawa K, Okunaka T, Yamamoto H, Tsuchida T, Usuda J, Kumasaka H, et al. Effectiveness of photodynamic therapy and Nd-YAG laser treatment of obstructed tracheobronchial malignancies. Diagn Ther Endosc. 1999;5:161–6.
11. Taber SW, Buschemeyer 3rd WC, Fingar VH, Wieman TJ. The treatment of malignant endobronchial obstruction with laser ablation. Surgery. 1999; 126:730–3. discussion 733–5.
12. Diaz-Jiménez JP, Martínez-Ballarín JE, Llunell A, Farrero E, Rodríguez A, Castro MJ. Efficacy and safety of photodynamic therapy versus NdYAG laser resection in NSCLC with airway obstruction. Eur Respir J. 1999;14:800–5.
13. Boxem TV, Muller M, Venmans B, Postmus P, Sutedja T. Nd-YAG laser vs bronchoscopic electrocautery for palliation of symptomatic airway obstruction: a cost-effectiveness study. Chest. 1999;116(4):1108–12.
14. Kato H, Cortese DA. Early detection of lung cancer by means of hematoporphyrin derivative fluorescence and laser photoradiation. Clin Chest Med. 1985;6: 237–53.
15. Lam S, Palcic B, McLean D, Hung J, Korbelik M, Profio AE. Detection of early lung cancer using low dose Photofrin II. Chest. 1990;97:333–7.
16. Sun J, Garfield DH, Lam B, Yan J, Gu A, Shen J, Han B. The value of autofluorescence bronchoscopy combined with white light bronchoscopy compared with white light alone in the diagnosis of intraepithelial neoplasia and invasive lung cancer: a metaanalysis. J Thorac Oncol. 2011;6(8):1336–44.
17. Lam S, MacAulay C, Hung J, LeRiche J, Profio AE, Palcic B. Detection of dysplasia and carcinoma in situ with a lung imaging fluorescence endoscope device. J Thorac Cardiovasc Surg. 1993;105(6):1035–40.
18. Häussinger K, Becker H, Stanzel F, Kreuzer A, Schmidt B, Strausz J, et al. Autofluorescence bronchoscopy with white light bronchoscopy compared with white light bronchoscopy alone for the detection of precancerous lesions: a European randomised controlled multicenter trial. Thorax. 2005;60(6):496–503.
19. Herth FJ, Ernst A, Becker HD. Autofluorescence bronchoscopy–a comparison of two systems (LIFE and D-Light). Respiration. 2003;70(4):395–8.
20. Furuse K, Fukuoka M, Kato H, Horai T, Kubota K, Kodama N, Kusunoki Y, Takifuji N, Okunaka T, Konaka C, et al. A prospective phase II study on photodynamic therapy with Photofrin II for centrally located early-stage lung cancer. The Japan Lung Cancer Photodynamic Therapy Study Group. J Clin Oncol. 1993;11:1852–7.
21. Kato H, Furukawa K, Sato M, Okunaka T, Kusunoki Y, Kawahara M, et al. Phase II clinical study of photodynamic therapy using mono-L-aspartyl chlorin e6 and diode laser for early superficial squamous cell carcinoma of the lung. Lung Cancer. 2003;42:103–11.
22. Wisnivesky JP, Yung RC, Mathur PN, Zulueta JJ. Diagnosis and treatment of bronchial intraepithelial neoplasia and early lung cancer of the central airways: diagnosis and management of lung cancer, 3rd

ed.: American College of Chest Physicians evidence-based clinical practice guidelines. Chest. 2013;143(5 Suppl):e263S–77.

23. Zaric B, Becker HD, Perin B, Jovelic A, Stojanovic G, Ilic MD, Eri Z, et al. Narrow band imaging videobronchoscopy improves assessment of lung cancer extension and influences therapeutic strategy. Jpn J Clin Oncol. 2009;39(10):657–63.
24. Simone 2nd CB, Friedberg JS, Glatstein E, Stevenson JP, Sterman DH, Hahn SM, Cengel KA. Photodynamic therapy for the treatment of non-small cell lung cancer. J Thorac Dis. 2012;4(1):63–75.

30 Photodynamic Diagnosis Mediated by 5-Aminolevulinic Acid for Urinary Bladder Cancer

Keiji Inoue, Hideo Fukuhara, and Taro Shuin

Urinary Bladder Cancer

Urinary bladder cancer is a common disease worldwide with more than 70,000 and 150,000 new cases diagnosed each year in the United States and Europe, respectively [1, 2]. In Japan, about 18,000 new cases are diagnosed, and the number of patients has recently been slowly increasing [3]. Bladder cancer is established risk factors include tobacco smoking, occupational exposure to aromatic amine carcinogens and Schistosoma haematobium infection [4, 5].

The standard treatment for non-muscle invasive bladder cancer (NMIBC) is the transurethral resection of bladder tumor (TURBT). More than 50,000 TURBTs for bladder cancer are performed each year in Japan [3]. TURBT facilitates conservation of the urinary bladder and a favorable prognosis. However, the incidence of postoperative intravesical recurrence is high, being a clinically serious problem. Multicentric development and intravesical dissemination are considered to be the causes of this postoperative intravesical recurrence, but, in addition, residual lesions endoscopically invisible during surgery, such as tiny lesions and flat lesions including dysplasia and carcinoma in situ (CIS), as well as concomitant flat lesions with raised tumor, are closely involved. The identification of these invisible lesions is essential to improve the diagnostic accuracy and therapeutic outcomes of NMIBC. Especially, CIS of urinary bladder is not only endoscopically invisible but also markedly atypical, and the prognosis is extremely poor. Therefore, photodynamic diagnosis (PDD) mediated by new photosensitive substance, 5-aminolevulinic acid (ALA) developed by Malik et al. in 1987 [6], has recently attracted attention as a new strategy for detection of these endoscopic invisible lesions in urinary bladder cancer.

Photodynamic Diagnosis (PDD) Mediated by 5-Aminolevulinic Acid (ALA) (ALA-PDD)

ALA has recently been attracting attention as a third-generation photosensitizer less toxic and more cancer-specific than previous photosensitive substances, such as first-generation hematoporphyrin mixtures and second-generation porphyrin derivatives. It is rapidly metabolized and excreted by normal cells, causing fewer adverse events and requiring no protection from

Electronic supplementary material: The online version of this chapter (doi:10.1007/978-3-319-15678-1_30) contains supplementary material, which is available to authorized users. Videos can also be accessed at http://link.springer.com/chapter/10.1007/978-3-319-15678-1_30.

K. Inoue, M.D., Ph.D. (✉) • H. Fukuhara, M.D., Ph.D.
T. Shuin, M.D., Ph.D.
Department of Urology, Kochi Medical School, Kohasu, Oko, Nankoku, Kochi 783-8505, Japan
e-mail: keiji@kochi-u.ac.jp; tsutomun@kochi-u.ac.jp

F.D. Dip et al. (eds.), *Fluorescence Imaging for Surgeons: Concepts and Applications*,
DOI 10.1007/978-3-319-15678-1_30, © Springer International Publishing Switzerland 2015

light after administration. Accordingly, the improvement of diagnostic and therapeutic accuracies is expected.

ALA is a natural amino acid that has existed in animals and plants for 3.6 billion years, and is a common precursor of hemoglobin and chlorophyll. ALA is an endogenous porphyrin synthesized from succinyl CoA and glycine, and is converted to a photoactive substance, protoporphyrin IX (PpIX), in mitochondria through several precursors in the cytoplasm. Intracellular iron is then inserted into PpIX by a catalyst, ferrochelatase, and metabolized to heme and bilirubin. Exogenously administered ALA is rapidly incorporated into the cytoplasm and biosynthesized into PpIX in mitochondria, similarly to endogenous ALA, but the negative feedback mechanism acts depending on the heme synthesis level in the metabolic process of PpIX in normal cells, and PpIX synthesis is the rate-limiting step. In contrast, in cancer cells, elevations of the activity levels of ATP-binding cassette (ABC) transporter and converting enzymes promote PpIX biosynthesis. Moreover, PpIX metabolism is inhibited due to enhanced transferrin receptor activity and reduced ferrochelatase activity, resulting in the marked progression of PpIX biosynthesis and accumulation. Actually, PpIX excessively accumulates in various cancer cells, and tumor selectivity is particularly high in the urinary tract epithelium [7]. Therefore, urothelial cancer (UC) of urinary bladder is a good indication for ALA-mediated PDD (ALA-PDD). These phenomena involved in PpIX biosynthesis and metabolism are considered to be due to the common biological characteristic of cancers: preference for anaerobic metabolism of various abnormal enzyme activities and cancers (Warburg effect). Since PpIX possesses photoactivity, it emits red fluorescence when it is excited by light irradiation at a specific wavelength, mainly visible blue light (375–445 nm), and cancer cells can be accurately identified by detecting the fluorescence. This is the mechanism of ALA-PDD. Therefore, ALA-PDD is the photodynamic technology based on the common biological characteristic of cancers, and is expected as a new diagnostic strategy for many cancers [8–10], especially the UC of urinary bladder [11–13].

An exclusive video camera system (TRICAM SL II 3CCD camera controller unit (CCU)/ TRICAM PDD 3CCD camera head (CH)), light source (D-Light C System (300 W xenon arc lamp)), and telescope (HOPKINS II PDD telescope 4 mm, 30°) of KARL STORZ GmbH&Co (Tuttlingen, Germany) are universally used for the ALA-PDD system. The light source system employs xenon light (output: 300 W), and the video camera system is equipped with a built-in blue light-penetrable, fluorescence-exciting band pass filter and long pass filter which blocks blue light and allows fluorescence observation. These filters enable instantaneous selection and switching of conventional white light and fluorescence-exciting blue light, readily facilitating observation of the lesions under the two light sources, and this is the most significant characteristic of this system. Recently, SPIES™ (Storz Professional Image Enhancement System) is a hot topic these days. It enables endoscopists to better identify important details and structures inside the body by modifying color contrast, including ALA-PDD. Compared to the current video system for PDD TRICAM (standard definition, PAL/NTSC, resolution ≧750 interlaced scan video signal, KARL STORZ Endoskope, Tuttlingen, Germany), the Image1 SPIES Camera System (Full HD resolution 1,080 × 1,920, progressive scan, KARL STORZ Endoskope, Tuttlingen, Germany) offers several image enhancement functions and is superior in brightness and resolution. Though there are not clinical studies yet, the sensitivity for red-colored fluorescence in PDD mode is increased and so a relative reduction in the false-positive rate is anticipated. Moreover, the improvement of visibility and clarity as well as the non-existence of delay in video image makes it even possible to resect tumors in PDD blue light autofluorescent mode. By using blue excitation light, the reflectance of hemoglobin is much less than under white light, which results in a better visibility especially in cases of turbid perfusion for irrigation due to inter-bladder bleedings. In view of this, PDD-TURB procedures so become safer and more reliable compared to conventional systems. Furthermore, the same camera head (IMAGE 1H3-Z FI SPIES) is also used for the indocyanine green fluorescence laparoscopy and so a highly utile piece of equipment.

The first clinical trial of PDD for bladder cancer adopting intravesical ALA administration was performed by Kriegmair et al. in 1994 [7]. Many clinical studies were performed thereafter, mainly in Europe, and it has been approved as medical care in 2005 in Europe but 2010 in the USA. In Japan, the first ALA-PDD for bladder cancer was performed in a clinical study in 2004 [11], and its safety and usefulness were confirmed [12, 13]. The advanced medical care in registered facilities permitted by the Ministry of Health, Labour and Welfare and also the investigator-initiated clinical trials supported by the Japan Medical Association Center for this application have already been finished in Japan; now, ALA is waiting for approval as a photosensitizer to diagnose urinary bladder cancer.

Clinical Results of ALA-PDD for Bladder Cancer

Diagnostic Accuracy of PDD

As described above, the fluorescent substance excessively accumulated specifically in cancer is excited by visible blue light irradiation, and cancer cells emitting red fluorescence are identified in PDD. ALA-PDD for bladder cancer was useful in detecting endoscopic invisible lesions such as tiny lesions (see Fig. 30.1a, b), flat lesions including dysplasia and CIS (see Fig. 30.2a, b), and concomitant flat lesions with raised tumor (see Fig. 30.3a, b and Video 30.1), avoiding an incorrect decision on adjuvant therapy. The first clinical study of ALA-PDD was reported by Kriegmair et al. [7], in which the diagnostic accuracy in 68 bladder cancer patients was 100.0 % sensitivity and 68.5 % specificity. Many clinical studies were performed mainly in Europe and United States, thereafter many clinical cases have been reported. In our clinical experience [13], the usefulness of PDD for the diagnosis of flat lesions was then shown in accumulated cases: 93.4 % sensitivity and 58.9 % specificity in 210 bladder cancer patients, and 72.1 % of flat lesions of CIS and dysplasia were detected only by PDD. Recently, meta-analysis and systematic review of ALA-PDD were published [14–17]. They showed higher sensitivity 92 % (versus WLE: 71 %) and lower specificity 57 % (versus WLE: 72 %), and higher additional detection rate in the diagnostic trials of ALA-PDD (see Table 30.1) [14–17].

Currently, ALA and ALA derivatives, such as an oral agent for brain tumors, Gliolan®, and an agent for intravesical administration, Hexvix®/Cysview®, have been approved as pharmaceuticals worldwide. Hexvix®/Cysview® was developed to increase the intratumor PpIX accumulation rate by converting water-soluble ALA to a fat-soluble ester compound as an intravesical

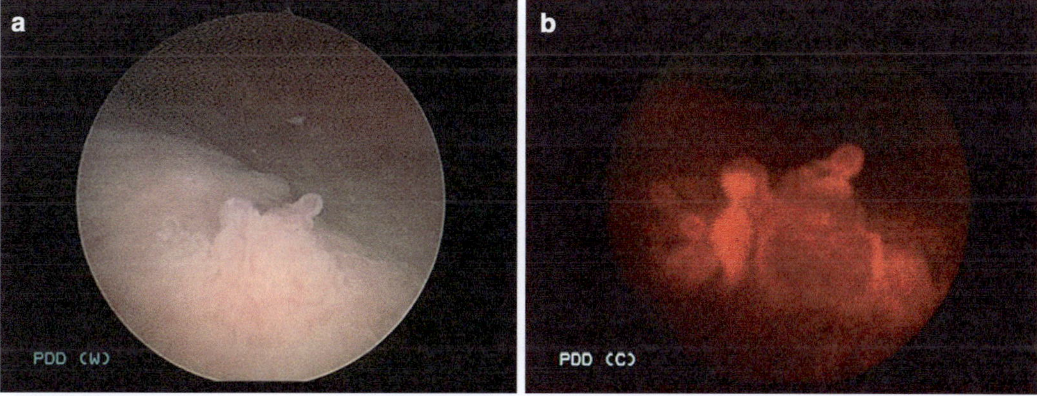

Fig. 30.1 (**a**) It is difficult to detect tiny lesion by conventional white light cystoscopy. (**b**) PDD cystoscopy prevents from overlooking of tiny lesion

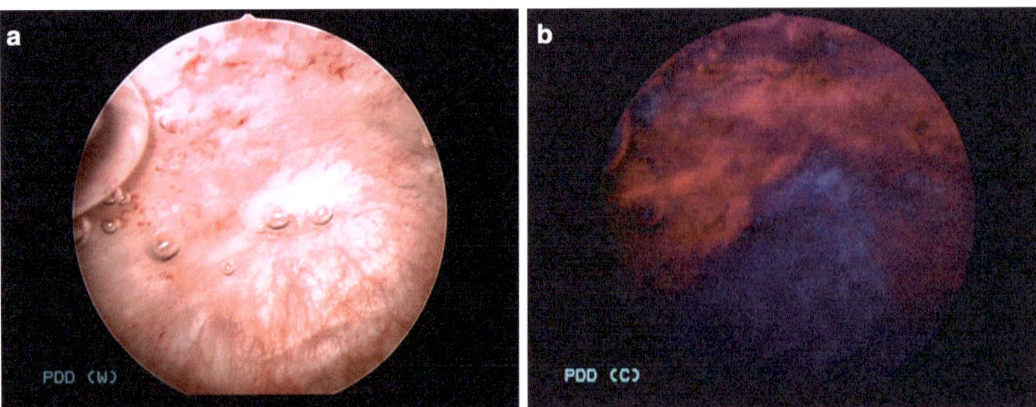

Fig. 30.2 (**a**) It is difficult to detect flat lesion by conventional white light cystoscopy. (**b**) PDD cystoscopy can adequately detect flat lesion such as dysplasia and carcinoma in situ

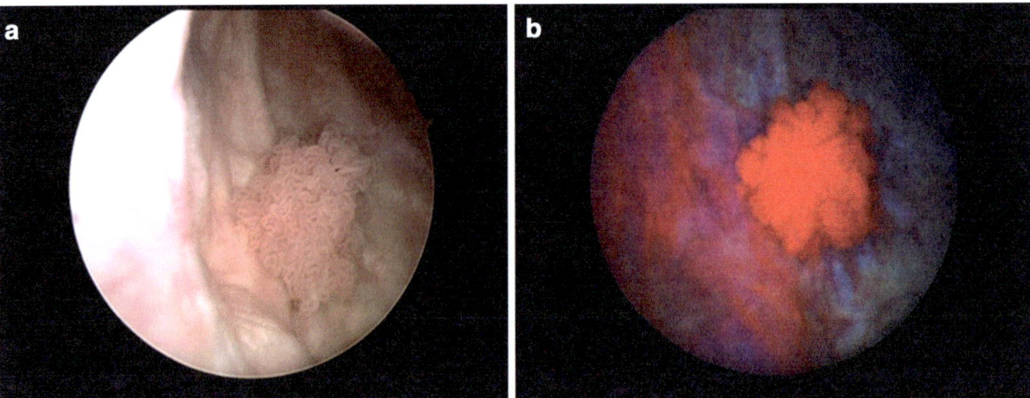

Fig. 30.3 (**a**) It is difficult to detect correct surgical margin by conventional white light cystoscopy. (**b**) PDD cystoscopy can confirm concomitant flat lesion with raised tumor

Table 30.1 Meta-analysis and systematic review of ALA-PDD

Author/year of publication (reference number)	Evidence synthesis and results			
	Diagnostic trials		Randomized trials	
		Number of assessed studies		Number of assessed studies
Kausch I, et al. 2010 [13]	Higher additional detection rate (all tumors: 20 %, CIS: 39 %)	12 studies	Less residual tumor (odds ratio: 0.28) and higher recurrent free survival (12–24 months)	5 studies
Mowatt G, et al. 2011 [14]	Higher sensitivity 92 % (versus WLE: 71 %) and lower specificity 57 % (versus WLE: 72 %)	27 studies	Less residual tumor (relative risk: 0.37) and longer recurrent-free survival (relative risk: 1.37)	4 studies
Burger M, et al. 2013 [15]	Higher additional detection rate (Ta: 14.7 %, CIS: 40.8 %)	10 studies	Higher recurrent-free survival (–12 months)	3 studies
Rink M, et al. 2013 [16]	Higher additional detection rate (Ta, T1: 7–29 %, CIS: 25–30 %)	32 studies	Lower residual tumor rate (20 %) and improve recurrent-free survival but not progression-free survival	13 studies

CIS carcinoma in situ, *WLC* white light examination

agent, and many favorable outcomes have been reported. However, its application cannot be expanded to systemic administration because it is a hexyl ester hydrochloride exhibiting strong cytotoxicity.

Transurethral Resection of a Bladder Tumor (TURBT) Navigated by ALA-PDD Fluorescence

ALA-PDD is not only for diagnosis, but also plays an assistive role for surgical resection by identifying the tumor range. As described above, the standard treatment for NMIBC is TURBT. In TURBT, additional resection under navigation of ALA-PDD fluorescence can avoid insufficient resection of tumors (see Video 30.2), reducing the rate of postoperative intravesical recurrence.

Regarding the recurrence-free survival rate, Denzinger et al. [18–20] compared the intravesical recurrence rate after TURBT over a long-term follow-up for a maximum of 8 years involving 191 bladder cancer patients. The recurrence-free survival rates after PDD-TURBT were significantly and consistently higher than that after conventional TURBT. They also performed multivariate analysis employing the Cox proportional-hazards model, in which the hazard ratio of PDD-TURBT was 0.29 (95 % confidence interval: 0.15–0.56, $p=0.0002$), showing that PDD-TURBT was an independent prognosis-improving factor related to intravesical recurrence. In our clinical experience [13], the recurrence-free survival rate over a median duration of follow-up of 22 months (0.2–68.7 months) in 99 non-muscle invasive bladder cancer patients was 86.9 % after 12 months, 74.7 % after 24 months, 69.7 % after 36 months, 67.7 % after 48 months, and 66.7 % after 60 months, showing that the rate was significantly increased compared to that after conventional white light TURBT. We also performed multivariate analysis employing the Cox proportional-hazards model. The hazard ratio of PDD-TURBT was 0.578 (95 % confidence interval: 0.371–0.888, $p=0.012$), showing that PDD-TURBT was an independent prognosis-improving factor related to intravesical recurrence. Moreover, meta-analysis and systematic review showed less residual tumor and improve recurrent-free survival but not progression-free survival in randomized trials of ALA-PDD (see Table 30.1) [14–17].

Adverse Events

In ALA-PDD, adverse events, such as phototoxic reactions, mainly photosensitive dermatitis, and liver disorder, induced by previous photosensitive substances, such as hematoporphyrin mixtures and porphyrin derivatives, is of the most serious concern. However, as described above, ALA is a natural amino acid contained in animals and plants, and it is less toxic, highly cancer-specific, and rapidly metabolized and excreted by normal cells. When ALA was systemically administered orally, the administered ALA was mostly eliminated from plasma after 6 h and excreted from the body after 24 h [21]; whereas, when ALA was administered into the bladder, the maximum plasma level of the predisposing cause of ALA-induced adverse events, protoporphyrin IX (PpIX), about 30 min after intravesical administration was about 1/100 of that after oral administration, and the half-life was also short (about 54 min) [22]. It is pharmacologically impossible for these conditions to cause phototoxic reactions, mainly photosensitive diseases, and liver disorder.

Actually, no photosensitive dermatitis developed, requiring no protection from light, in not only these pharmacological verifications but also an intervention study reported by Filbeck et al. [23], in which patients treated with intravesically applied ALA underwent experimental ultraviolet light irradiation, and a study involving the largest number of bladder cancer patients diagnosed by PDD using ALA (1,713 applications in 875 patients). In our clinical experience involving the 210 bladder cancer patients [13], although no special precaution, such as liver support and light shielding, was implemented throughout PDD, bladder irritation symptoms, such as pollakiuria and urgency, were observed after intravesical

ALA administration in 17.3 %, photosensitive diseases after oral ALA administration in 4.4 %, elevation of blood liver enzyme levels [Aspartate Aminotransferase (AST), Alanine Aminotransferase (ALT)] in 3.0 %, and nausea/vomiting in 3.0 %, but all of these events were mild and transient, and no serious adverse event occurred.

Conclusion

ALA-PDD is a clearly well-established navigation system of diagnosis for urinary bladder cancer, particularly CIS. Two major academic societies, European Association of Urology [24, 25] and American Urological Association [26], already showed the specific indications and also the general recommendations of ALA-PDD for urinary bladder cancer in their guidelines. In also Japan, it is hoped the establishment of fluorescence-guided navigation system based on ALA-PDD in urinary bladder cancer.

Moreover, photodynamic technology employing ALA is based on the common biological characteristic of cancers, which may be clinically applicable for not only bladder cancer [7, 11–20, 24–26], but also various other forms of cancer, such as prostate cancer [8, 9], renal cancer [10], brain tumors [27], pharyngo-laryngeal cancer [28], gastric cancer [29–31], colon cancer [32], rectal cancer [32], lung cancer [33], pleural malignancies [34], uterine cervical cancer [35], skin cancer [36], and spinal meningioma [37].

References

1. Siegel R, Naishadham D, Jemal A. Cancer statistics, 2013. CA Cancer J Clin. 2013;63(1):11–30.
2. Ferlay J, Steliarova-Foucher E, Lortet-Tieulent J, Rosso S, Coebergh JW, Comber H, et al. Cancer incidence and mortality patterns in Europe: estimates for 40 countries in 2012. Eur J Cancer. 2013;49:1374–403.
3. Matsuda A, Matsuda T, Shibata A, Katanoda K, Sobue T, Nishimoto H, et al. Cancer incidence and incidence rates in Japan in 2007: a study of 21 population-based cancer registries for the monitoring of cancer incidence in Japan (MCIJ) project. Jpn J Clin Oncol. 2013;43:328–36.
4. Cogliano VJ, Baan R, Straif K, Grosse Y, Lauby-Secretan B, El Ghissassi F, et al. Preventable exposures associated with human cancers. J Natl Cancer Inst. 2011;103:1827–39.
5. Ferrís J, Berbel O, Alonso-López J, Garcia J, Ortega JA. Environmental non-occupational risk factors associated with bladder cancer. Actas Urol Esp. 2013;37:579–86.
6. Malik Z, Lugaci H. Destruction of erythroleukaemic cells by photoactivation of endogenous porphyrins. Br J Cancer. 1987;56:589–95.
7. Kriegmair M, Baumgartner R, Knuechel R, Steinbach P, Ehsan A, Lumper W, et al. Fluorescence photodetection of neoplastic urothelial lesions following intravesical instillation of 5-aminolevulinic acid. Urology. 1994;44:836–41.
8. Fukuhara H, Inoue K, Kuno T, Kamei M, Shimamoto T, Fukata S, et al. Photodynamic diagnosis of positive margin during radical prostatectomy: preliminary experience with 5-aminolevulinic acid. Int J Urol. 2011;18:585–91.
9. Inoue K, Ashida S, Fukuhara H, Iiyama T, Miyamura M, Kurabayashi K, et al. Application of 5-aminolevulinic acid (ALA) mediated photodynamic diagnosis (PDD) to robot-assisted laparoscopic radical prostatectomy (RALP). Urology. 2013;82:1175–8.
10. Hoda MR, Popken G. Surgical outcomes of fluorescence-guided laparoscopic partial nephrectomy using 5-aminolevulinic acid-induced protoporphyrin IX. J Surg Res. 2009;154:220–5.
11. Inoue K, Karashima T, Kamada M, Kurabayashi A, Ohtsuki Y, Shuin T. Clinical experience with intravesical instillations of 5-aminolevulinic acid (5-ALA) for the photodynamic diagnosis using fluorescence cystoscopy for bladder cancer. Nippon Hinyokika Gakkai Zasshi. 2006;97:719–29.
12. Inoue K, Kuno T, Fukuhara H, Hamaguchi T, Fukata S, Karashima T, et al. Clinical experience with transurethral resection of bladder tumor (TUR-Bt) guided by photodynamic diagnosis (PDD). Nippon Hinyokika Gakkai Zasshi. 2009;100:661–70.
13. Inoue K, Fukuhara H, Shimamoto T, Kamada M, Iiyama T, Miyamura M, et al. Comparison between intravesical and oral administration of 5-aminolevulinic acid in the clinical benefit of photodynamic diagnosis for non-muscle invasive bladder cancer. Cancer. 2012;118:1062–74.
14. Kausch I, Sommerauer M, Montorsi F, Stenzl A, Jacqmin D, Jichlinski P, et al. Photodynamic diagnosis in non-muscle-invasive bladder cancer: a systematic review and cumulative analysis of prospective studies. Eur Urol. 2010;57:595–606.
15. Mowatt G, N'Dow J, Vale L, Nabi G, Boachie C, Cook JA, et al. Photodynamic diagnosis of bladder cancer compared with white light cystoscopy: systematic review and meta-analysis. Int J Technol Assess Health Care. 2011;27:3–10.
16. Burger M, Grossman HB, Droller M, Schmidbauer J, Hermann G, Drăgoescu O, et al. Photodynamic diagnosis of non-muscle-invasive bladder cancer with hexaminolevulinate cystoscopy: a meta-analysis of detection and recurrence based on raw data. Eur Urol. 2013;64:846–54.

17. Rink M, Babjuk M, Catto JW, Jichlinski P, Shariat SF, Stenzl A, et al. Hexyl aminolevulinate-guided fluorescence cystoscopy in the diagnosis and follow-up of patients with non-muscle-invasive bladder cancer: a critical review of the current literature. Eur Urol. 2013;64:624–38.
18. Zaak D, Hungerhuber E, Schneede P, Stepp H, Frimberger D, Corvin S, et al. Role of 5-aminolevulinic acid in the detection of urothelial premalignant lesions. Cancer. 2002;95:1234–8.
19. Hungerhuber E, Stepp H, Kriegmair M, Stief C, Hofstetter A, Hartmann A, et al. Seven years' experience with 5-aminolevulinic acid in detection of transitional cell carcinoma of the bladder. Urology. 2007;69(2):260–4.
20. Denzinger S, Burger M, Walter B, Knuechel R, Roessler W, Wieland WF, et al. Clinically relevant reduction in risk of recurrence of superficial bladder cancer using 5-aminolevulinic acidinduced fluorescence diagnosis: 8-year results of prospective randomized study. Urology. 2007;69:675–9.
21. Rick K, Sroka R, Stepp H, Kriegmair M, Huber RM, Jacob K, et al. Pharmacokinetics of 5-aminolevulinic acid-induced protoporphyrin IX in skin and blood. J Photochem Photobiol B. 1997;40:313–9.
22. Popken G, Schultze-Seemann W, Seiler KU, Birkel M, Wetterauer U. Intravesical administration of 5-aminolevulinic acid (5-ALA). Safety and pharmacokinetics of 5-ALA and its metabolite protoporphyrin IX. Eur J Clin Pharmacol. 2000;56:241–6.
23. Filbeck T, Wimmershoff MB, Pichlmeier U, Karrer S, Wieland WF, Szeimies RM, et al. No generalized skin phototoxicity after intravesical application of 5-aminolevulinic acid for fluorescence diagnosis of superficial bladder cancer. Urol Int. 2000;64:126–8.
24. Witjes JA, Redorta JP, Jacqmin D, Sofras F, Malmström PU, Riedl C, et al. Hexaminolevulinate-guided fluorescence cystoscopy in the diagnosis and follow-up of patients with non-muscle-invasive bladder cancer: review of the evidence and recommendations. Eur Urol. 2010;57:607–14.
25. European Association of Urology. Guidelines on Non-muscle-invasive Bladder Cancer (TaT1 and CIS). European Association of Urology Guidelines. 2013 edition, pp. 10–16.
26. Guideline for the management of nonmuscle invasive bladder cancer: (Stages Ta, T1, and Tis). Chapter 1: the management of bladder cancer: diagnosis and treatment recommendations. 2007 Update, pp. 14–15.
27. Friesen SA, Hjortland GO, Madsen SJ, Hirschberg H, Engebraten O, Nesland JM, et al. 5-Aminolevulinic acid-based photodynamic detection and therapy of brain tumors (review). Int J Oncol. 2002;21:577–82.
28. Csanády M, Kiss JG, Iván L, Jóri J, Czigner J. ALA (5-aminolevulinic acid)-induced protoporphyrin IX fluorescence in the endoscopic diagnostic and control of pharyngo-laryngeal cancer. Eur Arch Otorhinolaryngol. 2004;26:262–6.
29. Namikawa T, Inoue K, Uemura S, Shiga M, Maeda H, Kitagawa H, et al. Photodynamic diagnosis using 5-aminolevulinic acid during gastrectomy for gastric cancer. J Surg Oncol. 2014;109:213. doi:10.1002/jso.23487.
30. Kishi K, Fujiwara Y, Yano M, Inoue M, Miyashiro I, Motoori M, et al. Staging laparoscopy using ALA-mediated photodynamic diagnosis improves the detection of peritoneal metastases in advanced gastric cancer. J Surg Oncol. 2012;106:294–8.
31. Murayama Y, Ichikawa D, Koizumi N, Komatsu S, Shiozaki A, Kuriu Y, et al. Staging fluorescence laparoscopy for gastric cancer by using 5-aminolevulinic acid. Anticancer Res. 2012;32:5421–7.
32. Messmann H, Knüchel R, Endlicher E, Hauser T, Szeimies RM, Kullmann F, et al. Photodynamic diagnosis of gastrointestinal precancerous lesions after sensitization with 5-aminolevulinic acid. A pilot study. Dtsch Med Wochenschr. 1998;123:515–21.
33. Baumgartner R, Huber RM, Schulz H, Stepp H, Rick K, Gamarra F, et al. Inhalation of 5-aminolevulinic acid: a new technique for fluorescence detection of early stage lung cancer. J Photochem Photobiol B. 1996;36:169–74.
34. Baas P, Triesscheijn M, Burgers S, van Pel R, Stewart F, Aalders M. Fluorescence detection of pleural malignancies using 5-aminolaevulinic acid. Chest. 2006;129:718–24.
35. Andikyan V, Kronschnabl M, Hillemanns M, Wang X, Stepp H, Hillemanns P. Fluorescence diagnosis with 5-ALA thermogel of cervical intraepithelial neoplasia. Gynakol Geburtshilfliche Rundsch. 2004;44:31–7.
36. Fritsch C, Ruzicka T. Fluorescence diagnosis and photodynamic therapy in dermatology from experimental state to clinic standard methods. J Environ Pathol Toxicol Oncol. 2006;25:425–39.
37. Muroi C, Fandino J, Coluccia D, Berkmann S, Fathi AR, Landolt H. 5-Aminolevulinic acid fluorescence-guided surgery for spinal meningioma. World Neurosurg. 2013;80:223. pii: S1878-8750(12)01450-7.

Real-Time Assessment of Intestinal Perfusion by Fluorescence-Based Enhanced Reality

31

Michele Diana, Vincent Agnus, and Jacques Marescaux

The anastomotic vascular supply, following gastrointestinal resections, is determined at the time of surgery, since the development of collateral circulation to improve local perfusion is improbable within the first days [1]. Consequently, well-vascularized margins of the anastomotic site should be correctly identified during the surgical procedure in order to reduce the risk of leakage or stenosis and ensure optimal healing [2].

Accuracy of clinical evaluation to predict anastomotic complications, irrespective of the surgeon's experience, is globally low [3], and relies on subjective and highly variable features such as the color of the serosal surface. Implementation in the clinical routine of an objective and accurate method to assess intestinal blood supply could be desirable to decrease the rate of anastomotic complications directly ascribable to reduced perfusion.

M. Diana, M.D. (✉) • V. Agnus, PhD
Department of General, Digestive and Endocrine Surgery, IRCAD-IHU, Institute of Minimally Invasive Image-Guided Surgery, University of Strasbourg, 1, place de l'Hôpital, 67095 Strasbourg, France
e-mail: michele.diana@ircad.fr

J. Marescaux, MD, FACS (Hon) FRCS (Hon.), FJSES
Research Institute Against Cancers of the Digestive System (IRCAD), Institute of Image-Guided Surgery (IHU-Strasbourg), 1 place de l'Hôpital, Strasbourg 67091, France
e-mail: jacques.marescaux@ircad.fr

A variety of techniques have been proposed for the intraoperative quantitative evaluation of intestinal perfusion by targeting different parameters. Doppler ultrasound and laser Doppler flowmetry can measure the blood flow; pulse oxymetry informs on arterial oxygen saturation (SpO2); near-infrared spectroscopy (NIRS) and visible light spectroscopy (VLS) aim to assess hemoglobin oxygen saturation (StO2) [4]. However, these techniques have a low penetration in the clinical setting, as they are time-consuming, quite expensive, or operator-dependent, and mostly remain research tools.

Fluorescence videography is especially important. It measures the units of fluorescence emitted by an injected dye (indocyanine green or fluorescein) when illuminated by a laser [4]. The units of fluorescence are proportional to the degree of intestinal perfusion. Fluorescence videography has the advantage of being relatively rapid, safe, and easy to perform.

In 2006, Toens et al. validated Indocyanine (IC) View Videography in a rabbit model of bowel ischemia. After intravenous injection of indocyanine green, the regions of interest of the ischemic bowel were illuminated with a laser (0.16 W; wavelength 780 nm). Digital videos were recorded using a Digital Video Camcorder (IC View System, Pulsion Medical System, Munich, Germany). A specifically designed software (IC-Calc, Pulsion Medical) was used to calculate fluorescence intensity arising from the

F.D. Dip et al. (eds.), *Fluorescence Imaging for Surgeons: Concepts and Applications*, DOI 10.1007/978-3-319-15678-1_31,

exposed bowel. Fluorescence intensity is proportional to the amount of fluorescent dye diffused in the tissue, and is consequently a marker of tissue perfusion. The increment of fluorescence (pixel intensity) in regions of interest was measured as the curve steepness of light emission [5], which was converted by the IC-Calc software into a color code cartography, displaying the actual degree of perfusion of the exposed bowel on a screen. The IC View showed a high accuracy in assessing intestinal viability in this animal model. In 2010, using a similar setting, in a case-matched study involving 402 patients undergoing open surgery colorectal resections, Kudszus et al. found that indocyanine green laser fluorescence angiography could significantly reduce the rate of anastomotic leakage [6].

In 2011, Matsui et al. used the FLARE™ (Fluorescence-Assisted Resection and Exploration) imaging system, consisting in light-emitting diodes in the spectra of white light (400–650 nm) and NIR fluorescence excitation light (745–779 nm), to predict ischemic bowel in an experimental model of mesenteric ischemia in pigs and rats. The exposed bowel images were acquired using customary optics and could be merged with ICG fluorescence distribution images. The FLARE™ system demonstrated a better accuracy in predicting bowel survival when compared with clinical evaluation alone [7].

The use of such technology was initially limited to open surgery but with new tools becoming available, fluorescence-based perfusion evaluation has been extended to the minimally invasive approach [8].

Recently, Carus et al. [9] proposed the use of laparoscopic indocyanine green fluorescence angiography to intraoperatively assess the perfusion of anastomosis in a series of 45 patients undergoing colorectal resections. Authors used a near-infrared endoscope (D-Light P, Karl Storz®, Tuttlingen, Germany) able to detect the signal emitted by the fluorescent dye administered by intravenous injection. To confirm perfusion adequacy, Carus et al. used the degree of "relative blueness," as provided by computer analysis.

However, this analysis is "static" and does not take into account the over-time diffusion of the fluorescent dye, which can reach ischemic areas by retrograde flow and provide incorrect information on the exact perfusion value (Fig. 31.1). Additionally, the fluorescent intensity itself could not be used for comparisons between patients without the use of a "standard" reference calibration tool (which is a squared spot yielding a constant signal when illuminated by the near-infrared light that should be introduced into the abdomen of the patient during laparoscopy).

We aimed to fuse the concept of fluorescence videography with augmented reality (AR) to guide intestinal resection and assess vascular supply at the future anastomotic site.

AR is the process of superimposing virtual images of a patient onto real-time intraoperative images to enhance visual perception. Virtual images are created from preoperative DICOM images (CT-scan, Nuclear Magnetic Resonance or US) by 3D software manipulation in order to obtain a virtual clone of the patient [10–12].

In our method of assessment of bowel vascular supply, FLuorescence-based Enhanced Reality (FLER), virtual images of bowel perfusion (virtual vascular cartography) are generated from intra-operative live images based on fluorescence signals of indocyanine green.

To push the concept of fluorescence-guided evaluation of the future anastomotic site forward, our Research and Development department developed a dedicated image analyzer software (ER-PERFUSION, IRCAD, France) to obtain FLER [13].

FLER is a fluorescence videography system integrating a near-infrared endoscope (D-Light P, Karl Storz®, Tuttlingen, Germany) able to detect the fluorescence signal emitted by indocyanine green (ICG) and the ER-PERFUSION software, which can generate a virtual perfusion cartography based on fluorescence time-to-peak.

To assess time-to-peak in FLER, a dye injection is performed in real time while looking at the region of interest, in order to capture and record the "dynamics" of diffusion as well as the evolution of fluorescence. The virtual perfu-

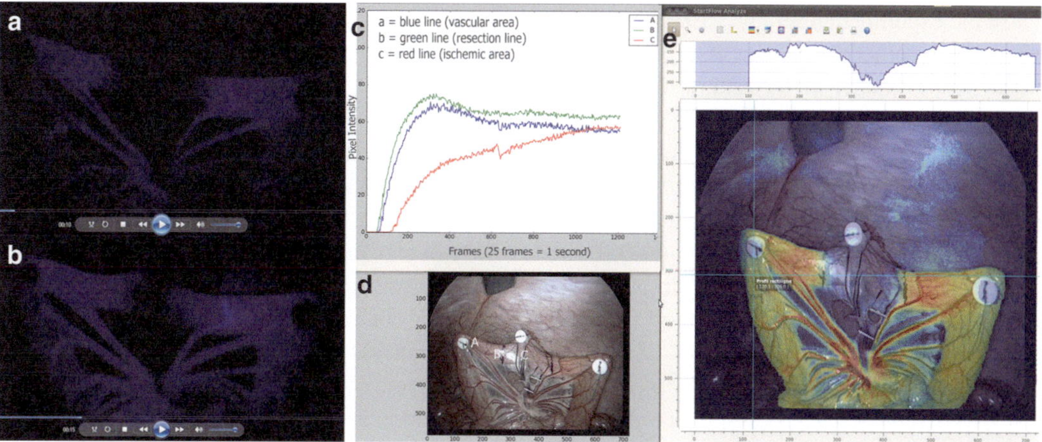

Fig. 31.1 Demonstration of retrograde refilling of the ischemic area by the fluorescent dye. (**a**) Five seconds recording after initial fluorescence signals arose from a partially ischemic bowel loop; (**b**) Five seconds later, the ischemic area was partly filled by retrograde flow (no vessels were apparent and the refilling was slow); (**c**) The graph shows the slope of pixel intensity evolution over time: the *blue* and the *green line* represent the evolution of fluorescence on the vascular area and resection line respectively, showing a very steep and fast increase; the *red line* is the fluorescence signal on the ischemic zone. Note that after some time, by retrograde filling, fluorescence intensity could reach similar values to those of vascular areas; (**d**) The bowel loop with some mesenteric vessels clipped to create a segmental ischemia: points *A*, *B*, *C* marked on the loop correspond to the vascular area, resection line and ischemic area as identified by fluorescence time-to-peak and corresponding to the slopes displayed on the (**c**) panel; (**e**) The image shows the enhanced reality obtained by using the steepness of the slopes (speed to reach maximum intensity) to create the virtual perfusion cartography which was overlapped with real-time image

Fig. 31.2 Schema of fluorescence-based enhanced reality. (**a**) A loop of small bowel was suspended and a window in the mesentery is made by sealing 3–4 vessels; (**b**) The fluorescence signal emitted by indocyanine green is captured by the near-infrared laparoscope (D-Light P, Karl Storz®, Tuttlingen, Germany); (**c**) A virtual perfusion cartography is generated using the ER-PERFUSION software, based on fluorescence time-to-peak and overlapped with laparoscopic images to obtain the FLER effect

sion cartography is created by averaging the signals over a 20–40 s video at the speed of 5–25 frames per second and attributing a color code based on the time required to reach the maximum intensity of each pixel. Each single pixel composing the virtual perfusion cartography is a dynamic image (2D + diffusion time) representing the average of several images (from 100 to 200). The virtual cartography can be overlapped with real-time laparoscopic images in order to obtain the enhanced reality effect (Fig. 31.2).

In the experimental setting, FLER showed the ability to exactly detect future resection lines in a laparoscopic model of mesenteric ischemia. In the initial experimental model of small bowel ischemia, the software was set to detect a reduction of fluorescence time-to-peak of 50 % and the areas identified by enhanced reality (ischemic, future resection lines, and vascular areas) showed significant and congruent changes in lactate levels, mitochondria respiratory rate, and metabolomic fingerprint of ischemia [13].

The use of the fluorescence time-to-peak slope has two advantages when compared to the use of the absolute value of "fluorescence intensity." First, time-to-peak is independent of the distance between the light source and the imaged area. This is not the case with fluorescence intensity, which is highly dependent on distance (a low perfused area may look more intensively fluorescent when observed closely and vice versa; a highly perfused area may look poorly fluorescent when observed from far away). We experimentally demonstrated the relationship between the light source-target distance and fluorescence intensity, using a customary electromagnetic tracking device (METRIS 3D) [14]. The METRIS 3D system is composed of a tube (1.2 m long and 2.2 mm in diameter) that fits into the operating channel of a conventional endoscope and contains seven 8 by 1 mm miniature electromagnetic probes distributed on its length. Electromagnetic probes are tracked by a commercially available magnetic tracking system (Aurora, Northern Digital Inc, Waterloo, Ontario, Canada). To track the position of the D-Light P laparoscope, the METRIS 3D system was attached onto the shaft of the laparoscope, and a 2 cm plastic tool was also fixed to establish a set distance. The laparoscope was then used to target the fluorescence calibrating tool in a dark environment, and was moved in and out several times (Fig. 31.3).

Secondly, time-to-peak allows for multiple and repetitive assessments, since the "noise" produced by the accumulation of fluorescent dye does not affect the steepness of the slope.

FLER: Development Steps and Future Works

After demonstrating the proof of the FLER concept, we worked in a stepwise fashion to validate the concept, to demonstrate the superior performances when compared to the clinical assessment, and to refine the software interface to be more user-friendly, effective and to add useful plugs-in.

Point 1: Clinical vs. FLER Evaluation in Segmental Sigmoid Ischemia

To establish the impact of FLER in the evaluation of perfusion at the anastomotic site, we used two models: first, a model of laparoscopic segmental sigmoid ischemia in 6 pigs. FLER could identify the ischemic area of the sigmoid colon and resection lines. FLER was compared to the clinical evaluation made by two blinded surgeons. A distance inferior to 1 cm between resection lines as defined by clinical evaluation and FLER was considered non-significant, which occurred in 50 % of cases (6 over 12 resection lines). In discordant cases, the FLER-guided marking found areas showing a better metabolic profile as demonstrated by a significantly lower level of capillary lactates when compared to those identified by clinical evaluation. Additionally, we used a probe-based Confocal Laser Endomicroscopy (CLE), the Cellvizio® system (Mauna Kea Technologies, France) to discriminate mucosal ischemic changes, to evaluate the transition zone from ischemic to vascular areas. Additionally, it was compared to the FLER system evaluation of the serosa. The probe can be inserted through conventional endoscopes to scan the intestinal

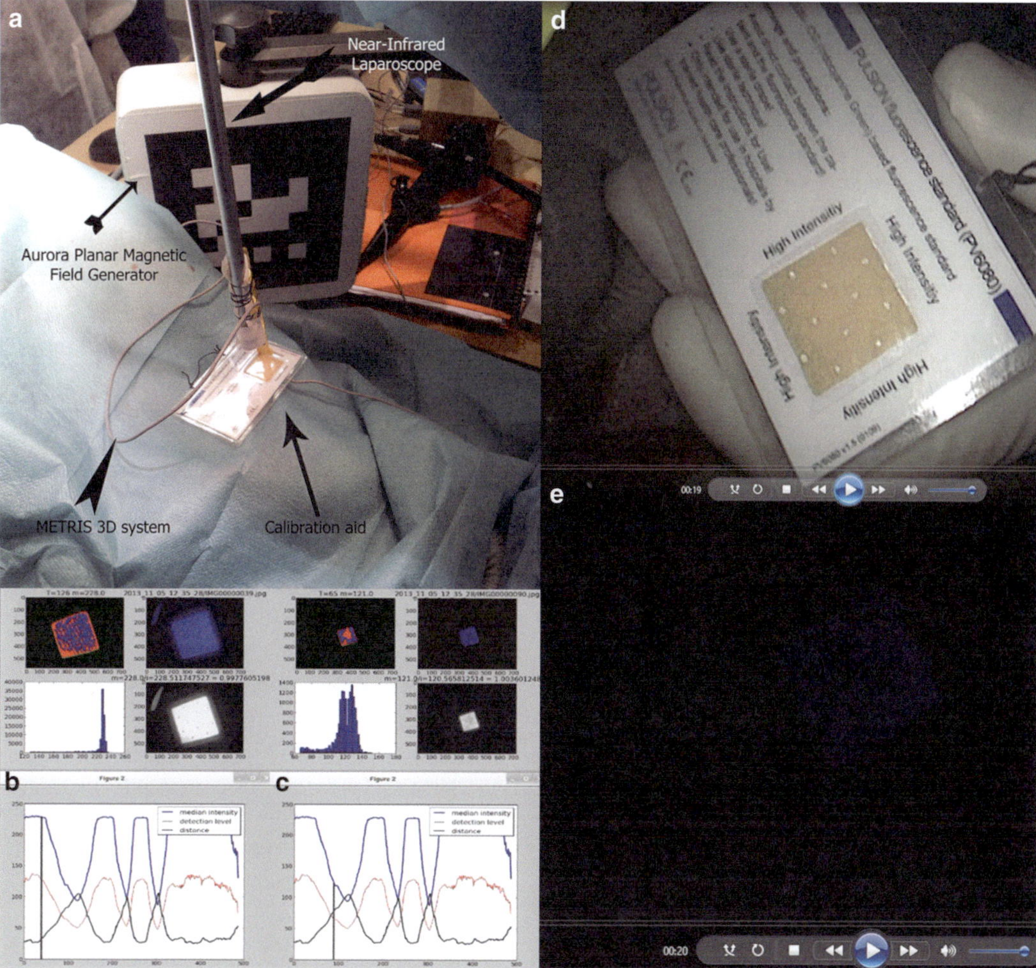

Fig. 31.3 Relationship between fluorescence intensity and distance. (**a**) The METRIS 3D system is composed of a tube (1.2 m long and 2.2 mm in diameter) which fits into the operating channel of a conventional endoscope and contains seven 8 by 1 mm miniature electromagnetic probes distributed on its length. Electromagnetic probes are tracked by a commercially available magnetic tracking system (Aurora, Northern Digital Inc., Waterloo, Ontario, Canada). To track the position of the D-Light P laparoscope and measure the distance between the tip of the scope and the fluorescence calibrating tool, the METRIS 3D system was attached onto the shaft of the laparoscope and a 2 cm plastic tool was also fixed to establish a set distance. In (**b**) and (**c**) the inverse relationship between fluorescence intensity and distance is shown: the *white square* represents the "standard" response of the calibration tool when illuminated with the near-infrared laparoscope. The shorter the distance (*black line*), the higher the intensity (*blue line*). (**d**) The fluorescence calibration tool is shown in white light, and (**e**) the same image is shown after switching the D-Light P laparoscope to near-infrared mode

mucosa providing information on the microcirculation giving capillary density or circulating blood cell velocity as well as the morphology of intestinal villi and crypts, demonstrating changes induced by the ischemic injury [15].

The inability to obtain a clear confocal image of enterocytes was an almost constant finding at the center of the ischemic area, secondary to the lack of penetration of fluorescein into the ischemic tissue. Moving toward future resection lines but still within the ischemic zone, a consistent change was the presence of enterocytes with a "target cell" image, with a clear acinar center. This was due to some fibrin deposits in the glandular lumen. This "target cell" image was not visible in vascular zones, in which the center of enterocytes was constantly black. The agreement between FLER and confocal evaluation was 91.6 % (Diana M et al. unpublished work).

Point 2: Ability to Perform Multiple Assessments and Comparison of Performance vs. Clinical Assessment Over Time

The next step was to determine the possibility to perform repeated injections and to evaluate the performances of FLER over time, since the accumulation of ICG from previous injections might alter the signal interpretation by increasing the noise given by a saturated background. To do so, we designed a model of small bowel ischemia involving 6 pigs. In each pig, a segmental ischemia was created in 3 small bowel loops (total of 18 loops). After 2, 4, and 6 h, the ischemic segments were evaluated by clinical assessment (by a second surgeon blinded to the extent of the mesenteric window) and by FLER, to determine presumed viable margins. Thanks to the use of time-to-peak, repeated assessments were possible up to 4 h using our setting, with an acceptable signal-to-background ratio. By evaluating time-to-peak (and not pixel intensity), the background can be zeroed to allow for an additional injection of dye and only the additional signal is interpreted to generate virtual perfusion cartography. In this model, Fluorescence-based Enhanced Reality allowed to determine the boundary between ischemic and vascularized zones even after repetitive assessments, with a better performance when compared to clinical evaluation (Diana M et al. unpublished work).

Point 3: Accuracy of the Software in a Survival Model of Bowel Resection and Anastomosis

The ER-PERFUSION software has been upgraded to (1) improve the interface, and (2) to improve efficacy by reducing the time to calculate virtual perfusion cartography as well as registration of cartography onto real images. In the latest version, only resection lines are projected onto the bowel with the enhanced reality effect. The accuracy of the software will be tested in a series of experimental bowel resections and anastomoses in a 2-week survival model (currently underway). Resection lines and subsequent anastomoses will be performed according to various degrees of perfusion, which will be determined by the ER-PERFUSION software.

Plug-in Application: Real-Time Tracking of FLER

The latest version of the software also integrates a tracking function which has been developed to allow for constant registration of virtual perfusion cartography obtained by fluorescence time-to-peak onto real images, even during bowel manipulation. To track and constantly register FLER, the virtual cartography is deformed by using an affine deformation created by pixel displacement estimated by a feature matching algorithm (BRIEF: binary robust independent elementary features) [16], in order to follow the motion of the regions of interest (Fig. 31.4).

To conclude, FLER is a promising technology to intraoperatively assess bowel perfusion and determine the future anastomotic site, as it is rapid and easy to use. In the experimental setting, FLER has shown a high accuracy to predict anastomotic healing and a superior performance when compared to clinical evaluation. Some developments are still underway to enhance the software's capabilities and ensure optimal transfer of this technology to the clinical setting.

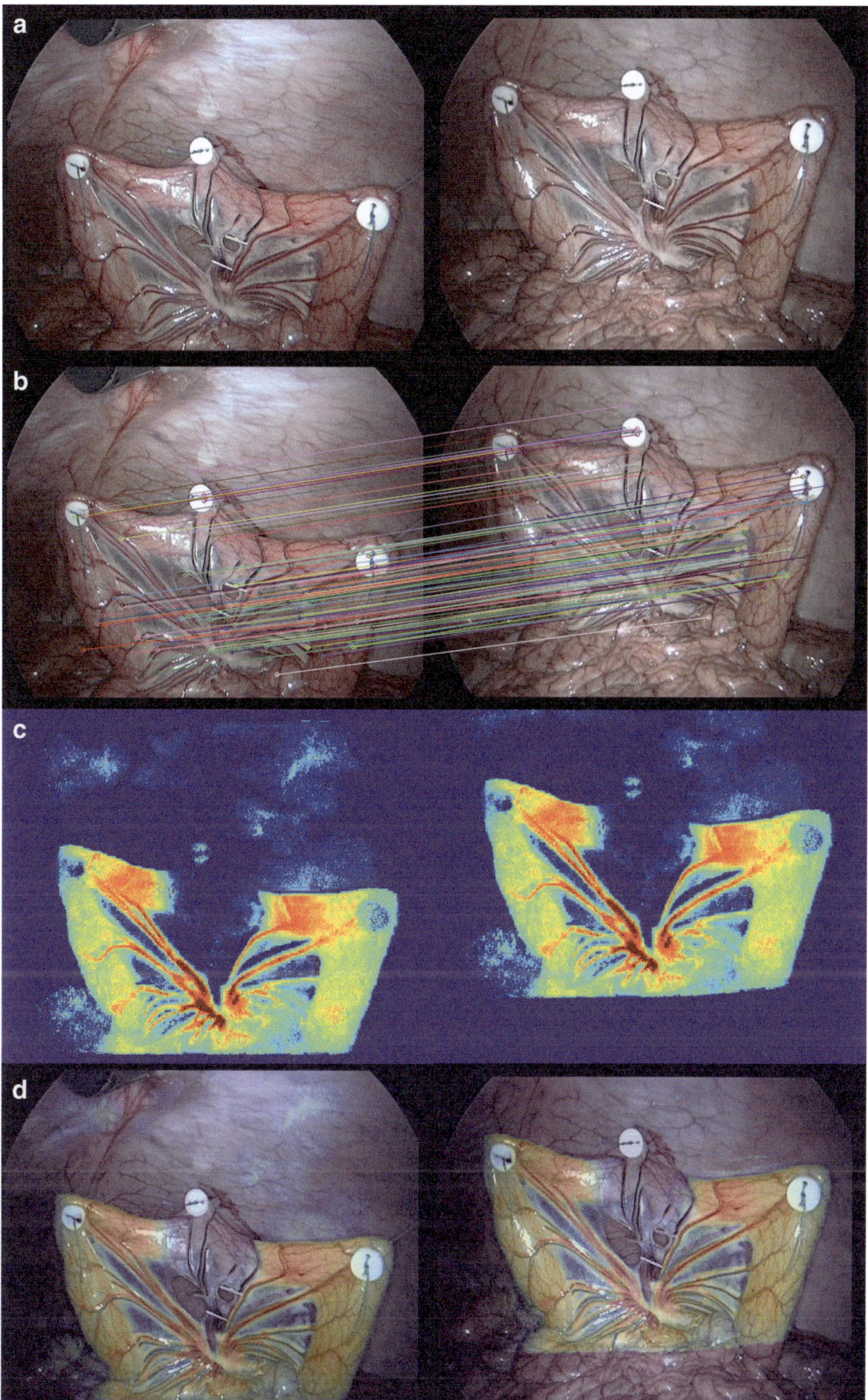

Fig. 31.4 Real-time tracking feature of FLER. (**a**) *Left*: the reference image; *Right*: the image after tilting the camera downward (producing a vertical down-up movement). (**b**) The two images are matched (feature matching algorithm), and (**c**) the same vectors of displacement are applied to the virtual perfusion cartography in order to (**d**) constantly register FLER on the laparoscopic image

References

1. Seike K, Koda K, Saito N, Oda K, Kosugi C, Shimizu K, et al. Laser Doppler assessment of the influence of division at the root of the inferior mesenteric artery on anastomotic blood flow in rectosigmoid cancer surgery. Int J Colorectal Dis. 2007;22(6):689–97.
2. Karliczek A, Benaron DA, Baas PC, Zeebregts CJ, Wiggers T, van Dam GM. Intraoperative assessment of microperfusion with visible light spectroscopy for prediction of anastomotic leakage in colorectal anastomoses. Colorectal Dis. 2010;12(10):1018–25.
3. Karliczek A, Harlaar NJ, Zeebregts CJ, Wiggers T, Baas PC, van Dam GM. Surgeons lack predictive accuracy for anastomotic leakage in gastrointestinal surgery. Int J Colorectal Dis. 2009;24(5):569–76.
4. Urbanavicius L, Pattyn P, de Putte DV, Venskutonis D. How to assess intestinal viability during surgery: a review of techniques. World J Gastrointest Surg. 2011;3(5):59–69.
5. Toens C, Krones CJ, Blum U, Fernandez V, Grommes J, Hoelzl F, et al. Validation of IC-VIEW fluorescence videography in a rabbit model of mesenteric ischaemia and reperfusion. Int J Colorectal Dis. 2006;21(4):332–8.
6. Kudszus S, Roesel C, Schachtrupp A, Hoer JJ. Intraoperative laser fluorescence angiography in colorectal surgery: a noninvasive analysis to reduce the rate of anastomotic leakage. Langenbecks Arch Surg. 2010;395(8):1025–30.
7. Matsui A, Winer JH, Laurence RG, Frangioni JV. Predicting the survival of experimental ischaemic small bowel using intraoperative near-infrared fluorescence angiography. Br J Surg. 2011;98(12):1725–34.
8. Cahill RA, Ris F, Mortensen NJ. Near-infrared laparoscopy for real-time intra-operative arterial and lymphatic perfusion imaging. Colorectal Dis. 2011;13 Suppl 7:12–7.
9. Carus T, Dammer R. Laparoscopic fluorescence angiography with indocyanine green to control the perfusion of gastrointestinal anastomoses intraoperatively. Surg Technol Int. 2012;22:27.
10. D'Agostino J, Diana M, Vix M, Soler L, Marescaux J. Three-dimensional virtual neck exploration before parathyroidectomy. NEJM. 2012;367(11):1072–3.
11. Marescaux J, Diana M, Soler L. Augmented reality and minimally invasive surgery. J Gastroenterol Hepatol Res. 2013;2(5):555–60.
12. Marzano E, Piardi T, Soler L, Diana M, Mutter D, Marescaux J, et al. Augmented reality-guided artery-first pancreatico-duodenectomy. J Gastrointest Surg. 2013;17(11):1980–3.
13. Diana M, Noll E, Diemunsch P, Dallemagne B, Benahmed MA, Agnus V, et al. Enhanced-reality video fluorescence: a real-time assessment of intestinal viability. Ann Surg. 2014;259.
14. Diana M, Wall J, Perretta S, Dallemagne B, Gonzales KD, Harrison MR, et al. Totally endoscopic magnetic enteral bypass by external guided Rendez-vous technique. Surg Innov. 2011;18(4):317–20.
15. Laemmel E, Genet M, Le Goualher G, Perchant A, Le Gargasson JF, Vicaut E. Fibered confocal fluorescence microscopy (Cell-viZio) facilitates extended imaging in the field of microcirculation. A comparison with intravital microscopy. J Vasc Res. 2004;41(5):400–11.
16. Calonder M, Lepetit V, Ozuysal M, Trzcinski T, Strecha C, Fua P. BRIEF: computing a local binary descriptor very fast. IEEE Trans Pattern Anal Mach Intell. 2012;34(7):1281–998.

Fluorescent Angiography for Flap Planning and Monitoring in Reconstructive Surgery

32

Georges Bettega, Clotilde Ochala, Marine Hitier, Cynthia Hamou, Stéphanie Guillermet, Pascal Gayet, and Jean-Luc Coll

Introduction

Reconstructive surgery has for objective to restore anatomical defects using autologous tissues called "flaps." A flap includes harvested tissues (skin, muscle, bone, etc. alone or combined) and their blood-supplying artery and veins (the pedicle). Blood supply is essential for flap survival. In case of major reconstruction, the flap is completely separated from the donor site and the vessels are anastomosed (reconnected) to the vessels of the recipient site. This type of flap is called a free-flap. Patency of the arterial and venous anastomosis is mandatory for free-flap survival. Close monitoring of the flap vascularization is important. The odds to salvage the flap depend on the rapidity of thrombectomy, in case of thrombosis (vascular obstruction).

Various types of flaps are available. Perforator flaps were developed to minimize donor-site morbidity. They are characterized by their plasticity and by the possibility to harvest large flaps on small pedicles (Fig. 32.1). They are more and more frequently indicated. More than 350 perforator vessels have been described [1]. Preoperative localization of the perforator vessels is essential for flap planning; but it is difficult because of the small diameter of these vessels (sometimes less than 1 mm).

Various methods have been used to determine the anatomy of the blood-supplying vessels [2], according to the type of flap. For example, the preoperative perforator mapping of a fibular flap or anterolateral thigh flap is often made by Doppler sonography [3]. But this mapping requires an experimented operator who is not necessarily the operating surgeon. Furthermore, the images obtained with this technique are not very contributive. CT angiography has also been used to map anterolateral thigh flaps or deep inferior epigastric flaps [4]. But it is an irradiating method that requires the injection of potentially allergenic iodine contrast agent. Its sensibility has not been determined yet.

Electronic supplementary material: The online version of this chapter (doi:10.1007/978-3-319-15678-1_32) contains supplementary material, which is available to authorized users. Videos can also be accessed at http://link.springer.com/chapter/10.1007/978-3-319-15678-1_32.

G. Bettega, M.D. Ph.D. (✉) • C. Ochala, M.D.
M. Hitier, M.D. • C. Hamou, M.D.
Plastic and Maxillofacial Surgery Department,
Grenoble University Hospital,
BP 217, Grenoble 38043, France
e-mail: gbettega@chu-grenoble.fr;
cochala@chu-grenoble.fr; mhitier@chu-grenoble.fr;
chamou@chu-grenoble.fr

S. Guillermet, Ph.D. • P. Gayet, Ph.D.
Fluoptics, BHT-Batiment 52 - 7 Parvais Louis Nell
CS 20050, 38040 Grenoble, France
e-mail: stephanie.guillermet@fluoptics.com;
pascal.gayet@fluoptics.com

J.-L. Coll, Ph.D.
INSERM–UJF U823, Institut Albert Bonniot,
University Joseph Fourier, Grenoble 38706, France
e-mail: jean-luc.coll@ujf-grenoble.fr

F.D. Dip et al. (eds.), *Fluorescence Imaging for Surgeons: Concepts and Applications*,
DOI 10.1007/978-3-319-15678-1_32, © Springer International Publishing Switzerland 2015

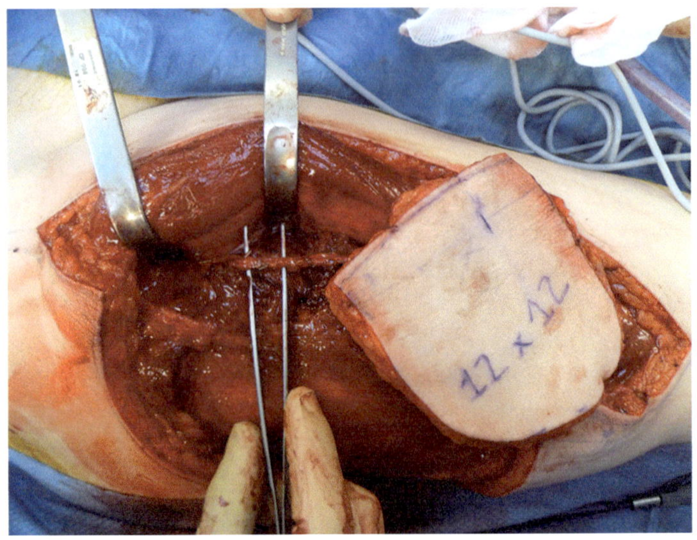

Fig. 32.1 A perforator flap

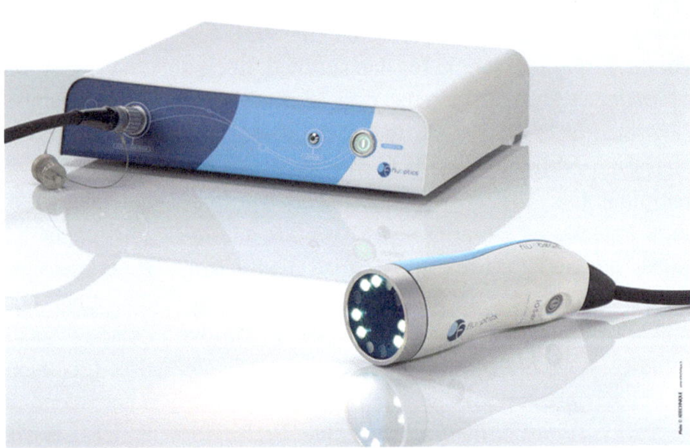

Fig. 32.2 The Fluobeam system for optical angiography

Indocyanine green fluorescence angiography (ICGFA) is used to explore superficial vascularization. It is particularly well adapted to evaluate skin perfusion. It should help to localize perforator vessels [5]. It has been also used for intraoperative assessment of flap perfusion and for postoperative flap monitoring [6]. We tested ICGFA in reconstructive surgery using the Fluobeam™ device (Fluoptics, Grenoble, France) for flap planning and flap monitoring (Fig. 32.2).

ICGFA for Flap Planning

The contribution of ICGFA for flap planning was assessed in animal studies. Giunta et al. correlated fluorescent perfusion index to flap necrosis in random pattern anterior abdominal rat flaps [7]. Lee et al. localized perforator vessels of epigastric flaps in pigs with near-infrared fluorescence imaging [8]. They found a 100 % correlation between the number of perforator vessels identified by fluorescence, x-ray angiography, and anatomic dissection. More recently, Onoda et al. compared perforator localization with computed tomography (MDCT), Doppler flowmetry, and ICGFA [9]. They concluded that "the identification of perforators was difficult with ICGFA in patients with a flap thickness >20 mm."

We performed a cadaveric study to confirm the feasibility and accuracy of ICGFA for flap planning, before its clinical use in humans. We also had for objective to define the ergonomics of a fluorescent system dedicated to reconstructive surgery. We explored anterolateral thigh flap perforators on 22 cadavers. Femoral vessels were exposed just below the inguinal ligament and catheterized.

In a first step, 20 mg of Monopeak Indocyanine Green (Infracyanine™, SERB laboratory, Paris, France) was injected and washed out with 20 ml of saline. The Fluobeam™ probe (Fluobeam FB800, Fluoptics, Grenoble, France) was fixed on an articulated arm, 20 cm above the thigh skin, and recording was started at the time of saline injection (Fig. 32.3). Perforator vessels appeared as fluorescent spots after a few seconds (Fig. 32.4). Those spots were marked on the skin and localized in an orthogonal landmark defined by the line joining the iliac crest to the patella and the perpendicular to its middle. Femoral vessels were washed out again with saline at the end of the fluorescence examination.

In a second step, the thigh was then detached above the iliac crest and below the knee. All the visible vessels on the distal cut were ligatured (especially the popliteal vessels) to prevent leaking during the injection of 20 ml of iodine solution (Omnipaque, 350 mg I/ml) for computed tomography angiography (CTA). The perforators identified by CTA were localized in the same landmark after reconstruction.

The last step was dissection of the thigh to expose the perforator vessels and to localize them, always in the same landmark. The coordinates of perforators given by CTA and ICGFA were compared with the coordinates given by dissection (gold standard); 97 perforator vessels were localized by ICGFA, 11 by CTA, and 76 by dissection. Unlike CTA, ICGFA localization was correlated to dissection; it was able to reveal even very tiny perforators. But it specificity was poor (14.7 %). ICGFA showed all the skin vascularization, it did not differentiate between surgically useful perforators and other superficial vessels. Nevertheless, we concluded that ICGFA would

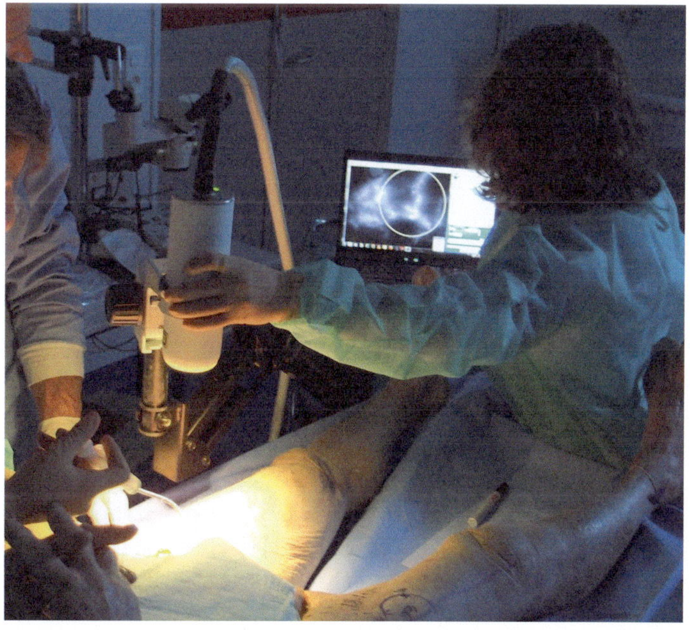

Fig. 32.3 Fluobeam™ system set-up to explore perforator vessels of the thigh

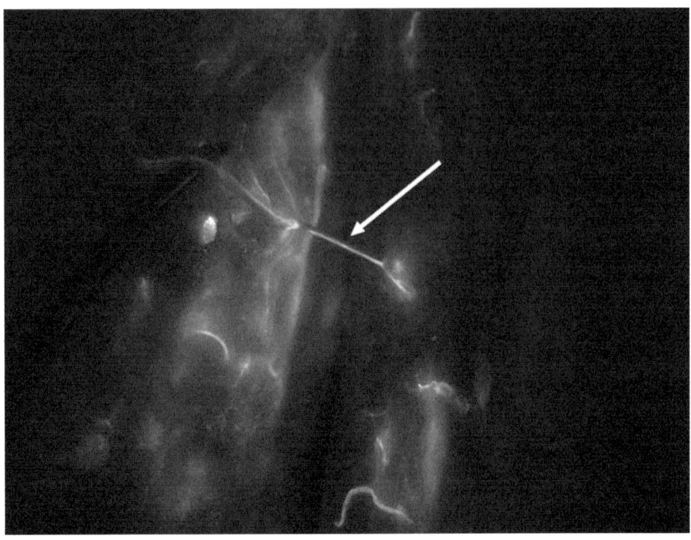

Fig. 32.4 Thigh. Perforator vessels appeared as fluorescent spots after a few seconds

help to plan perforator flaps, when combining its data with anatomical knowledge. This anatomical study also allowed validating the ergonomics to transfer this technology for clinical trials, and even in the operating room.

The goal of our clinical study was to validate the relevance of ICGFA for perforator flap planning, and to compare it with the gold standard (i.e. surgical dissection) and with routine investigation in our center. Twenty patients were included. Reconstruction was planned with one of the three following flaps: fibular flap, anterolateral thigh flap (ALT), or deep inferior epigastric flap (DIEP). Routine planning was made with echo-Doppler for fibular flaps and with CTA for ALT or DIEP in our department. The identified perforator vessels were localized in a specific landmark according to the flap (Fig. 32.5).

The day before surgery, 0.025 mg/kg of Infracyanine™ was injected intravenously and the skin area of the flap was explored with the Fluobeam™ system. Superficial vessels appeared as fluorescent spots and were marked with a pen on the skin and localized in the specific landmark (Video 32.1).

The position of perforator vessels was determined per-operatively in the same landmark after dissection. The three sets of localization coordinates (surgery, ICGFA, Doppler or CTA) were compared, surgical data being the reference.

ICGFA localization was accurate in a range of 0.90 cm ± 0.96, compared to anatomical localization. It was more accurate than echo-Doppler for fibular flaps (1.28 cm ± 1.01). But its specificity was poor (48 %); as in the cadaveric study, because it revealed all the superficial vascularization and not just the perforator vessels useful for surgery. Nevertheless, it was possible to differentiate and to identify the targeted vessels that first lit up in the area where they were supposed to be, according to surgical experience.

One of the difficulties with fluorescence angiography is the rapidity of skin illumination; the spots of the perforator vessels are very quickly drowned in the surrounding fluorescent noise, and it becomes hard to localize the perforator vessels accurately. A solution is to use a fluorescent ruler stitched on the skin, and to record the fluorescent data on video. The video can then be viewed image by image to improve optical localization (Video 32.2).

These data need to be confirmed with more patient studies. Nevertheless, ICGFA seems to be an accurate and safe method for perforator localization. It can be used directly by the surgeon, and does not require any irradiation or iodine contrast agent.

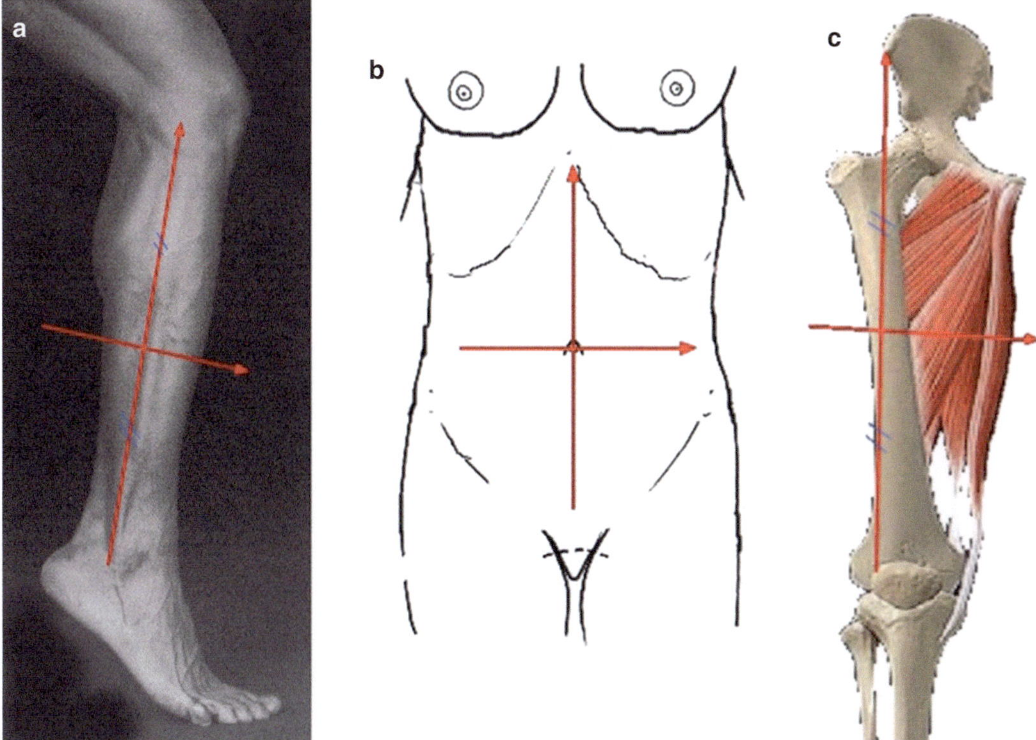

Fig. 32.5 The identified perforator vessels were localized in a specific landmark according to the flap

ICGFA for Flap Monitoring

The major complication of flap surgery is the obstruction of nutrient vessels (arterial and/or venous thrombosis). This can lead to flap failure (necrosis) with dramatic functional and esthetic consequences, if not treated rapidly. Flap monitoring is critical [10]. The most common technique for flap monitoring is clinical assessment. The surgeon monitors skin flap color, flap temperature, and capillary refill (ring test). In case of doubt, he can puncture the flap to verify bleeding, but this may harm the flap. This technique is very dependent on surgeon experience and signs of complication (thrombosis) appear late.

Various instrumental techniques have been developed for flap monitoring (implantable Doppler system, color duplex sonography, near-infrared spectroscopy, micro-dialysis, laser Doppler flowmetry, etc.) but none are used routinely. They are either very complex (even invasive) and expensive, or simple and equivalent to clinical assessment.

Few authors have studied flap monitoring by ICGFA [6, 8, 11]. Holm, using microscope-integrated indocyanine green near-infrared video-angiography, found that an intrinsic transit time (ITT: time needed for fluorescent dye to flow from arterial to venous anastomosis) superior to 50 s was correlated with early anastomotic complications [12]. The use of iterative ICG injection for flap monitoring has not been reported yet.

We first tested the Fluobeam™ system on abdominal rat flaps to identify relevant fluorescent parameters assessing flap perfusion. Both epigastric flaps were raised for each rat. The right one served as control. The pedicle of the left one was clamped (artery or vein) to simulate obstruction, and 0.025 mg/kg of Infracyanine™ was injected in the tail vein or in the jugular vein. Fluorescence imaging was performed in real-time (Fig. 32.6).

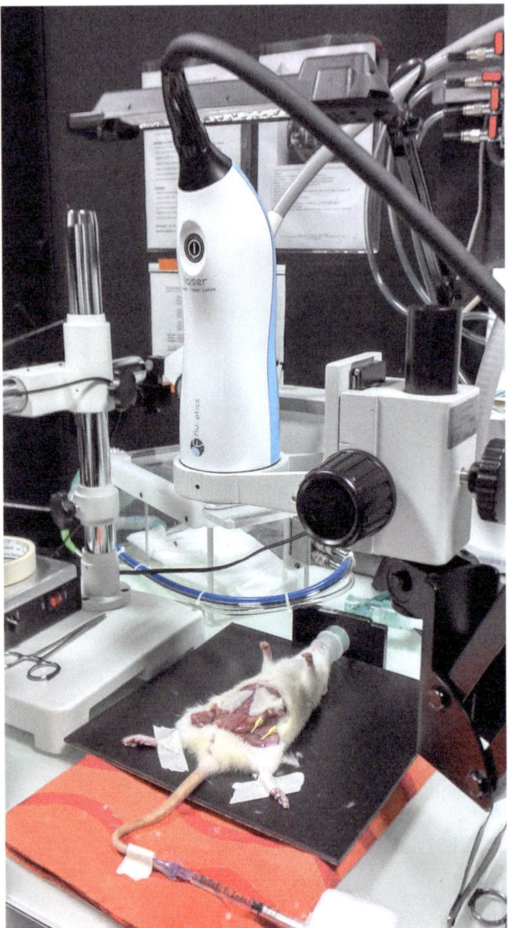

Fig. 32.6 The pedicle of the left one was clamped (artery or vein) to simulate obstruction, and 0.025 mg/kg of Infracyanine™ was injected in the tail vein or in the jugular vein. Fluorescence imaging was performed in real-time

Three parameters were calculated: the delay before onset of fluorescence in the flap after ICG injection, the rise speed of the fluorescent signal, and the maximum fluorescence intensity.

There was a clear difference in perfusion, after artery clamping, between right and left flap, the latter appearing dark (Fig. 32.7). There was a higher maximum and rise speed fluorescence in the control flap (Fig. 32.8). The delay before onset of fluorescence was the same. The flap also appeared dark after vein clamping, and the speed rise was slower on the left side (Fig. 32.9). The delay of onset of fluorescence was similar again.

These results seem to demonstrate that the rise speed of the fluorescent signal and the maximum fluorescence intensity could be good indicators to assess flap perfusion.

Our clinical experience was based on a clinical trial including 20 patients reconstructed by free flaps (the same patients included in the previous study dealing with flap planning). Ethical approval was obtained for the study. Each patient received iterative injections of 0.025 mg/kg Infracyanine™ after giving their informed consent. An injection was performed intraoperatively just after flap pedicle anastomosis to recipient vessels. The same injection was repeated every 6 h for 4 days, beginning 2 h after surgery. Each patient received 19 injections (including the preoperative one for flap planning). The parameters measured per-operatively were ITT and skin flap fluorescence average intensity (Video 32.3). The postoperative parameters were the same as for the animal experiment, i.e. the delay before onset of fluorescence in the flap after ICG injection, the rise speed of the fluorescent signal, and the maximum fluorescence intensity. There was no complication with the iterative ICG injections but two patients had a flap complication. A venous thrombosis was clinically identified in both cases 20 h after surgery requiring surgical revision of the anastomosis and both flaps were salvaged.

Another patient presented with partial necrosis of the skin paddle 14 days postoperatively. The skin paddle of the flap was not fluorescent per-operatively for these three patients.

Only one of the patients had an ITT over 40 s, the mean ITT of the group was 12.8 s.

The fluorescence intensity evolution, on postoperative monitoring, was clearly different for these patients.

Figure 32.10 illustrates the fluorescent curves for a non-complicated flap, Fig. 32.11 for one of the thrombotic flaps. In both cases of venous thrombosis, the fluorescence curves were abnormal after the first postoperative control without any clinical evidence of flap congestion. The clinical signs appeared after 20 h in both cases. Figure 32.12 shows the difference in fluorescent

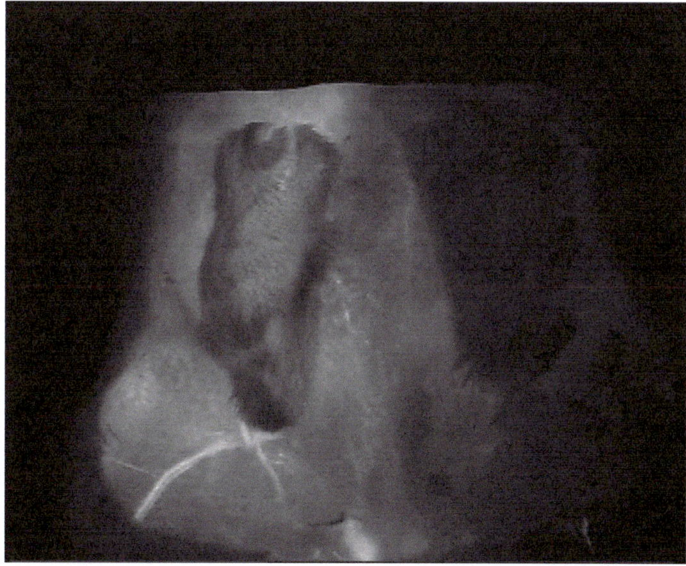

Fig. 32.7 There was a clear difference in perfusion, after artery clamping, between right and left flap, the latter appearing dark

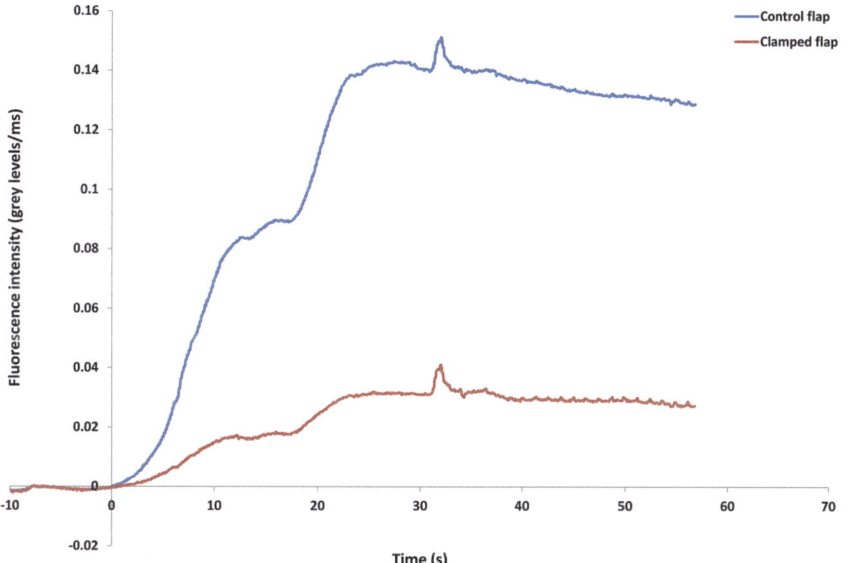

Fig. 32.8 Fluorescence intensity curve for each flap after ICG injection when the left epigastric artery is clamped

illumination between a well-vascularized flap and a thrombotic one.

Our data do not allow concluding on the relevance and the predictability of ITT for the flap outcome. It was longer only in one case of thrombosis. But the postoperative intensity of flap fluorescence and the dynamics of fluorescence seem to be good and early indicators of flap perfusion.

Conclusion

ICGFA seems to be a good tool for flap planning and flap monitoring. We need more experience to confirm the preliminary data. But this technology must be adapted to reconstructive surgery constraints. The surgeon needs a large field of vision to explore the entire area concerned by the planned

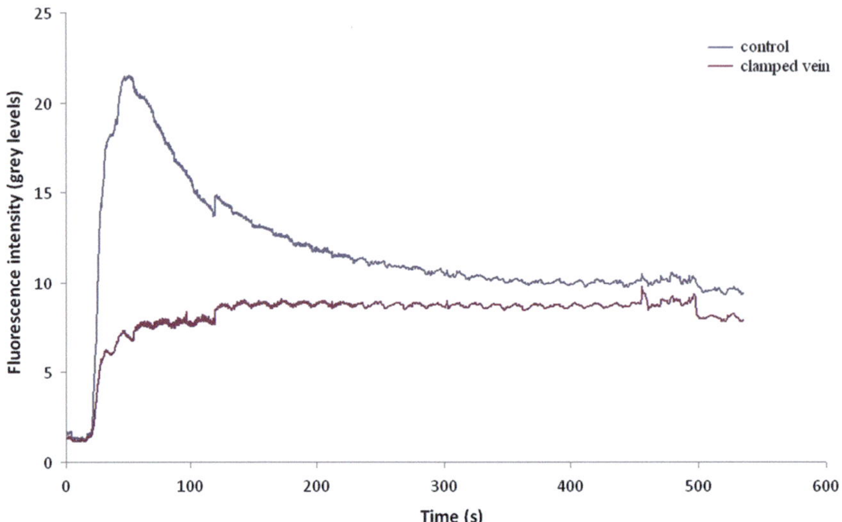

Fig. 32.9 Fluorescence intensity curve for each flap after ICG injection when the left epigastric vein is clamped

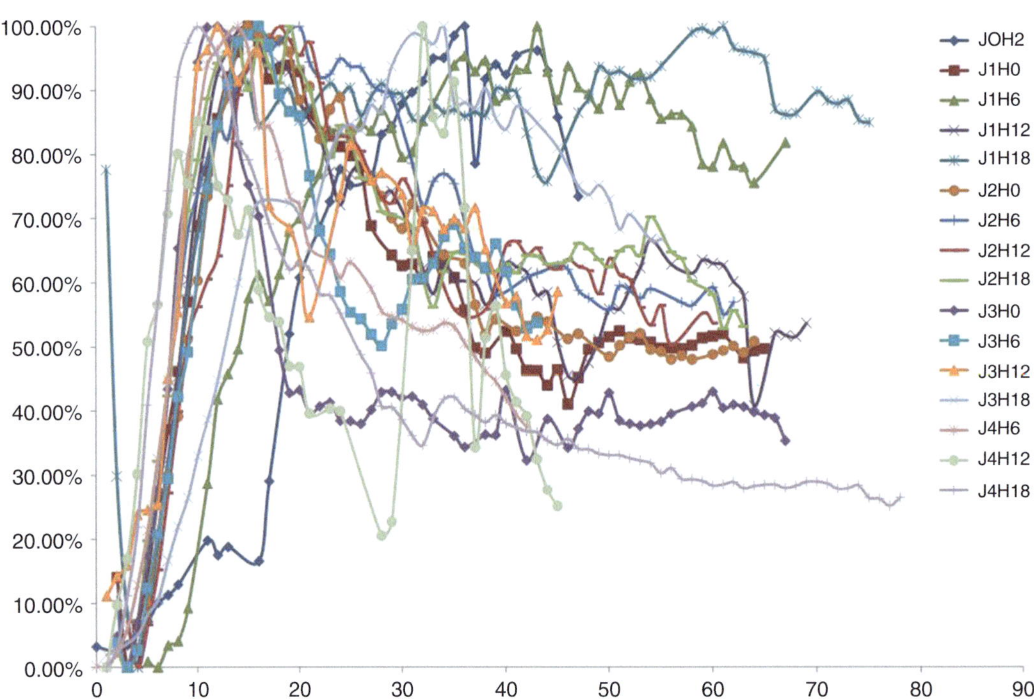

Fig. 32.10 Postoperative fluorescence intensity curves for a non-complicated flap during the 4 days of follow-up

flap. For example, it is important to visualize the lateral side of the leg in a range of 15–20 cm to identify a maximum of perforators, for a fibular flap. It is the same for an anterolateral thigh flap. Video recording with fluorescent tools is helpful to increase the accuracy of perforator localization. Conversely, the size of the field of exploration is not as important for flap monitoring, but the probe (laser light source and camera) must be small enough to allow exploration of deep areas (the oral

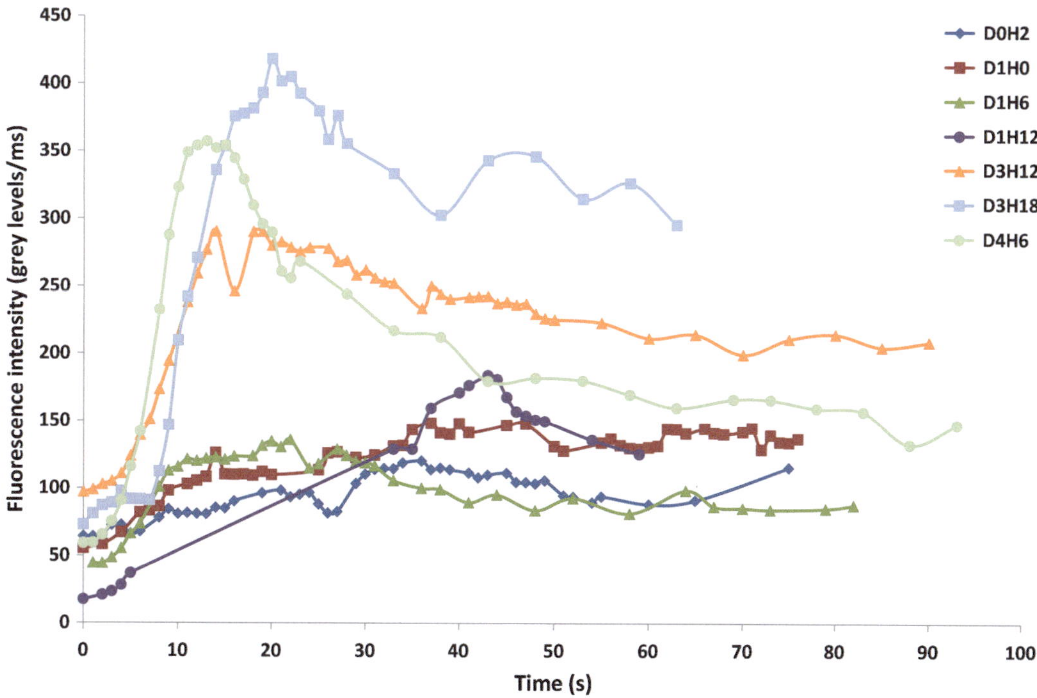

Fig. 32.11 Postoperative fluorescence intensity curves for a flap complicated by a venous thrombosis requiring surgical revision, 42 h after the first surgery. The number of curves was decreased to better read the figure. The dynamics of fluorescence were already abnormal at the first postoperative control (D0H2). Note the normalization of curves after revision (D3-D4)

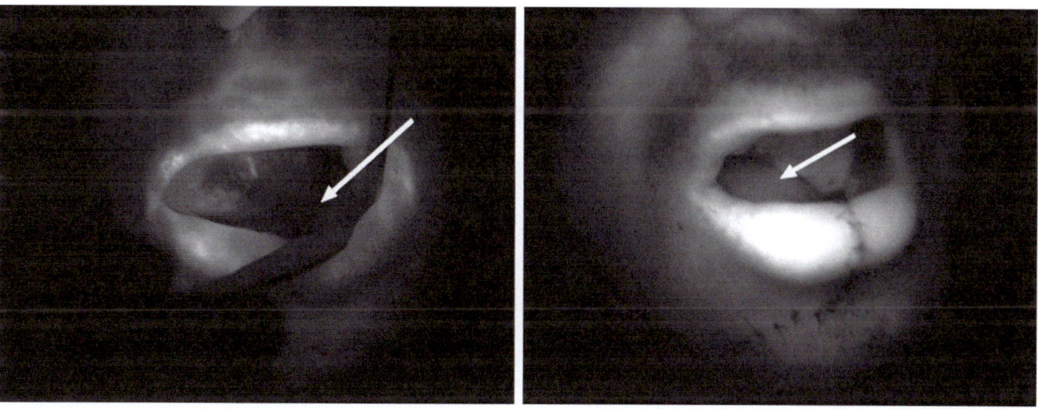

Fig. 32.12 Image of postoperative fluorescence in a fibular flap for mandibular reconstruction: on the *right side* for a non-complicated flap (good perfusion), on the *left side* for a flap complicated by a venous thrombosis

cavity in particular). Coupling the fluorescent probe with an endoscope could be a solution for flap monitoring in upper aero-digestive reconstruction. Betz previously tested this technique with a rigid endoscope in 2009 [13]. Fluostick™ (Fluoptics, Grenoble, France), a miniaturized version of the Fluobeam™, seems to be adapted to flap monitoring in the oral cavity.

Disclosure: The clinical trial mentioned in this article was supported by a grant from GRAVIT-innovation (Grenoble-Alpes) and the Grenoble University Hospital clinical research management.

References

1. Saint-Cyr M, Wong C, Schaverien M, Mojallal A, Rohrich RJ. The perforasome theory: vascular anatomy and clinical implication. Plast Reconstr Surg. 2009;124(5):1529–44.
2. Smit JM, Klein S, Werker PM. An overview of methods for vascular mapping in the planning of free flaps. J Plast Reconstr Aesth Surg. 2010;63(9):e674–82.
3. Iida H, Ohashi I, Kishimoto S, Umeda T, Hata Y. Preoperative assessment of anterolateral thigh flap cutaneous perforators by colour doppler flowmetry. Br J Plast Surg. 2003;56(1):21–5.
4. Rozen WM, Ashton MW, Stella DL, Ferris S, White DC, Phillips TJ, Taylor GI. Developments in perforator Imaging for the anterolateral thigh flap: CT angiography and CT-guided stereotaxy. Microsurgery. 2008;28(4):227–32.
5. Azuma R, Morimoto Y, Masumoto K. Detection of skin perforators by indocyanine green fluorescence nearly infrared angiography. Plast Reconstr Surg. 2008;122(4):1062–7.
6. Liu DZ, Mathes DW, Zenn MR, Neligan PC. The application of indocyanine green fluorescence angiography in plastic surgery. Microsurgery. 2011;27(6):355–64.
7. Giunta RE, Holzbach T, Taskov C, Holm PS, Brill T, Gansbacher B, et al. Prediction of flap necrosis with laser induced indocyanine green fluorescence in a rat model. Br J Plast Surg. 2005;58(5):695–701.
8. Lee BT, Matsui A, Hutteman M, Lin SJ, Winer JH, Laurence RG, et al. Intraoperative near-infrared fluorescence imaging in perforator flap reconstruction: current research and early clinical experience. J Reconstr Microsurg. 2010;26(1):59–65.
9. Onoda S, Azumi S, Hasegawa K, Kimata Y. Preoperative identification of perforator vessels by combining MDCT, doppler flowmetry, and ICG fluorescent angiography. Microsurgery. 2013;33(4):265–9.
10. Smit JM, Zeebregts CJ, Acosta R, Werker PM. Advancements in free flap monitoring in the last decade: a critical review. Plast Reconstr Surg. 2010;125(1):177–85.
11. Newman MI, Samson MC. The application of laser-assisted Indocyanine green fluorescent dye angiography in microsurgical breast reconstruction. J Reconstr Microsurg. 2009;25(1):21.
12. Holm C, Dornseifer U, Sturtz G, Basso G, Schuster T, Ninkovic M. The intrinsic transit time of free microvascular flaps: clinical and prognostic implications. Microsurgery. 2010;30(2):91–6.
13. Betz CS, Zhorzel S, Schachenmayr H, Stepp H, Matthias C, Hopper C, Harréus U. Endoscopic assessment of free flap perfusion in the upper aerodigestive tract using indocyanine green: a pilot study. J Plast Reconstr Aesthet Surg. 2013;66(5):667–74.

Fluorescein Detection of Myocardial Ischemia in an Experimental Model of Acute Coronary Occlusion

33

Fernando D. Dip, Alejandro Damonte, Gaston Quiche, Marcelo Damonte, Fernando M. Safdie, Nicolas Brozzi, Raul J. Rosenthal, and Pedro Ferraina

Introduction

Fluorescein has been studied and used since the nineteenth century. However, to our knowledge no studies have demonstrated the capability of fluorescein to evaluate and delineate infarcted myocardium after an ischemic event. The significant volume of distribution in largely vascularized tissues, such as the myocardium together with its low cost and minimal known adverse effects make fluorescein an ideal substrate for this experimental project [1].

We proposed to study the capacity of fluorescein to detect an injured area of myocardium secondary to an ischemic event. Therefore, fluorescein was intravenous administered in an in-vivo animal model. Ischemia was provoked via ligation of the second diagonal artery (Fig. 33.1). The individuals were scarified after the ischemic offense and the tissues harvested for analysis.

We hypothesize that fluorescein will provide a detailed anatomic demarcation of the ischemic myocardium and help us document tissue recovery after reperfusion. Considering that direct visualization of the injured myocardium does not allow clear identification of the compromised area, we believe that our study has a potential clinical application.

F.D. Dip, M.D., M.A.A.C. (✉)
Section of Minimally Invasive Surgery, Department of General and Vascular Surgery, The Bariatric and Metabolic Institute, Cleveland Clinic Florida, 2950 Cleveland Clinic Boulevard, Weston, FL 33331, USA

Oncological Surgical Division, Division of Surgical Research, Department of Surgery, Hospital de Clinicas Buenos Aires, University of Buenos Aires, Buenos Aires, Argentina
e-mail: fernandodip@gmail.com

A. Damonte, M.D. • G. Quiche, M.D.
Department of Surgery, Hospital de Clinicas Buenos Aires, University of Buenos Aires, Buenos Aires, Argentina

Division of Surgical Research, Hospital de Clinicas Buenos Aires, University of Buenos Aires, Buenos Aires, Argentina

M. Damonte, M.D.
Department of Surgery, Hospital de Clínicas, Jose de San Martin, University of Buenos Aires, Buenos Aires, Argentina

F.M. Safdie, M.D.
Department of General Surgery, Cleveland Clinic Florida, 2950 Cleveland Clinic Blvd., Weston, FL, USA

N. Brozzi, M.D.
Department of Cardiothoracic Surgery, Cleveland Clinic Florida, 2950 Cleveland Clinic Blvd., Weston, FL, USA

R.J. Rosenthal, M.D., F.A.C.S.
Section of Minimally Invasive Surgery, Department of General and Vascular Surgery, The Bariatric and Metabolic Institute, Cleveland Clinic Florida, 2950 Cleveland Clinic Boulevard, Weston, FL 33331, USA
e-mail: rosentr@ccf.org

P. Ferraina, M.D.
Hospital de Clinicas Jose de San Martin, Buenos Aires, Argentina

F.D. Dip et al. (eds.), *Fluorescence Imaging for Surgeons: Concepts and Applications*, DOI 10.1007/978-3-319-15678-1_33,

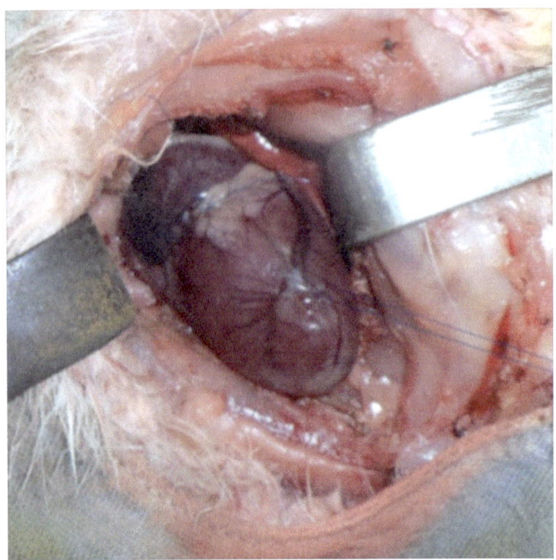

Fig. 33.1 Heart exposure and ligation of second diagonal artery

Although multiple authors have studied ischemia-induced cardiac injury in animal models, to our knowledge this is the first study demonstrating an in-vivo method for direct visualization of the injured myocardium [2].

Materials and Methods

All animal protocols were approved by the Institutional Animal Care and Use Committee (IACUC) at University of Buenos Aires. All studies were performed at Hospital de Clinicas "Jose de San Martin."

Fluorescein

Fluorescein is an organic compound, slightly soluble in water and alcohol. It is a fluorophore widely utilized as diagnostic tool in the field of ophthalmology and optometry. It has also been shown to be appropriate to evaluate hepatic function and liver blood flow. Originally described by Dr. Johann Fiedrich Wilhelm Adolf Baeyer (1835–1917) who was awarded with the Nobel Prize in biochemistry for his discovery. The fluorescence of this molecule is very intense; peak excitation occurs at 494 nm and peak emission at 521 nm (Fig. 33.2). Fluorescein has been used safely in humans in the past, but some side effects including nausea, vomiting, hives, hypotension, and rarely anaphylactic shock have been described in the past.

Ten New Zeeland ($N = 10$) rabbits were used for this study. Each rabbit served as its own control. General anesthesia was achieved using 35 mg/kg of ketamine in combination with 5 mg/kg of intramuscular xilacaine. The rabbits were clipped at the sternal and ear level. Intravenous access was achieved by placement of a 20-gauge angiocath in the area of the left pinna. A lateral tracheostomy was placed using a number 3 endotracheal tube to secure the airway. Pediatric ventilators were used. Propofol and fentanyl were utilized to achieve proper sedation and analgesia.

The surgical technique included: sternotomy, pericardiectomy, cardiac luxation, and ligation of the second diagonal artery with 6.0 polypropylene suture (Fig. 33.1).

The ischemic area was analyzed under the xenon and UV light before and after the administration of 0.01 mg/kg of sodium fluorescein. Blood samples were taken and quantification of cardiac enzymes including CK, CK-MB, LDH, and Troponin was done before and 90 min after the induced ischemia. Oxygen saturation, heart rate, respiratory rate, and EKG were monitored continuously.

All the animals were euthanized according to the IACUC UBA regulations. The heart and lungs were resected en bloc, and submitted to the pathology department.

Results

Ten New Zeeland ($N = 10$) rabbits with a mean average body weight of 3.1 ± 0.6 kg were used for this study. No perioperative mortality was noticed and all the rabbits were euthanized at the end of the study.

Every rabbit functioned as its own control. Initial EKG, HR, RR, and oxygen saturation were measured and used as baseline (BL) values. No abnormalities were found on these parameters confirming an adequate health status of the animals.

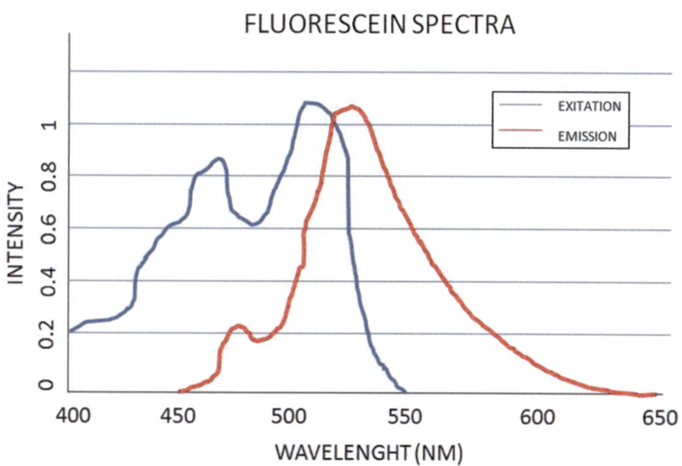

Fig. 33.2 Fluorescein spectrum

Table 33.1 Results

Cardiac enzymes	Baseline	Post-ischemia	*P* value
Troponin (ng/ml)	0.06±0.06	19.6±5.9	0.00000005
CPK (U/L)	1,072±121.7	359.5±95.7	0.00000002
CPK-MB (ng/ml)	0.89±0.42	3.89±1.9	0.00028593
LDH (U/L)	159.7±112.2	1,012±359.9	0.00000781

No significant difference was observed between the values for pre- and post-myocardial infarction in the categories of heart rate (145±5 bits/min) or respiratory rate (28±5 breaths/min). Consistently, no variation was appreciated in the saturation of oxygen post myocardial infarction (97±2 %).

On the contrary, an ST (1.8±0.65 mm) elevation was observed in the EKG tracing immediately after the ligation of the second diagonal artery.

Biochemical (enzymatic) analysis was performed at baseline and at 90 min after complete occlusion of the second diagonal artery. These were compared using Student's *t* test. Results were considered significant when the p value <0.05.

LDH (BL) 159.7±112.2 (U/L) vs. LDH (PI) 1,012±359.9 (U/L) ($p<0.0000781$).

CPK (BL) 1,072±121.7 (U/L) vs. CPK (PI) 359.5±95.7 (U/L) ($p<0.00000002$).

CPK-MB (BL) 0.89±0.42 (ng/ml) vs. CPK-MB (PI) 3.89±1.9 (ng/ml) ($p<0.00028593$).

Troponin (BL) 0.06±0.06 (ng/ml) vs. Troponin (PI) 19.6±5.9 (ng/ml) ($p<0.00000005$) (see Table 33.1).

The infarcted area was analyzed with the xenon and UV light before and after the administration of fluorescein 0.01 ml/kg via peripheral intravenous injection. Prior to the administration of fluorescein, the xenon and UV light failed to delineate the compromise-ischemic area (see Fig. 33.1).

Despite the injection of fluorescein, the infarcted area was still not demarcated when the specimen was observed under the xenon light. However, a significant contrast was observed between the well vascularized tissue and the infarcted area when evaluated under UV light (see Figs. 33.3a–c and 33.4). The area measured 0.722±0.39 cm^2.

Once the rabbits were euthanized, the specimens were collected and sent to the pathology laboratory for evaluation and analysis. Consistently, the ischemic area under the UV light demonstrated significant muscle fibers disorganization, inflammation, eosinophilia, and vacuolization of the cytoplasm (Figs. 33.5 and 33.6).

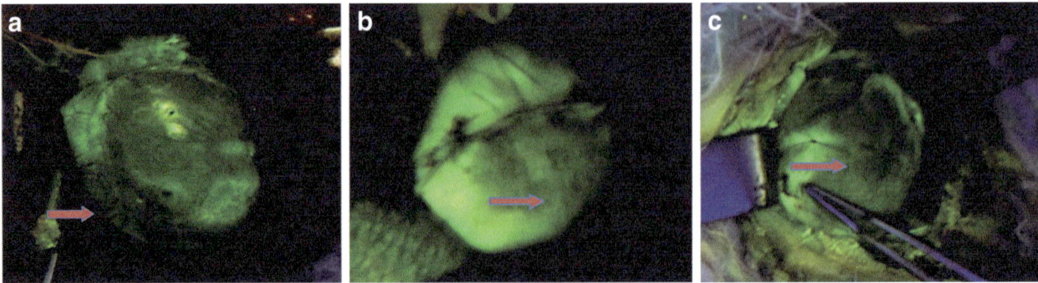

Fig. 33.3 Anterior (**a**), lateral (**b**) and inferior (**c**) view of the ischemic heart under UV light after the administration of fluorescein

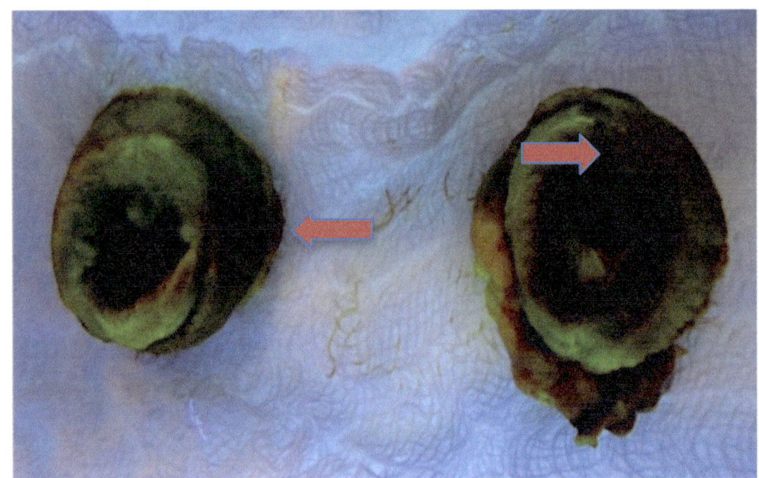

Fig. 33.4 Transverse view of the heart after administration of fluorescein. *Arrows* indicate the ischemic area. *Left* image Area 1 and *right* image Area 2

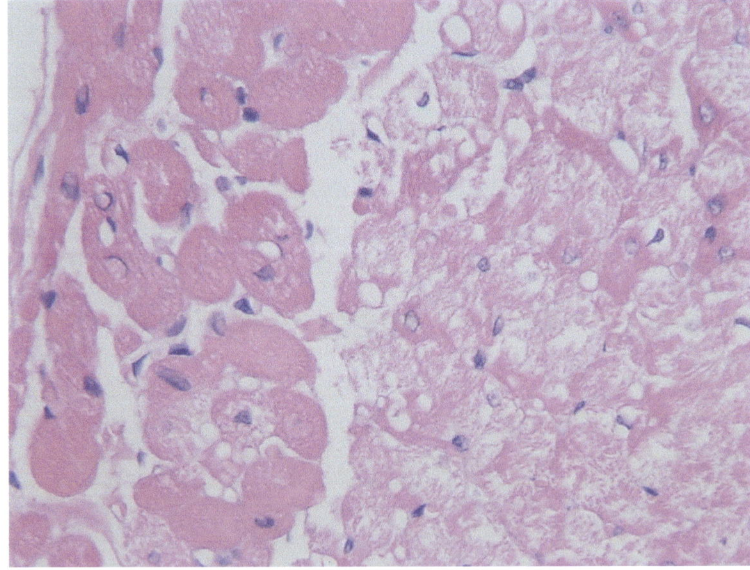

Fig. 33.5 Microscopic view of the ischemic tissue demonstrating muscle fibers disorganization, inflammation, eosinophilia, and vacuolization of the cytoplasm (Area 1)

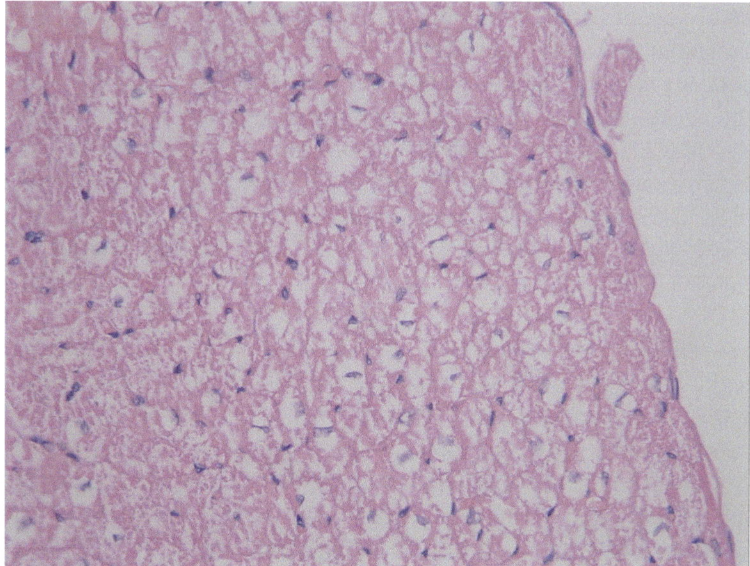

Fig. 33.6 Microscopic view of the ischemic tissue demonstrating muscle fibers disorganization, inflammation, eosinophilia, and vacuolization of the cytoplasm (Area 2)

Discussion

We present the use of sodium fluorescein as an aid to visualize the compromised myocardium after an acute ischemic event. The anatomical resemblance with the human heart, simple and quick access to the mediastinum, and the size of the coronary arteries make the rabbit a great animal model for this experiment.

After complete occlusion of the second diagonal artery by ligation of this vessel, there was an immediate increase on the cardiac enzymes including LDH, CPK, CPK-MB, and Troponin. Significant changes in the EKG trace included ST elevation in every individual confirming myocardial injury [3].

Currently, the most widely used methods to evaluate an ischemic area in the heart include the EKG, 2-D echocardiogram, and coronary angiogram. The EKG is a non-invasive, inexpensive modality. The 2-D echocardiogram can detect areas with hypo or complete akinesia of the ischemic area [4]. The coronary angiogram is an invasive and expensive method that allows an excellent evaluation of the vascular anatomy of the heart; it has the advantage of being diagnostic and therapeutic. The high sensitivity and specificity make it the gold standard for the diagnosis of ischemic cardiomyopathy. Nonetheless, the coronary angiogram does not allow a direct visualization of the ischemic tissue but rather the abrupt discontinuation of the blood flow [5]. Therefore, our interest is to describe a method that not only will allow a direct and accurate way to measure the affected myocardium after an ischemic event but rather will also allow the visualization of the recovered myocardium after revascularization procedures [6–8].

In our experimental model, we demonstrated in real time the safety, feasibility, and accuracy of the fluorescein in delineating the compromised area after an ischemic injury when analyzed under the UV light. Here, we also demonstrated how the previously affected myocardium recovers its capacity to absorb the fluorescein and express fluorescence once revascularization has occurred. This could potentially be used intraoperatively to evaluate the effectiveness of revascularization procedures such as coronary artery by-pass grafting (CABG) [1, 9].

Many different substances have been utilized in the medical industry. The indocynine green (ICG) has an absorption of 780 nm wavelength and can emit up to 830 nm of energy. ICG has been used to evaluate liver remnant function and currently is being used to identify intraoperatively the biliary system in order to prevent or reduce the risk of injuring the common bile duct during laparoscopic cholecystectomies [9–12]. ICG has been also used for visualization of the ureters during pelvic operations, to prevent inadvertent injuries. Methylene blue has been used to evaluate anastomotic leaks and to treat methemoglobinemia [10, 13]. We decided to use fluorescein and UV light since this possesses a larger quantum of energy which results in a greater fluorescent expression [1].

Due to its low molecular weight (332.306 g/mol) fluorescein diffuses through capillaries, rendering this technique not applicable to sentinel lymph node biopsies [14].

To confirm the unique characteristic of fluorescein to identify ischemic myocardium, we analyzed the infarcted area under regular light. We found no differences either before or after ligation of the second diagonal artery. On contrary, once we administered the fluorescein and analyzed the same tissue sample under 530 nm wavelength light, we were able to identify the ischemic area of myocardium distal to the previously ligated second diagonal artery. Since the emission field of fluorescein is detectable to the naked eyes, no filters were needed. This is one of the advantages of fluorescein over ICG [15].

Evaluation of the specimens by specialized pathologist confirmed that those areas that did not stain with the sodium fluorescein were actual areas that had suffered an ischemic event. Although there are softwares as the one described in the work of Diana et al. to perform a digital subtraction to calculate the exact compromised surface, we chose not to use it, since the objective of our work was to prove the principle that this method was safe, feasible, and had clinical applications. We hope that this first study serves as a platform for more complex studies that will provide a better understanding of the dynamics of ischemia and reperfusion processes as well as guide revascularization procedures of the heart [16, 17].

Conclusion

The use of sodium fluorescein via intravenous administration is safe, feasible, and was able to identify compromised myocardium after induced ischemia. The findings were supported by the changes noticed in the EKG tracing, biochemistry analysis and validated by histopathologic analysis of the samples. This study provides evidence to support the principle that sodium fluorescein has a potential application for in-vivo evaluation of ischemia and revascularization procedures.

Acknowledgements The authors thank Giselle Romero, M.D. Staff Pathology Department, Hospital de Clínicas, "Jose de San Martin" University of Buenos Aires.

Lic.Paulo Daniel Pascuini. Economist, University of Buenos Aires.

References

1. Valeur B. Molecular fluorescence: principles and applications, vol. 1. New York, NY: Wiley-VCH; 2001.
2. Podesser B, et al. Epicardial branches of the coronary arteries and their distribution in the rabbit heart: the rabbit heart as a model of regional ischemia. Anat Rec. 1997;247(4):521–7.
3. Kobayashi T, et al. Electrocardiograms corresponding to the development of myocardial infarction in anesthetized WHHLMI rabbits (Oryctolagus cuniculus), an animal model for familial hypercholesterolemia. Comp Med. 2012;62(5):409–18.
4. Jia C, Olafsson R, Kim K, Kolias TJ, Rubin JM, Weitzel WF, Witte RS, Huang SW, Richards MS, Deng CX, O'Donnell M. Two-dimensional strain imaging of controlled rabbit hearts. Ultrasound Med Biol. 2009;35(9):1488–501.
5. de Carvalho VB, Macruz R, Arie S, Martins JR, Pina RS, de Oliveira SA, Pileggi F, Décourt LV, Zerbini Ede J. Development of coronary atherosclerosis evaluated by cinecoronariography. Arq Bras Cardiol. 1980;34(6):431–9.
6. Soltesz EG, Laurence RG, De Grand AM, Cohn LH, Mihaljevic T, Frangioni JV. Image-guided quantification of cardioplegia delivery during cardiac surgery. Heart Surg Forum. 2007;10:E381–6.
7. Taggart DP, Choudhary B, Anastasiadis K, Abu-Omar Y, Balacumaraswami L, Pigott DW. Preliminary experience with a novel intraoperative fluorescence imaging technique to evaluate the patency of bypass grafts in total arterial revascularization. Ann Thorac Surg. 2003;75:870–3.

8. Nakayama A, et al. Functional near-infrared fluorescence imaging for cardiac surgery and targeted gene therapy. Mol Imaging. 2002;1(4):365–77.
9. Frangioni JV. In vivo near-infrared fluorescence imaging. Curr Opin Chem Biol. 2003;7:626–34.
10. De Grand AM, Frangioni JV. An operational near-infrared fluorescence imaging system prototype for large animal surgery. Technol Cancer Res Treat. 2003; 2:553.
11. Eiichi TMD, Shunsuke Ohnishi MD, Rita G, Laurence B. Real-time intraoperative ureteral guidance using invisible near-infrared fluorescence. J Urol. 2007; 178(5):2197–202.
12. Rangaraj AT, et al. Real-time visualization and quantification of retrograde cardioplegia delivery using near infrared fluorescent imaging. J Card Surg. 2008;23(6):701–8.
13. Nyarangi-Dix JN, Pahernik S, Bermejo JL, Prado L, Hohenfellner M. Significance of the intraoperative methylene blue test for postoperative evaluation of the vesicourethral anastomosis. Adv Urol. 2012;2012:702412.
14. van der Vorst JR, Schaafsma BE, Verbeek FP, Swijnenburg RJ, Hutteman M, Liefers GJ, van de Velde CJ, Frangioni JV, Vahrmeijer AL. Dose optimization for near-infrared fluorescence sentinel lymph node mapping in patients with melanoma. Br J Dermatol. 2013;168(1):93–8.
15. Ishizawa T, Fukushima N, Shibahara J, Masuda K, Tamura S, Aoki T, Hasegawa K, Beck Y, Fukayama M, Kokudo N. Real-time identification of liver cancers by using indocyanine green fluorescent imaging. Cancer. 2009;115(11):2491–504.
16. Diana M, Noll E, Diemunsch P, Dallemagne B, Benahmed MA, Agnus V, Soler L, Barry B, Namer IJ, Demartines N, Charles AL, Geny B, Marescaux J. Enhanced-reality video fluorescence. a real-time assessment of intestinal viability. Ann Surg. 2013;259:700.
17. Harrar N, Idali B, Moutaouakkil S, el Belhadji M, Zaghloul K, Amraoui A, Benaguida M. Anaphylactic shock caused by application of fluorescein on the ocular conjunctiva. Presse Med. 1996;25(32):1546–7. PMID 8952662 (in French).

Spatial and Temporal Monitoring of Perfusion with Indocyanine Green Videoangiography in the Rat Heart

34

Outi M. Villet, Jarmo T. Alander, and Ari L. Harjula

Abbreviations

DNIF	Dynamic near-infrared fluorescence (imaging)
DSA	Digital subtraction angiography
IA	Intra-arterial
ICG	Indocyanine green
ICGA	ICG angiography
IV	Intravenous
LAD	Left anterior descending
LED	Light-emitting diode
LITA	Left internal thoracic artery
NIR	Near-infrared
PCA	Principle component analysis
RBC	Red blood cell
RoI	Region of interest

Introduction

Indocyanine green (ICG) contrast agent has been used in clinical medicine since the 1960s. It is still the only near-infrared (NIR) fluorescence dye approved for clinical use. ICG is ideal for angiography, as it confines to the vascular compartment by binding to plasma proteins and is non-toxic. It does not have any known metabolites and is fast extracted by liver into bile allowing repeated use. ICG fluoresces at about 800 nm and longer near-infrared wavelengths. These wavelengths are located in the tissue optical window without much autofluorescence resulting in good signal-to-noise ratio even with quite modest optical and NIR imaging techniques.

ICG has recently gained a great deal of interest in several fields of surgery [1]. A comprehensive review was recently published by Alander et al. regarding ICG in intraoperational use [1]. Most studies have used ICG in visualization of blood circulation, i.e. in ICG angiography (ICGA). However, ICG can also be used in visualization of lymph nodes and even peripheral nerves [2]. In this study, we used ICG in evaluating perfusion dynamics in infarcted and healthy rat myocardium. In addition to basic visualization of blood circulation, dynamics of fluorescence were used to evaluate the outcome of a novel revascularization procedure—a modified Vineberg operation.

Among hundreds of ICGA publications there seems to be only a few that deal with dynamics of ICGA. Kang et al. have used ICGA dynamics analysis for measuring peripheral vascular insufficiency [3, 4]. Awano et al. have used ICGA dynamics analysis with ischemic stroke [5]. Chen et al. have used analysis of ICGA dynamics in

O.M. Villet, Ph.D. (✉) • A.L. Harjula, M.D./Ph.D.
Heart and Lung Center, HUCH,
Haartmaninkatu 4, Helsinki 00100, Finland
e-mail: outivillet@gmail.com; ari.harjula@hus.fi

J.T. Alander, Ph.D.
Department of Electrical and Energy Engineering,
University of Vaasa, Yliopistonranta 10,
65101 Vaasa, Finland
e-mail: jarmo.alander@uva.fi; jal@uwasa.fi

F.D. Dip et al. (eds.), *Fluorescence Imaging for Surgeons: Concepts and Applications*,
DOI 10.1007/978-3-319-15678-1_34,

cases of cerebral aneurysms and cerebral arteriovenous malformations [6]. Oda et al. analyzed the ICGA dynamics of 39 patients having 43 intracranial aneurysms that underwent microsurgical clipping [7]. Jonak et al. have measured ICGA dynamics when studying in vivo laser scanning microscopy of human skin [8].

Very recently, Kono et al. have compared digital subtraction angiography (DSA), intra-venous ICGA (IV-ICGA), and intra-arterial ICGA (IA-ICGA) in cerebral arteriovenous malformation surgery [9]. They compared the dynamics of observed fluorescence intensity between IA- and IV-ICGA. According to their study, the IA-ICGA seems to give better discrimination between the different vessels. This is due to much shorter clearance time period (about 15 s) of IA-ICGA than that of IV-ICGA (about 15 min). This means among other things that IA-ICGA can be applied several times during any vascular operation [10].

Holling et al. claim that dynamic ICGA provides better intraoperative understanding of arteriovenous fistulae [11]. They measured the rise time of parenchyma and transit time from artery to parenchyma before and after occlusion of fistulae. Their results revealed a significant change in flow dynamics in the draining vein after surgical occlusion. The analysis showed rise time and transit time to be significantly shorter, when comparing pre-occlusion with post-occlusion.

A more technical contribution to ICGA dynamics related to neovascularization in retina is given by Holmes et al. [12]. They give a mathematical model of the flow dynamics, which is further used in the image processing of videoangiography.

Murakami et al. have measured the time to peak for cerebral cavernous malfunctions by reviewing recorded video ICGA [13]. Woodard and Most have estimated the perfusion of facial flaps using ICGA dynamics measured by SPY [14]. Son et al. have recently used ImageJ for quantitative analysis of intraoperative ICGA videos for aneurysm surgery [15]. Mohajerani et al. have used Principal Component Analysis (PCA) in analysis of ICGA dynamics in patients having rheumatoid arthritis in hand joints [16]. Piper et al. have demonstrated whole-body skin perfusion monitoring for humans using Dynamic Near-Infrared Fluorescence (DNIF) imaging [17] using ICGA and image processing [18]. They have made the image segmentation by the time it takes each pixel to reach 50 % of its final intensity.

Ku and Choi have measured the dynamics of mice cerebral blood flow with ICG [19]. Amoozegar et al. have developed a method for small animal in vivo inner organ imaging using ICG fluorescence dynamics [20]. Mücke et al. have analyzed ICGA dynamics using a mice model [21].

Methods

Surgical Procedure

Rats were anesthetized, intubated, and ventilated during open chest procedure. Heart was exposed by median sternotomy. Myocardial infarction was induced by left anterior descending (LAD) artery ligation. To revascularize the infarcted myocardium, left internal thoracic artery (LITA) was extracted from chest wall and the distal open tip was tunneled into the infarcted left ventricular wall. Chest was closed and the rat was tended postsurgically.

Four weeks postsurgery rat chest was opened and the myocardium and LITA were imaged with fluorescence microscopy using the indocyanine green (ICG) dye (1 mg/kg diluted in saline) slowly injected into the left ventricle through apex. Pharmaceutically available form of dry ICG is stable at room temperature. As the dissolved ICG decomposes at room temperature, the solution had to be prepared just prior to injection. Dry ICG was dissolved in water and diluted with saline for isotonic solution.

Imaging Instrumentation

ICG imaging is a type of optical fluorescence imaging. The principle of ICG angiography (ICGA) is shown in Fig. 34.1. When ICG is used

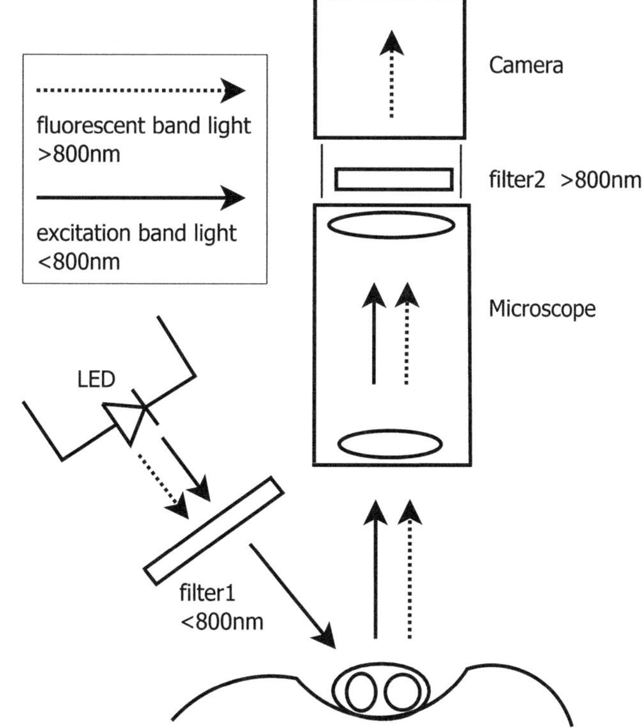

Fig. 34.1 Imaging setup: The imaging instrumentation consists of the following parts: LED light source (*left*), filter1 for filtering out wavelengths over 800 nm, the object of imaging (*bottom*), operation microscope, filter2 to filter out wavelengths under 800 nm, and the NIR-sensitive digital camera, which is further connected to a PC for image recording. The *solid arrows* denote light of wavelengths under 800 nm (excitation), while the *dotted arrows* denote light of wavelengths over 800 nm (fluorescence)

with an operational microscope, it closely resembles ordinary fluorescence microscopy and similar optical setup can be used as described in Fig. 34.1.

Tissue is illuminated by a near-infrared (NIR) light source giving excitation energy to ICG molecules. The optimum wavelength of the excitation is about 780 nm, while any radiation above 800 nm is filtered away, i.e. in the fluorescence band. Three main possible light source types can be used: halogen bulb + filter, LED + filter, and a NIR laser. In our setup we used LED + filter light source, as it is energy-efficient, economical and easy to control electronically. The LED spectrum somewhat overlaps the fluorescence spectra making the filter necessary for a LED-type light source (Fig. 34.1). The LED (LED780-66-60, Roithner Lasertechnik GmbH, Vienna, Austria) produced maximum 1.0W light power having spectral maximum at 780 nm and 30° light cone without lens. The LED was fed by a 500 mA constant current source.

Fluorescence was recorded by a NIR-sensitive camera (ORCA-05G, Hamamatsu, Japan) without mechanical shutter. The image format was 16-bit monochrome of which 14 bits were used by the camera. Thus the intensities are represented by integer values in the range [0,4095].

Without filters excitation light and fluorescence signals are summed at the sensor and are inseparable in the resulting image. To enable the separation of these signals both camera and light source filters are important for the imaging system. The excitation light should not contain any fluorescence wavelengths, as they should originate only from the fluorescing ICG. Thus a filter was placed in front of the camera on the microscope ocular to block the excitation light. The intensity of the fluorescence is much lower than that of the excitation light. Ideally, the two filters should divide the spectrum into two strictly non-overlapping bands. This can be best done using interference filter pairs separating the spectrum at about 800 nm (Chroma

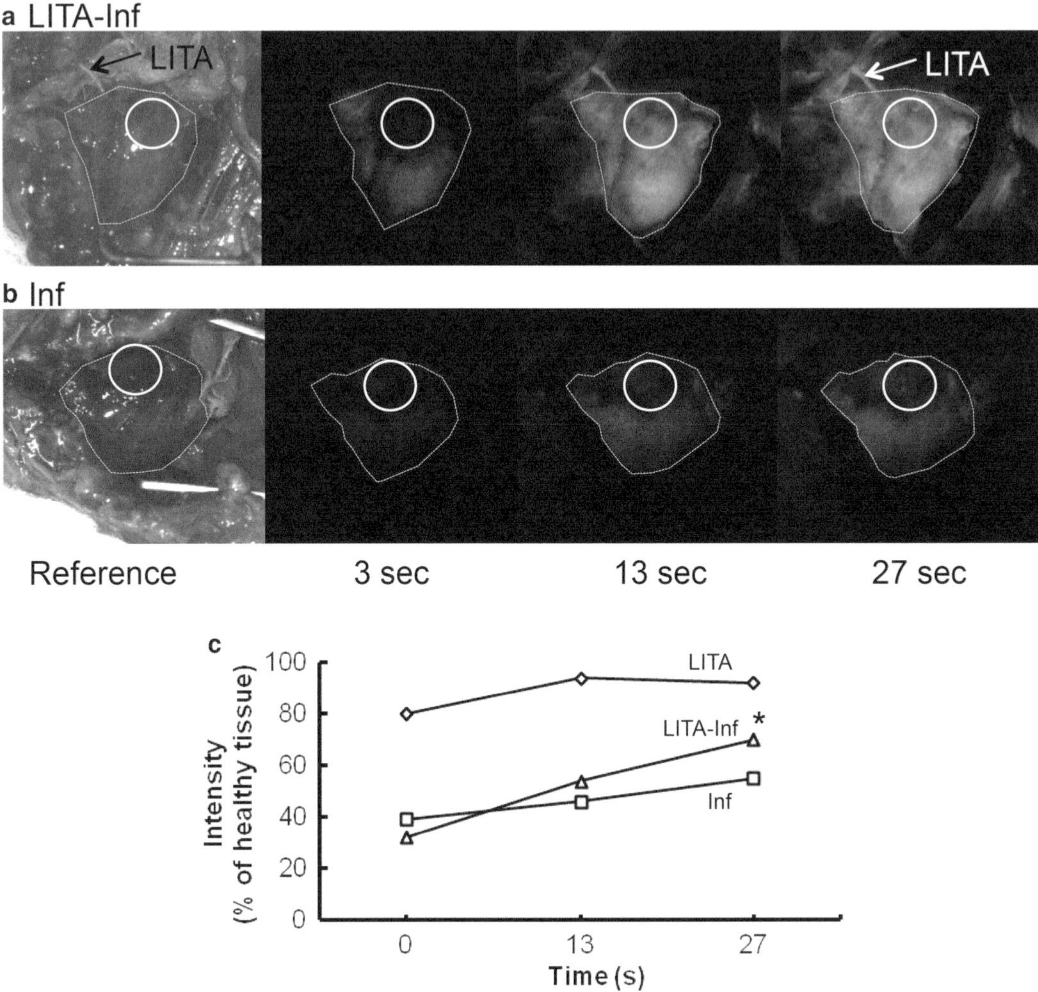

Fig. 34.2 ICG imaging of rat heart 4 weeks after myocardial infarction with (**a**) (LITA-Inf) and without (**b**) (Inf) LITA implantation. *Circle* (infarct area), *Dotted line* (heart outline), *Arrows* pointing at implanted LITA. (**c**) ICG intensity dynamics from three time points. Preliminary results imply that LITA implantation (LITA-Inf) increases perfusion in the infarcted area. $*p<0.05$ LITA-Inf vs. Inf, LITA

Technology, Brattleboro, VT, USA). In addition, a standard operational microscope (Wild 327733, Heerbrugg, Switzerland) was used to focus the image on the camera sensor and a PC for recording the ICGA videos via IEEE 1394 interface.

Visual reference images (Fig. 34.2, leftmost images) prior to fluorescence imaging were acquired with the same setup when the LED light source was replaced by an ordinary microscope broadband white light source while keeping the filter2 (Fig. 34.1) in front of the camera. ICGA videos were acquired with a rate of 3 frames per second with the exposure time of 0.05 s using HCImage software provided by Hamamatsu. During ICGA the operation room was darkened and all extra lighting was switched off. Analysis of the videos was done offline as described in the Image Processing section.

Image Processing

Image processing was done using ImageJ [22], for which a few plugins described next were implemented. The consecutive frames of the interesting time interval of the recorded ICGA video were collected into a directory, which was input to the ImageJ program as an image stack. In addition to the ICGA frames the stack may also contain dark frames (at the end of the stack, which is created according to the alphabetical order of the file names) allowing background subtraction at the RoIs. A file giving general parameters and coordinates of RoIs (Regions of Interests) is used by the implemented plugins. Currently, there are three ImageJ plugins for evaluating the dynamics:

- ICGA_Read_RoIs: reads RoI from a file into a table for the other plugins. Either a fixed number of RoIs is given for each frame, or one RoI set is copied to all frames for further editing by the next plugin.
- ICGA_Edit_RoIs: The stack is processed from the last one (most intense fluorescence) one RoI at a time allowing the user to move the RoI by a mouse. Updates the list of RoIs accordingly and finally outputs them to the log window, from where the list can be saved into a file using copy/paste for later use.
- ICGA_Measure_RoIs: scans the stack frame by frame and calculates the average intensity (signal) of the RoIs and the whole frame. If dark frames are included in the stack the corresponding average background (baseline) at the RoIs is subtracted from the average signal. Outputs the result to the log window for further processing by, e.g. Excel (copy/paste).

Microsoft Excel was used in further processing of RoI intensity data.

Regions of Interest

Beating heart is challenging for automatic image registration as distinctly defined landmarks are difficult to establish in fluorescence images. Therefore, RoIs were defined manually from the ICGA videos. Plugins were used to help in the manual definition of RoIs by giving default values for the coordinates of each RoI. Currently, the RoIs are square shaped—the edge length of which is given in the RoI-file among other parameters. Observe that the coordinates of the RoI give the center point of the RoI, not the upper left corner which is the default in ImageJ. This geometry allows straightforward RoI size tuning, if needed, by using only one parameter, the edge length.

Calibration

Video sequence can contain both black (dark current) and white (sensitivity) frames for calibration. Black frames were used to estimate average background intensity for each RoI. There can be any number of black frames at the end of the stack. Here, we used the average of ten black frames for background subtraction.

Analysis of Dynamics

The conventional method to display ICGA is to use absolute intensity. In order to be able to compare ICG fluorescence levels, the average intensities at RoIs were normalized by dividing the average intensity of the tissue region of interest (infracted myocardium) with the average intensity of the reference region (healthy myocardium in the same heart) and displaying the resulting values as percentages (Fig. 34.2c). Three frames (1 s) were pooled together to reduce noise.

In these experiments, we have used the following ways to display the dynamics extracted from the raw video frames:

- *A*: absolute intensity,
- *N*: normalization of intensity so that the intensities are approximately equal in the pseudo steady state, and
- *N*/*A*: *N* divided by the average intensity of the whole frame.

From the above information, we can evaluate several parameters characterizing dynamics like:

- T_0, start time: the time that the fluorescence starts.
- T_s, settling time: the time that is needed for the fluorescence to settle to a constant level. In practice there is no clear settling time and strictly speaking the concentration of ICG is constantly decreasing due to extraction by liver. Therefore, we have used T_s as the time

within which the fluorescence seems to be constant enough for most RoIs.

- I_s, settling intensity: the intensity that the RoI seems to reach after a reasonable settling time T_s.
- T_{50}, rise time to 50 %: the time from T_0 it takes the intensity to increase to the value $I_s/2$.

Results and Discussion

Our imaging setup and developed analyses system enabled us to perform ICGA in a small animal model with a fast moving object, the rat heart. The sensitivity and the resolution of the images acquired with the system seem to be adequate for at least qualitative in vivo visual monitoring of perfusion and quantitative perfusion estimation offline.

Our preliminary results showed patency of implanted LITA 4 weeks post operation (Fig. 34.2a). This was clearly visible from the ICGA videos as the ICG filled the implanted LITA. In addition, the direction of ICG flow toward the myocardium was observed. Preliminary studies combining LAD ligation with LITA implantation also demonstrated the direction of flow in the attached LITA. Interestingly, as the ICG is injected through apex into the left ventricle, firstly only the healthy tissue of the heart receives the dye leaving the infarct area dark. This is followed by LITA staining from the direction of the subclavian artery toward the myocardium followed by staining of the infarcted area. These preliminary studies imply the possibility of the implanted LITA to actually form collaterals in the damaged area and provide it with a totally new blood source (Fig. 34.2).

Our preliminary experiments with the imaging system showed a typical perfusion in a healthy myocardium and a darker poorly perfused area in a heart 4 weeks post LAD ligation (Fig. 34.2b). It is impossible to establish perfectly equal infarct sizes between animals with standard LAD ligation method. In general, a good average value for the infarct area is estimated by histological methods. However, to quantify the damaged area from the whole heart is time-consuming with standard histological methods. Thus the ICG imaging method could provide a rapid and effective tool to estimate the infracted area in the rat heart.

These studies were performed with spontaneously beating rat hearts. As the recording frequency is only 3 frames per second in our current system, it is not possible to acquire images exactly at the same time point in the cardiac cycle, though this was attempted. Therefore, some noise will be introduced in the perfusion measurements from individual time points. As we will perform a more through analyses of the dynamics data from all the time points this variation will become less significant.

Observe that the quantum efficiency of ICG is nonlinear, which means in practice that the estimation of ICG concentration based on the highest intensity values may contain large systematic error. Another source of small systematic error may be due to the temporal changes of the molecular environment of the ICG molecules. Another error source type is the absorption of the fluorescence back to the tissue, primarily to blood [22]. Here we have assumed that the concentrations are always small enough that these types of errors are negligible, however.

Even if the operating room was darkened, there was some ambient light present. Another source of background light was the light source. Even if it does not generate much light in the fluorescent band and even if that part of the source spectra was filtered, there is some small amount of light leaking through the filter also in the fluorescent band. The third source of the background intensity is the camera sensor that generates some amount of signal even in total darkness. The recorded black frames contain all these so that the subtraction of the average background should mostly eliminate the effect of these error sources from the analyzed frames.

Conclusions and Future

Here, we have applied our method in a revascularized infarcted rat myocardium. However, as the ICG is non-toxic and approved for clinical use, the proposed method could also be used for intraoperational perfusion monitoring of other perfusion sensitive organs such as brain or liver in humans. Additionally, the cost of ICGA is relatively small compared to other used imaging methods.

This imaging method allows rapid evaluation of tissue perfusion in a relatively small and moving area of interest. The sensitivity and the resolution of the images seem to be adequate for at least qualitative in vivo visual monitoring of perfusion and quantitative perfusion estimation offline.

The ICGA dynamics depends on the structure and dynamics of the cardiovascular system. In this work ICGA dynamics was monitored only by measuring the average intensity at three one second long time intervals (at 0, 13 and 27 s), which allowed us to compare perfusion at several regions of interest. However, a much more detailed monitoring of intensity and thus perfusion is possible both spatially and temporally, which could be used to assess as well local as organ level perfusion.

Our future activities include cardiovascular system identification based on the temporal and spatial ICGA intensity dynamics. The resulting system model could be used for, e.g. intensity prediction in perfusion estimation or assessment of the functionality of the cardiovascular system in general. Further studies will be needed to implement this method to online processing. In principle, using high-performance computing by, e.g. field programmable gate arrays (FPGA) the online processing could include the video processing. This would create totally new possibilities for the method in interactive intraoperational clinical use.

Acknowledgements Funding for this project was provided by The Finnish Academy of Science (A.H.), The Hospital District of Helsinki and Uusimaa EVO-grants (A.H.), Finnish Foundation for Cardiovascular Research (O.V.), Foundation of Emil Aaltonen (O.V.), Finnish Angiological Foundation (O.V.) and Bothnia-Atlantica/Field-NIRce (J.A). Dr. Tommi Pätilä and Dr. Yukiko Imanishi are acknowledged for study design and experimental help, respectively. MSc. Kari Kaunonen and MSc. Kari Koivuporras are acknowledged for constructing the LED light.

References

1. Alander JT, Kaartinen I, Laakso A, Patila T, Spillmann T, Tuchin VV, et al. A review of indocyanine green fluorescent imaging in surgery. Int J Biomed Imaging. 2012;2012:940585.
2. Morimoto D, Isu T, Kim K, Imai T, Yamazaki K, Matsumoto R, et al. Surgical treatment of superior cluneal nerve entrapment neuropathy. J Neurosurg Spine. 2013;19(1):71–5.
3. Kang Y, Lee J, Kwon K, Choi C. Application of novel dynamic optical imaging for evaluation of peripheral tissue perfusion. Int J Cardiol. 2010;145(3):e99–101.
4. Kang Y, Lee J, An Y, Jeon J, Choi C. Segmental analysis of indocyanine green pharmacokinetics for the reliable diagnosis of functional vascular insufficiency. J Biomed Opt. 2011;16(3):030504.
5. Awano T, Sakatani K, Yokose N, Kondo Y, Igarashi T, Hoshino T, et al. Intraoperative EC-IC bypass blood flow assessment with indocyanine green angiography in moyamoya and non-moyamoya ischemic stroke. World Neurosurg. 2010;73(6):668–74.
6. Chen SF, Kato Y, Oda J, Kumar A, Watabe T, Imizu S, et al. The application of intraoperative near-infrared indocyanine green videoangiography and analysis of fluorescence intensity in cerebrovascular surgery. Surg Neurol Int. 2011;2:42.
7. Oda J, Kato Y, Chen SF, Sodhiya P, Watabe T, Imizu S, et al. Intraoperative near-infrared indocyanine green-videoangiography (ICG-VA) and graphic analysis of fluorescence intensity in cerebral aneurysm surgery. J Clin Neurosci. 2011;18(8):1097–100.
8. Jonak C, Skvara H, Kunstfeld R, Trautinger F, Schmid JA. Intradermal indocyanine green for in vivo fluorescence laser scanning microscopy of human skin: a pilot study. PLoS One. 2011;6(8):e23972.
9. Kono K, Uka A, Mori M, Haga S, Hamada Y, Nagata S. Intra-arterial injection of indocyanine green in cerebral arteriovenous malformation surgery. Turk Neurosurg. 2013;23(5):676–9.
10. Yamamoto S, Kim P, Kurokawa R, Itoki K, Kawamoto S. Selective intraarterial injection of ICG for fluorescence angiography as a guide to extirpate perimedullary arteriovenous fistulas. Acta Neurochir (Wien). 2012;154(3):457–63.
11. Holling M, Brokinkel B, Ewelt C, Fischer BR, Stummer W. Dynamic ICG fluorescence provides better intraoperative understanding of arteriovenous fistulae. Neurosurgery. 2013;73(1 Suppl Operative): ons93–8. discussion ons99.
12. Holmes T, Invernizzi A, Larkin S, Staurenghi G. Dynamic indocyanine green angiography measurements. J Biomed Opt. 2012;17(11):116028.
13. Murakami K, Endo T, Tominaga T. An analysis of flow dynamics in cerebral cavernous malformation and orbital cavernous angioma using indocyanine green videoangiography. Acta Neurochir (Wien). 2012;154(7):1169–75.

14. Woodard CR, Most SP. Intraoperative angiography using laser-assisted indocyanine green imaging to map perfusion of forehead flaps. Arch Facial Plast Surg. 2012;14(4):263–9.
15. Son YJ, Kim JE, Park SB, Lee SH, Chung YS, Yang HJ. Quantitative analysis of intraoperative indocyanine green video angiography in aneurysm surgery. J Cerebrovasc Endovasc Neurosurg. 2013;15(2): 76–84.
16. Mohajerani P, Meier R, Noel PB, Rummeny EJ, Ntziachristos V. Spatiotemporal analysis for indocyanine green-aided imaging of rheumatoid arthritis in hand joints. J Biomed Opt. 2013;18(9):097004.
17. Hillman EM, Amoozegar CB, Wang T, McCaslin AF, Bouchard MB, Mansfield J, et al. In vivo optical imaging and dynamic contrast methods for biomedical research. Phil Trans A Math Phys Eng Sci. 2011;369(1955):4620–43.
18. Piper SK, Habermehl C, Schmitz CH, Kuebler WM, Obrig H, Steinbrink J, et al. Towards whole-body fluorescence imaging in humans. PLoS One. 2013;8(12): e83749.
19. Ku T, Choi C. Noninvasive optical measurement of cerebral blood flow in mice using molecular dynamics analysis of indocyanine green. PLoS One. 2012;7(10):e48383.
20. Amoozegar CB, Wang T, Bouchard MB, McCaslin AF, Blaner WS, Levenson RM, et al. Dynamic contrast-enhanced optical imaging of in vivo organ function. J Biomed Opt. 2012;17(9):96003–1.
21. Mucke T, Reeps C, Wolff KD, Mitchell DA, Fichter AM, Scholz M. Objective qualitative and quantitative assessment of blood flow with near-infrared angiography in microvascular anastomoses in the rat model. Microsurgery. 2013;33(4):287–96.
22. Burger W, Burge MJ. Digital image processing: an algorithmic introduction using java. Berlin: Springer; 2007.

Ureter Identification Using Methylene Blue and Fluorescein

35

Fernando D. Dip, Alejandro Daniel Moreira Grecco, David Nguyen, Luis Sarotto, Sandy Perrins, and Raul J. Rosenthal

Introduction

The intraoperative search for the ureters is a necessary step in almost all abdominopelvic surgeries secondary to the proximity of the ureters to the peritoneum [1]. Ureteral identification is a critical step in pelvic surgery in order to assure prevention of ureteral injury. Current literature notes ureteral injury occurs in 0.7–1.8 % of pelvic surgeries, and its occurrence has risen with the wide use of laparoscopic and robotic surgery [2–5]. Some procedures present a higher risk of inadvertent ureteral injury, specifically colorectal surgery, vascular surgery, and gynecological surgery [6]. Injury mechanisms include partial or total ligations, ureteral stricture, laceration, partial or total transection, flattening due to the clamp occlusion or thermal injury [7]. The key to diagnosing and managing a ureteric injury is to have a low threshold for suspecting its presence in the surgical field [8]. Importantly, intraoperative detection of such injuries can reduce morbidity and medical costs [9, 10].

Even though injury is infrequent, significant effort is required and OR time is spent during ureteral identification. Many maneuvers have been described to identify the ureters intraoperatively, from the use of vital dyes, such as indigo carmine or methylene blue, to the insertion of ureteral stents, lighted catheters and the use of radioscopy [9–11]. Some of these are invasive procedures, not free of their own respective complications, while others remain less effective. Radioscopy exposes the patient and caregivers to ionizing radiation. None of these methods provides real-time identification of structures. For example, identification of the ureters is usually performed during colonic resection after the medial to lateral dissection of the mesocolon and the Kelly maneuver is performed (ureteral com-

F.D. Dip, M.D., M.A.A.C. (✉)
Section of Minimally Invasive Surgery, Department of General and Vascular Surgery, The Bariatric and Metabolic Institute, Cleveland Clinic Florida, 2950 Cleveland Clinic Boulevard, Weston, FL 33331, USA

Oncological Surgical Division, Division of Surgical Research, Department of Surgery, Hospital de Clinicas Buenos Aires, University of Buenos Aires Buenos Aires, Argentina
e-mail: fernandodip@gmail.com

A.D. Moreira Grecco, M.D.
Department of Colorectal Surgery, Hospital de Clinicas Jose de San Martin, 2354 Cordoba Ave., 7th Floor, Office 1, Caba, Buenos Aires 1026, Argentina
e-mail: a_moreira_grecco@hotmail.com

D. Nguyen, M.D. • L. Sarotto, M.D.
Section of Surgery, Department of General Surgery, Hospital de Clinicas Jose de San Martin, 2351 Córdoba Ave., Ciudad Autónoma de Buenos Aires, Buenos Aires C112OAAR, Argentina

S. Perrins, M.D. • R.J. Rosenthal, M.D., F.A.C.S.
Section of Minimally Invasive Surgery, Department of General and Vascular Surgery, The Bariatric and Metabolic Institute, Cleveland Clinic Florida, 2950 Cleveland Clinic Boulevard, Weston, FL 33331, USA
e-mail: rosentr@ccf.org

F.D. Dip et al. (eds.), *Fluorescence Imaging for Surgeons: Concepts and Applications*, DOI 10.1007/978-3-319-15678-1_35, © Springer International Publishing Switzerland 2015

pression to evaluate peristaltic movement). Loss of three-dimensional view and tactile sensations during laparoscopy have increased ureteral injury and lesions [12–14]. The lack of ureteral identification has been described as a cause of conversion to open surgery during laparoscopic colorectal procedures resulting in significantly greater patient morbidity and higher operative cost [14, 15].

The use of fluorescence during pelvic surgery allows for real-time intraoperative ureteral visualization, and poses the theoretical advantage of reducing the time required for adequate dissection. Secondarily, real-time identification would prevent or lower the incidence of ureteral injury. An ideal imaging technique should be available during both open and minimally invasive surgeries. The ideal fluorescent dye for ureteral imaging would be intravenously injected at a low dose, yet provide sensitive ureteral imaging over an extended period of time.

Multimodal laparoscopic imaging systems possessing the capability for extended spectrum irradiation and visualization within a single, unified camera system are now available. The use of selective, discrete wavelength filters on laparoscopic cameras enables visualization of these structures by the surgeon [16]. These systems provide enhanced intracorporeal operative anatomic and dynamic perfusion assessment, which potentially improves patient outcome.

The near-infrared (NIR) spectrum is of great potential utility for the purposes of examining inducible fluorescence in the abdominopelvic tissue. Fluorescence can be achieved through administration of specific tracer agents, either directly into the circulation (e.g. for anastomotic perfusion assessment at the time of stapling), into the lymphatic system (e.g. for lymph basin road-mapping and/or focused target nodal assessment), or via renal injection/clearance for ureteral identification secondary to urinary excretion. It is likely that this technological capability will find a clear and common indication with colorectal, female pelvic, and general surgical departments worldwide in the near future [16].

Certainly, critical to ureteral identification is a fluorescent dye that is rapidly excreted in the urine. There are scattered reports of the use of fluorescence identification of the ureter in the literature [17–19]. More specifically, these report of the use of CW800-CA, methylene blue and fluorescein in detail. CW800-CA is not currently FDA-approved. Contrarily, methylene blue and fluorescein are FDA-approved and are in widespread use for specific surgical procedures such as sentinel lymph node identification and retinography.

Near-infrared radiation ($\lambda = 700–840$ nm) seems to have significant advantages over other wavelengths. Light in this range of wavelengths has demonstrated more tissue penetration while in use for endoscopic procedures and improved perfusion evaluation after bowel anastomosis [20].

IRDYE™800CW (CW800-CA)

The NIR wavelength penetrates deeply into the living tissues due to its low absorbance. A fluorescent dye reflective at this wavelength would help to identify a target anatomic structure. Tanaka et al. [18] reported the ureteral identification in animal models using IRDye™800CW (CW800-CA). This same group administered the dye intravenously and evaluated visualization between "white light" ($\lambda = 400–700$ nm) and NIR with a 725–775 nm light source in both 350 g rats and 35 kg porcine models. NIR fluorescence was visualized prior to dye injection in the ureters with the use of the NIR light. The authors demonstrated a maximum signal and peristaltic flow in the ureters at roughly 10 min after CW800-CA IV administration. One of the advantages of the fluorescent method described was the lack of ionizing radiation use.

Methylene Blue

Methylene blue (MB) is an FDA-approved dye that has been extensively used in for different clinical applications. Its light absorption is ~670 nm. One of its main advantages is its high threshold for adverse reactions, appearing only after a high dose of 5 mg/kg.

Matsui et al. reported a pig experience using methylene blue and a 670 nm laser diode light for fluorescence ureteral identification [17, 21, 22].

Approximately 10–30 min after MB administration, ureters were successfully identified secondary to a higher reflectivity with a 0.5 mg/kg dose of MB during conventional and laparoscopic surgery.

Matsui's group also established MB as an excellent IV dye due to its renal clearance and great ureteral fluorescent visualization. Adding furosemide to the animals increased the urine flow without increasing the fluorescent signal of the ureters. Similar results had been obtained using furosemide to increased radiological contrast-agent excretion during CT scans. It must be noted, however, that the lowest possible dose should be used due to MB side effects observed with doses higher than 5 mg/kg [23].

Verdeek et al. [19] reported the first human experience utilizing methylene blue for fluorescent in vivo ureteral visualization. Twelve patients who underwent lower abdominal surgery were included in their cohort. NIR fluorescence imaging was performed using the Mini-FLARE™ imaging system. In all patients, both ureters could be clearly visualized within 10 min after infusion of methylene blue in 0.25, 0.5, or 1 mg/kg doses. The signal lasted at least up to 60 min after injection. Higher dose groups were associated with a greater signal-to-background ratio despite the lack of statistical significance ($p=0.811$). Verdeek concluded that the use of methylene blue for ureteral fluorescent visualization is possible during pelvic surgery.

Indocyanine Green

The use of retrograde injection in the ureters with indocyanine green (ICG) has been described, achieving good identification of structures, but requiring ureteral intraoperative catheterization. Frangioni et al. [24] have described the possibility of visualizing ureters by fluorescence using CW 800-CA or methylene blue injected intravenously.

Tanaka reported his experience on ureteral visualization in a porcine model using retrograde injection of 10 μM of ICG into the ureters which resulted in immediate visualization of the ureters, even when embedded in surrounding tissue. In the case of ureteral injury, ICG leak can be visualized directly, in real-time, permitting immediate localization of the injury site

Lee at al. [25] described the use of retrograde or percutaneous injection of ICG during robotic surgery for ureteral stenosis in seven patients. Intraureteral administration of 25 mg/10 ml of ICG provides real-time ureteral stenosis localization and aids in ureteral repair when used in combination with the NIR modality of robotic surgery.

Fluorescein

Our group is working on the use of fluorescein in the intraoperative identification of the ureters. All procedures were performed at the Experimental Surgery Centre of the Jose de San Martin Clinical Hospital University of Buenos Aires. Animals were used in accordance with an approved institutional protocol.

Twenty male Wister rats of 250 g were administered a dose of 4 mg/kg of sodium fluorescein intravenously. After the administration of the contrast, the specimens were submitted to surgery. The retroperitoneum was observed with white xenon light and a 530 nm wavelength light (filtered Xenon Short Arc lamp, Karl Storz, Tuttlingen, Germany) alternatively. The images were acquired from a digital visualization device (AIDA Compact II System, Karl Storz, Tuttlingen, Germany). When exposing the retroperitoneum to xenon ("white") light, it was possible to visualize with the naked eye the location of the kidneys and the gallbladder without being able to determine the path or location of the ureters. The ureter and urine had no autofluorescence at the moment of laparotomy nor prior to the stimulation with the infrared light (Fig. 35.1).

We were able to demonstrate that ureters can be visualized using intravenous administration of fluorescein in rats. An acceptable intensity of fluorescence was observed using a 4 mg/kg dose. Excitation of the fluorescein was noted inside the ureters only, but not in the kidney parenchyma or other retroperitoneal structures. Some of the most notable advantages of fluorescein include ease of use, low cost, and low rate of side effects at a

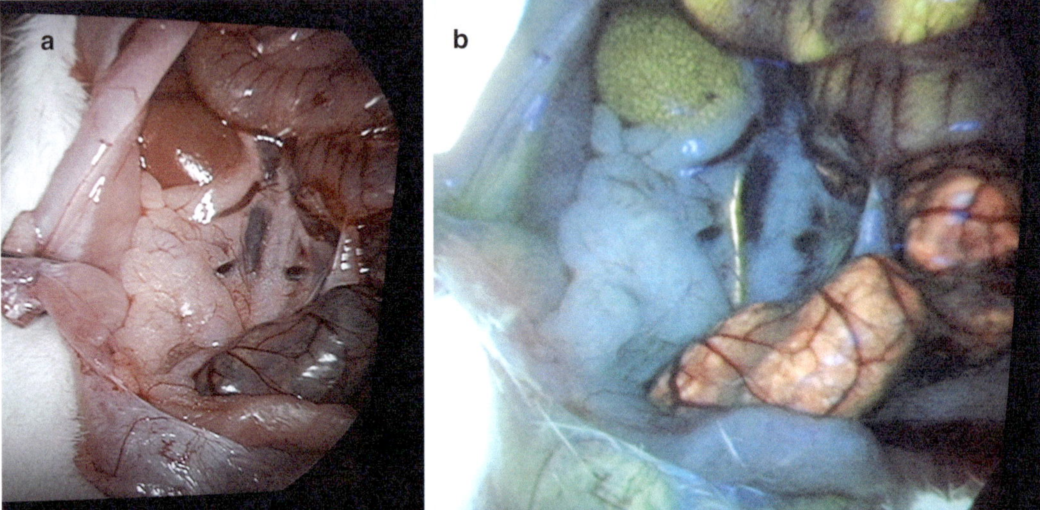

Fig. 35.1 (**a**) Exposed abdominal cavity with xenon light source, no auto-fluorescence is detected, ureters are not visible. (**b**) After an average of 8 min (subjects were evaluated in the range of 5–12 min), and employing a pedal to alternate between xenon "white light" and 530 nm wavelength light sources, it was possible to visualize ureteral fluorescence in all of the animals. The distribution and location of the ureters were noted directly, with a clear distinction from the nearby vascular structures. This is especially useful for the surgeon since it provides anatomic orientation, therefore enabling visualization of real-time ureteral peristaltic flow

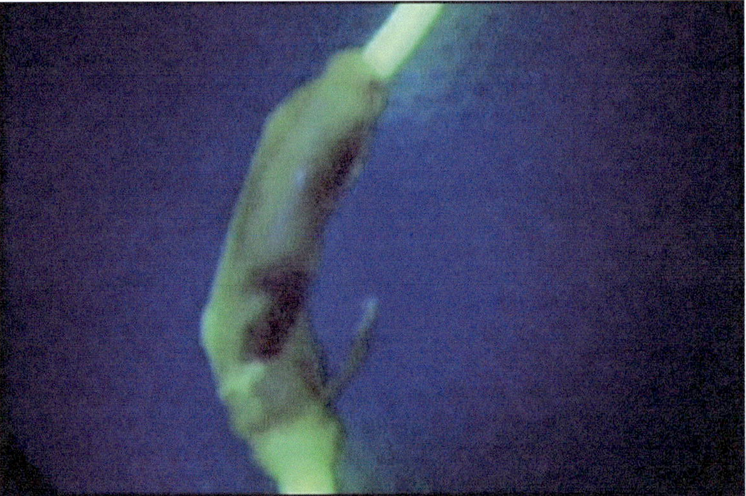

Fig. 35.2 Ex vivo fluorescence of human ureter after nephrectomy

relatively large dose. The incidence of adverse reactions is 0.6 %, which is mainly urticaria and in very few reported cases, post-injection anaphylactic shock in quick bolus [26–28].

Other reports have described the use of fluorescein during open surgery [29] (Fig. 35.2).

Conclusion

The use of a multimodal optical system that expands the surgeon's light spectrum of view can improve surgical performance making structures clearly visible during laparoscopic and open surgery. This allows for shorter procedural duration and improved prevention and incidence of ureteral injury associated with complex pelvic surgery. Optical imaging using invisible NIR fluorescent light has several advantages over currently available intraoperative techniques. First, visualization of the ureters does not require ionizing radiation, and uses only safe wavelengths of light for ample excitation. Secondly, because fluorescence emission is invisible to the human eye, the surgical field is not stained or changed in any way. The blue dyes that are currently used stain the surgical field and have relatively poor contrast. Thirdly, imaging can be performed in real-time (up to 15 frames per second) with the merged image from the color video and NIR fluorescent cameras providing anatomical landmarks that are easily identifiable. More work is needed to identify the optimal contrast agent and light wavelength with which it is most optimally visualized. Overall, this field of knowledge is of great interest and reports a great growing potential.

References

1. Park S, Pearle MS. Imaging for percutaneous renal access and management of renal calculi. Urol Clin North Am. 2006;33:353–64.
2. Ostrzenski A, Radolinski B, Ostrzenska KM. A review of laparoscopic ureteral injury in pelvic surgery. Obstet Gynecol Surv. 2003;58(12):794–9.
3. Bothwell WM, Cathcart KS, Bombardt PA. An online, column-switching high-performance liquid chromatographic procedure for the removal of probenecid from human plasma, serum, or urine in the quantitative determination of cefmetazole or cefoxitin. J Pharm Biomed Anal. 1989;7:987–95.
4. Kuno K, Menzin A, Kauder HH, Sison C, Gal D. Prophylactic ureteral catheterization in gynecologic surgery. Urology. 1998;52:1004–8.
5. Parpala-Spårman T, Paananen I, Santala M, Ohtonen P, Hellström P. Increasing numbers of ureteric injuries after the introduction of laparoscopic surgery. Scand J Urol Nephrol. 2008;42(5):422–7.
6. Wang AC. The techniques of trocar insertion and intraoperative rethrocystoscopy in tension-free vaginal taping: an experience of 600 cases. Acta Obstet Gynecol Scand. 2004;83:293–8.
7. Visco AG, Taber KH, Weidner AC, Barber MD, Myers ER. Cost-effectiveness of universal cystoscopy to identify ureteral injury at hysterectomy. Obstet Gynecol. 2001;97:685–92.
8. Abboudi H, Ahmed K, Royle J, Khan MS, Dasgupta P, N'Dow J. Ureteric injury: a challenging condition to diagnose and manage. Nat Rev Urol. 2013;10(2): 108–15.
9. Meirow D, Moriel EZ, Zilberman M, Farkas A. Evaluation and treatment of iatrogenic ureteral injuries during obstetric and gynecologic operations for nonmalignant conditions. J Am Coll Surg. 1994; 178:144–8.
10. Selzman AA, Spirnak JP. Iatrogenic ureteral injuries: a 20-year experience in treating 165 injuries. J Urol. 1996;155:878–81.
11. Redan JA, McCarus SD. Protect the ureters. JSLS. 2009;13(2):139–41.
12. Siddighi S, Carr KR. Lighted stents facilitate robotic-assisted laparoscopic ureterovaginal fistula repair. Int Urogynecol J. 2013;24(3):515–7.
13. Kusaka J, Matsumoto S, Hagiwara S, Koga H, Noguchi T. Use of perioperative ureteral stent in abdominal aortic aneurysm with retroperitoneal fibrosis - a report of two cases. Kr J Anesthesiol. 2012;63(1):76–9.
14. Palaniappa NC, Telem DA, Ranasinghe NE, Divino CM. Incidence of iatrogenic ureteral injury after laparoscopic colectomy. Arch Surg. 2012;147(3):267–71.
15. Casillas S, Delaney CP, Senagore AJ, Brady K, Fazio VW. Does conversion of a laparoscopic colectomy adversely affect patient outcome? Dis Colon Rectum. 2004;47(10):1680–5.
16. Cahill RA, Ris F, Mortensen NJ. Near-infrared laparoscopy for real-time intra-operative arterial and lymphatic perfusion imaging. Colorectal Dis. 2011;13 Suppl 7:12–7.
17. Matsui A, Tanaka E, Choi HS, Kianzad V, Gioux S, Lomnes SJ, Frangioni JV. Real-time, near-infrared, fluorescence-guided identification of the ureters using methylene blue. Surgery. 2010;148(1):78–86.
18. Tanaka E, Ohnishi S, Laurence RG, Choi HS, Humblet V, Frangioni JV. Real-time intraoperative ureteral guidance using invisible near-infrared fluorescence. J Urol. 2007;178(5):2197–202.
19. Verbeek FP, van der Vorst JR, Schaafsma BE, Swijnenburg RJ, Gaarenstroom KN, Elzevier HW, van de Velde CJ, Frangioni JV, Vahrmeijer AL. Intraoperative near infrared fluorescence guided identification of the ureters using low dose methylene blue: a first in human experience. J Urol. 2013;190(2):574–9.
20. Cahill RA, Mortensen NJ. Intraoperative augmented reality for laparoscopic colorectal surgery by intraoperative near-infrared fluorescence imaging and optical coherence tomography. Minerva Chir. 2010;65(4): 451–62.
21. Gioux S, Kianzad V, Ciocan R, Gupta S, Oketokoun R, Frangioni JV. High power, computer-controlled, LED-based light sources for fluorescence imaging and image-guided surgery. Mol Imaging. 2009;8: 156–65.

22. Troyan SL, Kianzad V, Gibbs-Strauss SL, Gioux S, Matsui A, Oketokoun R, et al. The FLARE™ intraoperative near-infrared fluorescence imaging system: a first-in-human clinical trial in breast cancer sentinel lymph node mapping. Ann Surg Oncol. 2009; 16:2943–52.
23. Khan MA, North AP, Chadwick DR. Prolonged postoperative altered mental status after methylene blue infusion during parathyroidectomy: a case report and review of the literature. Ann R Coll Surg Engl. 2007;89(2):W9–11.
24. Frangioni JV. In vivo near-infrared fluorescence imaging. Curr Opin Chem Biol. 2003;7:626–34.
25. Lee Z, Simhan J, Parker DC, Reilly C, Llukani E, Lee DI, Mydlo JH, Eun DD. Novel use of indocyanine green for intraoperative, real-time localization of ureteral stenosis during robot-assisted ureteroureterostomy. Urology. 2013;82:729–33.
26. Kwiterovich KA, Maguire MG, Murphy RP, Schachat AP, Bressler NM, Bressler SB, Fine SL. Frequency of adverse systemic reactions after fluorescein angiography. Results of a prospective study. Ophthalmology. 1991;98:1139–42.
27. Jennings BJ, Mathews DE. Adverse reactions during retinal fluorescein angiography. J Am Optom Assoc. 1994;65:465–71.
28. Gómez-Ulla F, Gutiérrez C, Seoane I. Severe anaphylactic reaction to orally administered fluorescein. Am J Ophthalmol. 1991;112:94.
29. Udshmadshuridze NS, Asikuri TO. Intra-operative imaging of the ureter with sodium fluorescein. Z Urol Nephrol. 1988;81(10):635–9.

Chymotrypsin Probe for Real-Time Visualization of Pancreatic Leak

36

Suguru Yamashita, Takeaki Ishizawa, Masayo Sakabe, Mako Kamiya, Yasuteru Urano, and Norihiro Kokudo

Introduction

Despite the increasing experience with pancreatic resections around the world—approximately 40,000 procedures per year in the United States [1, 2] and more than 11,000 in Japan [3]—postoperative pancreatic fistulas (PFs) still occur with an incidence of 10–25 % after pancreatojejunostomy and 30–50 % after distal pancreatectomy. PFs remain one of the most serious complications of digestive surgery and can lead to in-hospital death following severe infections and massive hemorrhage [4–7]. A major factor making it difficult to prevent PFs is that there has been no technique available that enables visualization of clear pancreatic juice leaking from the stump of the pancreas during surgery. Another problem in management after pancreatic surgery is that there has been no way to evaluate the severity of a postoperative PF based on protease activity in drained abdominal fluids. At present, postoperative PF is diagnosed according to amylase levels in drained fluids, as proposed by the International Study Group on Pancreatic Fistula (ISGPF) [8–12]. Amylase, a glycolytic enzyme, is theoretically a surrogate of the proteases that can cause tissue damage, leading to severe postoperative complications.

We have designed a fluorescent probe that allows us to visualize leakage of pancreatic juice during surgery and thus evaluate protease activity in drained fluids. The probe is activated by the chymotrypsin included in pancreatic juice and

Electronic supplementary material: The online version of this chapter (doi:10.1007/978-3-319-15678-1_36) contains supplementary material, which is available to authorized users. Videos can also be accessed at http://link.springer.com/chapter/10.1007/978-3-319-15678-1_36.

S. Yamashita, M.D.
Department of Surgery, Graduate School of Medicine, University of Tokyo, 7-3-1 Hongo, Bunkyo-ku, Tokyo 113-8655, Japan
e-mail: origin0304@yahoo.co.jp; take1438@gmail.com

T. Ishizawa, M.D., Ph.D., F.A.C.S.
Department of Gastroenterological Surgery, Cancer Institute Hospital, Japanese Foundation for Cancer Research, Koto-ku, Tokyo, Japan

Hepato-Biliary-Pancreatic Surgery Division, Department of Surgery, Graduate School of Medicine, University of Tokyo, 7-3-1 Hongo, Bunkyo-ku, Tokyo 113-8655, Japan
e-mail: tish-tky@umin.ac.jp

M. Sakabe, Ph.D. • M. Kamiya, Ph.D.
Y. Urano, Ph.D. (✉)
Laboratory of Chemical Biology and Molecular Imaging, Graduate School of Medicine, University of Tokyo, 7-3-1 Hongo, Bunkyo-ku, Tokyo 113-8655, Japan
e-mail: masayo_s_peeotu93@yahoo.co.jp; mkamiya@m.u-tokyo.ac.jp; uranokun@m.u-tokyo.ac.jp

N. Kokudo, M.D., Ph.D. (✉)
Hepato-Biliary-Pancreatic Surgery Division, Department of Surgery, Graduate School of Medicine, University of Tokyo, 7-3-1 Hongo, Bunkyo-ku, Tokyo, 113-8655, Japan
e-mail: KOKUDO-2SU@h.u-tokyo.ac.jp

F.D. Dip et al. (eds.), *Fluorescence Imaging for Surgeons: Concepts and Applications*, DOI 10.1007/978-3-319-15678-1_36, © Springer International Publishing Switzerland 2015

rapidly emits light. We describe the details of this fluorescent probe and its ability to identify pancreatic leak. We also show how it allows estimation of the risk of a postoperative pancreatic fistula.

Development of Chymotrypsin Probe for Visualization of Pancreatic Juice

The glutaryl phenylalanine hydroxymethyl rhodamine green (gPhe-HMRG) probe, a newly designed and synthesized compound, uses HMRG as the platform [13, 14]. HMRG is excited with long-wavelength light (around 490 nm) but emits a fluorescent signal with a peak wavelength around 520 nm only when it is hydrolyzed by chymotrypsin (Fig. 36.1) [14, 15]. Because pure pancreatic juice produced in humans normally contains not chymotrypsin but chymotrypsinogen [16], this trypsin-added gPhe-HMRG probe (chymotrypsin probe) can be used to detect pancreatic leaks.

Figure 36.2 shows the trends of the median fluorescence intensity (FI) obtained following addition of the chymotrypsin probe to drained fluid samples in previous series of pancreatic resection. The FI of pure pancreatic juice obtained using this probe increased with time and reached 0.59 (0.14–1.23) arbitrary units (a.u.) after 30 min of incubation. The FIs of intestinal fluids and abdominal fluids with the chymotrypsin probe

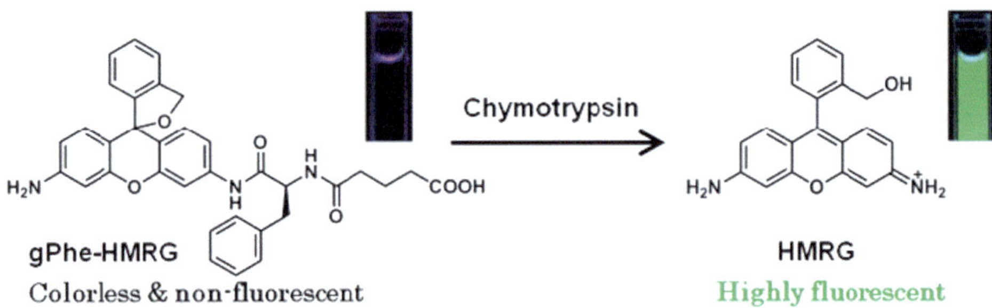

Fig. 36.1 Structures of glutaryl phenylalanine hydroxylmethyl rhodamine green (gPhe-HMRG). The gPhe-HMRG probe is activated with chymotrypsin secreted in pancreatic juice and emits HMRG fluorescence

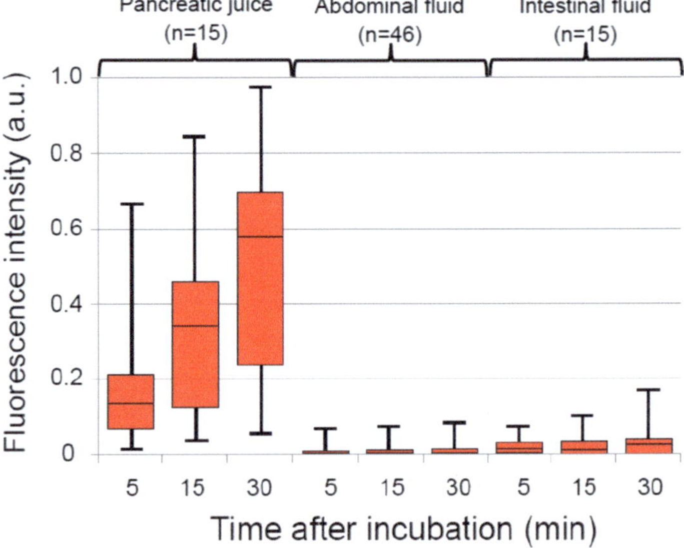

Fig. 36.2 Fluorescence intensity (FI) of drained fluid samples with the chymotrypsin probe according to the time after incubation. The FI increased with time in all fluid samples when the chymotrypsin probe yielded fluorescence almost exclusively in pure pancreatic juice (*red bars*). Median values (*horizontal lines in boxes*), interquartile range (*box*), and range (*error bars*) are shown in arbitrary units (a.u.)

were close to zero: 0.02 a.u. (−0.04 to +0.17 a.u.) and 0.001 (−0.01 to +0.07 a.u.), respectively. The FI of drained fluids for the chymotrypsin probe 30 min after incubation showed positive correlations with amylase levels in corresponding fluid samples (Spearman's rank-correlation test, $\rho=0.678$, $P<0.001$).

Application of Chymotrypsin Probe for Intraoperative Identification of Pancreatic Leak and Prediction of Postoperative PF

To identify leakage of pancreatic juice during surgery, the chymotrypsin probe is sprayed on a filter paper that is attached on the cut surface of the pancreas for 10 s following resection of the pancreatic specimen (administration of this probe into a patient's abdominal cavity has not yet been approved). Fluorescence images of leaking pancreatic juice on the pancreatic stump can be identified by naked-eye examination through light-blocking glasses (515 nm long-pass) under blue light illumination beginning 3 min after administration of the chymotrypsin probe (Fig. 36.3).

In our previous series consisting of 30 patients who underwent pancreatic resection [11], the fluorescence patterns of the pancreatic stump transcribed onto the filter paper were grossly classified into the following three types: fluorescent signal visualized only on the main pancreatic duct (MPD) stump (duct type, $n=16$), fluorescent signal visualized on the MPD stump and the surrounding

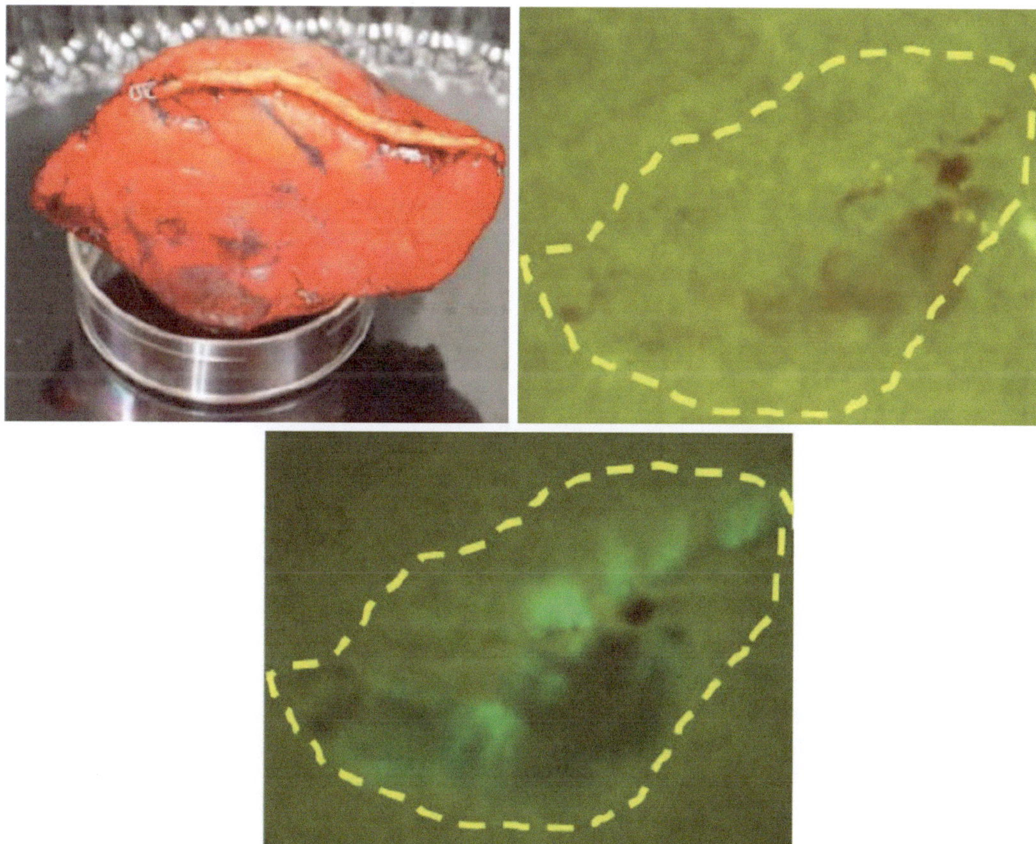

Fig. 36.3 Fluorescence imaging on a filter paper transcribing the pancreatic stump that had been closed with a surgical stapler during human distal pancreatectomy. *Left*. Gross appearance of the pancreatic stump in the resected specimen. *Right*. Appearance of the filter paper immediately after placement on the pancreatic stump (*top*). Fluorescence image of the filter paper obtained 1 min after spraying with the chymotrypsin probe (*bottom*). *Yellow dotted line* indicates the border of the pancreatic stump

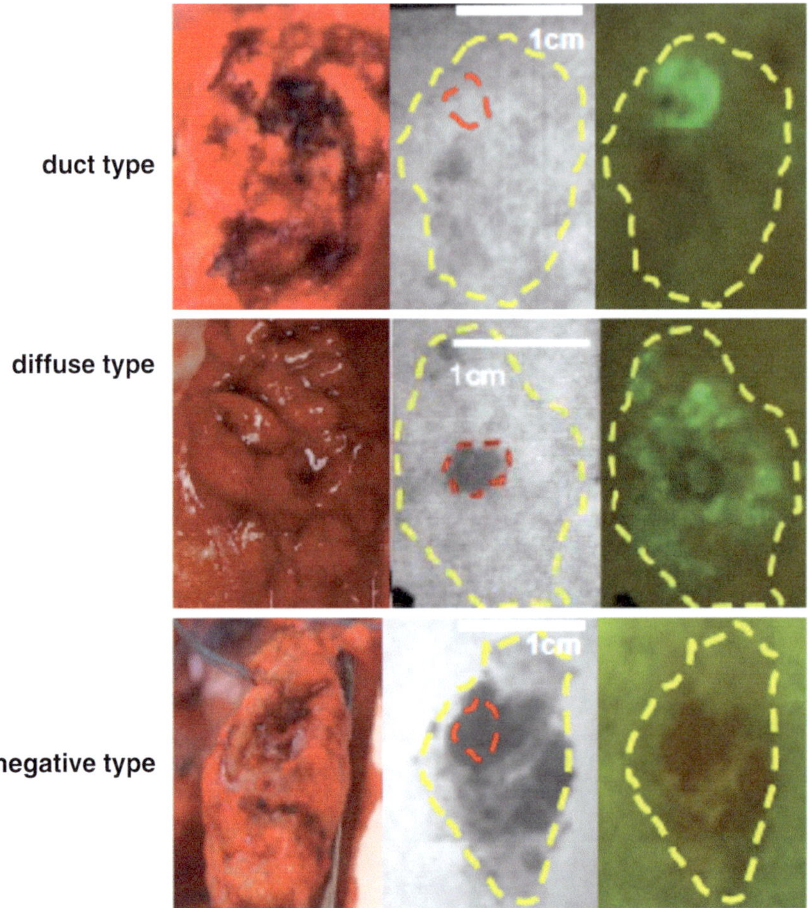

Fig. 36.4 Fluorescence patterns of the pancreatic stump transcribed onto filter papers. *Left*. Gross appearance of the pancreatic stump. *Middle*. Monochromatic appearance of the filter paper immediately after placement on the pancreatic stump. *Right*. Fluorescence images of the filter paper obtained 1 min after spraying with the chymotrypsin probe. *Red* and *yellow dotted lines* indicate the borders of the main pancreatic duct and pancreatic stump, respectively

pancreatic parenchyma (diffuse type, $n=7$), and no fluorescent signal (negative type, $n=7$) (Fig. 36.4). Postoperatively, the incidence of PFs was 57 % in 23 patients with fluorescent signals on the pancreatic stump (duct type and diffuse type), whereas none of the patients with negative fluorescence signals suffered from postoperative PF. In addition, severe PF (ISGPF grade C, $n=3$) developed only in patients with a diffuse-type fluorescence signal. These patients were deemed likely to be at high risk of postoperative PF because the lesion originated from the stump of the pancreatic parenchyma regardless of how accurately the pancreatojejunostomy was created. When FI of the pancreatic juice on filter paper was evaluated using the Maestro imaging system (CRi, Woburn, MA, USA), it correlated positively with the amylase levels in pancreatic juice collected from external pancreatostomy tubes on postoperative day 1 ($\rho=0.664$, $P<0.001$). These results suggest that the chymotrypsin probe during surgery enables detection of a pancreatic leak and estimation of the risk of postoperative PF. This knowledge may be helpful for surgeons regarding the need to undertake additional surgical procedures to close a pancreatic leak and to place (or omit) prophylactic abdominal drainage according to the risk of postoperative PF in each patient.

Preclinical Study for the Use of Chymotrypsin Probe in the Abdominal Cavity to Detect Pancreatic Leak During Surgery

The authors believe that, in future, the chymotrypsin probe should be sprayed directly on the pancreatic stump in the patient's abdominal cavity as it could enable surgeons to suture the point of the pancreatic leak with some accuracy. At present, this technique is being simulated using a swine model. In this model, the chymotrypsin probe is sprayed on the pancreatic stump within the abdominal cavity, following full mobilization and division of the pancreatic parenchyma with electrocautery or a surgical stapler. Fluorescence imaging using a FluorVivo imaging system (Indec, Santa Clara, CA, USA) with the blue-filter setting (excitation 470–490 nm; emission 515 nm longpass) then enables identification of the pancreatic leak from the MPD following division of the pancreas with electrocautery (Fig. 36.5a) or from a stapler line following closure of the pancreas

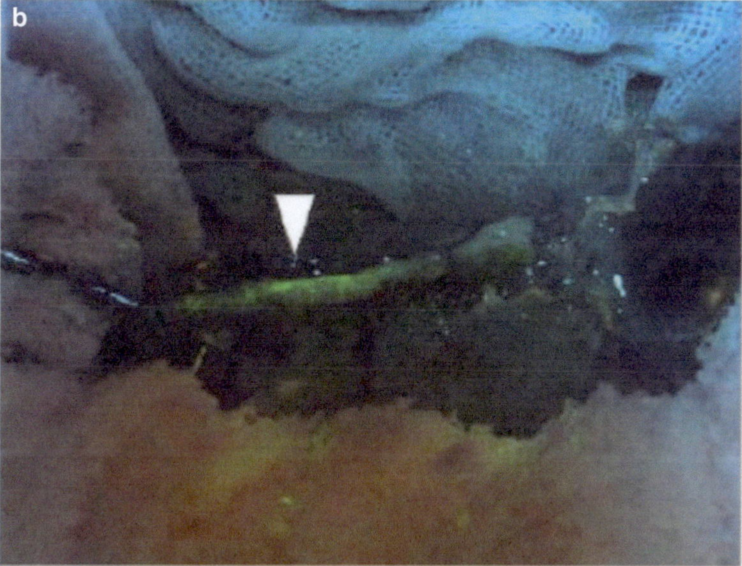

Fig. 36.5 Fluorescence images of the pancreatic stumps using chymotrypsin probe in a swine model. (**a**) Fluorescence imaging identified pancreatic leakage from the stump of the main pancreatic duct following division of the pancreatic body with electrocautery (*arrow*). (**b**) Fluorescence imaging visualized pancreatic leakage on a suture line of the pancreatic stump with a surgical stapler (*arrowheads*)

(Fig. 36.5b). Fluorescence of the leaking pancreatic juice is identifiable by naked-eye examination through light-blocking glasses (515 nm long-pass) beginning 1 min after administration of the chymotrypsin probe (see Video 36.1), which enables surgeons to make pinpoint sutures and confirm arrest of the pancreatic leak.

Conclusion

The chymotrypsin probe can identify the presence of a pancreatic leak on the pancreatic stump and estimate the risk of postoperative PF through rapid visualization of protease activity in pancreatic juice during surgery. This technique could enable surgeons to curtail the leak during surgery and apply appropriate management of prophylactic abdominal drainage according to the risk of developing a symptomatic PF postoperatively. However, preclinical safety trials are needed before this probe can be administered directly into a patient's abdominal cavity.

Financial Disclosure: The authors have no relevant affiliations or financial involvement with any organization or entity with a financial interest in or financial conflict with the subject matter or materials discussed in this manuscript. This includes employment, consultancies, honoraria, stock ownership or options, expert testimony, grants or patents received or pending, and royalties. No writing assistance was used in the production of this manuscript.

Funding/Support: This work was supported by grants from the Takeda Science Foundation, the Ministry of Education, Culture, Sports, Science, Technology of Japan, and the Ministry of Health, Labour, and Welfare of Japan, and Global Leader Program for Social Design and Management.

References

1. Finks JF, Osborne NH, Birkmeyer JD. Trends in hospital volume and operative mortality for high-risk surgery. N Engl J Med. 2011;364:2128–37.
2. McPhee JT, Hill JS, Whalen GF, Zayaruzny M, Litwin DE, Sullivan ME, Anderson FA, Tseng JF. Perioperative mortality for pancreatectomy: a national perspective. Ann Surg. 2007;246:246–53.
3. Ministry of Health, Labour and Welfare of Japan. Proceedings of Central Social Insurance Medical Council, 2008. Accessed 26 June 2009. http://www.mhlw.go.jp/shingi/2009/05/s0514-6.html
4. Yeo CJ, Cameron JL, Sohn TA, Lillemoe KD, Pitt HA, Talamini MA, Hruban RH, Ord SE, Sauter PK, Coleman J, Zahurak ML, Grochow LB, Abrams RA. Six hundred fifty consecutive pancreaticoduodenectomies in the 1990s: pathology, complications, and outcomes. Ann Surg. 1997;226:248–57.
5. Mckay A, Mackenzie S, Sutherland FR, Bathe OF, Doig C, Dort J, Vollmer Jr CM, Dixon E. Meta-analysis of pancreaticojejunostomy versus pancreaticogastrostomy reconstruction after pancreaticoduodenectomy. Br J Surg. 2006;93:929–36.
6. Fuks D, Piessen G, Huet E, Tavernier M, Zerbib P, Michot F, Scotte M, Triboulet JP, Mariette C, Chiche L, Salame E, Segol P, Pruvot FR, Mauvais F, Roman H, Verhaeghe P, Regimbeau JM. Life-threatening postoperative pancreatic fistula (grade C) after pancreaticoduodenectomy: incidence, prognosis, and risk factors. Am J Surg. 2009;197:702–9.
7. Birkmeyer JD, Siewers AE, Finlayson EVA, Stukel TA, Lucas FL, Batista I, Welch HG, Wennberg DE. Hospital volume and surgical mortality in the United States. N Engl J Med. 2002;346:1128–37.
8. Bassi C, Dervenis C, Butturini G, Fingerhut A, Yeo C, Izbicki J, Neoptolemos J, Sarr M, Traverso W, Buchler M, International study group on pancreatic fistula definition. Postoperative pancreatic fistula: an international study group (ISGPF) definition. Surgery. 2005;138:8–13.
9. Kawai M, Tani M, Terasawa H, Ina S, Hirono S, Nishioka R, Miyazawa M, Uchiyama K, Yamaue H. Early removal of prophylactic drains reduces the risk of intra-abdominal infections in patients with pancreatic head resection: prospective study for 104 consecutive patients. Ann Surg. 2006;244:1–7.
10. Molinari E, Bassi C, Salvia R, Butturini G, Crippa S, Talamini G, Falconi M, Pederzoli P. Amylase value in drains after pancreatic resection as predictive factor of postoperative pancreatic fistula: results of a prospective study in 137 patients. Ann Surg. 2007;246:281–7.
11. Shyr YM, Su CH, Wu CW, Lui WY. Does drainage fluid amylase reflect pancreatic leakage after pancreaticoduodenectomy? World J Surg. 2003;27:606–10.
12. Gebauer F, Kloth K, Tachezy M, Vashist YK, Cataldegirmen G, Izbicki JR, Bockhorn M. Options and limitations in applying the fistula classification by the International Study Group for Pancreatic Fistula. Ann Surg. 2012;256:130–8.
13. Urano Y, Asanuma D, Hama Y, Koyama Y, Barrett T, Kamiya M, Nagano T, Watanabe T, Hasegawa A, Choyke PL, Kobayashi H. Selective molecular imaging of viable cancer cells with pH-activatable fluorescence probes. Nat Med. 2009;15:104–9.

14. Urano Y, Sakabe M, Kosaka N, Ogawa M, Mitsunaga M, Asanuma D, Kamiya M, Young MR, Nagano T, Choyke PL, Kobayashi H. Rapid cancer detection by topically spraying a γ-glutamyltranspeptidase-activated fluorescent probe. Sci Transl Med. 2011;3:110–9.
15. Yamashita S, Sakabe M, Ishizawa T, Hasegawa K, Urano Y, Kokudo N. Visualization of the leakage of pancreatic juice using a chymotrypsin-activated fluorescent probe. Br J Surg. 2013;100:1220–8.
16. Whitcomb DC, Lowe ME. Human pancreatic digestive enzymes. Dig Dis Sci. 2007;52:1–17.

Miniaturized Clinical Imaging Device for Optical Surgery

37

Paul Dorval, Ihab Atallah, Gabriele Barabino, Maxime Henry, Michèle Keramidas, Fabien Stenard, Clément Milet, Christian Righini, Philippe Rizo, Stéphanie Guillermet, Véronique Josserand, and Jean-Luc Coll

Introduction

In recent years, a lot of improvements have been done concerning medical imaging. The specific technology of fluorescence imaging has shown a huge potential in oncology, vascular, and lymphatic related surgeries [1]. To be more specific, fluorescence is useful for sentinel lymph node in oncologic procedures (breast, skin, gastric, and colorectal cancers in majority), action on lymphedema, vascular visualization of free flaps in reconstructive surgery, tumor resection, and general vascular mapping [2–4]. Indocyanine green (ICG) and methylene blue are FDA-approved probes for fluorescence procedures [5, 6]. Other probes such as Patent Blue V, approved in Europe, have shown a potential for fluorescence image-guided surgeries [7].

A breakthrough is still needed in the design of fluorescence imaging devices to make them easy to use for everyday surgery. Such procedures as the resection of peritoneal carcinomatosis using fluorescence imaging have been successfully experimented in preclinical model [8]. Nevertheless, the size of existing instrument does not permit to translate it to humans. Most of the available devices for clinical investigations or everyday procedures are too bulky to image the sides of cavities or specific organs [9–11].

The miniaturization of fluorescence imaging technologies is the purpose of some studies and prototypes have been developed for specific applications. Head and neck cancers are one of the fields where improvement of the footprint of fluorescence imaging system is really needed [12–14]. Some other miniaturizations of fluorescence technology projects have been explored, especially for in vivo microscopy applications [15, 16]. Finally, developments for preclinical

P. Dorval • P. Rizo, Ph.D. • S. Guillermet, Ph.D.
Fluoptics, 7 Parvis Louis Neel CS20050, Grenoble, France
e-mail: paul.dorval@fluoptics.com; philippe.rizo@fluoptics.com; stephanie.guillermet@fluopics.com

I. Atallah, M.D. • C. Righini, M.D., Ph.D.
INSERM-UJF U823, Institut Albert Bonniot, CHU-Grenoble, University Joseph Fourier, Grenoble 38706, France
e-mail: iatallah@CHU-grenoble.fr; crighini@chu-grenoble.fr

G. Barabino, M.D.
Hopital nord CHU Saint Etienne, Avenue Albert Raimond, Saint Etienne 42270, France
e-mail: g.barabino77@gmail.com

M. Henry • M. Keramidas • C. Milet • V. Josserand, Ph.D. • J.-L. Coll, Ph.D. (✉)
INSERM-UJF U823, Institut Albert Bonniot, University Joseph Fourier, Grenoble 38706, France
e-mail: maxime.henry@uif.grenoble.fr; michellekeramidas@hotmail.com; clement.milet@gmail.com; veronique.josserand@uif-grenoble.fr; Jean-luc.coll@ujf-grenoble.fr

F. Stenard, M.D.
Groupe Hospitalier Mutualiste de Grenoble, 8-12 rue du Dr. Calmette, CS 10043, Grenoble 38028, France
e-mail: f.stenard@ghm-grenoble.fr

F.D. Dip et al. (eds.), *Fluorescence Imaging for Surgeons: Concepts and Applications*,
DOI 10.1007/978-3-319-15678-1_37,

procedures or clinical studies only have been achieved, but none have reached an approval yet [17–20].

The purpose of this study is to show the development and the first results of a CE-marked miniaturized, handheld, fluorescence image-guided surgery system named Fluostick™.

Materials and Methods

The goal of FIGS systems is to image fluorescent probes previously injected to the patient. To achieve this, the system is able to send specific excitation light correlated to the probes imaged and collect the emission light produced. The FIGS system mostly consists of an imaging head linked to one or several control boxes including hardware and software to acquire and display images. Due to their size and weight, most of the available systems on the market are fixed to a mechanical arm and are able to image the surgical field from the top only. FIGS systems are medical imaging devices operating in the sterile area of the theatre. None of the existing systems are sterile or autoclavable. Thus, a sterile cover and sterile shields are needed to isolate the FIGS system from the sterile area. As an imaging device, the sterile cover of the fluorescence system has to have good optical performance. The most common probes approved for surgery emit light in a spectrum region comprised between 650 and 900 nm. The ambient light of the theatre, as well as the shadowless surgical light, contains this type of emission. It is particularly true for xenon- or tungsten-type light. Therefore, to perform good fluorescence acquisitions, the light of the operating room must be turned off. Some FIGS systems provide an additional filtered white light which helps the surgeon to perform his or her gesture but will not disrupt the acquisition of fluorescence emission light. In a miniaturized system, we will mostly consider the miniaturization of the terminal head of the system, also called the optical head. Indeed, it is the only part of the FIGS miniaturized system which will enter the sterile zone and the working field of the surgeon.

As said before, the FIGS system has to provide light to excite the fluorescent probe and should furnish additional white light to help the surgeon to perform his or her gesture. There are several ways to generate light. In the medical field, for powerful white light emission, the LED technologies are more and more used instead of older and less reliable techniques such as xenon lamps. Recent advances in the LED field permit to generate a huge amount of light with tiny and low-powered system. The ways to generate excitation light are different due to its characteristics. Indeed, excitation light is narrowed to a small part of the spectrum because of the specificities of fluorescent probes. Moreover, the excitation light has to be quite powerful to be efficient for fluorescence imaging purpose. Using laser technologies is the easiest way to produce these types of light. Nevertheless, the use of filtered LED is not prohibited and can be found in numerous FIGS products. The great advantage of laser technology for miniaturization purpose in comparison to LED is that it can be easily coupled to an optic fiber and deported to a remote box.

After generating the excitation light, the second functionality of a FIGS system is to be able to collect the few amount of light emitted by the fluorescent probes. The whole acquisition chain has to be optimized for the emission spectrum of the considered probe. We will focus on the most common fluorescent probe, the ICG, a product approved for lymphatic and vascular procedures. The emission spectrum of ICG starts at 700 nm and ends at 850 nm (near-infrared light, NIR). So, in order to perform a good acquisition, all the parts of this chain must be optimized for the NIR region of the spectrum. The particularity of a fluorescence imaging system in comparison to a simple camera is its ability to discriminate fluorescence emission to other types of incoming light. To do so, filters are placed in front of the lens of the system. These filters are critical compounds of the system and have to be able to isolate fluorescence emission light from the powerful excitation and white lights which illuminate the working field.

The filtered light emitted by the fluorescent probe first meets the optical lens of the system.

The selection of a lens impacts in many ways the performances and the maneuverability of a FIGS system. The lens has to be able to collect the maximum of the incoming photons from the surgical field. Several parameters are important to consider. First, the F-number is the ability of the lens to collect light. The minus the F-number, the greater the amount of light collected by the lens. Then, the focal length determines the field of view of the lens. The combination of these two parameters, knowing the working distance, will define the ability of the system to collect light from a determined field of view. The other parameter that can be evaluated from these parameters is the depth of field of view of the system. This is particularly important for a miniaturized system, supposed to be handheld. Augmenting the depth of view will help the surgeon to focus on the regions of interest. Considering miniaturization purposes, the size of the lens is important to take into account, knowing that, for a constant imaged circle, the smaller the F-number, the larger the size of the lens. Behind the lens, the imaging sensor is another important part of the system. The low quantity of emitted light and its spectrum characteristics made the work hard enough for common imaging sensors. Two main technologies currently coexist. One, called charge-coupled devices (CCD), is particularly adapted to low signal and high dynamic acquisition. The other technology, named CMOS Active Pixel Sensors (CMOS-APS), is adapted to small and low-powered devices. A lot of improvements have been achieved concerning the CMOS-APS technology recently, especially about noise and overall sensitivity. Other technologies exist, as electron-multiplying CDD or other image intensifier approaches, but lead to significant challenge in terms of integration, noise, and resolution degradation. In order to select a sensor for fluorescence imaging purpose, several characteristics have to be evaluated. The quantum yield of a sensor expresses its ability to detect light at a specific wavelength. The well capacity of a sensor's pixel is directly linked to the dynamic of signal acquired. Another aspect which must be examined is the overall noise of a sensor. This aspect is important to consider because very low light signals will have to be acquired in fluorescence imaging.

For miniaturization purposes, it is also important to consider the electronics linked to a sensor. A native numeric sensor will need important electronics to acquire the signal and send it through a digital connection (USB or Ethernet for instance). Analog sensors have the advantage to necessitate few electronics integrated directly behind the sensor. Numeric sensors and cameras require more space to integrate. Some of the electronics could be deported from the terminal head to the control boxes but signals have to be sent very properly through at least the 3 m of cable required between the sterile and the non-sterile zone.

Results

The Fluostick™, a Fluorescence Imaging System

Currently existing FIGS instruments approved for operating room deal with all the constraints explained here except one, the footprint. Indeed, current devices address vascular applications where the size of the instrument, fixed to a mechanical arm, does not really matter. In these applications a wide field should be imaged. Nevertheless, in lymphatic or some cancer-related surgeries (lung cancers for instance) we should be able to image the sides of the cavities or organs. None of the existing instruments are able to do this. We identified a real need for surgeons to use smaller and lighter fluorescence imaging devices. Nevertheless, it is important to notice that the primary goal of such a development is to design an effective, approved, and reliable fluorescence imaging device for surgery. This means that the device will have to detect the quantity of ICG down to 5 pmol, an accurate fluorescence detection level, as good as existing FIGS system. The second design goal is to improve the size and the maneuverability of the fluorescence imaging system and develop a real handheld system for surgery. Some existing systems for other modalities have been taken as

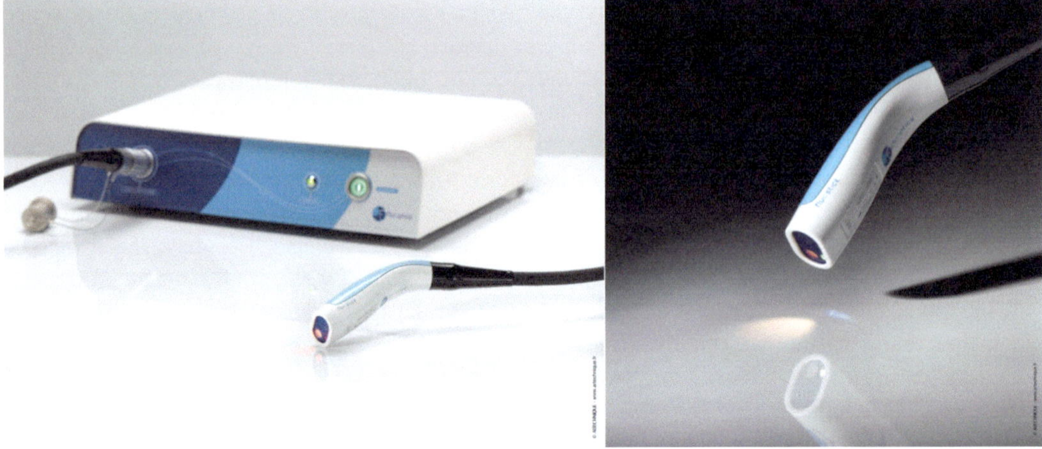

Fig. 37.1 The Fluostick™, optical head and control box

references, ultrasound and gamma detection systems. By reducing the size of fluorescence imaging devices, the aim is to promote fluorescence imaging surgery for applications where size matters.

The system is shown in Fig. 37.1. The system is composed of a control box linked to the miniaturized optical head through a custom cable. In the operating room, the camera head and the cable are fitted to a sterile drape cover with an optical windowed tip. The control box is set on a table at 1.5 m from the surgical field, in the non-sterile zone of the theatre. The control box is linked to a computer thanks to an Ethernet cable, and a specific software drives the system, displays the images live, and is able to save them.

We have seen previously that, in order to reach an accurate fluorescence performance, the whole optical acquisition chain must be optimized for NIR light. Filtering is a critical part of the system. The filters used to collect the fluorescence signal are custom-made and particularly optimized for ICG. Behind the emission filters, the optical lens of the system is NIR coated to avoid reflection and to collect the maximum of incoming signal. A compromise is found between the F-number of the lens and the focal length to preserve the depth of field of our imaging system. It is particularly important considering the fact that the system is handheld and not sealed to a mechanical arm. The CCD sensor selected for the camera is NIR optimized. With a 1/3″ in. size and a definition of 752 × 576 px, it offers a generous pixel size of 6.5 µm and a high well capacity directly linked to a good dynamic range. The excitation is provided by a laser located in the electrical box. An optical fiber drives it to the head of the system. At the tip of the system, the fiber is expanded by an optical combination and spread quite uniformly and with no speckle to the imaged field. The optimized optical combination makes of Fluostick™ a class 1 laser system. At a 10 cm working distance, the laser illumination is 12.5 mW/cm² at 750 nm. At a 5 cm working distance, the laser illumination goes up to 25 mW/cm² at 750 nm, which is very comfortable for real-time and fast ICG visualization. In most of the cases, a FIGS system does not authorize the surgeon to keep the surgical shadowless light on, which could contain NIR light, during NIR acquisitions. The white light accessory of the Fluostick™ is able to provide a high-quality white light with a 4,000 lux power at 10 cm working distance and a very high color rendering index of 93. The aim is to be able to display a wide range of values of ICG concentrations on a single exposition. The analog signal, out of the optical head, is digitalized in the control box by high-range electronics before being sent to the computer. At the end, the system is able to image properly a quantity of ICG down to 5 pmol of product in solution with a very low background.

Fig. 37.2 Images of three different quantities of ICG, from left to right, 5, 10, and 100 pmol

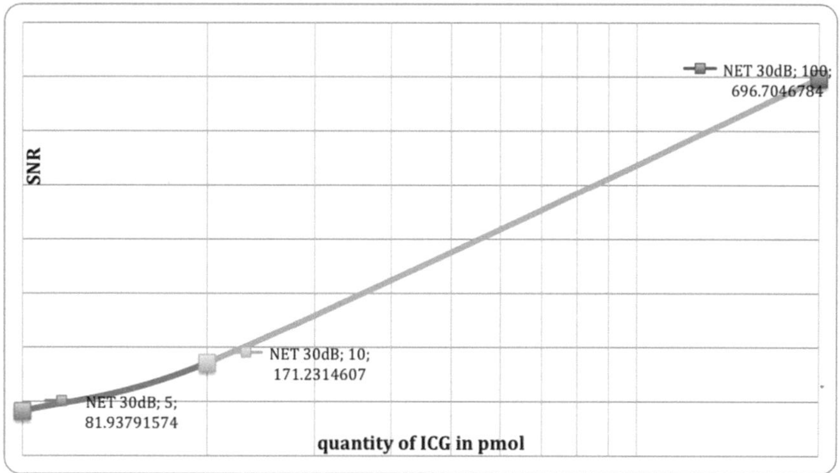

Fig. 37.3 Signal-to-noise ratio for different quantities of ICG

Figure 37.2 shows the results of imaging three different 10 μL drops of ICG.

The concentrations of the product are 0.5, 1, and 10 μM which correspond to 5, 10, and 100 pmol of ICG, respectively. Thanks to a very low background, the values of signal-to-noise ratio (SNR) are high (see Fig. 37.3). The spatial resolution of the image is around 70 μm. Figure 37.4 shows an image of an USAF1951 resolution chart acquired with the Fluostick™. In this picture, to the well-displayed group 2, element 6 corresponds to a resolution of 70.1 μm.

We showed previously that the Fluostick™ has good fluorescence imaging performances. The purpose of this study is to generate the smallest footprint possible with the best fluorescence performance. The initial specification was to develop a handheld instrument. The size constraints were to build the system in a footprint not bigger than 800 mm² with a length of 150 mm. Because the Fluostick™ is considered a handheld system, its frame rate should not be reasonably less than 25 frames per second. We can conclude that the system will collect the fluorescence signal with a 20 ms maximum exposition time. The three main technical choices made to reduce the size of the optical head were a fiber optical deported laser excitation; a single but powerful and high-quality LED for white light excitation; and a 1/3″ in. CCD analog sensor with the minimum of embedded electronics to interface it.

After investigation, the oblong format front section has been chosen for being the best compromise to fit the sensor, the laser excitation, and the white light illumination together. The oblong section fits a rectangle of 34 × 24 mm. Figure 37.5 displays the inside of the optical head and the disposition of the components. A curved shape has been given to the optical head with the purpose to be easier to handle. Moreover, the

Fig. 37.4 Image of an USAF1951 resolution chart acquired with the Fluostick™

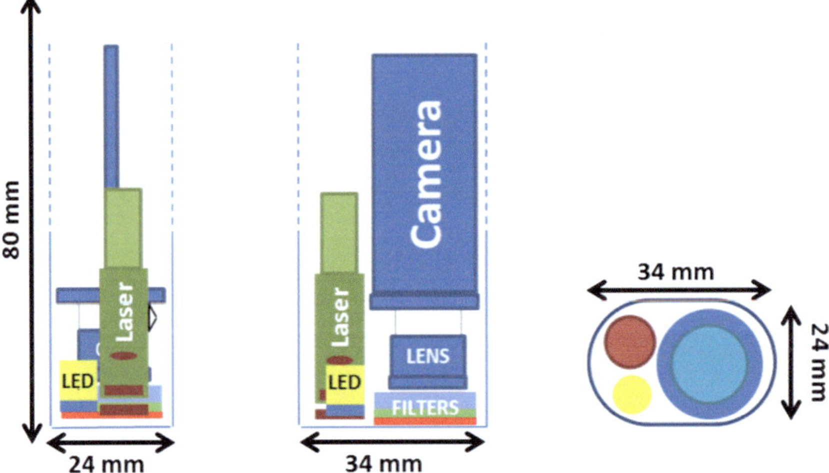

Fig. 37.5 Schematics, disposition of the components

profiles of the system have been thought to fit the hand. A triangle form is designed to help the prehension of the optical head. Some cross sections of a 3D model of the optical head are shown in Fig. 37.6. The system is no longer than 140 mm.

The entire design is thought to make the surgeon able to grip the system and acquire images in every situation. Table 37.1 presents several parameters of the Fluostick™ system.

The Fluostick™, Improves Head and Neck Cancer Resection in a Preclinical Orthotopic Model

We used the Fluostick™ to guide head and neck squamous cell carcinoma (HNSCC) resection in an optimized orthotopic animal model for head and neck cancer. We performed a systemic administration of Angiostamp™ 800, an RGD-based probe that targets αvβ3 integrin, in nude

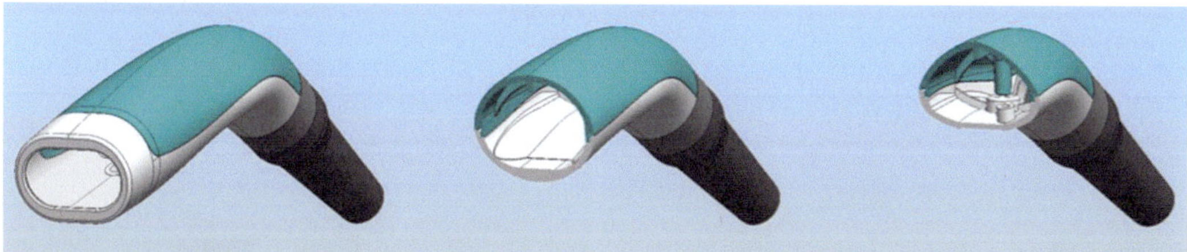

Fig. 37.6 Cross sections of the Fluostick™ optical head

Table 37.1 Fluostick™ review of performance

Footprint	34×24 mm
Length	140 mm
Weight	125 g
Working distance	50–100 mm
Field of view	40×30 mm at 100 mm working distance
Depth of field	15 mm
ICG sensitivity	0.5 pmol
Fluorescence excitation	2.5 mW/cm^2 at 50 mm working distance
F-number	1.8
Optical resolution	70 μm
Sensor definition	756×576 px
Sensor characteristics	High NIR sensitivity monochrome CCD, 25 fps
White light intensity	4,000/ux
White light temperature	4,000 K
White color rendering index	93
Dynamic range	8 bits

mice presenting orthotopic HNSCC tumors developed after intraoral implantation of tumor fragments obtained from subcutaneous tumors derived from a human HNSCC. These tumors had a positive expression of αvβ3 integrin. Tumor resection was performed with and without the aid of Fluostick™. NIR optical imaging-guided surgery using Fluostick™ helped to detect fluorescent cancer residues that could remain unidentified if resection was done exclusively under visual guidance (Fig. 37.7). These residues were measured and analyzed microscopically and we found that Fluostick™ could detect fluorescent cancer foci as small as 185 μm. The detection of these residues had a positive impact on the recurrence-free survival rate of mice in comparison with mice which underwent tumor resection without the aid of Fluostick™. This preclinical stage is an important step before testing Fluostick™ in HNSCC resection in humans.

Colorectal Peritoneal Carcinomatosis

The data from the French registry (FRANCIM network) show that the incidence of colorectal cancer (CRC) has been increasing during the last 30 years. Unfortunately, more than half of patients diagnosed with colorectal cancers present at diagnosis, or will present during their monitoring, visceral metastases (mainly liver) and peritoneal carcinomatosis (PC). Like isolated liver metastases, peritoneal carcinomatosis can be considered as a "regional stage" of metastatic disease. PC secondary to colorectal cancer is characterized by the development of solid tumor deposits on the peritoneal surface. Tumor implantation and growth may lead to invasion of any organ or structure covered by the peritoneum. This peritoneal dissemination can appear in 30–40 % of patients [21, 22]. The natural history of the disease is always fatal, with a median survival of 6 months [23]. However, since the early 1990s, several teams have conducted phase I–II studies to assess the interest of the combination of intraperitoneal chemotherapy with hyperthermia to cytoreductive surgery in the therapeutic management of this pathology. Only complete surgical cytoreduction associated with chemotherapy plus or minus intraperitoneal hyperthermia immediately provides a 5-year survival of 30–48 % [24–26]. The diagnosis of the initial tumor mass is an important prognostic factor of the disease itself. Curability is more easily obtained for limited carcinomatosis, carefully extirpated by the surgeon with a wide peritumoral tissue rather than bulky carcinomatosis

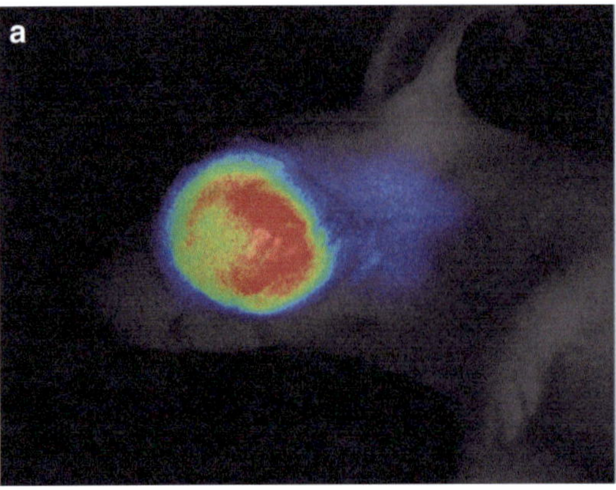

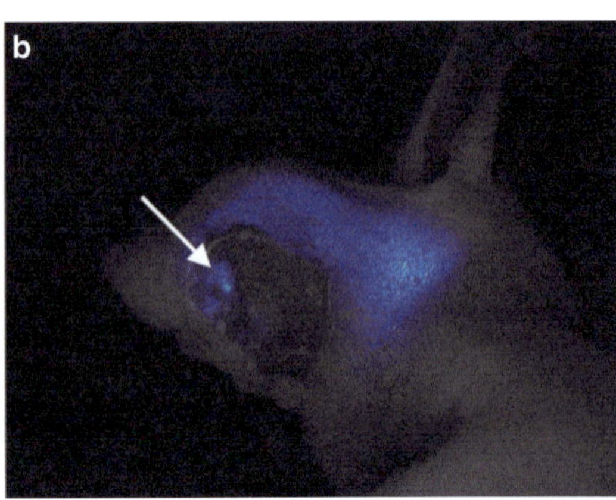

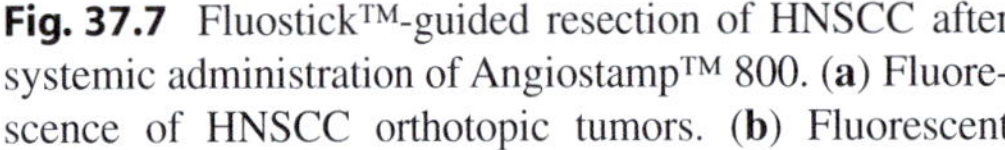

Fig. 37.7 Fluostick™-guided resection of HNSCC after systemic administration of Angiostamp™ 800. (**a**) Fluorescence of HNSCC orthotopic tumors. (**b**) Fluorescent residues (*arrow*) unintentionally left behind after total macroscopic resection of the tumor

which requires complex resections. The classical concept of R0 resection is no longer appropriate in the context of peritoneal surgery and was replaced by CC-0 (macroscopically complete cytoreduction) [27]. In addition, the tumor volume is closely related to potential metastatic disease which determines the progression-free survival. The "Peritoneal Cancer Index" (PCI) of Sugarbaker is a tool for precise quantitative assessment for carcinomatosis of gastrointestinal tumor origin. It is based on the distribution of tumor nodules and their dimension [27–29]. The peritoneal cavity is divided into 9 regions and 4 dials for the hail, 13 regions in total (Fig. 37.8).

The best indications of surgery are associated with debulking surgery satisfactorily (downstaging procedure), which allows a macroscopically complete cytoreduction, leaving only the nodules less than 5 mm, equivalent to a carcinomatosis score 1 in the PCI classification [30]. At present, the difficulty of the most complete surgical resection can result in being able to view a maximum of cancerous lesions, some of which are not visible. Optical fluorescence imaging in the near infrared can improve the intraoperative detection of small dispersed cancerous lesions of the peritoneal cavity, after they are labeled with ICG injected systemically the day before the operation. We are currently investigating how efficiently ICG can help the detection of peritoneal carcinomatosis of colorectal origin in human patients. However, since this application has not been clinically validated yet, it is not possible to rely on the ICG staining for taking the decision to remove or not a suspicious area. That is why we are currently performing a study in which the nodules are removed first under accepted procedures that do not rely on fluorescence, and the extracted tissues are then submitted to an ex vivo analysis of their fluorescent content using Fluostick before being characterized by the pathologists. The preliminary results are very encouraging (Fig. 37.9). The form of the Fluostick is well designed to fit the surgeon's needs. The images show a fine detection of millimetric lesions with a satisfying sensibility of detection although this will have to be formerly determined when the study will be completed.

Discussion

Fluorescence image-guided surgery shows more and more its important potential. It helps surgery to perform faster, better, and less expensively than current existing standards.

The main objective of this study was to illustrate the development and the design of a miniaturized FIGS system and to display sample results of its preclinical and clinical evaluations. The performances of the Fluostick™ are similar to those currently demonstrated by existing instruments.

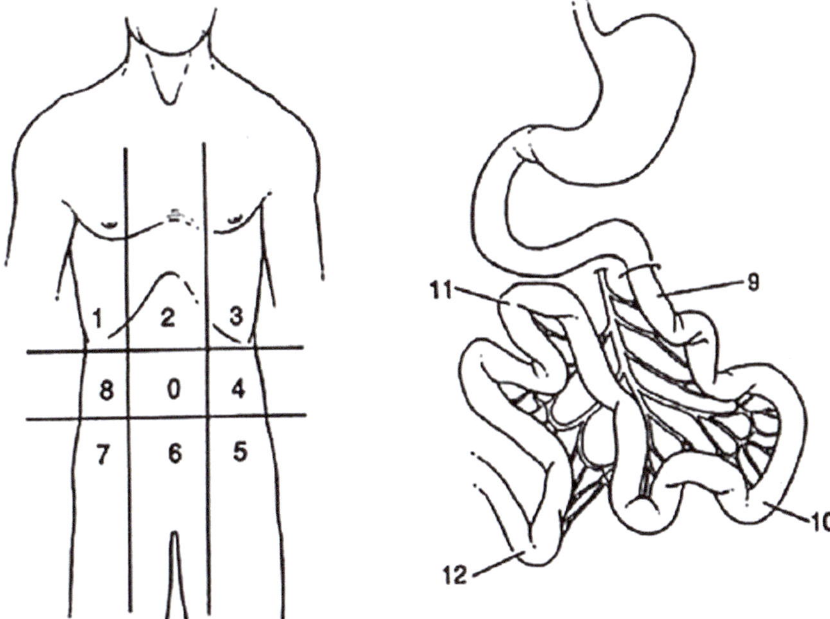

Fig. 37.8 Peritoneal cancer index of Sugarbaker (PCI)

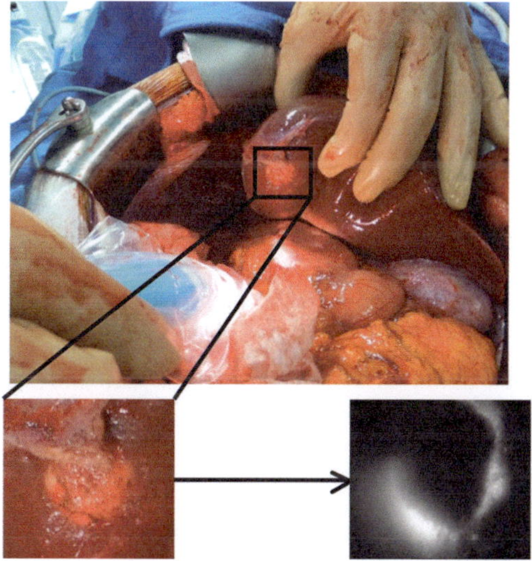

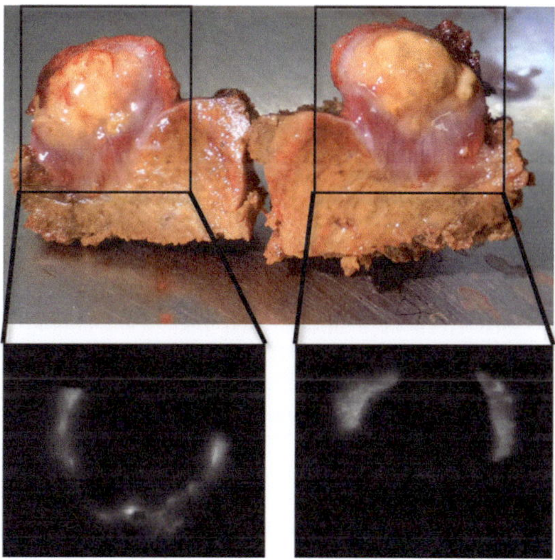

Fig. 37.9 Surgery of hepatic metastasis of adenocarcinoma of the left colon assisted by fluorescence imaging. The circumference of the metastasis is made fluorescent through the injection of Infracyanine[R] (*left panel*). Fluorescence is detected with a Fluostick. After surgical removal, the periphery of the peritoneal metastasis is found fluorescent (*right panel*)

The system has been successfully validated in the operating room and showed its potential for numerous applications such as liver tumor resection or vascular related surgeries. The choices we made for design purpose, such as the curved shape or the sensor technology, will help us in developing new solutions for minimally invasive surgery and other modalities where fluorescence imaging is not or less developed.

References

1. Taruttis A, Ntziachristos V. Translational optical imaging. Am J Roentgenol. 2012;199:263–71.
2. Rasmussen JC, Tan IC, Marshall MV, Adams KE, Kwon S, Fife CE, Maus EA, Smith LA, Covington KR, Sevick-Muraca EM. Human lymphatic architecture and dynamic transport imaged using near-infrared fluorescence. Transl Oncol. 2010;3:362–72.
3. Verbeek FP, van der Vorst JR, Schaafsma BE, Hutteman M, Bonsing BA, van Leeuwen FW, Frangioni JV, van de Velde CJ, Swijnenburg RJ, Vahrmeijer AL. Image-guided hepatopancreatobiliary surgery using near-infrared fluorescent light. J Hepatobiliary Pancreat Sci. 2012;19:626–37.
4. Handa T, Katare RG, Nishimori H, Wariishi S, Fukutomi T, Yamamoto M, Sasaguri S, Sato T. New device for intraoperative graft assessment: HyperEye charge-coupled device camera system. Gen Thorac Cardiovasc Surg. 2010;58:68–77.
5. Polom K, Murawa D, Rho YS, Nowaczyk P, Hunerbein M, Murawa P. Current trends and emerging future of indocyanine green usage in surgery and oncology: a literature review. Cancer. 2011;17:4812–22.
6. Schaafsma BE, Mieog JS, Hutteman M, van der Vorst JR, Kuppen PJ, Lowik CW, Frangioni JV, van de Velde CJ, Vahrmeijer AL. The clinical use of indocyanine green as a near-infrared fluorescent contrast agent for image-guided oncologic surgery. J Surg Oncol. 2011;104:323–32.
7. Tellier F, Steibel J, Chabrier R, Ble FX, Tubaldo H, Rasata R, Chambron J, Duportail G, Simon H, Rodier JF, Poulet P. Sentinel lymph nodes fluorescence detection and imaging using Patent Blue V bound to human serum albumin. Biomed Opt Express. 2012;3:2306–16.
8. Keramidas M, Josserand V, Righini CA, Wenk C, Faure C, Coll JL. Intraoperative near-infrared image-guided surgery for peritoneal carcinomatosis in a preclinical experimental model. Br J Surg. 2010;97:737–43.
9. Troyan SL, Kianzad V, Gibbs-Strauss SL, Gioux S, Matsui A, Oketokoun R, Ngo L, Khamene A, Azar F, Frangioni JV. The FLARE intraoperative near-infrared fluorescence imaging system: a first-in-human clinical trial in breast cancer sentinel lymph node mapping. Ann Surg Oncol. 2009;16:2943–52.
10. Mieog JS, Troyan SL, Hutteman M, Donohoe KJ, van der Vorst JR, Stockdale A, Liefers GJ, Choi HS, Gibbs-Strauss SL, Putter H, Gioux S, Kuppen PJ, Ashitate Y, Lowik CW, Smit VT, Oketokoun R, Ngo LH, van de Velde CJ, Frangioni JV, Vahrmeijer AL. Toward optimization of imaging system and lymphatic tracer for near-infrared fluorescent sentinel lymph node mapping in breast cancer. Ann Surg Oncol. 2011;18:2483–91.
11. Themelis G, Yoo JS, Soh KS, Schulz R, Ntziachristos V. Real-time intraoperative fluorescence imaging system using light-absorption correction. J Biomed Opt. 2009;14:064012.
12. Lane PM, Gilhuly T, Whitehead P, Zeng H, Poh CF, Ng S, Williams PM, Zhang L, Rosin MP, MacAulay CE. Simple device for the direct visualization of oral-cavity tissue fluorescence. J Biomed Opt. 2006;11:024006.
13. Poh CF, Ng SP, Williams PM, Zhang L, Laronde DM, Lane P, Macaulay C, Rosin MP. Direct fluorescence visualization of clinically occult high-risk oral premalignant disease using a simple hand-held device. Head Neck. 2007;29:71–6.
14. Roblyer D, Richards-Kortum R, Sokolov K, El-Naggar AK, Williams MD, Kurachi C, Gillenwater AM. Multispectral optical imaging device for in vivo detection of oral neoplasia. J Biomed Opt. 2008;13:024019.
15. Ghosh KK, Burns LD, Cocker ED, Nimmerjahn A, Ziv Y, Gamal AE, Schnitzer MJ. Miniaturized integration of a fluorescence microscope. Nat Methods. 2011;8:871–8.
16. Shin D, Pierce MC, Gillenwater AM, Williams MD, Richards-Kortum RR. A fiber-optic fluorescence microscope using a consumer-grade digital camera for in vivo cellular imaging. PLoS One. 2010;5:e11218.
17. Kakareka JW, McCann TE, Kosaka Mitsunaga NM, Morgan NY, Pohida TJ, Choyke PI, Kobayashi H. A portable fluorescence camera for testing surgical specimens in the operating room: description and early evaluation. Mol Imaging Biol. 2011;13:862–7.
18. Gioux S, Coutard JG, Berger M, Grateau H, Josserand V, Keramidas M, Righini C, Coll JL, Dinten JM. FluoSTIC: miniaturized fluorescence image-guided surgery system. J Biomed Opt. 2012;17:106014.
19. Wang X, Bhaumik S, Li Q, Staudinger VP, Yazdanfar S. Compact instrument for fluorescence image-guided surgery. J Biomed Opt. 2010;15:020509.
20. Liu Y, Bauer AQ, Akers WJ, Sudlow G, Liang K, Shen D, Berezin MY, Culver JP, Achilefu S. Hands-free, wireless goggles for near-infrared fluorescence and real-time image-guided surgery. Surgery. 2011;149:689–98.
21. Chu DZ, Lang NP, Thompson C, Osteen PK, Westbrook KC. Peritoneal carcinomatosis in nongynecologic malignancy. A prospective study of prognostic factors. Cancer. 1989;63:364–7.

22. Sadeghi B, Arvieux C, Glehen O, Beaujard AC, Rivoire M, Baulieux J, Fontaumard E, Brachet A, Caillot JL, Faure JL, Porcheron J, Peix JL, Francois Y, Vignal J, Gilly FN. Peritoneal carcinomatosis from non-gynecologic malignancies: results of the EVOCAPE 1 multicentric prospective study. Cancer. 2000;88:358–63.
23. Jayne DG, Fook S, Loi C, Seow-Choen F. Peritoneal carcinomatosis from colorectal cancer. Br J Surg. 2002;89:1545–50.
24. Glehen O, Kwiatkowski F, Sugarbaker PH, Elias D, Levine EA, De Simone M, Barone R, Yonemura Y, Cavaliere F, Quenet F, Gutman M, Tentes AA, Lorimier G, Bernard JL, Bereder JM, Porcheron J, Gomez-Portilla A, Shen P, Deraco M, Rat P. Cytoreductive surgery combined with perioperative intraperitoneal chemotherapy for the management of peritoneal carcinomatosis from colorectal cancer: a multi-institutional study. J Clin Oncol. 2004;22: 3284–92.
25. da Silva RG, Sugarbaker PH. Analysis of prognostic factors in seventy patients having a complete cytoreduction plus perioperative intraperitoneal chemotherapy for carcinomatosis from colorectal cancer. J Am Coll Surg. 2006;203:878–86.
26. Verwaal VJ, van Ruth S, Witkamp A, Boot H, van Slooten G, Zoetmulder FA. Long-term survival of peritoneal carcinomatosis of colorectal origin. Ann Surg Oncol. 2005;12:65–71.
27. Begossi G, Gonzalez-Moreno S, Ortega-Perez G, Fon LJ, Sugarbaker PH. Cytoreduction and intraperitoneal chemotherapy for the management of peritoneal carcinomatosis, sarcomatosis and mesothelioma. Eur J Surg Oncol. 2002;28:80–7.
28. Dromain C, Leboulleux S, Auperin A, Goere D, Malka D, Lumbroso J, Schumberger M, Sigal MR, Elias D. Staging of peritoneal carcinomatosis: enhanced CT vs PET/CT. Abdom Imaging. 2008;33:87–93.
29. Sugarbaker PH. Intraperitoneal chemotherapy and cytoreductive surgery for the prevention and treatment of peritoneal carcinomatosis and sarcomatosis. Semin Surg Oncol. 1998;14:254–61.
30. Esquivel J. Cytoreductive surgery for peritoneal malignancies – development of standards of care for the community. Surg Oncol Clin N Am. 2007;16: 653–66.

Index

F.D. Dip et al. (eds.), *Fluorescence Imaging for Surgeons: Concepts and Applications*, DOI 10.1007/978-3-319-15678-1, © Springer International Publishing Switzerland 2015

Q

R

S